This book describes recent advances in the application of chaos theory to classical scattering and nonequilibrium statistical mechanics generally, and to transport by deterministic diffusion in particular.

The author presents the basic tools of dynamical systems theory, such as dynamical instability, topological analysis, periodic-orbit methods, Liouvillian dynamics, dynamical randomness and large-deviation formalism. These tools are applied to chaotic scattering and to transport in systems near equilibrium and maintained out of equilibrium. Chaotic scattering is illustrated with disk scatterers and with examples of unimolecular chemical reactions and then generalized to transport in spatially extented systems. Transport and chaotic properties are inter-connected. The phenomenological laws of hydrodynamics and of irreversible thermodynamics are derived from first principles defined in dynamical systems theory. The foundations of statistical mechanics are discussed on the basis of the hypothesis of microscopic chaos.

This book will be of interest to researchers interested in chaos, dynamical systems, chaotic scattering and statistical mechanics in theoretical, computational and mathematical physics and also in theoretical chemistry.

Cambridge Nonlinear Science Series 9

Chaos, scattering and statistical mechanics

TITLES IN PRINT IN THIS SERIES

**G. M. Zaslavsky, R. Z. Sagdeev,
D. A. Usikov and A. A. Chernikov**

1. Weak chaos and quasi-regular patterns

J. L. McCauley

2. Chaos, dynamics and fractals: an algorithmic approach to deterministic chaos

K. Nakamura

3. Quantum chaos: a new paradigm for nonlinear dynamics

C. Beck and F. Schlögl

4. Thermodynamics of chaotic systems: an introduction

P. Meakin

5. Fractals, scaling and growth far from equilibrium

R. Badii and A. Politi

6. Complexity – hierarchical structures and scaling in physics

H. Kantz and T. Schreiber

7. Nonlinear time series analysis

T. Bohr, M. H. Jensen, G. Paladin and A. Vulpiani

8. Dynamical systems approach to turbulence

P. Gaspard

9. Chaos, scattering and statistical mechanics

Chaos, scattering and statistical mechanics

Pierre Gaspard
Université Libre de Bruxelles
Faculté des Sciences
Center for Nonlinear Phenomena and Complex Systems

CAMBRIDGE
UNIVERSITY PRESS

CAMBRIDGE UNIVERSITY PRESS
Cambridge, New York, Melbourne, Madrid, Cape Town, Singapore, São Paulo

Cambridge University Press
The Edinburgh Building, Cambridge CB2 2RU, UK

Published in the United States of America by Cambridge University Press, New York

www.cambridge.org
Information on this title: www.cambridge.org/9780521395113

First published 1998
This digitally printed first paperback version 2005

A catalogue record for this publication is available from the British Library

Library of Congress Cataloguing in Publication data

Gaspard, Pierre, 1959–
Chaos, scattering and statistical mechanics / Pierre Gaspard.
p. cm. – (Cambridge nonlinear science series; 9)
Includes bibliographical references and index.
ISBN 0 521 39511 9
1. Statistical mechanics–Congresses. 2. Chaotic behavior in systems–Congresses.
3. Scattering (Physics)–Congresses.
I. Title. II. Series.
QC174.7.G37 1998
530.13′01′1857–dc21 97-22890 CIP

ISBN-13 978-0-521-39511-3 hardback
ISBN-10 0-521-39511-9 hardback

ISBN-13 978-0-521-01825-8 paperback
ISBN-10 0-521-01825-0 paperback

Contents

Preface

Today, there is a growing interest in understanding the role of chaos in nonequilibrium statistical mechanics. Although ergodic theory has been one of the seeds of modern dynamical systems theory, it is only recently that new methods have been developed – especially, in periodic-orbit theory – in order to quantitatively characterize the microscopic chaos as well as the intrinsic rates of decay or relaxation of statistical ensembles of trajectories. One of these intrinsic rates is the escape rate associated with the so-called fractal repeller which plays a central role in chaotic scattering. During recent years, chaotic scattering has been discovered in many different fields, from celestial mechanics and hydrodynamics to atomic, molecular, mesoscopic, and nuclear physics. In molecular systems, chaotic scattering provides a classical and statistical understanding of chemical reactions. Chaotic scattering is also closely related to transport processes like diffusion or viscosity. In this way, relationships can be established between the transport coefficients and the characteristic quantities of microscopic chaos, such as the Lyapunov exponents, the Kolmogorov–Sinai entropy, or the fractal dimensions. These results and their developments shed new light on nonequilibrium statistical mechanics and the problem of irreversibility.

The aim of the present book is to describe the theory of chaotic scattering and this new approach to nonequilibrium statistical mechanics starting from the principles of dynamical systems theory and from the hypothesis of microscopic chaos. For lack of space and time, the book only contains results on classical dynamical systems, although many fascinating and closely connected results have also been obtained in the context of quantum dynamics.

This book is the outcome of several collaborations which I have had with my colleagues:

My deepest gratitude goes to G. Nicolis of the Université Libre de Bruxelles (Brussels). This work would not have been possible without his continuous support and encouragement for the past fifteen years.

It is my great pleasure to thank S. A. Rice who is at the origin of this work. Indeed, I started studying the three-disk scatterer as a model of unimolecular reaction during my postdoctoral stay of two years in his department at the University of Chicago.

This work owes much to J. R. Dorfman of the University of Maryland (College Park). I would like to express here both my gratitude and admiration.

P. Cvitanović of the Niels Bohr Institute (Copenhagen) has greatly determined the development of this work. I thank him for fruitful contacts since 1983 and for his support in the completion of this work.

It is another pleasure to thank X.-J. Wang (Boston) who will find in these pages the results and continuation of our past collaboration.

I am also particularly indebted to S. Tasaki (Kyoto) for most enlightening discussions during our collaboration on Liouvillian dynamics.

I would also like to thank warmly my other collaborators J.-P. Eckmann (Geneva), K. Nakamura (Osaka), D. Alonso (La Laguna), F. Baras (Brussels), I. Burghardt (Bonn), T. Okuda (Osaka), A. Provata (Greece), T. Schreiber (Germany), S. Tersigni (USA).

I also wish to express my thanks to I. Prigogine, Director of the Solvay Institutes, as well as to R. Balescu (Brussels) who have had considerable influence on this work. Moreover, I gratefully acknowledge the fruitful discussions, suggestions, or support received from J.-P. Boon (Brussels), J. Bricmont (Louvain-la-Neuve), L. A. Bunimovich (Atlanta), G. Casati (Milano), Ph. Choquard (Lausanne), E. G. D. Cohen (New York), C. P. Dettmann (Sydney), D. Driebe (Austin), Y. Elskens (Marseille), H. H. Hasegawa (Austin), M. Herman (Brussels), L. Kadanoff (Chicago), R. Kapral (Toronto), J. Lebowitz (Rutgers), C. Maes (Leuven), M. Malek Mansour (Brussels), M. Mareschal (Brussels), G. P. Morriss (Kensington), D. Ruelle (Paris), U. Smilansky (Rehovot), T. Tél (Budapest), J.-W. Turner (Brussels). My best thanks also to my other colleagues of the Center for Nonlinear Phenomena and Complex Systems at ULB and also to Mr P. Kinet, Mrs I. Saverino, and Mrs S. Wellens.

I thank the Université Libre de Bruxelles (ULB), its Faculty of Sciences, and its Department of Physics for continuous and active support. The present work is based on lectures given in the Physics Department of ULB during the years 1991–92 and 1992–93.

I would like to thank the Fonds National de la Recherche Scientifique (FNRS Belgium) for continuous financial support. This research has also been supported, in part, by the European Commission under the programs 'Science' and

'Human Capital and Mobility', by the Belgian Federal Government under the program 'Pôles d'Attraction Interuniversitaire', by the Communauté française de Belgique under contract No. ARC 93/98-166, by the Banque Nationale de Belgique.

Pierre Gaspard

Introduction

The idea that gases are disordered or amorphous states of matter is old. Actually, the word *gas* was created from the Greek word *chaos* by Joan-Baptista van Helmont (1577–1644). This Flemish physician and chemist born in Brussels was the first to distinguish different kinds of gases thanks to the experimental method and he also invented an air thermoscope which was the precursor of the modern thermometer. He was contemporary with Bacon (1561–1626), Galileo (1564–1642), Kepler (1571–1630), Descartes (1596–1650), Torricelli (1608–1647), as well as with the famous painter Rubens (1577–1640). His son published his work *Ortus medicinae, id est initia phisicare inaudita* at Amsterdam in 1648 (Farber 1961).

During the XIXth century, the spatial disorder of gases and of matter in general was quantitatively characterized with the concept of entropy per unit volume. However, the idea of dynamical chaos, i.e., of temporal disorder in physical systems like gases is more recent as it results from a long sequence of observations and works which extends throughout the XXth century with the development of statistical mechanics.

Today, we may say that statistical mechanics and kinetic theory are among the greatest successes of modern science. Since Maxwell and Boltzmann, macroscopic properties of matter can be explained in terms of the motion of atoms and molecules composing matter. In particular, transport properties like diffusion, viscosity, or heat conductivity can be predicted in terms of the parameters of the microscopic Hamiltonians, which are the masses of the atoms and molecules, and the coupling constants of their interaction (Maxwell 1890, Boltzmann 1896).

Since Boltzmann, many kinetic and master equations have been derived from microscopic Hamiltonians. Methods have been developed to derive effective macroscopic equations either directly from the microscopic Hamiltonian or, more often, from the intermediate kinetic equations, leading to hydrodynamic phenomenological equations such as the Navier–Stokes equation and its generalizations, and that for all the possible phases of matter like gases, liquids, solids, liquid crystals, plasmas, quantum fluids, radiative-transfer systems, magnetic systems, reaction-diffusion systems,... Most of these phenomenological equations (except the nondissipative ones like wave equations and nonlinear integrable equations) have irreversible terms due to transport and other dissipative processes. The works by Onsager (1931) and, later, by Callen and Welton (1951) led to the results of Green (1951–60), Kubo (1957), and others that the transport and dissipative coefficients can be obtained as integrals of the time correlation functions. In this way, the methods of nonequilibrium statistical mechanics have provided a foundation for hydrodynamics and other continuous-media theories as well as to nonequilibrium thermodynamics. It is in thermodynamics that the second law has been formulated which plays so important a role in our understanding of dissipative structures (Glansdorff and Prigogine 1971, Nicolis and Prigogine 1977) as well as of all engines in nature from steam engines down to molecular engines and other bioenergetic processes.

In spite of its great success, nonequilibrium statistical mechanics suffers from a lack of firm foundations especially concerning the intrinsic dynamical properties which may justify the introduction of irreversible phenomenological equations. The paradox of irreversibility is well known, that the macroscopic phenomenological equations are irreversible while Hamiltonian equations are reversible. Such paradoxes question the very existence of dissipative processes in spite of their fundamental importance. Boltzmann and Gibbs (1902) certainly opened the way to the resolution of such paradoxes by their introduction of statistical ensembles of trajectories in phase space. However, it was not until very recently that suitable mathematical tools were introduced for the description of the time evolution of such statistical ensembles.

Tremendous advances have been carried out in kinetic theory since Boltzmann's derivation of his famous equation. Major efforts have been devoted to the extension of kinetic theory to dense gases since Enskog (1921). Methods have been developed for systematic derivations of kinetic equations based, in particular, on the reduced n-particle distribution functions. A hierarchy of coupled equations ruling the time evolution of these reduced distributions was derived by Yvon (1935), Bogoliubov (1946), Born and Green (1946), and Kirkwood (1946). Kinetic equations were derived by truncating this BBGKY hierarchy, leading to expansions of the transport properties in terms of the particle density. However, such expansions turned out to be nonanalytic in the particle density and, in the same context, the time correlation functions of typical fluids were found to have long-time tails (Alder and Wainwright 1969; Dorfman and Cohen 1970, 1972,

1975; Ernst et al. 1970; Pomeau and Résibois 1975). These results showed that dynamical effects can progressively overwhelm the mere Boltzmannian kinetics as the density increases.

Other systems with hard-core interactions have also been studied like the Lorentz gas models of mixtures of light and infinitely heavy particles (van Beijeren 1982). Moreover, a kinetic approach to chemically reacting systems was pioneered by Prigogine (1949).

Since the work by Pauli (1928), kinetic equations have also been obtained by perturbative methods for weakly coupled systems. Uehling and Uhlenbeck (1933) derived a quantum analogue of Boltzmann's equation. Van Hove (1955–1959) systematizes the derivation of quantum master equations in the weak-coupling limit. Brout and Prigogine (1956) developed a different approach based on perturbative expansions in action-angle variables around integrable systems. Kinetic equations were also derived for plasmas by Landau (1936), Vlassov (1938), and more recently, by Balescu (1960) and Lenard (1960) who included the Debye screening of Coulomb interaction in the kinetic description. Many other systems have been and continue to be studied in kinetic theory.[1]

Each kinetic equation necessarily involves a stochastic assumption such as Boltzmann's *Stosszahlansatz*, which is introduced by some truncation of the evolution equation either in the low density limit or in the weak coupling limit. In recent works, in particular, by Lebowitz and Spohn, scaling limits have been used to justify more rigorously these stochastic assumptions (Spohn 1980). Nevertheless it is difficult to admit that such important phenomena as transport and reaction processes do not have a more fundamental justification in terms of the intrinsic properties of the underlying microscopic dynamics. Recently, adequate constructions have been proposed in the theory of chaotic dynamical systems, which overcome the previous difficulties caused by the stochastic assumptions.

The way to the modern theory of dynamical systems has followed a long history during the XXth century since the early works in ergodic theory by Koopman (1931), Birkhoff (1931), von Neumann (1932), and others. Many consequences of dynamical instability, especially, on the topology of trajectories have been discovered since the pioneering work of Poincaré (1892) and Lyapunov (1907). Several simple dynamical systems were invented and studied like the geodesic flows on negative-curvature surfaces by Hadamard (1898), the billiards by Birkhoff (1927), and the baker map[2] by Seidel (1933). Homoclinic orbits and symbolic dynamics have been central concepts in these slow developments by Birkhoff (1927), Hopf (1937), Hedlund (1939), Smale (1980b), and Shilnikov (1970), among others. The advent of modern computers has opened an era of systematic exploration of phase-space structures since the works by Lorenz,

1. For further information see Cohen (1962), Balescu (1975), and Spohn (1980).
2. So called in Arnold and Avez (1968).

Hénon and Heiles (Lorenz 1963, 1993; Hénon and Heiles 1964; Hénon 1969, 1976).

In the late thirties and early forties, visionaries like Wiener (1948), Ulam and von Neumann (1947) paved the way to the fusion of dynamical systems theory with probability theory. In the late forties and early fifties, the connection between ergodic theory and Shannon's information theory (1949) was a turning point which led to the widespread recognition that deterministic dynamical systems may be as random as a coin tossing process, following the work by Kolmogorov (1958, 1959), Sinai (1959), Ornstein (1974), and others. A precise characterization of dynamical randomness – and thus of today's chaos – was introduced with the Kolmogorov–Sinai (KS) entropy per unit time. This notion opened the way to a statistical (large-deviation) formalism of dynamical chaos, elaborated in particular by Bowen and Ruelle (1975) from the analogy between the time randomness of chaotic trajectories and the space randomness of a system configuration in equilibrium statistical mechanics. In this context, Mandelbrot (1982) furthermore introduced the concept of fractal and powerful methods were developed to calculate fractal dimensions of natural objects.

These advances in ergodic theory and in dynamical systems theory have provided important contributions to the foundation of statistical mechanics. Inspired by Krylov (1944, 1979), Sinai (1970) proved the ergodic hypothesis for the two-dimensional two-disk system. Since then, the ergodicity of several larger hard-ball systems has also been proved (Szász 1996). Moreover, Bunimovich and Sinai (1980) proved the existence of a positive and finite diffusion coefficient in the periodic Lorentz gas with a finite horizon. More recently, Bunimovich and Spohn (1996) proved the existence of a positive and finite viscosity coefficient for a system of two disks in elastic collisions.

In these recent works, the property which turns out to play the fundamental role is the dynamical instability of trajectories as characterized by a positive Lyapunov exponent. This dynamical instability is at the origin of a hyperbolic structure which covers the phase space with segments of stable and unstable manifolds. On the one hand, these invariant sets form a Hopf chain along which different trajectories may be connected in phase space, which allows to prove transitivity and hence ergodicity. On the other hand, the dynamical instability is at the origin of a loss of correlation in time between initially close phase-space cells. This loss of correlations yields the property of mixing and guarantees a rapid decay of the time correlation functions. Consequently, the dynamical fluctuations around time averages turn out to be Gaussian distributed so that transport processes may be normal. The dynamical instability is also at the origin of dynamical randomness and hence of chaos in these systems.

We should nevertheless remark that the relevance of ergodicity and of mixing to kinetic theory is arguable. Ergodicity is defined by the equality of time averages with phase-space averages. This property is very important for the determination of the transitive invariant sets of trajectories, which yields the decomposition of phase space into ergodic components. However, it is not very relevant for kinetic and transport properties because no time scale can be defined using ergodicity alone. Moreover, ergodicity is both very common and very rare. For instance, a quasiperiodic motion with incommensurable frequencies in an integrable system is already ergodic although transport does not exist in integrable systems. In contrast, a high-density hard-sphere gas near close packing is not ergodic because the spheres are blocked except for small fluctuations. Nevertheless, these small fluctuations are chaotic, which may be enough to guarantee the existence of relaxation rates in this solid phase. In the same order of ideas, infinite systems of harmonic oscillators or of free particles like the ideal gases and solids are ergodic and even mixing. However, the decay of time correlations is not fast enough for normal transport to exist. Similarly, transport is anomalous in integrable solitonic nonlinear systems like the Toda lattice even if it is mixing. We conclude here that ergodicity and mixing are too weak properties to justify transport properties although they are important necessary conditions.

In contrast, dynamical instability results generically in the formation of homoclinic orbits and of topological chaos. Therefore, the analytic constants of motion are broken in these homoclinic tangles and transport through phase space may occur along the chaotic zones (Bensimon and Kadanoff 1984, MacKay et al. 1984, Dana et al. 1989, Meiss 1992, Wiggins 1992). Accordingly, we understand that dynamical instability can be at the origin of transport processes and we should expect a connection between chaos and transport properties.

In order to explain such a connection, let us first start from generalities concerning chaos. It is well known that dynamical instability induces a sensitivity to initial conditions such that two trajectories initially very close separate exponentially before the distance between them becomes so large that they follow completely different paths. This sensitivity to initial conditions limits the possible predictions on the trajectories because the initial condition is only known with a given precision $\epsilon_{\text{initial}}$. For a rate of separation of nearby trajectories given by the maximum Lyapunov exponent λ_{max}, the error between the predicted and the actual trajectories grows as $\epsilon_t \simeq \epsilon_{\text{initial}} \exp(\lambda_{\text{max}} t)$. After a finite time, the error becomes larger than the precision ϵ_{final} allowed to remain predictive, which defines the so-called Lyapunov time: $t_{\text{Lyap}} \simeq (1/\lambda_{\text{max}}) \ln(\epsilon_{\text{final}}/\epsilon_{\text{initial}})$. Given initial and final precisions, long-time predictions are thus always limited. Beyond the Lyapunov time, the time evolution of the system can only be described in statistical terms. A statistical description is thus required, especially for all the properties related to the approach to the ergodic invariant measure. In classical atomic or molecular systems, the ergodic invariant measure is precisely the state which defines the thermodynamic equilibrium. Therefore, we may expect that the properties of relaxation toward equilibrium should find a statistical description in chaotic systems. Chaos provides here a natural justification for the introduction of statistical ensembles. Since chaos may already appear in systems with only two degrees of freedom (with a three-dimensional flow) statistical mechanics is justified already for small classical systems. This is in contrast with the common expectation that the accumulation of many particles is required for the foundation of statistical mechanics. The presence of many particles is certainly an important feature of macroscopic systems with $N_{\text{Avogadro}} \sim 10^{23}$ particles but this large number is not a sufficient condition for transport properties, in particular. The ideal gases and solids constitute counterexamples to this supposition.

On the contrary, dynamical instability and chaos define intrinsic time scales of the dynamics of a system. Moreover, they are very elementary properties of trajectories. Indeed, dynamical instability is a property of linear stability. In classical Hamiltonian systems, the equations of motion are defined from the first variation of the action, which is Hamilton's famous action principle of classical mechanics. Linear stability comes next since it is obtained by the second variation of the action, as known since Jacobi (DeWitt-Morette et al. 1979).

The positivity of Lyapunov exponents is a very common property in natural systems as shown in Table 1, which presents several values obtained either experimentally or numerically. In bounded systems, the KS entropy per unit time which characterizes dynamical randomness is known to be given by the sum of positive Lyapunov exponents according to Pesin's formula:

$$h_{\text{KS}} = \sum_{\lambda_i > 0} \lambda_i \,, \tag{0.1}$$

(Eckmann and Ruelle 1985). The known examples of chaos at the macroscopic level present Lyapunov exponents and KS entropies of the order of 0.01–100 digits/sec. In celestial mechanics, chaotic systems have a much lower power of dynamical randomness. For instance, Hyperion has a Lyapunov exponent of the order of 10^{-7} digits/sec, while Mars and the Solar System would become dynamically unstable on much longer time scales.

On the other hand, chaos takes place on a much shorter time scale in the classical motion of atoms and molecules in matter. A typical Lyapunov exponent in a gas can be estimated by a reasoning first carried out by Krylov (1944, 1979). The particles are supposed to undergo elastic collisions like hard spheres of mass m and diameter d (see Fig. 1). In a fluid of density ρ and temperature T, the mean free path is equal to $\ell \sim 1/(\rho\pi d^2)$, while the mean velocity is $\bar{v} = \sqrt{3k_{\text{B}}T/m}$. The impact parameter b at a binary collision is defined as the distance between the incoming trajectory and the parallel line passing through the centre of mass. The velocity angle φ_1 after the collision is

$$\varphi_1 = \pi - 2 \arcsin \frac{2b}{d} . \qquad (0.2)$$

Because of the mean separation ℓ between the successive collisions, a difference

Table 1 Maximum Lyapunov exponent and KS entropy per unit time in typical systems.

	$\lambda_{\max}^{-1}$ [sec/digit]	$\lambda_{\max}$ [digit/sec]	h_{KS} [digit/sec]
Solar System[(a)]	$3.6\ 10^{14}$	$2.8\ 10^{-15}$	–
Pluto[(b)]	$1.5\ 10^{15}$	$7\ 10^{-16}$	–
Obliquity of Mars[(c)]	$1–4\ 10^{14}$	$1–3\ 10^{-15}$	–
Rotation of Hyperion (satellite of Saturn)[(d)]	$8.5\ 10^{6}$	$1.2\ 10^{-7}$	–
Chemical chaotic oscillations[(e)]	880	0.0011	0.0011
Hydrodynamic chaotic oscillations[(f)]	5.4	0.19	0.26
1 cm^3 of argon at room temperature[(g)]	10^{-10}	10^{10}	10^{29}
1 cm^3 of argon at triple point[(h)]	10^{-15}	10^{15}	$1.4\ 10^{34}$

The Lyapunov exponents are here expressed in the units of digits/sec defined in relation to $10^{\lambda_{\text{digit}} t}$. We also find the units of bits/sec with $2^{\lambda_{\text{bit}} t}$, and of nats/sec with $e^{\lambda_{\text{nat}} t}$ which is most often used. The digit/sec is convenient to visualize the Lyapunov exponent as an error growth along the digits of the phase-space variables of the system.

(*a*) Laskar (1989) observed numerically a Lyapunov time of 5 10^6 y/nat.

(*b*) Sussman and Wisdom (1988) observed numerically a Lyapunov time of 20 10^6 y/nat.

(*c*) Touma and Wisdom (1993) observed numerically a Lyapunov time of 1–5 10^6 y/nat.

(*d*) Wisdom et al. (1984).

(*e*) Roux et al. (1983) reported an experimental value of 0.3 ± 0.1 nats/period in the Belousov–Zhabotinskii reaction for oscillations of mean period of 114 sec.

(*f*) Eckmann et al. (1986) reported an experimental spectrum of three non-negative Lyapunov exponents $\{\lambda^{(k)}\} = \{0.43, 0.17, 0.0\}$ nat/sec (with an error of 5%) for the Rayleigh–Bénard convection in silicon oil at Pr = 30, $\text{Ra}/\text{Ra}_c = 180$ in a cell of height $h = 1$ cm and of horizontal sizes $l_x = 4$ cm, $l_y = 1$ cm.

(*g*) See below Eqs. (0.6)–(0.7).

(*h*) Numerical result by Posch and Hoover (1988, 1989).

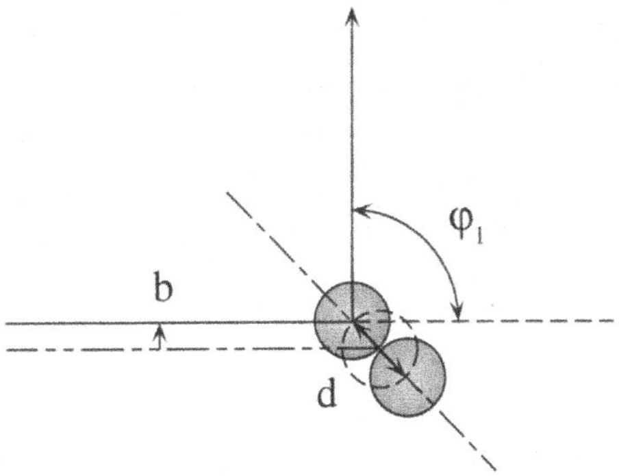

Figure 1 Geometry of an elastic collision between two hard spheres.

$\delta\varphi_0$ on the velocity angle at the previous collision grows up to a difference $\delta b = \ell\ \delta\varphi_0$ in the impact parameter before the current collision. The velocity angle after the collision is modified by

$$\delta\varphi_1 = \frac{4\ell}{d|\sin(\varphi_1/2)|}\ \delta\varphi_0 \sim \frac{4\ell}{d}\ \delta\varphi_0 . \tag{0.3}$$

After n collisions, the difference in the velocity angle will thus have increased as

$$\delta\varphi_n \sim \left(\frac{4\ell}{d}\right)^n \delta\varphi_0 . \tag{0.4}$$

This increase occurs during a time $T_n \sim n\ell/\bar{v}$ so that the maximum Lyapunov exponent

$$\lambda_{\text{max}} = \lim_{n\to\infty} \frac{1}{T_n} \ln\left|\frac{\delta\varphi_n}{\delta\varphi_0}\right| , \tag{0.5}$$

is estimated as

$$\lambda_{\text{max}} \sim \frac{\bar{v}}{\ell} \ln\frac{4\ell}{d} \sim \rho\ \pi\ d^2 \sqrt{\frac{3k_{\text{B}}T}{m}}\ \ln\frac{4}{\rho\pi d^3} , \tag{0.6}$$

in nats/sec. For a gas at room temperature and atmospheric pressure, we find

$$\lambda_{\text{max}} \sim 10^{10}\ \text{digits/sec} . \tag{0.7}$$

The Lyapunov time scale is thus of the order of the mean intercollisional time in the fluid, which is the kinetic time scale. This short time scale is to be compared with the much longer hydrodynamic time scale, which is arbitrarily long for hydrodynamic fluctuations of arbitrarily long wavelengths. This difference of time scales determines to a large extent the relationship between microscopic chaos and the transport properties. By the way, we should remark here that the macroscopic chaos takes place on the hydrodynamic time scale fixed by some nonequilibrium conditions, so that the microscopic and the macroscopic chaos are very different phenomena.

We notice that a many-particle system like a gas has several different unstable directions and so many positive Lyapunov exponents. In Hamiltonian systems, the Lyapunov exponents form pairs of positive and negative exponents of equal absolute values $(+\lambda_i, -\lambda_i)$ with $i = 1, 2, \ldots, Nd$ and $\lambda_i \geq 0$. For a d-dimensional system of N particles in a box with periodic boundary conditions, energy and the total linear momentum are conserved so that $2(d+1)$ Lyapunov exponents vanishes. Therefore, we expect in general $i_{\text{max}} = Nd - d - 1$ positive Lyapunov exponents. In the thermodynamic limit $N \to \infty$, the positive Lyapunov exponents may form a quasicontinuous spectrum given by

$$\lambda_i = \lambda_{\text{max}}\ f\left(\frac{i}{i_{\text{max}}}\right) , \tag{0.8}$$

where $f(x)$ is a dimensionless function such that $f(1) = 1$ (Sinai 1996). This

Lyapunov spectrum has been numerically observed in the one-dimensional Fermi–Pasta–Ulam chain with $f(x) \simeq x$ (Livi et al. 1986), in a three-dimensional system of particles with a repulsive Lennard-Jones interaction in fluid and solid phases (Posch and Hoover 1988, 1989; Morriss 1989), as well as in the hard-sphere gas (Dellago et al. 1995, 1996).

The KS entropy is the sum of all the positive Lyapunov exponents so that it can be estimated as

$$h_{\rm KS} \;=\; \sum_{\lambda_i>0} \lambda_i \;\simeq\; \int_0^{i_{\rm max}} \lambda_i \, di \;=\; C \; i_{\rm max} \, \lambda_{\rm max} \; , \tag{0.9}$$

with $C = \int_0^1 f(x)dx$, which is extensive by this argument. An intensive entropy per unit time and unit volume would thus have the following estimation for a three-dimensional fluid

$$h^{\rm (time,space)} \;=\; 3 \; C \; \rho \; \lambda_{\rm max} \;\sim\; \rho^2 \; \pi \; d^2 \sqrt{\frac{3k_{\rm B}T}{m}} \; \ln \frac{4}{\rho \pi d^3} \; , \tag{0.10}$$

which takes the value (Gaspard and Nicolis 1985)

$$h^{\rm (time,space)} \;\sim\; 10^{29} \; {\rm digits/sec \; cm^3} \; , \tag{0.11}$$

for a dilute gas like argon at room temperature and pressure. Systematic evaluations of this entropy have recently been obtained for dilute gases in equilibrium (van Beijeren et al. 1997). For argon near its critical point at 48 atmospheres and 150 K, this entropy takes the larger value (Posch and Hoover 1988, 1989)

$$h^{\rm (time,space)} \;\simeq\; 1.4 \;\times\; 10^{34} \; {\rm digits/sec \; cm^3} \; . \tag{0.12}$$

The entropy per unit time and unit volume characterizes the randomness which occurs not only along the time axis but also in position space. Since this quantity is positive such systems may be said to develop spatio-temporal chaos. We notice that the dynamical randomness developed by the classical motion of atoms and molecules in matter is gigantic. This randomness acts like a reservoir to animate stochastically many different transport and relaxation processes. In particular, the Brownian motion is such a process in which the microscopic chaos drives the erratic motion of a colloidal particle in suspension in a fluid. The dynamical randomness of the Brownian particle may be evaluated and shown to be very small. Nevertheless, this time randomness provides evidence of the microscopic chaos, as we shall argue later.

The purpose of the present work is to describe the relationship between chaos and transport properties. Indeed, the previous arguments suggest that such relationships exist and are fundamental for nonequilibrium statistical mechanics. Precise relationships can be established at different levels.

Firstly, transport processes like diffusion can be conceived in terms of chaotic scattering, which leads to the so-called escape-rate formulas relating the diffusion coefficient to the Lyapunov exponents and the KS entropy or the fractal dimensions, characterizing chaos in scattering systems (Gaspard and Nicolis 1990). This result was generalized to the other transport and reaction-rate coefficients by Dorfman and Gaspard (1995).

Secondly, in the same context, hydrodynamic modes of diffusion and nonequilibrium steady states can be calculated directly in terms of the Liouvillian dynamics in phase space. The dispersion relations of the dissipative processes turn out to be given as classical (Pollicott–Ruelle) resonances or generalized eigenvalues of the Liouvillian dynamics of statistical ensembles (Gaspard 1996). In some systems, the diffusion coefficient can even be calculated in terms of the unstable periodic orbits which are dense in chaos. This new approach to nonequilibrium statistical mechanics establishes a direct connection between the phase-space dynamics and the hydrodynamic laws. In particular, this approach has recently led, in a simple model of diffusion, to the derivation of the entropy production expected from irreversible thermodynamics (Gaspard 1997b).

The work is organized as follows. The first chapters describe the general methods of dynamical systems theory which are thereafter applied to spatially extended systems of relevance to diffusion and to the other transport processes. Chapter 1 is concerned with the definition of dynamical systems in phase space and the linear stability (or instability) of their trajectories. The Lyapunov exponents are introduced among other quantities characterizing the linear stability. Chapter 2 shows how dynamical instability may result in topological chaos. The time evolution of statistical ensembles, the so-called Liouvillian dynamics, is described in Chapter 3 where the periodic-orbit theory is exposed. Concepts like Frobenius–Perron operator, trace formulas, zeta functions, Pollicott–Ruelle resonances, generalized eigenvalue problem are introduced. Probabilistic chaos and the large-deviation formalism are presented in Chapter 4 where Pesin's formula is generalized to scattering systems and where fractal dimensions are defined. These methods are applied to chaotic scattering in Chapter 5. The scattering theory of transport is developed in Chapter 6 where the escape-rate formulas are derived and the relationship between the escape rate and the thermostatted-system approach is discussed. The hydrodynamic modes of diffusion are constructed in Chapter 7 on the basis of the Liouvillian dynamics of spatially extended systems. Chapter 8 contains the construction of the nonequilibrium steady states as well as a derivation of the entropy production of irreversible thermodynamics in the case of diffusion in the multibaker map. The characterization of stochastic processes with the ε-entropy per unit time is given in Chapter 9 where we also discuss the possible observation of microscopic chaos in the laboratory. Conclusions and perspectives are drawn in Chapter 10.

Our purpose is therefore to show that hydrodynamics and nonequilibrium thermodynamics can in principle be directly derived from the Liouvillian dynamics ruling the time evolution of statistical ensembles of trajectories in phase space, and that without tempering the basic Hamiltonian dynamics. Relaxation toward thermodynamic equilibrium is described in terms of the hydrodynamic modes which are constructed by generalized eigenvalue problems as previously considered in particular by Balescu (1960) and by Kadanoff and Swift (1968). The obtained results provide counterexamples to some assumptions of the phenomenological and kinetic theories, in particular, that hydrodynamic modes and nonequilibrium steady states can be represented by functions. Instead, they appear as distributions and under certain circumstances as singular measures with some fractal-like properties. This singular character turns out to play a fundamental role in the connection with nonequilibrium thermodynamics. In this way, the present work provides the foundation for an approach to nonequilibrium statistical mechanics based on the hypothesis of microscopic chaos that many-particle systems typically have a positive Kolmogorov–Sinai entropy per unit time.

Chapter 1

Dynamical systems and their linear stability

1.1 Dynamics in phase space

1.1.1 The group of time evolutions

The time evolution of many natural systems is ruled by ordinary differential equations which are autonomous if there is no time-dependent forcing. In this respect, we consider a dynamical system described by a set of first-order differential equations

$$\dot{\mathbf{X}} = \mathbf{F}(\mathbf{X}) , \tag{1.1}$$

where the dot denotes a derivative with respect to the time. Higher-order differential equations which are local in time can be reduced in general to the form (1.1).

The trajectory is represented in the space $\mathcal{M} \subseteq \mathbb{R}^M$ of the variables $\mathbf{X}$, which is called the *phase space*. Phase-space volumes may be preserved if the divergence of the vector field $\mathbf{F}$ vanishes: $\mathrm{div}\mathbf{F} = 0$. Otherwise, the volumes may expand or contract. In dissipative systems, they usually contract at least globally over a long time interval in view of the overall stability of these systems.

The equations (1.1) induce a one-parameter group, called the *flow*,

$$\mathbf{X} = \mathbf{\Phi}^t \mathbf{X}_0 . \tag{1.2}$$

which is in general a nonlinear function of the time t and of the initial conditions $\mathbf{X}_0$. The flow defines a group because:

$$\mathbf{\Phi}^{t+\tau} = \mathbf{\Phi}^t \circ \mathbf{\Phi}^\tau . \tag{1.3}$$

The evolution of phase-space volumes is controlled by the Jacobian determinant of the transformation (1.2):

$$|\det \partial_{\mathbf{X}} \mathbf{\Phi}^t| = \exp \int_0^t \mathrm{div}\mathbf{F} \, d\tau . \tag{1.4}$$

If the volumes are preserved, the Jacobian determinant remains equal to unity: $|\det \partial_{\mathbf{X}} \mathbf{\Phi}^t| = 1$.

1.1.2 The Poincaré map

A very powerful tool for the analysis of the phase-space geometry is the surface of section introduced by Poincaré (1892). A (hyper)surface $\mathscr{P}(\mathbf{X}) = 0$ is defined in phase space in such a way that it intersects transversally (if possible) all the trajectories of interest. Since the flow is deterministic, each point of intersection of a trajectory (1.2) is uniquely determined by the previous intersection, which defines the Poincaré map $\boldsymbol{\phi}$. Moreover, the successive intersections do not occur at equal times (i.e., they are not isochronic in general except for periodically driven systems). Therefore, there exists another function called the first-return time function which determines the successive intersection times $\{t_n\}_{n=-\infty}^{+\infty}$. In this way, the flow may be reduced to the map

$$\begin{aligned} \mathbf{x}_{n+1} &= \boldsymbol{\phi}(\mathbf{x}_n) , \\ t_{n+1} &= t_n + T(\mathbf{x}_n) , \end{aligned} \tag{1.5}$$

where $\mathbf{x} \in \mathbb{R}^{M-1}$ are some coordinates which are intrinsic to the surface of section. Reciprocally, a flow may be reconstructed from the map (1.5), which is then equivalent to the original flow (1.2). Indeed, each phase-space point $\mathbf{X}$ may be represented by the coordinate $\mathbf{x}$ of the previous intersection and by the time $\tau \in [0, T(\mathbf{x})[$ elapsed since the last intersection. A so-called *suspended flow* is thus defined by

$$\tilde{\mathbf{\Phi}}^t(\mathbf{x}, \tau) = \left[\boldsymbol{\phi}^n \mathbf{x}, \tau + t - \sum_{j=0}^{n-1} T(\boldsymbol{\phi}^j \mathbf{x}) \right] ,$$

$$\text{for} \quad 0 \le \tau + t - \sum_{j=0}^{n-1} T(\boldsymbol{\phi}^j \mathbf{x}) < T(\boldsymbol{\phi}^n \mathbf{x}) . \tag{1.6}$$

If we know the correspondence between both systems of coordinates, $\mathbf{X} = \mathbf{G}(\mathbf{x}, \tau)$, then we have equivalence between both the representations (1.2) and (1.6) of the flow: $\mathbf{\Phi}^t = \mathbf{G} \circ \tilde{\mathbf{\Phi}}^t \circ \mathbf{G}^{-1}$.

There are many examples of dynamical systems which are either dissipative, time-reversible-but-not-volume-preserving, volume-preserving, Hamiltonian, symplectic, mechanical like billiards,... which have been invented since Newton. Let us here briefly recall some properties of Hamiltonian systems and of billiards.

1.1.3 Hamiltonian systems

This special class of systems appears in conservative problems of mechanics as in celestial mechanics but also in the statistical mechanics ruling the motion of atoms and molecules in matter. A mechanical system of N particles is modelled with a Hamiltonian function $H(\mathbf{q}, \mathbf{p})$ where $\mathbf{X} = (\mathbf{q}, \mathbf{p})$ is a $2Nd$-dimensional vector space with $\mathbf{q} = (\mathbf{q}_1, \mathbf{q}_2, \ldots, \mathbf{q}_N)$, $\mathbf{q}_i$ being the (d-dimensional) position of particle i, and $\mathbf{p} = (\mathbf{p}_1, \mathbf{p}_2, \ldots, \mathbf{p}_N)$ with $\mathbf{p}_i$ the (d-dimensional) momentum of particle i. For nonrelativistic particles of mass m, the Hamiltonian function has the form

$$H(\mathbf{q}, \mathbf{p}) = \sum_{i=1}^{N} \frac{\mathbf{p}_i^2}{2m} + V(\mathbf{q}_1, \ldots \mathbf{q}_N) \tag{1.7}$$

where $V(\mathbf{q})$ is the potential energy of interaction between the particles. Hamilton's equations of motion are

$$\dot{\mathbf{q}} = \frac{\partial H(\mathbf{q}, \mathbf{p})}{\partial \mathbf{p}}, \quad \dot{\mathbf{p}} = -\frac{\partial H(\mathbf{q}, \mathbf{p})}{\partial \mathbf{q}}, \tag{1.8}$$

which are of the form (1.1) since

$$\dot{\mathbf{X}} = \mathbf{F}(\mathbf{X}) = \mathbf{\Sigma} \cdot \partial_{\mathbf{X}} H, \tag{1.9}$$

in terms of the fundamental matrix of the symplectic structure, i.e.,

$$\mathbf{\Sigma} = \begin{pmatrix} 0 & 1 \\ -1 & 0 \end{pmatrix}. \tag{1.10}$$

The number $f = Nd$ of pairs of conjugate variables (q_i, p_i) is called the number of degrees of freedom. The total number of canonical variables, i.e., the phase-space dimension, is thus always the double of the number of degrees of freedom: $2f$. We will suppose either that the boundaries of the system are hard walls of infinite mass, or that periodic boundary conditions are applied, or else that the phase space is unbounded as in scattering systems. In each case, the total energy is conserved. In systems without infinitely massive walls, the total momentum is also conserved. We will always consider a system at one fixed energy E, and therefore the $2f$-dimensional phase space reduces to the $(2f-1)$-dimensional constant energy surface defined by the condition that $H = E$ and called the *energy shell.*

1.1.4 Billiards

These other conservative systems are defined by a Hamiltonian like (1.7) for the motion of particles between obstacles on which the phase-space point $\mathbf{X} = (\mathbf{q}, \mathbf{p})$ undergoes elastic bounces. Examples of billiards are the Lorentz-type billiards where the obstacles are hard disks or spheres fixed in space and onto which independent point particles undergo elastic collisions. The obstacles may form a finite scatterer, a periodic lattice, or be randomly distributed. Most often, the motion is free between the obstacles so that $V = 0$ in Eq. (1.7). The velocity is defined by $\mathbf{v} = \mathbf{p}/m$. At each elastic collision, the velocity changes according to the collision rule

$$\mathbf{v}^{(+)} = \mathbf{v}^{(-)} - 2(\mathbf{n} \cdot \mathbf{v}^{(-)})\, \mathbf{n} , \tag{1.11}$$

where $\mathbf{v}^{(-)}$ and $\mathbf{v}^{(+)}$ denote the velocity before and after the collision, while $\mathbf{n}(\mathbf{q})$ is a unit vector perpendicular to the obstacle at the point of impact $\mathbf{q}$. Since energy is conserved between and at collisions, the magnitude $\|\mathbf{v}\|$ of the velocity is also conserved so that the time of flight is equal to the length of the trajectory on the energy shell corresponding to a unit velocity $\|\mathbf{v}\| = 1$.

A related billiard is the hard-sphere gas in which spherical particles of mass m and diameter d undergo mutual elastic collisions. The positions of the particles are restricted by the conditions $\|\mathbf{q}_i - \mathbf{q}_j\| > d$. The same collision rule as (1.11) is obeyed in which $\mathbf{v} = (\mathbf{v}_1, \ldots, \mathbf{v}_N)$ denotes the velocities of all the particles and $\mathbf{n} = (0, \ldots, 0, \epsilon_{ij}/\sqrt{2}, 0, \ldots, 0, -\epsilon_{ij}/\sqrt{2}, 0, \ldots, 0)$ is used for a binary collision between the particles No. i and No. j, where ϵ_{ij} is the unit vector in the direction of the line joining their centres at the time of collision.

Other types of billiards exist with some external fields, with moving boundaries, or with other collision rules.

1.2 Linear stability and the tangent space

1.2.1 The fundamental matrix

An important characterization of the trajectories is given by their stability and, in particular, by their linear stability which controls the way infinitesimal perturbations evolve with time. These perturbations can be calculated by integration of a set of coupled equations, one for the trajectory that passes through the point $\mathbf{X}_0$ and one for the trajectory that deviates by an infinitesimal amount $\delta\mathbf{X}$ from the reference trajectory, $\mathbf{X}_t = \mathbf{\Phi}^t\mathbf{X}_0$,

$$\delta\dot{\mathbf{X}} = \frac{\partial\mathbf{F}(\mathbf{X})}{\partial\mathbf{X}} \cdot \delta\mathbf{X}\,, \tag{1.12}$$

to linear order in $\delta\mathbf{X}$. These infinitesimal vectors $\delta\mathbf{X}$ belong to a linear vector space which is tangent to the phase space $\mathscr{M}$ at each phase point $\mathbf{X}$ and which is called the *tangent space* and denoted by $\mathscr{T}\mathscr{M}(\mathbf{X})$.

Since Eq. (1.12) is linear, all of its solutions can be expressed as

$$\delta\mathbf{X}_t = \frac{\partial\mathbf{\Phi}^t(\mathbf{X}_0)}{\partial\mathbf{X}_0} \cdot \delta\mathbf{X}_0 = \mathsf{M}(t,\mathbf{X}_0)\cdot\delta\mathbf{X}_0\,. \tag{1.13}$$

In Eq. (1.13), we have set $\mathbf{X}_0$ and $\delta\mathbf{X}_0$ to be the value of $\mathbf{X}$ and $\delta\mathbf{X}$ at $t = 0$, and $\mathsf{M}(t,\mathbf{X}_0)$ is the fundamental matrix given by

$$\mathsf{M}(t,\mathbf{X}_0) = \mathbb{T}\,\exp\int_0^t \frac{\partial\mathbf{F}}{\partial\mathbf{X}}(\mathbf{\Phi}^\tau\mathbf{X}_0)\,d\tau \tag{1.14}$$

where $\mathbb{T}\exp\int_0^t(\cdot)d\tau$ denotes a time-ordered exponential, i.e., a Dyson series. The fundamental matrix thus obeys the evolution equation

$$\dot{\mathsf{M}}(t,\mathbf{X}_0) = \frac{\partial\mathbf{F}(\mathbf{\Phi}^t\mathbf{X}_0)}{\partial\mathbf{X}_0}\cdot\mathsf{M}(t,\mathbf{X}_0)\,. \tag{1.15}$$

We mention that the construction of the fundamental matrix requires the simultaneous integration of both Eqs. (1.1) and (1.15).

1.2.2 Lyapunov exponents

The infinitesimal perturbation $\delta\mathbf{X}$ may grow at most exponentially, which is evidence for the sensitivity to initial conditions. This exponential growth can be characterized by the Lyapunov exponent associated with an arbitrary tangent vector $\bar{\mathbf{e}}$,

$$\lambda(\mathbf{X},\bar{\mathbf{e}}) = \lim_{t\to\infty}\frac{1}{t}\ln\|\mathsf{M}(t,\mathbf{X})\cdot\bar{\mathbf{e}}\|\,. \tag{1.16}$$

A theorem by Oseledets (1968) shows that the limit (1.16) takes its value from a discrete set called the spectrum of Lyapunov exponents satisfying

$$\lambda^{(1)}(\mathbf{X}) > \lambda^{(2)}(\mathbf{X}) > \cdots > \lambda^{(r)}(\mathbf{X})\,, \tag{1.17}$$

with multiplicities

$$m^{(1)}(\mathbf{X}), m^{(2)}(\mathbf{X}), \ldots, m^{(r)}(\mathbf{X})\,, \tag{1.18}$$

which sum up to the dimension of the tangent space $M = \dim\mathcal{T}\mathcal{M} = \dim\mathcal{M}$

$$M = \sum_{k=1}^{r} m^{(k)}(\mathbf{X}) \ . \tag{1.19}$$

The value $\lambda^{(k)}(\mathbf{X})$ is obtained from Eq. (1.16) when the tangent vector $\bar{\mathbf{e}}$ belongs to the subset $\mathcal{V}^{(k)}(\mathbf{X})\backslash\ \mathcal{V}^{(k+1)}(\mathbf{X})$ (that is, the set of points in $\mathcal{V}^{(k)}$ but not in $\mathcal{V}^{(k+1)}$), where $\{\mathcal{V}^{(k)}(\mathbf{X})\}_{k=1}^{r}$ are nested linear subspaces of the tangent space

$$\mathcal{T}\mathcal{M}(\mathbf{X}) \ = \ \mathcal{V}^{(1)}(\mathbf{X}) \supset \mathcal{V}^{(2)}(\mathbf{X}) \supset \cdots \supset \mathcal{V}^{(r)}(\mathbf{X}) \ , \tag{1.20}$$

such that $m^{(k)}(\mathbf{X}) = \dim[\mathcal{V}^{(k)}(\mathbf{X})] - \dim[\mathcal{V}^{(k+1)}(\mathbf{X})]$.

For continuous-time systems, the Lyapunov exponent associated with the direction of the flow always vanishes for trajectories which do not approach any stationary point. This result is obtained by differentiating Eq. (1.1) with respect to time to observe that $\delta\mathbf{X} = \varepsilon\dot{\mathbf{X}} = \varepsilon\mathbf{F}(\mathbf{X})$ is always a solution of Eq. (1.12) so that

$$\dot{\mathbf{X}}_t \ = \ \mathsf{M}(t, \mathbf{X}_0) \cdot \dot{\mathbf{X}}_0 \ . \tag{1.21}$$

Therefore, for the vector $\bar{\mathbf{e}} = \dot{\mathbf{X}}_0 / \|\dot{\mathbf{X}}_0\|$ taken in the flow direction, the Lyapunov exponent (1.16) becomes

$$\begin{aligned} \lambda(\mathbf{X}, \bar{\mathbf{e}}) \ &= \ \lim_{t\to\infty} \ \frac{1}{t} \ \ln\left(\|\dot{\mathbf{X}}_t\| / \|\dot{\mathbf{X}}_0\|\right) \\ &= \ \lim_{t\to\infty} \ \frac{1}{t} \left[\ln\|\mathbf{F}(\mathbf{X}_t)\| \ - \ \ln\|\mathbf{F}(\mathbf{X}_0)\|\right] \ = \ 0 \ , \end{aligned} \tag{1.22}$$

if $\mathbf{F}(\mathbf{X}_t) \neq 0$. *Q.E.D.*

1.2.3 Decomposition of the tangent space into orthogonal directions

A method of calculation of the Lyapunov exponents can be obtained by considering the distance between a reference trajectory $\mathbf{X}_t$ and its infinitesimal perturbation $\mathbf{X}_t + \delta\mathbf{X}_t$

$$\|\delta\mathbf{X}_t\|^2 = \delta\mathbf{X}_0^{\mathrm{T}} \cdot \mathsf{M}^{\mathrm{T}}(t, \mathbf{X}_0) \cdot \mathsf{M}(t, \mathbf{X}_0) \cdot \delta\mathbf{X}_0 \ , \tag{1.23}$$

where the superscript T denotes a transpose.

Equation (1.23) defines a positive-definite quadratic form given by the real symmetric matrix, $\mathsf{M}^{\mathrm{T}} \cdot \mathsf{M}$, which can thus be diagonalized in terms of its real eigenvalues $\{\sigma_j(t, \mathbf{X}_0)\}$, and orthonormal eigenvectors $\{\mathbf{u}_j(t, \mathbf{X}_0)\}$ as

$$\mathsf{M}^{\mathrm{T}}(t, \mathbf{X}_0) \cdot \mathsf{M}(t, \mathbf{X}_0) \ = \ \sum_{j=1}^{M} \ \mathbf{u}_j(t, \mathbf{X}_0) \ \sigma_j(t, \mathbf{X}_0) \ \mathbf{u}_j^{\mathrm{T}}(t, \mathbf{X}_0) \ . \tag{1.24}$$

Because the matrix (1.24) is real symmetric, the eigenvectors span the M-dimensional space $\mathscr{T}\mathscr{M}(\mathbf{X}_0)$ tangent to the phase space at $\mathbf{X}_0$. The local Lyapunov exponents of the trajectory at initial point $\mathbf{X}_0$ may now be defined as

$$\lambda_j(\mathbf{X}_0) = \lim_{t\to\infty} \frac{1}{2t} \ln \sigma_j(t,\mathbf{X}_0) , \tag{1.25}$$

giving exponents which are equivalent to Eq. (1.16). Depending on the sign of the Lyapunov exponent $\lambda_j(\mathbf{X}_0)$, the corresponding directions are stable ($\lambda_j < 0$), or unstable ($\lambda_j > 0$), or neutral ($\lambda_j = 0$). The neutral directions include the direction of the flow (and if we wished to include it, the direction perpendicular to the energy surface). The different Lyapunov exponents can then be ordered according to Eq. (1.17).

With the previous results, we can show that the sum of all the Lyapunov exponents is directly related to the change of volumes in phase space according to

$$\sum_j \lambda_j(\mathbf{X}) = \lim_{t\to\infty} \frac{1}{t} \int_0^t (\text{div } \mathbf{F})(\mathbf{\Phi}^\tau \mathbf{X})\, d\tau , \tag{1.26}$$

which vanishes for conservative flows. This result follows by summing Eq. (1.25) over all the Lyapunov exponents to get

$$\sum_j \lambda_j(\mathbf{X}) = \lim_{t\to\infty} \frac{1}{2t} \ln \prod_j \sigma_j(t,\mathbf{X}) . \tag{1.27}$$

The product of all the eigenvalues σ_j of the quadratic form (1.24) is equal to its determinant

$$\prod_j \sigma_j = \det \mathsf{M}^{\mathrm{T}} \cdot \mathsf{M} = |\det \mathsf{M}|^2 . \tag{1.28}$$

Since the determinant of the fundamental matrix M is given by the Jacobian determinant of the flow (1.2) according to (1.13), Eq. (1.4) can here be used to obtain (1.26). *Q.E.D.*

The linear stability of periodic orbits can be characterized by the eigenvalues Λ_j of the fundamental matrix $\mathsf{M}(T)$ where T is the period of the orbit. Indeed, the Λ_j are related to the preceding eigenvalues by $\sigma_j = |\Lambda_j|^2$, so that

$$\lambda_j = \frac{1}{T} \ln |\Lambda_j|. \tag{1.29}$$

It is worth noting that to determine the Lyapunov exponents for non-periodic points $\mathbf{X}_0$, one should, in general determine the Lyapunov exponents from the eigenvalues of the symmetric, positive definite form $\mathsf{M}^{\mathrm{T}}(t,\mathbf{X}_0)\cdot\mathsf{M}(t,\mathbf{X}_0)$ but for a point $\mathbf{X}$ on a periodic orbit, one may determine the Lyapunov exponents by computing the eigenvalues of $\mathsf{M}(T,\mathbf{X})$ directly.

1.2.4 Homological decomposition of the multiplicative cocycle

The Multiplicative Ergodic Theorem of Oseledets (1968) allows us to clearly identify the locally stable and unstable directions. This is accomplished through the use of the property that the matrix $\mathsf{M}(t, \mathbf{X})$ is a *multiplicative cocycle.* That is, $\mathsf{M}(t, \mathbf{X})$ satisfies the important group-type relation

$$\mathsf{M}(t+\tau, \mathbf{X}) = \mathsf{M}(t, \mathbf{\Phi}^{\tau}\mathbf{X}) \cdot \mathsf{M}(\tau, \mathbf{X}) , \tag{1.30}$$

for any positive time τ. This is just a restatement of the time evolution property of the solution of Eq. (1.1). Next we define a Lyapunov homology as a local, linear transformation between the cocycle M and another cocycle $\tilde{\mathsf{M}}$ of the form

$$\mathsf{M}(t, \mathbf{X}) = \mathsf{R}(\mathbf{\Phi}^{t}\mathbf{X}) \cdot \tilde{\mathsf{M}}(t, \mathbf{X}) \cdot \mathsf{R}^{-1}(\mathbf{X}) , \tag{1.31}$$

such that the transformation matrices R are not exponentially growing. That is, such that

$$\lim_{t\to\infty} \frac{1}{t} \ln \|\mathsf{R}(\mathbf{\Phi}^{t}\mathbf{X})\| = 0 . \tag{1.32}$$

One can easily check that the cocycle M satisfies Eq. (1.30) provided that the cocycle $\tilde{\mathsf{M}}$ does also. The purpose of the Lyapunov homology is to find conditions under which the cocycle M might be reduced to a diagonal cocycle $\tilde{\mathsf{M}}$ such that

$$\mathsf{M}(t, \mathbf{X}) = \sum_i \mathbf{e}_i(\mathbf{\Phi}^{t}\mathbf{X}) \, \Lambda_i(t, \mathbf{X}) \, \mathbf{f}_i^{\mathrm{T}}(\mathbf{X}) , \tag{1.33}$$

where $\mathbf{e}_i(\mathbf{X})$ are vectors with components $e_{i,k}(\mathbf{X}) = R_{ki}(\mathbf{X})$ and $\mathbf{f}_i$ are vectors with components $f_{i,k}(\mathbf{X}) = [\mathsf{R}^{-1}(\mathbf{X})]_{ik}$. These vectors form a set of biorthogonal pairs satisfying

$$\sum_i \mathbf{e}_i \, \mathbf{f}_i^{\mathrm{T}} = \mathbf{1} , \tag{1.34}$$

and

$$\mathbf{f}_i^{\mathrm{T}} \cdot \mathbf{e}_j = \delta_{ij} , \tag{1.35}$$

at each phase-space point $\mathbf{X}$. We note that the orthonormal eigenvectors $\mathbf{u}_j(t, \mathbf{X}_0)$ are not directly related to the directions $\{\mathbf{e}_i(\mathbf{X}_0)\}$ which are not orthogonal. The decomposition (1.33) of the fundamental matrix is also referred to as the Mather spectrum (Pesin 1989).

The function $\Lambda_i(t, \mathbf{X})$ is called the *stretching factor* of the direction $\mathbf{e}_i(\mathbf{X})$. It must also satisfy a multiplicative cocycle relation

$$\Lambda_i(t+\tau, \mathbf{X}) = \Lambda_i(t, \mathbf{\Phi}^{\tau}\mathbf{X}) \, \Lambda_i(\tau, \mathbf{X}) . \tag{1.36}$$

We emphasize that this construction is not a diagonalization of the matrix M in the usual sense because the vector $\mathbf{f}_i$ is evaluated at the initial point $\mathbf{X}$ while the vector $\mathbf{e}_i$ is evaluated at the final point $\mathbf{\Phi}^t\mathbf{X}$. It is also important to note that $\mathbf{e}_i(\mathbf{\Phi}^t\mathbf{X})$ and $\mathbf{f}_i(\mathbf{X})$ are not, in general, mutually orthogonal.

When the matrix M is applied to one of the direction vectors $\mathbf{e}_i$ one obtains

$$\mathsf{M}(t,\mathbf{X}) \cdot \mathbf{e}_i(\mathbf{X}) = \Lambda_i(t,\mathbf{X})\,\mathbf{e}_i(\mathbf{\Phi}^t\mathbf{X}) . \tag{1.37}$$

According to the Lyapunov condition (1.32), the vector $\mathbf{e}_i(\mathbf{\Phi}^t\mathbf{X})$ is not growing exponentially, so that the entire exponential growth on the right-hand side of Eq. (1.37) is contained in the function $\Lambda_i(t,\mathbf{X})$. Therefore the Lyapunov exponent associated with the direction $\bar{\mathbf{e}} = \mathbf{e}_i(\mathbf{X})$ is

$$\lambda_i(\mathbf{X}) \equiv \lambda(\mathbf{X},\mathbf{e}_i) = \lim_{t\to\infty} \frac{1}{t} \ln |\Lambda_i(t,\mathbf{X})| , \tag{1.38}$$

according to (1.16) and (1.33). We conclude that the Lyapunov exponents can equivalently be defined by the stretching factors Λ_i. The different directions $\{\mathbf{e}_i(\mathbf{X})\}$ of the tangent space $\mathscr{T}\mathscr{M}(\mathbf{X})$ can be ordered as in Eqs. (1.17)–(1.20). In particular, a linear subspace $\mathscr{V}^{(k)}(\mathbf{X})$ of Eq. (1.20) is spanned by the set of unit vectors $\{\mathbf{e}_i(\mathbf{X})\}_{i\in I^{(k)}}$ corresponding to the Lyapunov exponents which satisfy $\lambda_i(\mathbf{X}) \leq \lambda^{(k)}(\mathbf{X})$ for $i \in I^{(k)}$.

As a result of these considerations, we can establish directions of expansion and of contraction in the tangent space at the point $\mathbf{X}$, which can be expressed as a direct sum of three linear subspaces

$$\mathscr{T}\mathscr{M}(\mathbf{X}) = \mathscr{E}_u(\mathbf{X}) \oplus \mathscr{E}_0(\mathbf{X}) \oplus \mathscr{E}_s(\mathbf{X}) , \tag{1.39}$$

spanned by the unstable, the neutral, and the stable directions, respectively, and obtained by combining together the linear subspaces $\{\mathbf{e}_i(\mathbf{X})\}$ with $\lambda_i > 0$, $\lambda_i = 0$, and $\lambda_i < 0$, respectively.

1.2.5 The local stretching rates

We notice that the local Lyapunov exponents are quantities that are constant on half trajectories $\mathbf{\Phi}^t\mathbf{X}$ with $t \in [0,+\infty[$ or $]-\infty,0]$ in the sense that $\lambda_i(\mathbf{X}) = \lambda_i(\mathbf{\Phi}^\tau\mathbf{X})$ for $\tau \in \mathbb{R}^\pm$, which follows from the cocycle property and the definition (1.16). Our purpose now is to introduce the local quantities that underlie the local Lyapunov exponents and could be varying functions along each trajectory. With this aim, we differentiate Eq. (1.37) with respect to time to get an equation of evolution for the stretching factors $\Lambda_i(t,\mathbf{X})$. Using the evolution equation

(1.15) of the fundamental matrix and the biorthogonality and completeness relations (1.34)–(1.35), we obtain

$$\dot{\Lambda}_i(t,\mathbf{X}) = \chi_i(\mathbf{\Phi}^t\mathbf{X})\,\Lambda_i(t,\mathbf{X}) , \tag{1.40}$$

or

$$\Lambda_i(t,\mathbf{X}) = \exp \int_0^t \chi_i(\mathbf{\Phi}^\tau\mathbf{X})\,d\tau , \tag{1.41}$$

where we introduce the *local stretching rate* in the direction $\mathbf{e}_i(\mathbf{X})$ by

$$\chi_i(\mathbf{X}) = \mathbf{f}_i^{\mathrm{T}}(\mathbf{X}) \cdot \left\{ \frac{\partial \mathbf{F}}{\partial \mathbf{X}}(\mathbf{X}) \cdot \mathbf{e}_i(\mathbf{X}) - \left[\mathbf{F}(\mathbf{X}) \cdot \frac{\partial}{\partial \mathbf{X}}\right] \cdot \mathbf{e}_i(\mathbf{X}) \right\} . \tag{1.42}$$

Such local quantities have been considered, in particular, by Eckhardt and Yao (1993a). The rates (1.42) are defined locally at each phase-space point. Accordingly, the local Lyapunov exponents are given by

$$\lambda_i(\mathbf{X}) = \lim_{t\to\infty} \frac{1}{t} \int_0^t \chi_i(\mathbf{\Phi}^\tau\mathbf{X})\,d\tau . \tag{1.43}$$

We remark that, for linear vector fields $\mathbf{F}(\mathbf{X}) = \mathsf{L} \cdot \mathbf{X}$, the first term of Eq. (1.42) directly provides the local Lyapunov exponents λ_i.

The homological decomposition (1.33) requires in general the knowledge of the vector fields $\{\mathbf{e}_i(\mathbf{X})\}_{i=1}^M$, which are not known except for the direction of the flow: $\mathbf{e}_j(\mathbf{X}) \propto \mathbf{F}(\mathbf{X})$. Thanks to the Poincaré surface of section, the homological decomposition of the flow can be constructed from a homological decomposition of the Poincaré mapping by propagating the vector fields from the surface of section into the bulk of phase space. In a surface of section, the different directions can be obtained locally by iterating the Poincaré map ϕ together with its Jacobian matrix either forward or backward in time. In this problem, the following equations are useful which give the vector fields as an eigenvalue problem:

$$(\partial_{\mathbf{X}}\mathbf{F}) \cdot \mathbf{e}_i - (\mathbf{F} \cdot \partial_{\mathbf{X}})\mathbf{e}_i = \chi_i\,\mathbf{e}_i , \tag{1.44}$$

$$\left(\partial_{\mathbf{X}}\mathbf{F}\right)^{\mathrm{T}} \cdot \mathbf{f}_i - (\mathbf{F} \cdot \partial_{\mathbf{X}})\mathbf{f}_i = \chi_i\,\mathbf{f}_i , \tag{1.45}$$

which hold at each point $\mathbf{X}$. Equation (1.44) can be rewritten as

$$[\mathbf{e}_i, \mathbf{F}] = \chi_i\,\mathbf{e}_i , \tag{1.46}$$

in terms of the Poisson bracket of two vector fields, $\mathbf{A}(\mathbf{X})$ and $\mathbf{B}(\mathbf{X})$, which is defined by a new vector field of components

$$[\mathbf{A},\mathbf{B}]_n = \sum_{m=1}^{M} \left(A_m \frac{\partial B_n}{\partial X_m} - B_m \frac{\partial A_n}{\partial X_m} \right) . \tag{1.47}$$

Accordingly, the vector fields $\{\mathbf{e}_i\}$ of the homological decomposition (1.33) of the fundamental matrix generate a Lie algebra (Arnold 1978). Let us remark that these vector fields can be rescaled according to: $\mathbf{e}_i(\mathbf{X}) \to \mathbf{e}_i(\mathbf{X}) \exp[\alpha(\mathbf{X})]$, under the condition that the scalar field $\alpha(\mathbf{X})$ is finite. The adjoint vector fields and the local stretching rates undergo a related rescaling, which leaves invariant the local Lyapunov exponents.

The linear stability analysis shows therefore that all of the Lyapunov exponents are defined in terms of the local stretching rates $\chi_i(\mathbf{X})$. Their sum

$$u(\mathbf{X}) = \sum_{\lambda_i>0} \chi_i(\mathbf{X}) , \tag{1.48}$$

defines at each phase-space point $\mathbf{X}$ the quantity we call the *local dispersion rate*. This quantity plays a central role in the following considerations. Let us note that some of the local stretching factors may be locally negative, although the corresponding local Lyapunov exponent is positive so that the sum should extend over all the positive Lyapunov exponents.

1.2.6 Stable and unstable manifolds

The *local stable and unstable manifolds* $W_s(\mathbf{X}_0)$ and $W_u(\mathbf{X}_0)$, respectively, associated with the point $\mathbf{X}_0$ of a trajectory, $\xi = \cup_{-\infty<t<+\infty} \mathbf{\Phi}^t \mathbf{X}_0$, play a central role in dynamical systems theory. These local manifolds are defined by

$$W_s(\mathbf{X}_0) = \{\mathbf{X} \in \mathcal{M} : \|\mathbf{\Phi}^t\mathbf{X} - \mathbf{\Phi}^t\mathbf{X}_0\| \to 0 \quad \text{for} \quad t \to +\infty\} \tag{1.49}$$

and

$$W_u(\mathbf{X}_0) = \{\mathbf{X} \in \mathcal{M} : \|\mathbf{\Phi}^t\mathbf{X} - \mathbf{\Phi}^t\mathbf{X}_0\| \to 0 \quad \text{for} \quad t \to -\infty\} \tag{1.50}$$

where $\|\cdot\|$ denotes a distance which we may take to be a Riemannian metric distance on the constant energy surface. The local stable and unstable manifolds are sets which extend globally in phase space. The local stable and unstable manifolds are tangent respectively to the stable and unstable directions given by the corresponding vector fields $\{\mathbf{e}_i(\mathbf{X}_0)\}$. The union of all the local stable or unstable manifolds of all the points of a trajectory $W_{s,u}(\xi) = \cup_{-\infty<t<+\infty} W_{s,u}(\mathbf{\Phi}^t\mathbf{X}_0)$ are invariant under the time evolution and are called the *global stable or unstable manifolds* or *invariant manifolds*.

Heteroclinic orbits are orbits belonging to the intersection between the stable and unstable manifolds of different trajectories: $W_s(\xi) \cap W_u(\xi')$. *Homoclinic orbits* are orbits belonging to the intersection of the stable and unstable manifolds of the same trajectory: $W_s(\xi) \cap W_u(\xi)$ (Poincaré 1892).

In the following chapter, we shall analyze the topology of trajectories in phase space and some consequences of hyperbolicity.

1.3 Linear stability of Hamiltonian systems

1.3.1 Symplectic dynamics and the pairing rule of the Lyapunov exponents

Since Hamiltonian systems (1.9) define symplectic systems, the fundamental matrix $\mathsf{M}(t, \mathbf{X})$ obeys the relations

$$\mathsf{M}^{\mathrm{T}} \cdot \boldsymbol{\Sigma} \cdot \mathsf{M} = \boldsymbol{\Sigma} \qquad \text{and} \qquad \mathsf{M} \cdot \boldsymbol{\Sigma} \cdot \mathsf{M}^{\mathrm{T}} = \boldsymbol{\Sigma} \, , \tag{1.51}$$

with the matrix $\boldsymbol{\Sigma}$ given by (1.10). As a result of these relations, if σ is an eigenvalue of $\mathsf{M}^{\mathrm{T}} \cdot \mathsf{M}$ then so is σ^{-1}. Accordingly, to each stable direction, there corresponds an unstable direction, and *vice versa.* Their Lyapunov exponents are respectively $-\lambda_i$ and $+\lambda_i$. Applying the conditions (1.51) to the homological decomposition (1.33), we obtain the so-called *pairing rule* according to which the vector fields of (1.33) and the associated local stretching rates (1.42) come in pairs

$$\{\chi(\mathbf{X}), \mathbf{e}(\mathbf{X}), \mathbf{f}(\mathbf{X})\} \;\leftrightarrow\; \{-\chi(\mathbf{X}), \boldsymbol{\Sigma} \cdot \mathbf{f}(\mathbf{X}), \boldsymbol{\Sigma} \cdot \mathbf{e}(\mathbf{X})\} \, , \tag{1.52}$$

with $\Lambda(t, \mathbf{X}) \leftrightarrow 1/\Lambda(t, \mathbf{X})$.

This pairing rule also implies that the sum of all of the Lyapunov exponents must be zero for conservative Hamiltonian systems and that phase-space volumes are preserved by the dynamics.

1.3.2 Vanishing Lyapunov exponents and the Lie group of continuous symmetries

Continuous symmetries have important consequences on the structure of linear stability in general systems. In Hamiltonian systems, continuous symmetries are known to imply the existence of constants of motion. Therefore, there is no sensitivity to initial conditions in some directions of the tangent space for which the Lyapunov exponents vanish. It turns out that the different constants of motion yield different vector fields $\mathbf{e}_j$ with corresponding $\chi_j = 0$, as we show below.

Continuous symmetries are defined by transformations $\mathbf{G}_\alpha(\mathbf{X})$ acting on the phase space $\mathscr{M}$ which depend on continuous parameters α_j. These symmetry transformations define a Lie group whose infinitesimal generators are vector fields obtained by differentiating the transformations with respect to the parameters

$$\mathbf{e}_j(\mathbf{X}) = \frac{\partial \mathbf{G}_0(\mathbf{X})}{\partial \alpha_j} . \tag{1.53}$$

Because the group transformations leave invariant the vector field (1.1) of the dynamical system, the vector fields (1.53) obey Eq. (1.44) with a vanishing local stretching factor $\chi_j(\mathbf{X}) = 0$

$$\frac{\partial \mathbf{F}}{\partial \mathbf{X}} \cdot \mathbf{e}_j - \frac{\partial \mathbf{e}_j}{\partial \mathbf{X}} \cdot \mathbf{F} = 0 . \tag{1.54}$$

Therefore, the corresponding Lyapunov exponents vanish.

On the other hand, Noether's theorem implies that constants of motions $\{C_j\}$ are associated with continuous symmetries in time-independent Hamiltonian systems (Arnold 1978). Their Poisson bracket with the Hamiltonian vanishes

$$\{H, C_j\} \equiv \frac{\partial H}{\partial \mathbf{X}} \cdot \mathbf{\Sigma} \cdot \frac{\partial C_j}{\partial \mathbf{X}} = 0 . \tag{1.55}$$

Differentiating Eq. (1.55) with respect to the phase-space variables $\mathbf{X}$, we obtain Eq. (1.54) if we do the identification

$$\mathbf{e}_j = \mathbf{\Sigma} \cdot \frac{\partial C_j}{\partial \mathbf{X}} . \tag{1.56}$$

This result shows how the constants of motion can be related to the vector fields of the homological decomposition (1.33). Comparing with (1.9), we notice that the constants of motion can be considered as Hamiltonians to generate the continuous symmetries taking for time the parameter α_j.

Moreover, the pairing rule implies the existence of another direction k associated with the direction j and with an adjoint vector given by the gradient fields of the constant of motion C_j

$$\mathbf{f}_k = \frac{\partial C_j}{\partial \mathbf{X}} . \tag{1.57}$$

The stretching factor of this direction also vanishes, $\chi_k = 0$.

A particular example of such constants of motion is the Hamiltonian itself which generates the flow and which corresponds to the symmetry under time translations. Consequently, two Lyapunov exponents always vanish in Hamiltonian systems: the one associated with the direction of the flow $\mathbf{F} = \mathbf{\Sigma} \cdot \partial_{\mathbf{X}} H$, as is the case for all continuous-time systems, and the one associated with the direction $\partial_{\mathbf{X}} H$ outside the energy shell $H = E$.

When the group of continuous symmetries of the system has several parameters, we have to choose mutually independent constants of motion $\{C_j\}_{j=1}^{c}$ where the first one is the Hamiltonian itself $C_1 = H$. The number of vanishing Lyapunov exponents is then equal to $2c$. For instance, in the case of a rotationally invariant system, the three components of angular momentum (L_x, L_y, L_z) are conserved. However, the three components of angular momentum are not independent since $\{L_x, L_y\} = L_z$. Therefore, we must consider for instance the three constants of motion H, L^2, and L_z which are independent since $\{H, L^2\} = \{H, L_z\} = \{L^2, L_z\} = 0$. Accordingly, there are three independent vector fields (1.56). Moreover, three other vector fields with vanishing Lyapunov exponents may be obtained by the pairing rule. Consequently, the number of vanishing Lyapunov exponents is equal to 6.

This discussion also shows that the relations (1.44)–(1.47) between the vector fields of the homological decomposition extends the Lie-group structure of the constants of motion into a larger structure including the directions with nonvanishing Lyapunov exponents.

1.3.3 Hamilton–Jacobi equation and the curvature of the wavefront

It is well-known that Hamilton's equations of motion can be deduced from a variational principle based on the action (Goldstein 1950; Arnold 1978, 1988)

$$W(\mathbf{q}, \mathbf{q}_0, T) = \int_0^T \mathbf{p}\cdot d\mathbf{q} - H dt . \tag{1.58}$$

The equations of motion result from the first variation of the action: $\delta W = 0$. On the other hand, the properties of linear stability can be studied at the level of the second variation of the action, $\delta^2 W$, which defines a Sturm–Liouville operator of second order in time called the Jacobi–Hill operator (DeWitt-Morette et al. 1979, Gaspard et al. 1995b). The solutions of this Jacobi–Hill operator are known as the Jacobi fields which are vector fields in the projection of the tangent space onto the position space. The Jacobi fields provide detailed knowledge of the properties of linear stability of a periodic orbit, which is essentially equivalent to the knowledge of the vector fields of the homological decomposition. The Jacobi fields allow the definition of a topological property of trajectories like the famous Morse index characterizing the winding of nearby trajectories around a given trajectory. We refer the reader to the literature concerning this question (Gutzwiller 1990, Arnold and Novikov 1990).

Here, we would like to obtain a method to calculate the local dispersion rate (1.48) in Hamiltonian systems. Our starting point is the Hamilton–Jacobi equation (Arnold 1978)

$$\partial_t W + H\left(\mathbf{q}, \partial_{\mathbf{q}} W\right) = 0 . \tag{1.59}$$

which rules the time evolution of a wavefront formed by the projection onto the position space of the current points of a family of trajectories. This wavefront is given in terms of the action (1.58) by the surface

$$W(\mathbf{q}, \mathbf{q}_0; t) = \text{constant}, \tag{1.60}$$

defined in position space $\mathbf{q}$. The normal to the wavefront is the momentum

$$\mathbf{p} = \partial_{\mathbf{q}} W, \tag{1.61}$$

of the trajectory passing by the current position $\mathbf{q}$.

The linear stability of the trajectory can be characterized by the variation of the momentum when the current position is perturbed on the front. This variation defines the curvature of the wavefront in terms of the matrix C as

$$\delta\mathbf{p} = \frac{\partial^2 W}{\partial \mathbf{q}^2} \cdot \delta\mathbf{q} = \mathsf{C}(t) \cdot \delta\mathbf{q}, \tag{1.62}$$

which is the second fundamental form of the wavefront up to a scaling factor (Vattay 1994). The trajectories of the wavefront are unstable if the momentum is diverging, i.e., if the wavefront appears concave with respect to the initial conditions. Such a dynamical instability can be characterized by looking at the time evolution of an infinitesimal volume sustained by perturbations $\delta\mathbf{q}_t$ on the current wavefront, issued from perturbations $\delta\mathbf{q}_0$ on the initial wavefront. Since the perturbation $\delta\mathbf{p}_0$ on the initial momentum is related to the perturbation on the initial position $\delta\mathbf{q}_0$ by

$$\delta\mathbf{p}_0 = \partial^2_{\mathbf{q}} W(\mathbf{q}, \mathbf{q}_0; t=0) \cdot \delta\mathbf{q}_0, \tag{1.63}$$

and because the initial momentum is given by

$$\mathbf{p}_0 = -\partial_{\mathbf{q}_0} W, \tag{1.64}$$

we find that the ratio between the final and the initial volumes is

$$|\det \frac{\partial \mathbf{q}_t}{\partial \mathbf{q}_0}| = \frac{|\det \frac{\partial \mathbf{p}_0}{\partial \mathbf{q}_0}|}{|\det \frac{\partial \mathbf{p}_0}{\partial \mathbf{q}_t}|} = \frac{|\det \mathsf{C}(0)|}{|\det \mathsf{D}(t)|}, \tag{1.65}$$

in terms of the Jacobian matrix

$$\mathsf{D} = \frac{\partial \mathbf{p}_0}{\partial \mathbf{q}} = -\frac{\partial^2 W}{\partial \mathbf{q}_0 \partial \mathbf{q}} \quad \text{or} \quad D_{ij} = -\frac{\partial^2 W}{\partial q_{0i} \partial q_j}. \tag{1.66}$$

The determinant of the Jacobian matrix (1.66) is known as the Morette–Van Hove determinant

$$D(\mathbf{q}, \mathbf{q}_0, t) = \det \frac{\partial \mathbf{p}_0}{\partial \mathbf{q}} = \det\left(-\frac{\partial^2 W}{\partial \mathbf{q} \partial \mathbf{q}_0}\right), \tag{1.67}$$

(Morette 1951, Van Hove 1951, Choquard and Steiner 1996). The absolute value of this determinant may be interpreted as the probability density in position space of a statistical ensemble of trajectories with uniformly distributed initial momenta $\mathbf{p}_0$ according to the Lebesgue measure $d^f p_0$, because $d^f p_0 = |D| d^f q$. The Morette–Van Hove determinant obeys the continuity equation

$$\partial_t D + \partial_{\mathbf{q}} \cdot (D\dot{\mathbf{q}}) = 0 , \tag{1.68}$$

where $\dot{\mathbf{q}} = \partial_{\mathbf{p}} H(\mathbf{q}, \partial_{\mathbf{q}} W)$, which guarantees the conservation of probability in position space. Introducing the total time derivative along a trajectory as

$$\frac{d}{dt} = \partial_t + \dot{\mathbf{q}} \cdot \partial_{\mathbf{q}} , \tag{1.69}$$

the Morette–Van Hove determinant varies along a trajectory according to

$$\frac{d}{dt} \ln D = - \operatorname{tr} \left(\frac{\partial^2 H}{\partial \mathbf{q} \partial \mathbf{p}} + \frac{\partial^2 H}{\partial \mathbf{p}^2} \cdot \mathsf{C} \right) , \tag{1.70}$$

in terms of the matrix (1.62) of the second fundamental form of the wavefront. Equation (1.70) shows that the time evolution of the Morette–Van Hove determinant is known if we find the equation of motion for the matrix $\mathsf{C}(t)$. The time evolution of both the curvature matrix (1.62) and the Jacobian matrix (1.66) is obtained by differentiation of the Hamilton–Jacobi equation, which leads to

$$\dot{\mathsf{C}} = - \mathsf{C} \cdot \frac{\partial^2 H}{\partial \mathbf{p}^2} \cdot \mathsf{C} - \mathsf{C} \cdot \frac{\partial^2 H}{\partial \mathbf{p} \partial \mathbf{q}} - \frac{\partial^2 H}{\partial \mathbf{q} \partial \mathbf{p}} \cdot \mathsf{C} - \frac{\partial^2 H}{\partial \mathbf{q}^2} , \tag{1.71}$$

$$\dot{\mathsf{D}} = - \mathsf{D} \cdot \frac{\partial^2 H}{\partial \mathbf{p}^2} \cdot \mathsf{C} - \mathsf{D} \cdot \frac{\partial^2 H}{\partial \mathbf{p} \partial \mathbf{q}} . \tag{1.72}$$

Equation (1.71) for the wavefront curvature is an equation of Ricatti type which may diverge like $\mathsf{C}(t) \sim (t - t_c)^{-1}$ at the so-called caustic points where the Morette–Van Hove determinant diverges. This divergence requires some regularization in order to integrate Eq. (1.71). If this solution is known, then the volume ratio (1.65) is given by

$$|\det \frac{\partial \mathbf{q}_t}{\partial \mathbf{q}_0}| = |D(0)^{-1} \det \mathsf{C}(0)| \exp \int_0^t \operatorname{tr} \left(\frac{\partial^2 H}{\partial \mathbf{q} \partial \mathbf{p}} + \frac{\partial^2 H}{\partial \mathbf{p}^2} \cdot \mathsf{C} \right) d\tau . \tag{1.73}$$

Equation (1.73) allows us to identify the local dispersion rate corresponding to the positive and some vanishing Lyapunov exponents as

$$u(\mathbf{q}, \mathbf{p}) = \sum_{\lambda_i \geq 0} \chi_i(\mathbf{q}, \mathbf{p}) = \operatorname{tr} \left[\frac{\partial^2 H}{\partial \mathbf{q} \partial \mathbf{p}} + \frac{\partial^2 H}{\partial \mathbf{p}^2} \cdot \mathsf{C}(\mathbf{q}, \mathbf{p}) \right] , \tag{1.74}$$

where the curvature matrix C is evaluated over the past half-trajectory passing by the current point $(\mathbf{q}, \mathbf{p})$ with an initial condition in the remote past: $\mathbf{\Phi}^t(\mathbf{q}, \mathbf{p})$ with $t \in] - \infty, 0]$.

Dynamical instability arises if the dispersion rate is positive everywhere in phase space. However, in most systems, the dispersion rate is positive only on average, in which cases dynamical instability originates from a mechanism of parametric instability (Pettini 1993).

We notice that, contrary to Eq. (1.48), the local stretching rates coming from some vanishing Lyapunov exponents may contribute to the local dispersion rate (1.74) because we have not eliminated the neutral directions which are either parallel to the flow or perpendicular to the energy shell. This elimination can be carried out as followed.

1.3.4 Elimination of the neutral directions

A change of coordinates can be carried out in order to eliminate the neutral directions corresponding to vanishing Lyapunov exponents, which reduces the fundamental matrix from the size $2f \times 2f$ to the size $(2f-2)\times(2f-2)$ (Eckhardt and Wintgen 1991). The smaller fundamental matrix is only concerned with linear perturbations in some Poincaré surface of section in a given energy shell and is called the *monodromy matrix*. This change of coordinates has to be specified only by the matrix transforming the tangent vectors from the old coordinates $\delta\mathbf{X} = (\delta\mathbf{q}, \delta\mathbf{p})$ to the tangent vector $\boldsymbol{\varepsilon}$ in the new coordinates:

$$\delta\mathbf{X}(t) = \mathsf{M}(t)\cdot\delta\mathbf{X}(0) = \mathsf{R}(t)\cdot\boldsymbol{\varepsilon}(t) = \mathsf{M}(t)\cdot\mathsf{R}(0)\cdot\boldsymbol{\varepsilon}(0)\,. \tag{1.75}$$

The matrix R is composed of the known vectors $\mathbf{e}_j$ associated with the vanishing Lyapunov exponents plus extra vectors to fill the matrix. These vectors must be linearly independent. The transformation (1.75) is nothing other than a Lyapunov homology (1.31). If the matrix R is invertible and nonsingular the Lyapunov exponents are unchanged under such a transformation.

In the new coordinates, the tangent vector evolves in time according to

$$\boldsymbol{\varepsilon}(t) = \mathsf{R}^{-1}(t)\cdot\mathsf{M}(t)\cdot\mathsf{R}(0)\cdot\boldsymbol{\varepsilon}(0)\,. \tag{1.76}$$

The new vector components are thus governed by the equation

$$\dot{\boldsymbol{\varepsilon}} = \mathsf{L}'\cdot\boldsymbol{\varepsilon}\,, \tag{1.77}$$

where

$$\mathsf{L}' = \mathsf{R}^{-1}\cdot(\mathsf{L}\cdot\mathsf{R} - \dot{\mathsf{R}})\,, \tag{1.78}$$

with the notation $\mathsf{L} = \partial_{\mathbf{X}}\mathbf{F}(\mathbf{X}_t)$ for the Jacobian of the vector field (1.1).

If we exploit only energy conservation, the first vectors may be chosen as: $\mathsf{R} = (\mathbf{F}, \boldsymbol{\Sigma}\cdot\mathbf{F}, \cdots)$, where $\mathbf{F}$ denotes the vector field of the flow. In this case, the components of the tangent vector are $\boldsymbol{\varepsilon} = (\varepsilon_\parallel, \varepsilon_E, \tilde{\boldsymbol{\varepsilon}})$, where $\varepsilon_\parallel$ is the component parallel to the flow while ε_E is the component outside the energy shell. The linearized matrix (1.78) has then the form

$$\mathsf{L}' = \begin{pmatrix} 0 & * & * & \cdots & * \\ 0 & 0 & 0 & \cdots & 0 \\ 0 & * & & & \\ \vdots & \vdots & & \mathsf{l} & \\ 0 & * & & & \end{pmatrix} , \tag{1.79}$$

where the stars denote nonvanishing elements which do not need to be specified. The form of the matrix (1.79) implies that, if the initial perturbation is within the energy shell, i.e., $\varepsilon_E(0) = 0$, it remains there for all times, $\varepsilon_E(t) = 0$. Therefore, the remaining components $\tilde{\boldsymbol{\varepsilon}}$ evolve in time independently of the two first components. This reasoning generalizes to the case of several constants of motion. We can introduce a reduced fundamental matrix m associated with the $2(f-1)$ remaining components which obeys

$$\dot{\mathsf{m}} = \mathsf{l} \cdot \mathsf{m} , \tag{1.80}$$

and which defines the monodromy matrix. The nontrivial Lyapunov exponents are thus given by

$$\lambda(\mathbf{X}, \tilde{\boldsymbol{\varepsilon}}) = \lim_{t\to\infty} \frac{1}{t} \ln \|\mathsf{m}(t, \mathbf{X}) \cdot \tilde{\boldsymbol{\varepsilon}}\| . \tag{1.81}$$

We notice that the monodromy matrix $\mathsf{m}(t, \mathbf{X})$ of a trajectory between two passages in a Poincaré surface of section is related to the Jacobian matrix $\partial_{\mathbf{x}}\boldsymbol{\phi}$ of the Poincaré map (1.5) in a given energy shell.

1.3.5 The local stretching rate in $f = 2$ hyperbolic Hamiltonian systems

For Hamiltonian systems $H = \mathbf{p}^2/2 + V$ with two degrees of freedom ($f = 2$), Eq. (1.12) for the time evolution of the tangent vectors writes

$$\delta\dot{\mathbf{X}} = \begin{pmatrix} 0 & 0 & 1 & 0 \\ 0 & 0 & 0 & 1 \\ -v_{11} & -v_{12} & 0 & 0 \\ -v_{21} & -v_{22} & 0 & 0 \end{pmatrix} \cdot \delta\mathbf{X} , \text{ with } \delta\mathbf{X} = \begin{pmatrix} \delta q_1 \\ \delta q_2 \\ \delta p_1 \\ \delta p_2 \end{pmatrix} , \tag{1.82}$$

with the notation $v_{ij} = \partial^2 V/\partial q_i \partial q_j$.

The tangent vector can be decomposed onto a set of four mutually orthogonal vectors according to

$$\delta\mathbf{X}(t) = \varepsilon_{\|}\mathbf{e}_{\|} + \varepsilon_E \mathbf{e}_E + \varepsilon_q \mathbf{e}_q + \varepsilon_p \mathbf{e}_p , \tag{1.83}$$

using the following matrix

$$\mathsf{R} = \begin{pmatrix} \dot{q}_1 & -\dot{p}_1/r^2 & -\dot{q}_2/r & -\dot{p}_2/r \\ \dot{q}_2 & -\dot{p}_2/r^2 & \dot{q}_1/r & \dot{p}_1/r \\ \dot{p}_1 & \dot{q}_1/r^2 & \dot{p}_2/r & -\dot{q}_2/r \\ \dot{p}_2 & \dot{q}_2/r^2 & -\dot{p}_1/r & \dot{q}_1/r \end{pmatrix}, \tag{1.84}$$

with $r = \|\dot{\mathbf{X}}\| = (\dot{q}_1^2 + \dot{q}_2^2 + \dot{p}_1^2 + \dot{p}_2^2)^{1/2}$ (Eckhardt and Wintgen 1991). The first column is the vector $\mathbf{e}_\parallel$ which is parallel to the flow while the second column is the vector $\mathbf{e}_E$ which points outside the energy shell and is proportional to the matrix $\mathbf{\Sigma}$ applied to the first column. The third column is chosen as a vector $\mathbf{e}_q$ to be perpendicular to the first column without exchanging positions with momenta. Similarly, the fourth column $\mathbf{e}_p$ is proportional to $\mathbf{\Sigma}$ applied to the second column. The matrix of the linearized system is here

$$\mathsf{L}' = \begin{pmatrix} 0 & e & -d & f \\ 0 & 0 & 0 & 0 \\ 0 & f & b & c \\ 0 & d & -a & -b \end{pmatrix}, \tag{1.85}$$

where the elements are obtained according to Eq. (1.78) by a symbolic manipulation program

$$a = \left[(\dot{q}_1^2 + \dot{q}_2^2)(v_{11} + v_{22}) + 2(\dot{p}_1^2 + \dot{p}_2^2)\right]/r^2, \tag{1.86}$$

$$b = \left[\dot{q}_1\dot{p}_1(v_{11} - 1) + \dot{q}_2\dot{p}_2(v_{22} - 1) - (\dot{q}_1\dot{p}_2 + \dot{q}_2\dot{p}_1)v_{12}\right]/r^2, \tag{1.87}$$

$$c = \Big[(\dot{q}_1^2 + \dot{p}_2^2)(v_{11} + 1) + (\dot{q}_2^2 + \dot{p}_1^2)(v_{22} + 1) + 2(\dot{q}_1\dot{q}_2 - \dot{p}_1\dot{p}_2)v_{12}\Big]/r^2, \tag{1.88}$$

$$d = (\dot{q}_1\dot{p}_2 - \dot{q}_2\dot{p}_1)(v_{11} + v_{22} - 2)/r^{3/2}, \tag{1.89}$$

$$e = \Big[(\dot{q}_1^2 - \dot{p}_1^2)(1 - v_{11}) + (\dot{q}_2^2 - \dot{p}_2^2)(1 - v_{22}) + 2(\dot{p}_1\dot{p}_2 - \dot{q}_1\dot{q}_2)v_{12}\Big]/r^2, \tag{1.90}$$

$$f = \Big[(\dot{q}_1\dot{q}_2 + \dot{p}_1\dot{p}_2)(v_{11} - v_{22}) - (\dot{q}_1^2 + \dot{p}_1^2 - \dot{q}_2^2 - \dot{p}_2^2)v_{12}\Big]/r^{3/2}. \tag{1.91}$$

When $\varepsilon_E = 0$, we observe from the matrix (1.85) that the components ε_q and ε_p form a closed system. The component $\varepsilon_\parallel$ is driven by the three others so that it remains constant if the three other components are initially set to zero, as expected for the direction parallel to the flow. The nontrivial Lyapunov exponents are therefore determined by the behaviour of the two components ε_q

and ε_p so that the linear stability problem is reduced to the integration of the 2×2 matrix system (1.80) with

$$\mathsf{l} = \begin{pmatrix} b & c \\ -a & -b \end{pmatrix} , \tag{1.92}$$

along the orbit.

A further reduction is possible to obtain the unique positive Lyapunov exponent in terms of the integral

$$\lambda(\mathbf{X}) = \lim_{T\to\infty} \frac{1}{T} \int_0^T \chi(\mathbf{\Phi}^t\mathbf{X})\, dt , \tag{1.93}$$

where χ is the corresponding local stretching rate. Indeed, in an unstable system, a tangent vector like $\tilde{\boldsymbol{\varepsilon}}$ solution of

$$\dot{\tilde{\boldsymbol{\varepsilon}}} = \mathsf{l} \cdot \tilde{\boldsymbol{\varepsilon}} , \tag{1.94}$$

tends to increase under forward time evolution at a rate equal to the positive Lyapunov exponent. The increase occurs for almost all initial conditions except those taken in the contracting direction of the tangent space. It is thus reasonable to assume that a tangent vector of the form

$$\tilde{\boldsymbol{\varepsilon}} = \begin{pmatrix} \chi \\ 1 \end{pmatrix} \varepsilon , \tag{1.95}$$

would not fall in the contracting direction which is only a line in the two-dimensional tangent space. Accordingly, we should expect that the coefficient χ will increase exponentially at the average rate of the positive Lyapunov exponent. Therefore, we can identify the coefficient χ as the local dispersion rate. Inserting the assumption (1.95) in Eq. (1.94) with the matrix (1.92) and eliminating ε , we obtain the Ricatti-type equation

$$\dot{\chi} = a\,\chi^2 + 2\,b\,\chi + c , \tag{1.96}$$

in terms of the coefficients (1.86)–(1.88).[1] By this method, the positive Lyapunov exponent can be practically obtained in Hamiltonian systems with two degrees of freedom. A similar method has been proposed by Pollner and Vattay (1996).

1.4 Linear stability in billiards

After presenting the theory of linear stability for a general system, we shall specialize to the case of conservative billiards defined by the free Hamiltonian

1. Singularities like $\chi(t) \sim (t - t_c)^{-1}$ could be avoided in the integration of such Ricatti-type equations by integration with respect to a regularized time defined by $d\tau = (\chi^2 + 1)dt$ where the term in χ^2 avoids a problem at $|\chi| = \infty$ and the term in 1 at $\chi = 0$. We remark that the integration of the system (1.80) with (1.92) does not require this regularization.

$H = (m/2)\mathbf{v}^2$ and the collision rule (1.11) on the border $\partial\mathcal{Q}$ of the domain of variation of the position $\mathbf{q} \in \mathcal{Q}$. For a point particle in elastic collision on a sphere of radius a, this domain is determined by the condition that $\|\mathbf{q}\| \geq a$. In general, the border of the billiard has for its equation $\sigma(\mathbf{q}) = 0$ with $\sigma(\mathbf{q}) > 0$ for $\mathbf{q} \in \mathcal{Q}$.

1.4.1 Some definitions

At each point of the border, a vector is defined which is perpendicular to the border and interior to the domain: $\mathbf{n}(\mathbf{q}) = \partial_{\mathbf{q}}\sigma/\|\partial_{\mathbf{q}}\sigma\|$. We denote by $\mathfrak{I}$ the subspace which is tangent to the border $\partial\mathcal{Q}$ and thus perpendicular to $\mathbf{n}(\mathbf{q})$. For the following discussion, we also need to define the linear subspaces $\mathfrak{I}^{(\pm)}$ which are perpendicular to the velocities $\mathbf{v}^{(\pm)}$ before and after each collision. Following Sinai (1979) (see also Sinai and Chernov 1987; Chernov 1991, 1994), we introduce the transformations:

(a) The projection of $\mathfrak{I}^{(+)}$ onto $\mathfrak{I}^{(-)}$, parallel to $\mathbf{n}$,

$$\mathsf{U}: \quad \mathfrak{I}^{(+)} \to \mathfrak{I}^{(-)}, \qquad \text{given by} \qquad \mathsf{U} = 1 - 2\,\mathbf{n}\,\mathbf{n}^{\mathrm{T}}, \tag{1.97}$$

which takes a vector in the hyperplane $\mathfrak{I}^{(+)}$ and projects it onto the hyperplane $\mathfrak{I}^{(-)}$ such that the difference between the two vectors is parallel to $\mathbf{n}$, and that the components of the two vectors in the directions perpendicular to $\mathbf{n}$ are equal. The superscript T denotes a transposed vector (or operator) and 1 is a unit operator. The transformation U is an invertible, orthogonal transformation given by an orthogonal matrix in the f-dimensional position space

$$\mathsf{U} = \mathsf{U}^{\mathrm{T}} = \mathsf{U}^{-1} = 1 - 2\,\mathbf{n}\,\mathbf{n}^{\mathrm{T}}, \tag{1.98}$$

with determinant $\det \mathsf{U} = -1$.

(b) The projection of $\mathfrak{I}^{(-)}$ onto $\mathfrak{I}$ parallel to $\mathbf{v}^{(-)}$

$$\mathsf{V}: \quad \mathfrak{I}^{(-)} \to \mathfrak{I}, \tag{1.99}$$

is given by the matrix

$$\mathsf{V} = 1 - \frac{\mathbf{v}^{(-)}\,\mathbf{n}^{\mathrm{T}}}{\mathbf{v}^{(-)}\cdot\mathbf{n}}\,. \tag{1.100}$$

(c) The projection of $\mathfrak{I}$ onto $\mathfrak{I}^{(-)}$, parallel to $\mathbf{n}$

$$\mathsf{V}^{\mathrm{T}}: \mathfrak{I} \to \mathfrak{I}^{(-)} \tag{1.101}$$

is given by the transpose of the matrix V, Eq. (1.100).

1.4.2 The second fundamental form

The convexity of the hypersurface $\sigma(\mathbf{q}) = 0$ determines the defocusing character of the collisions. The convexity is characterized by an operator known as the second fundamental form,[2] and which gives the variation of the normal vector with respect to a variation of a point on the hypersurface. That is, we define the second fundamental form $\mathsf{K}(\mathbf{q})$ by

$$\mathbf{n}(\mathbf{q}+\delta\mathbf{q}) - \mathbf{n}(\mathbf{q}) = \mathsf{K}(\mathbf{q})\cdot\delta\mathbf{q} \tag{1.102}$$

where both $\mathbf{q}$ and $\mathbf{q}+\delta\mathbf{q}$ belong to the hypersurface. Consequently, the variation $\delta\mathbf{q}$ is perpendicular to the normal, $\mathbf{n}\cdot\delta\mathbf{q} = 0$. Similarly, $\mathbf{n}^2 = 1$ implies that $\mathbf{n}\cdot\delta\mathbf{n} = 0$. As a result, the second fundamental form maps vectors, $\delta\mathbf{q}$, in the tangent plane onto vectors $\delta\mathbf{n}$ in the tangent plane,

$$\mathsf{K} : \mathfrak{I} \to \mathfrak{I}, \tag{1.103}$$

which is symmetric. In order to see the meaning of the sign of the second fundamental form consider the inner product

$$\delta\mathbf{q}^{\mathrm{T}} \cdot \delta\mathbf{n} \;=\; \delta\mathbf{q}^{\mathrm{T}} \cdot \mathsf{K} \cdot \delta\mathbf{q}\,. \tag{1.104}$$

For surfaces that are defocusing, this inner product satisfies $\delta\mathbf{q}^{\mathrm{T}}\cdot\delta\mathbf{n} > 0$, while for focusing surfaces $\delta\mathbf{q}^{\mathrm{T}}\cdot\delta\mathbf{n} < 0$, as may be immediately checked by considering a sphere in three dimensions and constructing this inner product for both outward ($\delta\mathbf{q}^{\mathrm{T}}\cdot\delta\mathbf{n} > 0$) and inward ($\delta\mathbf{q}^{\mathrm{T}}\cdot\delta\mathbf{n} < 0$) normals. In the case of outward normals, $\mathsf{K} = \frac{1}{a}\,\mathsf{1}_{\perp}$, where a is the radius of the sphere, $\mathsf{1}_{\perp} = \mathbf{e}_\theta\mathbf{e}_\theta^{\mathrm{T}} + \mathbf{e}_\phi\mathbf{e}_\phi^{\mathrm{T}}$ is the unit operator in the tangent plane to the sphere at point $\mathbf{q}$, and $\mathbf{e}_\theta$ and $\mathbf{e}_\phi$ are mutually orthogonal unit vectors in this tangent plane.

If the operator (1.103) is nonnegative, $\mathsf{K} \geq 0$, the collisions are thus defocusing or neutral, and they may lead to a dynamical instability. In the case that $\mathsf{K} \leq 0$, the collisions are focusing or neutral, but, as illustrated by the stadium billiard (Bunimovich 1979), the collisions do not necessarily lead to a dynamically stable situation because an excess of focusing can also induce a dynamical instability by a mechanism similar to the aforementioned parametric instability (Pettini 1993).

1.4.3 The tangent space of the billiard

To develop a method to compute dynamical quantities for a hard-sphere gas we need to consider infinitesimal perturbations of some reference trajectory. The

2. The first fundamental form determines the Riemannian distance between two nearby points on the surface (Kreyszig 1991).

perturbations are best represented in a local frame of coordinates with one axis parallel to the velocity $\mathbf{v}$

$$\delta\mathbf{q} = \delta q_{\parallel}\mathbf{e}_{\parallel} + \delta q_{\perp 1}\mathbf{e}_{\perp 1} + \delta q_{\perp 2}\mathbf{e}_{\perp 2} + \cdots + \delta q_{\perp f-1}\mathbf{e}_{\perp f-1} \tag{1.105}$$

where $\mathbf{e}_{\parallel} = \mathbf{v}$ with $\mathbf{v}^2 = 1$. An equation similar to Eq. (1.105) is obtained for the infinitesimal variation in velocity with respect to the reference trajectory. We remark that the condition $\mathbf{v}^2 = 1$ implies that the velocity perturbation is necessarily perpendicular to $\mathbf{v}$: $\mathbf{v}\cdot\delta\mathbf{v} = 0$. Therefore we have $\delta v_{\parallel} = 0$. Similarly we can set $\delta q_{\parallel} = 0$, otherwise the perturbation can be assigned to another position along the trajectory. Consequently, the infinitesimal perturbation belongs to a linear space of dimension $2(f-1)$.

Our purpose is to obtain the time evolution of the infinitesimal perturbation over a time interval T in order to define the monodromy matrix $\mathsf{m}(T)$ as

$$\begin{pmatrix} \delta\mathbf{q}_{\perp}(T) \\ \delta\mathbf{v}_{\perp}(T) \end{pmatrix} = \mathsf{m}(T)\cdot \begin{pmatrix} \delta\mathbf{q}_{\perp}(0) \\ \delta\mathbf{v}_{\perp}(0) \end{pmatrix} . \tag{1.106}$$

The time evolution is composed of collisions and free flights between the collisions. We shall first determine the monodromy matrix for a free flight and then the matrix for a collision.

1.4.4 Free flight

During a free flight the velocity vector is constant although the position changes according to

$$\mathbf{q}(t) = \mathbf{v}(0)\, t + \mathbf{q}(0) , \tag{1.107}$$

where $[\mathbf{q}(0), \mathbf{v}(0)]$ are the position and velocity after the previous collision. As a direct consequence, before the next collision

$$\begin{pmatrix} \delta\mathbf{q}_{\perp}(t) \\ \delta\mathbf{v}_{\perp}(t) \end{pmatrix} = \begin{pmatrix} \delta\mathbf{q}_{\perp}(0) + t\,\delta\mathbf{v}_{\perp}(0) \\ \delta\mathbf{v}_{\perp}(0) \end{pmatrix} , \tag{1.108}$$

where t is the time of flight between two consecutive collisions, which is also equal to the distance between impact points since the velocity $\mathbf{v}$ is normalized to unity. The monodromy matrix for a free flight is therefore

$$\mathsf{m}_{\text{free flight}} = \begin{pmatrix} \mathsf{1} & t\,\mathsf{1} \\ \mathsf{0} & \mathsf{1} \end{pmatrix} . \tag{1.109}$$

1.4.5 Collision

If the collision illustrated in Fig. 1.1 is perturbed, the position and velocity just before the collision are displaced by $(\delta\mathbf{q}_\perp^{(-)}, \delta\mathbf{v}_\perp^{(-)})$ which both belong to the subspace $\mathfrak{I}^{(-)}$ perpendicular to the incident velocity $\mathbf{v}^{(-)}$. As a consequence, the collision of the perturbed trajectory does not occur at the collision point $\mathbf{q}$ of the reference trajectory but at the nearby point $\mathbf{q}+\delta\mathbf{q}$ on the hypersurface. We note that the perturbation $\delta\mathbf{q}$ of the impact point belongs to the tangent subspace $\mathfrak{I}$ and is determined from the perturbed position $\delta\mathbf{q}_\perp^{(-)}$ by the projection from $\mathfrak{I}^{(-)}$ onto $\mathfrak{I}$ parallel to the incident velocity $\mathbf{v}^{(-)}$, that is by Eq. (1.100)

$$\delta\mathbf{q} = \mathsf{V}\cdot\delta\mathbf{q}_\perp^{(-)}. \tag{1.110}$$

The perturbed trajectory after the collision is issued from the point $\mathbf{q}+\delta\mathbf{q}$ with a velocity $\mathbf{v}^{(+)}+\delta\mathbf{v}^{(+)}$ given by the collision rule Eq. (1.11). The intersection of this trajectory with the tangent space $\mathfrak{I}$ defines the perturbation in position $\delta\mathbf{q}_\perp^{(+)}$ of the outgoing trajectory. Therefore the outgoing perturbation is given by the projection of the perturbation Eq. (1.110) of the impact point from the tangent subspace $\mathfrak{I}$ onto the subspace $\mathfrak{I}^{(+)}$ parallel to the outgoing velocity $\mathbf{v}^{(+)}$. In Fig. 1.1, we observe that the composition of the two successive projections, $\mathfrak{I}^{(-)}\to\mathfrak{I}\to\mathfrak{I}^{(+)}$ is the projection from $\mathfrak{I}^{(-)}$ onto $\mathfrak{I}^{(+)}$ parallel to the normal vector $\mathbf{n}$ which is the inverse U^{-1} of the isometry Eqs. (1.97)–(1.98),

$$\delta\mathbf{q}_\perp^{(+)} = \mathsf{U}^{-1}\cdot\delta\mathbf{q}_\perp^{(-)}. \tag{1.111}$$

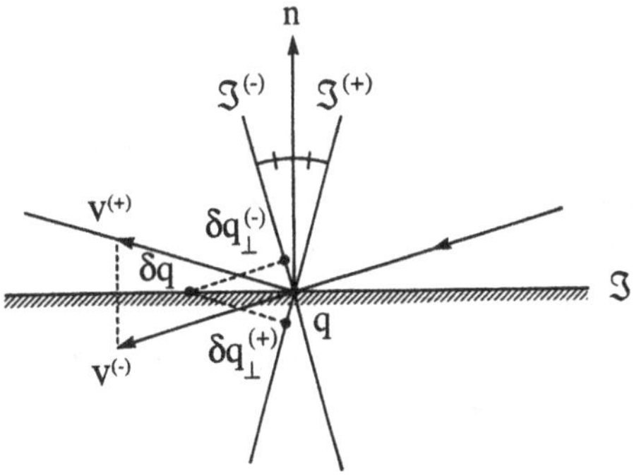

Figure 1.1 Geometry of an elastic collision. $\mathbf{n}$ is the vector normal to the hypersurface of the billiard. $\mathbf{v}^{(\pm)}$ are respectively the velocities after and before the collision. $\mathfrak{I}$ is the tangent subspace. $\mathfrak{I}^{(\pm)}$ are the subspaces perpendicular to the outgoing and ingoing velocities. $\delta\mathbf{q}^{(\pm)}$ are the infinitesimal perturbations in position after and before collision. $\delta\mathbf{q}$ is the infinitesimal perturbation of the impact point $\mathbf{q}$. The curvature of the collision hypersurface is not represented for the following reason. The hypersurface curvature plays a role only in the infinitesimal perturbations in velocities $\delta\mathbf{v}^{(\pm)}$ which are straightforwardly derived from the collision rule (1.11) by analysis.

At the perturbed impact point, the normal vector is no longer identical to $\mathbf{n}$ but is perturbed as determined by the second fundamental form, Eq. (1.102)–(1.103). Accordingly, the velocity perturbation after the collision can be obtained in a straightforward way by differentiating the collision rule Eq. (1.11)

$$\begin{aligned}\delta\mathbf{v}_\perp^{(+)} &= \delta\mathbf{v}_\perp^{(-)} - 2(\mathbf{n}\cdot\delta\mathbf{v}_\perp^{(-)})\mathbf{n} \\ &\quad - 2(\mathbf{n}\cdot\mathbf{v}^{(-)})\mathsf{K}\cdot\delta\mathbf{q} - 2(\mathbf{v}^{(-)}\cdot\mathsf{K}\cdot\delta\mathbf{q})\mathbf{n}\end{aligned} \tag{1.112}$$

where $\delta\mathbf{q}$ is the perturbation of the impact point, Eq. (1.110). Defining the velocity angle ϕ by

$$\cos\phi = \mathbf{n}\cdot\mathbf{v}^{(+)} = -\mathbf{n}\cdot\mathbf{v}^{(-)} \tag{1.113}$$

and using Eqs. (1.98) and (1.110), we find that Eq. (1.112) becomes

$$\delta\mathbf{v}_\perp^{(+)} = \mathsf{U}^{-1}\cdot\left[\delta\mathbf{v}_\perp^{(-)} + (2\cos\phi)\,\mathsf{V}^{\mathrm{T}}\cdot\mathsf{K}\cdot\mathsf{V}\cdot\delta\mathbf{q}_\perp^{(-)}\right] . \tag{1.114}$$

According to Eqs. (1.111) and (1.114), the monodromy matrix of a collision is

$$\begin{aligned}\mathsf{m}_{\text{collision}} &= \begin{pmatrix} \mathsf{U}^{-1} & 0 \\ (2\cos\phi)\mathsf{U}^{-1}\cdot\mathsf{V}^{\mathrm{T}}\cdot\mathsf{K}\cdot\mathsf{V} & \mathsf{U}^{-1} \end{pmatrix} \\ &= \begin{pmatrix} \mathsf{U}^{-1} & 0 \\ 0 & \mathsf{U}^{-1} \end{pmatrix} \cdot \begin{pmatrix} 1 & 0 \\ (2\cos\phi)\mathsf{V}^{\mathrm{T}}\cdot\mathsf{K}\cdot\mathsf{V} & 1 \end{pmatrix} .\end{aligned} \tag{1.115}$$

1.4.6 Expanding and contracting horospheres

The local Lyapunov exponents are determined by the rate of dispersion of trajectories in the vicinity of the reference trajectory. The dispersion is controlled by the "horosphere" which is a local sphere tangent to a wavefront of trajectories accompanying the reference trajectory and issued from a common initial position in the past. The wavefront is expanding so that we are concerned with an expanding horosphere which is nothing other than the local unstable manifold.

The expanding wavefront has a local curvature which is characterized by the second fundamental operator B_u defined by

$$\delta\mathbf{v}_\perp = \frac{d}{dt}\delta\mathbf{q}_\perp = \mathsf{B}_u\cdot\delta\mathbf{q}_\perp . \tag{1.116}$$

Accordingly, the unstable subspace has the following parametric representation

$$\begin{pmatrix} \delta\mathbf{q}_\perp \\ \delta\mathbf{v}_\perp \end{pmatrix} = \begin{pmatrix} \delta\mathbf{q}_\perp \\ \mathsf{B}_u\cdot\delta\mathbf{q}_\perp \end{pmatrix} , \tag{1.117}$$

in the tangent space. A similar representation holds for the stable subspace.

Let us consider a trajectory from an initial condition $\mathbf{X} = (\mathbf{q}, \mathbf{v})$. Collisions occur at the impact points $\mathbf{X}_n$ and times t_n with $n \in \mathbb{Z}$. We denote by $\tau_n = t_{n+1} - t_n$ the time between the collisions n and $n+1$. Furthermore, $\mathsf{B}_u^{(-)}(n)$ denotes the second fundamental operator of the horosphere immediately before the nth collision, i.e., at the end of the preceding free flight between the collisions n and $n-1$. On the other hand, $\mathsf{B}_u^{(+)}(n)$ denotes the fundamental operator immediately after the nth collision. $\mathsf{B}_u(t)$ denotes the fundamental operator during a free flight.

We assume that the second fundamental operator of the expanding horosphere is fixed at some collision in the remote past and look for the operator at the initial condition. The operator is successively modified by the free flights and collisions according to the monodromy matrices Eqs. (1.109) and (1.115). The second fundamental operator B'_u is related to the one before the monodromy matrix m by

$$\begin{pmatrix} \delta\mathbf{q}'_\perp \\ \delta\mathbf{v}'_\perp \end{pmatrix} = \begin{pmatrix} \delta\mathbf{q}'_\perp \\ \mathsf{B}'_u \cdot \delta\mathbf{q}'_\perp \end{pmatrix} = \mathsf{m} \cdot \begin{pmatrix} \delta\mathbf{q}_\perp \\ \mathsf{B}_u \cdot \delta\mathbf{q}_\perp \end{pmatrix} \tag{1.118}$$

which is solved by eliminating $\delta\mathbf{q}_\perp$ and $\delta\mathbf{q}'_\perp$ between both lines.

Applying this equation to the monodromy matrix Eq. (1.109) for a free flight, we find

$$\mathsf{B}_u^{(-)}(n+1) = \left[\tau_n \mathsf{1} + \mathsf{B}_u^{(+)}(n)^{-1}\right]^{-1} \qquad \text{(free flight)}. \tag{1.119}$$

For a collision, we have from Eq. (1.115)

$$\mathsf{B}_u^{(+)}(n) = \mathsf{U}_n^{-1} \cdot \left[(2\cos\phi_n)\mathsf{V}_n^{\mathrm{T}} \cdot \mathsf{K}_n \cdot \mathsf{V}_n + \mathsf{B}_u^{(-)}(n)\right] \cdot \mathsf{U}_n \quad \text{(collision)}. \tag{1.120}$$

Combining Eqs. (1.119) and (1.120) for successive backward collisions, we obtain Sinai's continued-fraction expression for the matrix:

$$\mathsf{B}_u(t) = \cfrac{\mathsf{1}}{\tau\mathsf{1} + \mathsf{U}_n^{-1} \cdot \left(\mathsf{W}_n + \cfrac{\mathsf{1}}{\tau_{n-1}\mathsf{1} + \mathsf{U}_{n-1}^{-1} \cdot \left(\mathsf{W}_{n-1} + \cfrac{\mathsf{1}}{\tau_{n-2}\mathsf{1} + \cdots}\right)^{-1} \cdot \mathsf{U}_{n-1}}\right)^{-1} \cdot \mathsf{U}_n} \tag{1.121}$$

where

$$\mathsf{W}_n = (2\cos\phi_n)\, \mathsf{V}_n^{\mathrm{T}} \cdot \mathsf{K}_n \cdot \mathsf{V}_n \,, \tag{1.122}$$

$\tau = t - t_n$, and $t_{n+1} > t > t_n$. Because $\mathsf{K}_n \geq 0$, and $\cos\phi_n \geq 0$, the matrix $\mathsf{B}_u(t) \geq 0$ so that the noncontracting character is maintained during the whole time evolution in the case of a hard-sphere gas. Let us emphasize that the operator (1.121) is defined locally for each initial condition $\mathbf{X}$ and can be obtained by integrating backward the trajectory $\mathbf{\Phi}^t\mathbf{X}$ for $-\infty < t < 0$ to determine the successive past collisions and the corresponding quantities appearing in Eq. (1.121).

To see the connection between the second fundamental operator and the Lyapunov exponents we proceed as follows. We use the fact that, between collisions, the quantity $\delta\mathbf{q}_\perp$ develops as

$$\begin{aligned}\delta\mathbf{q}_\perp(t_n) &= \delta\mathbf{q}_\perp(t_{n-1}+\tau_{n-1}) \\ &= \delta\mathbf{q}_\perp(t_{n-1})+\tau_{n-1}\delta\mathbf{v}_\perp(t_{n-1}) \\ &= \left[1+\tau_{n-1}\mathsf{B}_u^{(+)}(n-1)\right]\cdot\delta\mathbf{q}_\perp(t_{n-1})\,, \end{aligned}\tag{1.123}$$

where the quantities are the values immediately after the collisions, as given by Eq. (1.120). After a sequence of n collisions, $\delta\mathbf{q}_\perp(t)$ is given by

$$\delta\mathbf{q}_\perp(t_n) = \left[1+\tau_{n-1}\mathsf{B}_u^{(+)}(n-1)\right]\cdot\left[1+\tau_{n-2}\mathsf{B}_u^{(+)}(n-2)\right]\cdots\left[1+\tau_0\mathsf{B}_u^{(+)}(0)\right]\cdot\delta\mathbf{q}_\perp(0)\,, \tag{1.124}$$

with $t_0=0$. If there is an exponential separation of trajectories we would expect that for long times $\|\delta\mathbf{q}_\perp(T)\| \sim (\exp\lambda^{(1)}T)\|\delta\mathbf{q}(0)\|$. Similar behaviours hold for k-volumes supported by infinitesimal vectors $\delta\mathbf{q}_\perp(T)$ with $k\le f-1$. The exponential growth rate of an infinitesimal $(f-1)$-volume would give the sum of positive Lyapunov exponents as

$$\begin{aligned}\sum_{\lambda_i>0}\lambda_i &= \lim_{N\to\infty}\frac{1}{T_N}\ln\left|\det\left[1+\tau_{N-1}\mathsf{B}_u^{(+)}(N-1)\right]\cdots\left[1+\tau_0\mathsf{B}_u^{(+)}(0)\right]\right| \\ &= \lim_{N\to\infty}\frac{1}{T_N}\sum_{n=0}^{N-1}\ln\left|\det\left[1+\tau_n\mathsf{B}_u^{(+)}(n)\right]\right| \\ &= \lim_{N\to\infty}\frac{1}{T_N}\sum_{n=0}^{N-1}\int_0^{\tau_n}d\tau\,\mathrm{tr}\left[\tau 1+\mathsf{B}_u^{(+)}(n)^{-1}\right]^{-1}\,, \end{aligned}\tag{1.125}$$

where T_N is the time during which N collisions occurred. Now by using Eq. (1.119) in the form

$$\mathsf{B}_u(t_n+\tau)=\mathsf{B}_u^{(+)}(n)\cdot\left[1+\tau\mathsf{B}_u^{(+)}(n)\right]^{-1}\,, \tag{1.126}$$

we readily find that

$$\sum_{\lambda_i>0}\lambda_i=\lim_{T\to\infty}\frac{1}{T}\int_0^T dt\,\mathrm{tr}\,\mathsf{B}_u(t)\,. \tag{1.127}$$

We can therefore identify the local dispersion rate for billiards as

$$u(\mathbf{X})=\sum_{\lambda_i>0}\chi_i(\mathbf{X})=\mathrm{tr}\,\mathsf{B}_u(\mathbf{X})\,. \tag{1.128}$$

The form Eq. (1.128) plays for billiards the role of Eqs. (1.48) or (1.74) for smooth dynamical systems. With respect to smooth Hamiltonian systems, a simplification for billiard systems comes from the fact that the local Lyapunov exponents are given directly in terms of quantities which can be constructed from successive collisions. Numerical calculations of the sum of positive Lyapunov exponents for billiard systems can be carried out efficiently and quickly using Eq. (1.125). It is worth noting that the individual Lyapunov exponents may also be calculated by using the second fundamental operator (1.121). Such a method has recently been applied to a numerical evaluation of the Lyapunov exponents in the hard-sphere gas at different densities by Dellago et al. (1995, 1996).

1.4.7 Two-dimensional hard-disk billiards ($f = 2$)

We shall here consider a billiard system consisting of one moving particle in a two-dimensional domain made of hard-disk scatterers of radius a. The collision dynamics on the disks is defocusing which induces a dynamical instability of the trajectories in such systems. Examples of such systems are the disk scatterers, the Sinai billiard, and the Lorentz gases. In such systems, one can apply the above method as follows.

We introduce a coordinate system which accompanies the trajectory in its motion. The x-axis is parallel to the velocity while the y-axis is perpendicular. The infinitesimal perturbations are always pointing along the y-axis. The space $\mathfrak{I}^{(-)}$ is parallel to the y-axis before the collision while the space $\mathfrak{I}^{(+)}$ is parallel to the y-axis after the collision.

The operator (1.97) is the projection from $\mathfrak{I}^{(+)}$ to $\mathfrak{I}^{(-)}$ parallel to the normal $\mathbf{n}$ and establishes the relation between the perturbations of the position before and after the collision, $\delta q_\perp^{(\pm)}$. We observe that the collision induces a reversal of the perturbation in position: $\delta q_\perp^{(+)} = -\delta q_\perp^{(-)}$. Accordingly, we have that

$$\mathsf{U} = -1 . \tag{1.129}$$

This result also follows from the fact that this operator has a determinant equal to -1 and is a scalar.

The operator (1.99) is the projection from $\mathfrak{I}^{(-)}$ to $\mathfrak{I}$ parallel to the ingoing velocity $\mathbf{v}^{(-)}$. If we denote by ϕ the angle between the normal and the incident ray, the geometry of collision shows that $\delta q_\perp^{(-)} = \delta q_\perp \cos\phi$ (see Fig. 1.1). By furthermore noting that the operator V^{T} is the projection from $\mathfrak{I}$ to $\mathfrak{I}^{(-)}$ parallel to the normal $\mathbf{n}$, we have that

$$\mathsf{V} = \mathsf{V}^{\mathrm{T}} = \frac{1}{\cos\phi} . \tag{1.130}$$

Since the curvature of the disks is given by the scalar $\mathsf{K} = 1/a$ we find the relation

$$\mathsf{W} = (2\cos\phi)\mathsf{V}^{\mathrm{T}}\cdot\mathsf{K}\cdot\mathsf{V} = \frac{2}{a\cos\phi} . \tag{1.131}$$

Substituting the previous results into the monodromy matrices of a free flight (1.109) and of a collision (1.115), we obtain

$$\mathsf{m}_{\text{free flight}} = \begin{pmatrix} 1 & \tau \\ 0 & 1 \end{pmatrix} , \tag{1.132}$$

$$\mathsf{m}_{\text{collision}} = \begin{pmatrix} -1 & 0 \\ -\frac{2}{a\cos\phi} & -1 \end{pmatrix} . \tag{1.133}$$

We emphasize that the monodromy matrix of a collision is evaluated at the point of impact, i.e., the angle ϕ is evaluated at the collision. The monodromy matrix associated with a succession of collisions and of free flights can thus be obtained by the multiplication of the corresponding monodromy matrices.

In the two-dimensional case, there is one positive Lyapunov exponent so that the expanding and contracting horospheres reduced to horocircles characterized by a scalar Gaussian curvature which is given by the continued fraction, Eq. (1.121). Around the initial position $\mathbf{q}_0 = (x_0, y_0)$, the wavefront is circular and its radius grows linearly with time t (or equivalently with the path length), so that

$$B_u(t) = \frac{1}{t} , \qquad \text{for} \qquad 0 < t < \tau_0 . \tag{1.134}$$

Therefore, the curvature of the wavefront diverges at the origin. At the first impact, the curvature is modified according to the geometry of the collision. After the nth collision, the wavefront is locally equivalent to a circle of radius ρ_n. During the free motion to the next collision, the radius of this circle continues to grow linearly with the path length (or the time) so that

$$B_u(t) = \frac{1}{\tau + \rho_n} , \qquad \text{for} \qquad 0 < \tau < \tau_n , \tag{1.135}$$

where $\tau_n = \|\mathbf{q}_{n+1} - \mathbf{q}_n\|$ is the distance or the time between the nth and the $(n+1)^{\text{th}}$ collisions and $t = \tau_0 + \tau_1 + \ldots + \tau_{N-1} + \tau$. We observe that the curvature of the horocircle decreases between each collision, as depicted in Fig. 1.2.

Differentiating the collision rule (1.11) with respect to variations in the positions and velocities of the wavefront particles at the nth collision, the wavefront curvatures before and after the collision on a disk of radius a are related by

$$\frac{1}{\rho_n} = B_u^{(+)} = B_u^{(-)} + \frac{2}{a\cos\phi_n} , \tag{1.136}$$

where ϕ_n is the angle between the outgoing velocity of the reference orbit and the normal at the point of impact. Since $-\pi/2 \leq \phi \leq +\pi/2$, we have that $\cos\phi \geq 0$. Consequently, the curvature increases at each collision which is due to the defocusing dynamics of collisions on disks.

Combining the results (1.135) and (1.136), we obtain that the curvature of the wavefront between the nth and $(n+1)^{\text{th}}$ collisions is given by the continuous fraction

$$B_u(t) = \cfrac{1}{\tau + \cfrac{1}{\cfrac{2}{a\cos\phi_n} + \cfrac{1}{\tau_{n-1} + \cfrac{1}{\cfrac{2}{a\cos\phi_{n-1}} + \cfrac{1}{\tau_{n-2} + \ldots}}}}} \tag{1.137}$$

where $0 < \tau < \tau_n$ and $t = \tau_0 + \tau_1 + \ldots + \tau_{N-1} + \tau$. Equation (1.137) is the two-dimensional analogue of Eqs. (1.121)–(1.122), and it was originally derived by Sinai (1970) (see also Gallavotti and Ornstein 1974, Bunimovich 1979).

The local curvature (1.137) of the expanding wavefront provides the rate of separation between the reference orbit and nearby orbits. Accordingly, the Lyapunov exponent of the orbit of initial condition $\mathbf{X}_0$ is given by

$$\lambda(\mathbf{X}_0) = \lim_{N\to\infty} \frac{1}{T_N} \sum_{n=1}^{N} \ln |1 + \tau_n B_u^{(+)}(\phi_n, \tau_{n-1}, \ldots, \tau_0)| , \tag{1.138}$$

where

$$B_u^{(+)}(\phi_n, \tau_{n-1}, \ldots, \tau_0) = \frac{2}{a\cos\phi_n} + \cfrac{1}{\tau_{n-1} + \cfrac{1}{B_u^{(+)}(\phi_{n-1}, \tau_{n-2}, \ldots, \tau_0)}} , \tag{1.139}$$

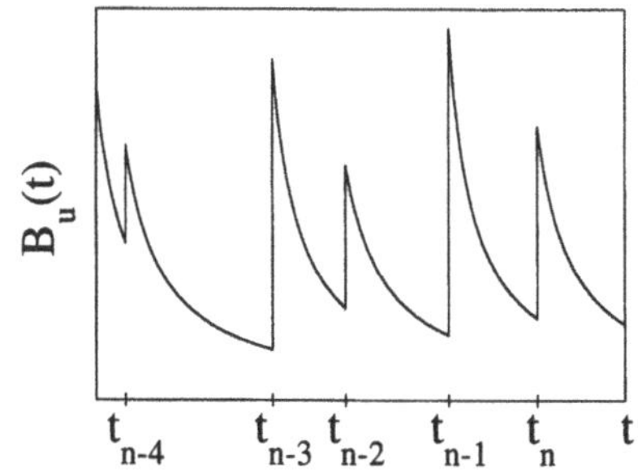

Figure 1.2 Typical behaviour of the curvature (1.137) of an expanding wavefront (horocircle) in a defocusing billiard.

with

$$B_u^{(+)}(\phi_1, \tau_0) = \frac{2}{a \cos \phi_1} + \frac{1}{\tau_0} , \tag{1.140}$$

and $T_N = \tau_1 + \ldots + \tau_N$. The stretching factor of a segment of trajectory over a time T_N can be estimated as the product

$$\Lambda(T_N, \mathbf{X}_0) \simeq \prod_{n=1}^{N} [1 + \tau_n \, B_u^{(+)}(\phi_n, \tau_{n-1}, \ldots, \tau_0)] . \tag{1.141}$$

We observe that the absolute value of each factor in (1.141) is greater than one so that we can expect an exponential growth of $|\Lambda|$ with N. Therefore, if the orbit has infinitely many collisions occurring with a finite mean free path $\bar{\tau}$ between collisions, the Lyapunov exponent is positive. Such two-dimensional billiards composed of hard disks are therefore dynamically unstable or hyperbolic (see Chapter 2).

Chapter 2

Topological chaos

2.1 Topology of trajectories in phase space

2.1.1 Phase portrait and invariant set

The geometric analysis of dynamical systems is carried out using the *phase portrait* which is the graphic representation of the trajectories in phase space (Nicolis 1995). This is a powerful method for low-dimensional dynamical systems where the phase portrait can be visualized as a two- or three-dimensional plot. The principle of construction of phase portraits is based on Cauchy's unicity theorem which has the consequence that trajectories do not cross each other in phase space. This constraint introduces a notion of parallelism of trajectories according to which certain configurations of trajectories are topologically excluded, which restricts considerably the types of possible phase portraits, especially, in two-dimensional flows.

One of the main focuses in the analysis of phase portraits is sets of points $\mathscr{I} \subseteq \mathscr{M}$ which are invariant under the time evolution, i.e.,

$$\textit{Invariant set}: \quad \mathbf{\Phi}^t \mathscr{I} = \mathscr{I}, \qquad \forall\, t. \tag{2.1}$$

The invariant sets – which are necessarily composed of complete trajectories – may have different stability and recurrent properties. The purpose of topological dynamics is to properly define such properties (Birkhoff 1927, Abraham and Mardsen 1978, Smale 1980b, Ruelle 1989c, Devaney 1990).

2.1.2 Critical orbits: stationary points and periodic orbits

The simplest invariant sets of a flow are the *stationary points* at which all the variables of the system remain constant in time:

$$\textit{Stationary point}: \quad \mathbf{\Phi}^t \mathbf{X}_s = \mathbf{X}_s, \quad \forall\, t. \tag{2.2}$$

In a flow, such a trajectory is the solution of the differential equations (1.1) because the vector field of the flow vanishes at the stationary point: $\dot{\mathbf{X}}_s = \mathbf{F}(\mathbf{X}_s) = 0$.

Other simple trajectories are the *periodic orbits* or *cycles*, which are solutions of the evolution law such that:

$$\textit{Periodic orbit}: \quad \exists\, T: \quad \mathbf{\Phi}^{t+T}\mathbf{X}_p = \mathbf{\Phi}^t \mathbf{X}_p, \quad \forall\, t. \tag{2.3}$$

In the case of a nontrivial periodic orbit, there still exist countably many real numbers T for which (2.3) holds, which are the repetitions rT_p (with $r \in \mathbb{Z}$) of the smallest nontrivial period T_p called the prime period (or simply, the period). In continuous-time systems, the periodic orbits form closed curves in phase space, which close on themselves for the first time after the prime period but also at repetitions of the prime period. As we shall see, it is important to distinguish between a prime periodic orbit and its repetitions.

The stationary points and the periodic orbits are called the *critical elements* of the flow. The set of all critical elements of the vector field is denoted by $\Gamma_{\mathbf{\Phi}}$. A critical element is said to be *isolated* in the phase space $\mathcal{M}$ if there is no distinct critical element of the same type arbitrarily close to it.

2.1.3 Nonwandering sets

The concept of wandering orbits in phase space is due to Birkhoff (1927). A point $\mathbf{X}$ is said to be *wandering* if there is a neighbourhood U of $\mathbf{X}$ such that $\cup_{|t|>t_0} \mathbf{\Phi}^t U \cap U = \emptyset$ for some positive time t_0. The wandering points form an invariant open subset of $\mathcal{M}$.

The complementary set is closed and is called the set $\Omega_{\mathbf{\Phi}}$ of the *nonwandering points*. A point $\mathbf{X}$ is nonwandering if, for all neighbourhoods U of $\mathbf{X}$ and for all $t_0 \geq 0$, there exists a time $t > t_0$ such that $U \cap \mathbf{\Phi}^t U$ is nonempty.

Examples of nonwandering orbits are the stationary points and the periodic orbits so that $\Gamma_{\mathbf{\Phi}} \subset \Omega_{\mathbf{\Phi}}$. But the nonwandering sets may also contain nonperiodic orbits.

In nonconservative systems, most of the trajectories tend to a subset of zero Lebesgue measure which is an attractor. In this case, most of the trajectories never come back nearby their initial conditions which are therefore wandering points. Accordingly, most points in the phase space of dissipative systems are nonrecurrent.

In contrast, recurrence is an essential aspect of conservative systems as shown by the celebrated:

Poincaré recurrence theorem. Let $\mathbf{\Phi}^t$ *be a volume-preserving flow leaving invariant a bounded domain* $\mathscr{D}$ *of the phase space* $\mathscr{M}$*. Then, in each neighbourhood* U *of some point* $\mathbf{Y}$ *of the domain* $\mathscr{D}$*, there exists a point* $\mathbf{X}$ *which comes back in the neighbourhood* U *for some time* $t > 0$*, i.e.* $\mathbf{\Phi}^t\mathbf{X} \in U$.

(For a proof, see Arnold 1978). We conclude that a bounded invariant domain $\mathscr{D}$ of a conservative flow belongs to its nonwandering set $\Omega_{\mathbf{\Phi}}$. This behaviour is in contrast with nonconservative systems where the nonwandering set is in general smaller than the phase space. However, this result does not extend to open conservative systems of scattering type where no bounded domain $\mathscr{D}$ is relevant.

2.1.4 Locally maximal invariant sets and global stability

In order to distinguish simple invariant sets like the critical orbits from larger ones like nonwandering sets, the following notion may be introduced (Alekseev and Yakobson 1981):

An invariant set $\mathscr{I}$ *is locally maximal iff there exists a neighbourhood* $U \supset \mathscr{I}$ *such that, for any invariant set* $\mathscr{I}'$*, the condition* $\mathscr{I} \subseteq \mathscr{I}' \subseteq U$ *implies* $\mathscr{I} = \mathscr{I}'$.

As a direct consequence, a locally maximal invariant set can be obtained from the time evolution of the neighbourhood U as

$$\mathscr{I} = \cap_{-\infty<t<+\infty} \mathbf{\Phi}^t U , \tag{2.4}$$

because the right-hand member defines an invariant set $\mathscr{I}'$ which should coincide with $\mathscr{I}$ by the above definition.

In some cases, the invariant set can be obtained from the sole forward time evolution:

$$\text{\textit{The invariant set is attracting iff}}: \quad \mathscr{I} = \cap_{t>0} \mathbf{\Phi}^t U , \tag{2.5}$$

for a so-called *fundamental neighbourhood* U such that

$$\mathbf{\Phi}^t U \subset U , \quad \text{for } t > 0 . \tag{2.6}$$

By reversing time, we have

$$\text{\textit{The invariant set is anti-attracting iff}}: \quad \mathscr{I} = \cap_{t<0} \mathbf{\Phi}^t U , \tag{2.7}$$

for a fundamental neighbourhood U such that

$$\mathbf{\Phi}^t U \subset U , \quad \text{for } t < 0 . \tag{2.8}$$

An invariant set is said to be *repelling* if it is not attracting. We notice that an anti-attracting set is repelling in all the phase-space directions so that we can also speak of an absolutely repelling set. However, most of the invariant sets are saddle-type repelling sets which are attracting in some directions and repelling in others, so that both limits $t \to \pm\infty$ are required to localize them in phase space.

2.1.5 Dense orbits and transitivity

We say that the orbit $\mathbf{\Phi}^t\mathbf{X}$ is *dense* in some invariant subset of the phase space ($\mathscr{I} \subset \mathscr{M}$) if the closure (Cl) of the set ξ of points composing the orbit is identical with the set $\mathscr{I}$

$$\mathrm{Cl}(\xi) = \mathrm{Cl}(\cup_{-\infty<t<+\infty}\mathbf{\Phi}^t\mathbf{X}) = \mathscr{I} . \tag{2.9}$$

A dense orbit is a trajectory which passes arbitrarily close to each point of $\mathscr{I}$.

If there exists an orbit which is dense in some subset $\mathscr{I}$ we say that the flow is *transitive* in $\mathscr{I}$. If there is an orbit which is dense in the whole phase space $\mathscr{M}$ the flow $\mathbf{\Phi}^t$ is said to be transitive, simply. A locally maximal invariant set which is transitive is made up of one piece and is therefore *indecomposable* into smaller locally maximal invariant sets.

The transitivity of a dynamically unstable system can be investigated by considering a web made of pieces of stable and unstable manifolds. If two points $\mathbf{X}$ and $\mathbf{Y}$ can be topologically connected with a sequence of intersecting pieces of successive stable and unstable manifolds then there exists an orbit which connects arbitrarily small neighbourhoods of both points. Such a sequence of alternating stable and unstable manifolds is called a Hopf chain (Hedlund 1939, Hopf 1939). If a Hopf chain exists between any two points in phase space then the system is transitive and therefore ergodic [according to the ergodic theorem of Birkhoff (1931)]. Sinai (1970, 1979) has generalized the construction by considering Cantor sets made of pieces of stable or unstable manifolds to prove transitivity in two-dimensional billiards. The result was extended to higher-dimensional billiards by Krámli et al. (1992).

2.1.6 Attractors, anti-attractors, and repellers

If transitivity does not hold in the whole phase space or, more generally, in the whole nonwandering set, this latter should be decomposed into smaller transitive invariant sets which are locally maximal: $\Omega = \cup_i \mathscr{I}_i$. This construction is known as the decomposition into ergodic components (Eckmann and Ruelle 1985).

After this decomposition, the different locally maximal invariant sets are of one of the following types:

A set $\mathscr{I}$ is respectively (a) an attractor, (b) an anti-attractor, (c) a repeller iff:

1. *$\mathscr{I}$ is a closed and compact set;*
2. *$\mathscr{I}$ is a locally maximal invariant set which is respectively (a) attracting, (b) anti-attracting, (c) repelling;*
3. *$\mathscr{I}$ is indecomposable.*

The *basin of attraction B* of an attractor $\mathscr{I}$ consists of all the initial conditions $\mathbf{X} \in \mathscr{M}$ which are attracted to $\mathscr{I}$ when $t \to +\infty$, i.e.

$$B = \cup_{t<0} \, \mathbf{\Phi}^t U \, , \tag{2.10}$$

if U is a fundamental neighbourhood of $\mathscr{I}$. In dissipative systems, the knowledge of the basins of attraction provides a complete description of the global stability of the system since the attracting set of each trajectory is then known. Different neighbouring basins of attraction are in general separated by repelling invariant sets called *basin boundaries* (Eckmann 1981, Grebogi et al. 1983, Ott 1993).

2.2 Hyperbolicity

2.2.1 Definition

We shall here make assumptions on the stability properties of the invariant subsets we are interested in.

An invariant subset $\mathscr{I} \subset \mathscr{M}$ is said to be *hyperbolic* if:

1. $\forall \mathbf{X} \in \mathscr{I}$*, the tangent space of a trajectory can be decomposed as*

$$\mathscr{T}\mathscr{M}(\mathbf{X}) = \mathscr{E}_{\mathrm{u}}(\mathbf{X}) \oplus \{\mathbf{F}(\mathbf{X})\} \oplus \mathscr{E}_{\mathrm{s}}(\mathbf{X}) \, , \tag{2.11}$$

into the stable and unstable subspaces together with a centre subspace containing only the direction of the flow.

2. *There exist a positive constant κ and a positive function $C(\mathbf{X}, \mathbf{X}')$ such that,* $\forall t \geq 0$*,* $\forall \mathbf{e}_{\mathrm{s}} \in \mathscr{E}_{\mathrm{s}}(\mathbf{X})$*,* $\forall \mathbf{e}_{\mathrm{u}} \in \mathscr{E}_{\mathrm{u}}(\mathbf{X})$*, and* $\forall \mathbf{X} \in \mathscr{I}$*, we have*

$$\begin{aligned} \|\mathsf{M}(t,\mathbf{X}) \cdot \mathbf{e}_{\mathrm{s}}\| &\leq C(\mathbf{X}, \Phi^t\mathbf{X})^{-1} \exp(-\kappa t) \, \|\mathbf{e}_{\mathrm{s}}\| \, , \\ \|\mathsf{M}(t,\mathbf{X}) \cdot \mathbf{e}_{\mathrm{u}}\| &\geq C(\mathbf{X}, \Phi^t\mathbf{X}) \exp(+\kappa t) \, \|\mathbf{e}_{\mathrm{u}}\| \, , \\ \gamma_{\mathrm{us}}(\Phi^t\mathbf{X}) &\geq C(\mathbf{X}, \Phi^t\mathbf{X}) \, \gamma_{\mathrm{us}}(\mathbf{X}) \, , \end{aligned} \tag{2.12}$$

where $\gamma_{\mathrm{us}}(\mathbf{X})$ denotes the angle between the stable and unstable linear subspaces.

A system is hyperbolic if it has a single hyperbolic invariant subset. We notice that the last condition of (2.12) means that the angle between the stable and unstable linear subspaces never vanishes. It is important to remark that the preceding definition is invariant under smooth changes of coordinates which do not modify the conditions (2.12). Since the direction of the flow is always neutral it must be excluded from the stable and unstable subspaces in (2.11). However, it is imperative that the definition requires that the neutral subspace does not include a direction other than that of the flow. Otherwise, the system would be nonhyperbolic as for a periodic orbit of centre type in a flow. If globally conserved quantities exist, hyperbolicity is possible but for the system restricted to a single hypersurface where all the conserved quantities are constant.

The invariant subset $\mathscr{I}$ is said to be *continuously hyperbolic* if, moreover:

3. *The stable and unstable linear subspaces* $\mathscr{E}_s(\mathbf{X})$ *and* $\mathscr{E}_u(\mathbf{X})$ *continuously vary with* $\mathbf{X}$.

In a hyperbolic system, all the periodic orbits have nonvanishing Lyapunov exponents (except for the only zero exponents associated with the direction of flow). Hence, the condition of hyperbolicity implies that the spectrum of Lyapunov exponents is

$$\lambda^{(1)}(\mathbf{X}) > \ldots > \lambda^{(\nu)}(\mathbf{X}) \geq \kappa > \lambda^{(\nu+1)}(\mathbf{X}) = 0 > -\kappa \geq \lambda^{(\nu+2)}(\mathbf{X}) > \ldots > \lambda^{(r)}(\mathbf{X}) , \tag{2.13}$$

for $\mathbf{X} \in \mathscr{I}$. The system is said to be *uniformly hyperbolic* if it is hyperbolic and if the Lyapunov spectrum is the same for all the trajectories of the invariant set

$$\lambda^{(k)}(\mathbf{X}) = \lambda^{(k)} , \qquad \forall\, \mathbf{X} \in \mathscr{I} . \tag{2.14}$$

Otherwise, the system is *nonuniformly hyperbolic* which is by far the most common situation.

When a system is continuously hyperbolic the stable and unstable manifolds of its trajectories extend without rupture in phase space. The stable and unstable manifolds are obtained by some kind of integration along the stable and unstable directions in the tangent space constructed in Chapter 1. The invariant manifolds are differentiable if these directions vary continuously with position. In bounded hyperbolic billiards where trajectories tangent to the obstacles may exist, the decomposition of the tangent space into stable and unstable directions exists but does not vary continuously, which results in ruptures of the stable and unstable manifolds (Sinai 1970, 1979).

The system is *nonhyperbolic* if there is a set of periodic orbits for which all of the Lyapunov exponents are zero. In nonhyperbolic systems, the Lyapunov exponents of the different orbits accumulate in general at zero so that $\kappa = 0$

in Eq. (2.13). We will consider systems of both hyperbolic and nonhyperbolic types, although there are only a few general statements that can be made about nonhyperbolic systems.

A hyperbolic repeller has stable and unstable manifolds which are made of all the stable or unstable manifolds of each orbit: $W_{s,u}(\mathscr{I}) = \cup_{\xi\in\mathscr{I}} W_{s,u}(\xi)$. The stable and unstable manifolds, $W_{s,u}(\xi)$ and $W_{s,u}(\mathscr{I})$, are invariant sets of the system. The transitive nonwandering subset $\mathscr{I}$ is always at the intersection of its stable and unstable manifolds

$$\mathscr{I} \;=\; W_s(\mathscr{I}) \,\cap\, W_u(\mathscr{I}) \,. \tag{2.15}$$

An attractor coincides with its unstable manifolds, i.e.,

$$\mathscr{I} \textit{ is attractor } \Rightarrow \mathscr{I} \;=\; W_u(\mathscr{I}) \,, \tag{2.16}$$

which results from (2.15) by observing that the closure of the stable manifolds forms the basin of attraction and, hence, contains all the trajectories of a fundamental neighbourhood of the attractor: $\mathscr{I} \subset U \subset \mathrm{Cl}[W_s(\mathscr{I})] = B$.

2.2.2 Escape-time functions

The dynamics in the neighbourhood of locally maximal nonwandering sets $\mathscr{I}$ can be characterized by the time taken by trajectories to escape from a neighbourhood U in the future or the past

$$T_U^{(+)}(\mathbf{X}) \;=\; \mathrm{Max}\{T>0: \quad \mathbf{\Phi}^t\mathbf{X}\in U \quad \forall t\in[0,T[\} \,, \tag{2.17}$$

$$T_U^{(-)}(\mathbf{X}) \;=\; \mathrm{Min}\{T<0: \quad \mathbf{\Phi}^t\mathbf{X}\in U \quad \forall t\in]T,0]\} \,, \tag{2.18}$$

for $\mathbf{X}\in U$.

If U is the fundamental neighbourhood of an attractor, then $T_U^{(+)}(\mathbf{X}) = +\infty$ for every point $\mathbf{X}\in U$. If the attractor is hyperbolic, $T_U^{(-)}(\mathbf{X})$ is finite for every point except those on the unstable manifold of the attractor where: $T_U^{(-)}(\mathbf{X}) = -\infty$ for $\mathbf{X}\in W_u(\mathscr{I})$.

A similar result with the exchanges $+\leftrightarrow-$ and $u\leftrightarrow s$ holds for anti-attractors by reversing time.

For a hyperbolic repeller, the forward time $T_U^{(+)}(\mathbf{X})$ to escape from U is finite except on the stable manifold of the repeller: $T_U^{(+)}(\mathbf{X}) = +\infty$ for $\mathbf{X}\in W_s(\mathscr{I})$. The backward time of escape is finite except on the unstable manifold where: $T_U^{(-)}(\mathbf{X}) = -\infty$ for $\mathbf{X}\in W_u(\mathscr{I})$. Therefore, the function $T_U^{(+)}(\mathbf{X}) + |T_U^{(-)}(\mathbf{X})|$ is finite in $U\backslash[W_s(\mathscr{I})\cup W_u(\mathscr{I})]$. This function is very useful to construct numerically repellers and their stable and unstable manifolds (Burghardt and Gaspard 1995).

By the way, we may use the escape-time functions to construct approximations to the stable or unstable manifolds of the repeller. For instance, the set of initial conditions for which the escape time is larger than a predetermined time $T > 0$ contains all of the trajectories still inside the neighbourhood U at the time T, i.e.

$$\Upsilon_U^{(+)}(T) \equiv \{\mathbf{X} \in U : \quad T < T_U^{(+)}(\mathbf{X})\} = \bigcap_{0<t<T} \mathbf{\Phi}^{-t} U . \tag{2.19}$$

In the long-time limit, this set contains the trapped trajectories of the repeller and its stable manifold, restricted to U:

$$\lim_{T\to\infty} \Upsilon_U^{(+)}(T) = \mathrm{Cl}[W_s(\mathscr{I})] \cap U . \tag{2.20}$$

In a similar way, we can define a set of initial conditions which had already arrived in U before some time $-T$ in the past

$$\Upsilon_U^{(-)}(T) = \{\mathbf{X} \in U : \quad T_U^{(-)}(\mathbf{X}) < -T\} = \bigcap_{-T<t<0} \mathbf{\Phi}^{-t} U . \tag{2.21}$$

In analogy with Eq. (2.20), we have that

$$\lim_{T\to\infty} \Upsilon_U^{(-)}(T) = \mathrm{Cl}[W_u(\mathscr{I})] \cap U . \tag{2.22}$$

The intersection of the two sets contains all of the trajectories which are inside the region U over the time interval $-T < t < +T$. The repeller is thus defined as the following set of points

$$\lim_{T\to\infty} \Upsilon_U^{(-)}(T) \cap \Upsilon_U^{(+)}(T) = \lim_{T\to\infty} \bigcap_{-T<t<+T} \mathbf{\Phi}^{-t} U = \mathscr{I} . \tag{2.23}$$

It is instructive to consider this construction for simple systems such as a two-dimensional Smale horseshoe map of the unit square (Smale 1980b, Guckenheimer and Holmes 1983, Tél 1990). One sees that for this map the sets $\Upsilon_U^{(+)}(T)$ and $\Upsilon_U^{(-)}(T)$ are thin strips parallel to the stable and unstable directions, respectively, each becoming the product of a Cantor set with a one-dimensional interval as $T \to \infty$. The repeller is the intersection of these two sets, and is a Cantor set which can be coded as a bi-infinite sequence of zeros and ones. The dynamics on the repeller is then isomorphic to the left-Bernoulli shift on these bi-infinite sequences. In this map, the repeller is hyperbolic, fractal and chaotic, which is a typical situation. Most remarkably, such repellers are very common in Hamiltonian systems of classical scattering theory.

2.2.3 Anosov and Axiom-A systems

Certain classes of dynamical systems have been the object of special attention for their relative simplicity in spite of their hyperbolic structure (Anosov 1967, Smale 1980b, Pesin 1989).

A system is called an *Axiom-A system* if:

1. *The nonwandering set* Ω_Φ *is continuously hyperbolic and compact.*
2. *The periodic orbits are dense in* Ω_Φ.

A system is called an *Anosov system* if *it is an Axiom-A system such that its whole phase* $\mathcal{M}$ *consists of nonwandering points:* $\Omega_\Phi = \mathcal{M}$.

We notice that, in the case of mappings, the fixed points are considered as periodic orbits of period one. In the case of flows, the nonwandering set Ω_Φ may contain stationary points. It is then required that the closure of the periodic orbits and of the stationary points are disjoint subsets and that the condition that the periodic orbits are dense holds only on the invariant subsets composed of the periodic orbits. For Axiom-A systems, Smale (1980b) has proved that the nonwandering set decomposes into a finite union of locally maximal nonwandering subsets which are transitive.

One remarkable property of the Anosov and Axiom-A systems is their structural stability (Abraham and Mardsen 1978). If the vector field or the mapping is slightly perturbed by a smooth enough perturbation, the trajectories of the perturbed system remain in one-to-one topological correspondence with the trajectories of the original system. The concept of structural stability lies on the Grobman-Hartman theorem thanks to which a homeomorphism can be constructed between the original system and the linearized one under the condition of hyperbolicity for fixed points in maps or stationary points in flows (Guckenheimer and Holmes 1983). Both the original and the perturbed systems may be linearized by two homeomorphisms which can be combined to apply the perturbed system onto the original one. By a generalization of these considerations, continuously hyperbolic systems can be shown to be structurally stable under the preceding assumptions by Anosov and Smale. In this sense, these systems turn out to be robust under slight perturbations, even if they are dynamically unstable with positive Lyapunov exponents. This argument has been very important in the historical development of the field of chaotic dynamics because it shows that such behaviours are not exceptional.

Today, however, the argument has been weakened by the discovery of many typical systems of great physical relevance which are not continuously hyperbolic and not structurally stable. This is the case already for billiards like the finite-horizon Lorentz gases which are still hyperbolic but not continuously hyperbolic. In these billiards, all the periodic orbits are unstable of saddle type but the topology of trajectories is not robust because some trajectories are tangent to the obstacles. On the other hand, most dynamical systems appear nonhyperbolic due to orbits in which all the Lyapunov exponents vanish. Nonhyperbolicity has often its origin in homoclinic tangencies between some stable and unstable manifolds. These homoclinic tangencies are known to generate infinitely many periodic attractors in some rare perturbations of the vector field, the so-called Newhouse phenomenon (Guckenheimer and Holmes 1983, Wang 1990); although, in many other perturbations of the vector field, a single chaotic attractor exists which is nonhyperbolic in the sense that Lyapunov exponents are arbitrarily close to zero. A similar phenomenon arises in Hamiltonian systems.

Nevertheless, systems of statistical mechanics with many degrees of freedom and at relatively high temperature numerically present a complete spectrum of positive Lyapunov exponents, which suggests that the system behaves like a hyperbolic system (see Introduction). In this regard, we can formulate the hypothesis that hyperbolic vector fields of symplectic type exist arbitrarily close to the vector field of a typical Hamiltonian system with *many* particles. A similar hypothesis has been proposed recently in the case of the so-called thermostatted systems as the Chaotic Hypothesis which supposes that a many-body thermostatted system can be approximated for all relevant properties by an Anosov system (Gallavotti and Cohen 1995a, 1995b). Let us mention that Smale (1980a) has previously formulated closely connected hypotheses and has discussed their relevance to the ergodic hypothesis.

2.3 Markov partition and symbolic dynamics in hyperbolic systems

We suppose that the dynamical system is hyperbolic and that, furthermore, we have a suitable surface of section so that the flow reduces to the Poincaré map (1.5). The hyperbolicity of the flow implies the hyperbolicity of the map as well. For a map, hyperbolicity is defined as before except that we do not need to consider the direction of the flow in (2.11). The stable and unstable manifolds $W_{s,u}(\mathbf{x})$ can then be constructed in the surface of section.

2.3.1 Partitioning phase space with stable and unstable manifolds

The idea of the Markov partition is to partition the phase space into curvilinear rectangles made out of segments of stable and unstable manifolds, which are intrinsic to the dynamics in phase space (Bowen 1975).

The cells of the partition $\mathscr{C} = \{C_1, C_2, \ldots, C_L\}$ are closed: $C_i = \mathrm{Cl}(C_i)$. The border of each cell $\partial C_i = \mathrm{Cl}(C_i) - \mathrm{Int}(C_i)$ (where Int denotes the interior) is composed of segments of the stable and unstable manifolds which are denoted by $\partial C_i = \partial_s C_i \cup \partial_u C_i$. A similar decomposition applies to the borders of the partition itself: $\partial\mathscr{C} = \partial_u\mathscr{C} \cup \partial_s\mathscr{C}$. Moreover, it is required that, for any pair of points of a cell, their stable and unstable manifolds intersect inside the rectangle

$$\forall\ \mathbf{x},\ \mathbf{x}' \in C_i\ , \qquad W_s(\mathbf{x}) \cap W_u(\mathbf{x}') \in C_i\ . \tag{2.24}$$

In order to have a Markov-type property on the visit of trajectories in successive cells, the cells must satisfy specific conditions on the way the images of the cells intersect with the cells themselves, according to the definition:

Definition. A Markov partition is a finite cover of the invariant set $\mathscr{I}$ by curvilinear rectangles $\mathscr{C} = \{C_1, C_2, \ldots, C_L\}$ such that:

1. $\mathrm{Int}C_i \cap \mathrm{Int}C_j = \emptyset,\ \forall\ i \neq j.$
2. *If* $\mathbf{x} \in \mathrm{Int}C_i$ *and* $\boldsymbol{\phi}(\mathbf{x}) \in \mathrm{Int}C_j$, *then*

$$\boldsymbol{\phi}\left[W_u(\mathbf{x}) \cap C_i\right] \supset W_u(\boldsymbol{\phi}\mathbf{x}) \cap C_j\ , \tag{2.25}$$

$$\boldsymbol{\phi}\left[W_s(\mathbf{x}) \cap C_i\right] \subset W_s(\boldsymbol{\phi}\mathbf{x}) \cap C_j\ . \tag{2.26}$$

This last condition can be rewritten in the form

$$\boldsymbol{\phi}^{-1}(\partial_u\mathscr{C}) \subset \partial_u\mathscr{C}\ , \tag{2.27}$$

$$\boldsymbol{\phi}(\partial_s\mathscr{C}) \subset \partial_s\mathscr{C}\ , \tag{2.28}$$

which expresses that the images of the stable sides of the partition are contained in the stable sides and that the unstable sides are contained in their images.

Markov partitions of arbitrarily small diameters exist in continuously hyperbolic systems according to a theorem by Bowen (1975).

2.3.2 Symbolic dynamics and shift

Once a partition has been constructed, symbolic sequences can be assigned to each trajectory and a symbolic dynamics can be constructed (Smale 1980b, Alekseev and Yakobson 1981).

The idea of a symbolic dynamics is to establish – if possible – a one-to-one correspondence between the trajectories of the system and sequences of symbols taken in an alphabet, $\omega_i \in \mathscr{A} = \{1, 2, 3, \ldots, L\}$, which is composed of a finite or countably infinite number of integers, or of alphabetical or nonalphabetical symbols in one-to-one correspondence with the cells of the partition $\mathscr{C} = \{C_1, C_2, \ldots, C_L\}$. We shall denote by $|\mathscr{A}|$ the number of elements in the set $\mathscr{A}$. If the map ϕ is invertible, a correspondence is established with bi-infinite sequences

$$\mathbf{x} \in \mathscr{I} \overset{\Theta}{\longleftrightarrow} \omega_{.} = \ldots \omega_{-3}\omega_{-2}\omega_{-1} \cdot \omega_0\omega_1\omega_2\omega_3 \ldots \in \Sigma \,, \tag{2.29}$$

where the dot denotes the origin of time when the initial condition is given. The symbolic sequences are simply infinite for noninvertible maps.

The transformation Θ is established by considering the intersection of the images of all the cells corresponding to the bi-infinite sequence:

$$\Theta(\omega_{.}) = \cap_{n=-\infty}^{+\infty} \phi^{-n}(C_{\omega_n}) = \mathbf{x} \,. \tag{2.30}$$

The one-to-one correspondence Θ between the invariant set $\mathscr{I}$ and the symbolic space Σ implies that the dynamics ϕ induces a dynamics on the symbolic space Σ according to

$$\phi \circ \Theta = \Theta \circ \sigma \,. \tag{2.31}$$

The map of this symbolic dynamics is called a *shift* because it acts like a shift to the left on the symbolic sequences

$$\sigma(\omega_{.}) = \sigma\Big(\{\omega_n\}_{n\in\mathbb{Z}}\Big) = \{\omega_{n+1}\}_{n\in\mathbb{Z}} \,. \tag{2.32}$$

A few observations are in order here about the symbolic representation of the orbits:

If an initial condition $\mathbf{x}$ is associated with a dotted symbolic sequence $\omega_{.}$, the same symbolic sequence without its dot denotes the whole trajectory

$$\xi = \cup_{-\infty<n<+\infty}\phi^n\mathbf{x} \overset{\Theta}{\longleftrightarrow} \omega = \ldots \omega_0\omega_1\omega_2 \ldots \omega_n\omega_{n+1} \ldots \,. \tag{2.33}$$

Periodic orbits are represented by periodic sequences like

$$\begin{aligned} \omega &= \ldots \underbrace{\omega_1\omega_2\ldots\omega_{n_p}}_{=p}\underbrace{\omega_1\omega_2\ldots\omega_{n_p}}_{=p}\underbrace{\omega_1\omega_2\ldots\omega_{n_p}}_{=p}\ldots \,, \\ &= [\omega_1\omega_2\ldots\omega_{n_p}]^\infty = p^\infty \,, \end{aligned} \tag{2.34}$$

defined by the infinite repetition of the word or pattern $p = \omega_1\omega_2\ldots\omega_{n_p}$ of minimal length. The orbit is therefore of prime period n_p in the number of symbols. A point of the periodic orbit is represented by a dotted sequence so that the periodic orbit has n_p distinct intersection points in the surface of section.

If $p = \omega_1\omega_2\ldots\omega_{n_p}$ and $q = \omega'_1\omega'_2\ldots\omega'_{n_q}$ are the patterns of two distinct periodic orbits p^∞ and q^∞, a heteroclinic orbit joining both of them is represented by

$$p^\infty\omega_1\omega_2\ldots\omega_{n-1}\omega_n q^\infty \,, \qquad \text{(heteroclinic orbit)} \,, \tag{2.35}$$

where the word $\omega_1 \ldots \omega_n$ characterizing the heteroclinic orbit is distinct from p or q. On the other hand, a homoclinic orbit associated with the periodic orbit p^∞ is represented by

$$p^\infty \omega_1 \omega_2 \ldots \omega_{n-1} \omega_n p^\infty \,, \qquad \text{(homoclinic orbit)} \,, \tag{2.36}$$

where the intermediate word is neither empty nor identical to p (otherwise the orbit is identical to p^∞).

2.3.3 Markov topological shift

The remarkable property of a Markov partition is that it generates a symbolic dynamics which is a topological Markov chain. For such a symbolic dynamics, the symbols follow each other with constraint only on pairs of successive symbols in the sequences (2.29). Such a constraint can be represented by a transition matrix, $A_{\omega\omega'}$, containing zeros and ones with a number of lines and columns equal to the number of symbols in the alphabet, such that

$$A_{\omega_n \omega_{n+1}} = \begin{cases} 1 \text{ if } \omega_n \omega_{n+1} \text{ is allowed} \,, \\ 0 \text{ otherwise} \,, \end{cases} \tag{2.37}$$

for all $\omega_n, \ \omega_{n+1} \in \mathscr{A}$. This transition matrix defines a grammatical rule in the composition of the words and sequences on the alphabet $\mathscr{A}$.

The symbolic dynamics defined on the alphabet $\mathscr{A}$ with the transition matrix A

$$\Sigma(\mathscr{A}, \mathsf{A}, \mathbb{Z}) = \{\omega_. : \omega_n \in \mathscr{A} \,, \ A_{\omega_n \omega_{n+1}} = 1 \,, \ \forall n \in \mathbb{Z}\} \,, \tag{2.38}$$

is called a *topological Markov shift* or *subshift of finite type.*

A topological Markov shift may be represented by the oriented graph of transitions corresponding to the transition matrix. The symbols of the alphabet are the vertices of this graph. The transitions between the symbols are represented by oriented edges between the vertices.

A special case of topological Markov shifts is the topological Bernoulli shifts for which the transition matrix is full of ones, $A_{\omega\omega'} = 1 \ \forall \omega, \omega'$, as is the case in the Smale horseshoe.

A transition matrix A can be associated with the previous Markov partition $\mathscr{C}$ according to

$$A_{\omega\omega'} = \begin{cases} 1 & \text{if} \quad \mathrm{Int} C_\omega \cap \phi^{-1}(\mathrm{Int} C_{\omega'}) \neq \emptyset \,, \\ 0 & \text{otherwise} \,. \end{cases} \tag{2.39}$$

The remarkable feature of the Markov condition (2.25)–(2.26) is that the symbolic dynamics is automatically a Markov shift:

$$\forall\, \omega_{\cdot} \in \Sigma(\mathscr{A}, \mathsf{A}), \quad \exists\, \mathbf{x} \in \mathscr{I} \; : \\ \mathbf{x} \in \cap_{n\in\mathbb{Z}}\, \phi^{-n}(\mathrm{Int}C_{\omega_n}) \,. \tag{2.40}$$

This important theorem by Sinai (1968) and Bowen (1975) is proved as follows.

Let us first consider the set

$$\mathscr{C}_s(\omega_{\cdot}) = \cap_{n=0}^{\infty}\, \phi^{-n}(\mathrm{Int}\ C_{\omega_n}) \,, \tag{2.41}$$

which is contained inside the cell C_{ω_0}. The set $\mathrm{Int}C_{\omega_0} \cap \phi^{-1}(\mathrm{Int}C_{\omega_1})$ with $A_{\omega_0\omega_1} = 1$ is composed of one or several curvilinear rectangles contained in C_{ω_0}. According to (2.25) or (2.27), the unstable sides of these rectangles are contained in the unstable sides of the rectangle C_{ω_0} as illustrated in Fig. 2.1a. According to (2.26), we have successively

$$\forall\, \mathbf{x} \in \mathscr{C}_s(\omega_{\cdot}) \; : \\ W_s(\mathbf{x}) \cap C_{\omega_0} \subset W_s(\mathbf{x}) \cap \phi^{-1}C_{\omega_1} \subset \ldots \subset W_s(\mathbf{x}) \cap \phi^{-n}C_{\omega_n} \; \cdots \,, \tag{2.42}$$

which means that the segment of the stable manifold of $\mathbf{x}$ in the cell C_{ω_0} extends from one unstable side to the other unstable side of C_{ω_0} *without rupture*, as shown in Fig. 2.1a. The set (2.41) is thus a set of segments of stable manifolds extending inside C_{ω_0} from one of its unstable sides to the other without rupture. Therefore, all the possible trajectories allowed by the Markov shift will be found inside this set.

Similarly, the set

$$\mathscr{C}_u(\omega_{\cdot}) = \cap_{n=0}^{\infty}\, \phi^{n}(\mathrm{Int}\ C_{\omega_{-n}}) \,, \tag{2.43}$$

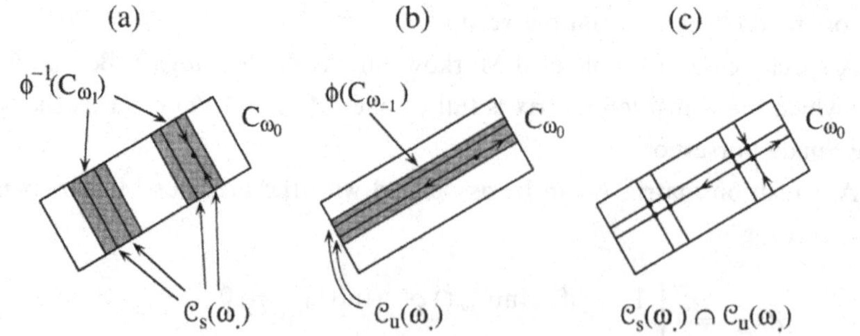

Figure 2.1 (a), (b), and (c) Formation of the sets $\mathscr{C}_u(\omega_{\cdot})$ and $\mathscr{C}_s(\omega_{\cdot})$ inside the cell C_{ω_0} at the basis of the theorem (2.40).

is a set of segments of unstable manifolds extending inside C_{ω_0} from one of its stable sides to its other again without rupture (see Fig. 2.1b).

Because the sets $\mathscr{C}_s(\omega_.)$ and $\mathscr{C}_u(\omega_.)$ are not empty, because the stable and unstable manifolds intersect transversally in hyperbolic systems, and because there is no rupture in the segments composing these sets, there exists at least one point at their intersection

$$\exists \, \mathbf{x} \in \mathscr{C}_s(\omega_.) \cap \mathscr{C}_u(\omega_.) \, , \tag{2.44}$$

which is the statement (2.40). *Q.E.D.*

This theorem shows the role of the definition of the Markov partition to guarantee the existence of at least one point corresponding to each sequence $\omega_.$ with *only* binary constraints expressed by the transition matrix (2.39).

In continuous hyperbolic systems, the stable and unstable manifolds extend without rupture over the whole phase space so that cells of finite sizes can be constructed. As a consequence, Markov partitions with a finite number L of cells exist. However, in bounded hyperbolic billiards, the stable and unstable manifolds are broken into pieces by the singularities so that the Markov partition contains cells of arbitrarily small sizes in such systems and the partition has countably many cells (Bunimovich and Sinai 1980).

We remark that the symbolic sequence is unique if the cells of the partition are not overlapping ($C_i \cap C_j = \emptyset$). It is possible to construct such a nonoverlapping Markov partition for hyperbolic repellers with escape. However, this is no longer the case in bounded hyperbolic systems where the cells cannot avoid overlapping because we need to cover the whole phase space. In this bounded case, the correspondence with the symbolic sequences is one-to-one only for trajectories which never pass through the border of the partition, i.e., for the trajectories of $\mathscr{I} \setminus \cup_{n \in \mathbb{Z}} \Phi^n(\partial \mathscr{C})$. This is not a problem for measure-theoretic properties because the ambiguity of the correspondence only occurs on the border of the partition which is of zero Lebesgue measure. But this does become a problem for counting periodic orbits. We shall come back on this point later.

2.4 Topological entropy

Our aim here is to characterize quantitatively the topological dynamics of general hyperbolic or nonhyperbolic systems. In this regard, we construct some coarse graining of phase space and we count how the elements of the coarse graining proliferate under the action of time evolution.

2.4.1 Nets and separated subsets

The problem of counting the elements of a set up to a precision has been clarified by Kolmogorov and Tikhomirov (1959) and later applied to topological dynamics by Bowen (1975) and Walters (1981). The idea is to consider a time-dependent distance between trajectories in phase space:

$$d_T(\mathbf{X},\mathbf{Y}) \;=\; \mathrm{Max}_{-T\leq t\leq +T}\; \|\mathbf{\Phi}^t\mathbf{X} \;-\; \mathbf{\Phi}^t\mathbf{Y}\| \;, \tag{2.45}$$

where $\|\cdot\|$ is a standard distance in the phase space itself. This time-dependent distance allows us to distinguish trajectories of the system.

A subset $\mathscr{K} \subseteq \mathscr{I}$ is then approximated up to a given precision ϵ by trajectories which are at a distance ϵ from each other. This is made precise with the following definitions:

1. *A subset $\mathscr{R}(\epsilon, T, \mathscr{K})$ is an (ϵ, T)-net or (ϵ, T)-spanning set for the set $\mathscr{K}$ if each point $\mathbf{X}$ of $\mathscr{K}$ is at a distance less than ϵ from a point $\mathbf{Y}$ of $\mathscr{R}$, i.e., $\forall \mathbf{X} \in \mathscr{K}$, $\exists \mathbf{Y} \in \mathscr{R}$: $d_T(\mathbf{X},\mathbf{Y}) \leq \epsilon$.*

2. *A subset $\mathscr{S}(\epsilon, T, \mathscr{K})$ of $\mathscr{K}$ is (ϵ, T)-separated if all the points of $\mathscr{S}$ are separated by a distance greater than ϵ: $d_T(\mathbf{X},\mathbf{Y}) > \epsilon$, $\forall \mathbf{X} \neq \mathbf{Y} \in \mathscr{S}$.*

The extremal numbers of elements contained in these counting devices are defined by

$$r(\epsilon, T, \mathscr{K}) \;=\; \mathrm{Min}_{\mathscr{R}}\; |\mathscr{R}(\epsilon, T, \mathscr{K})| \;, \tag{2.46}$$

$$s(\epsilon, T, \mathscr{K}) \;=\; \mathrm{Max}_{\mathscr{S}}\; |\mathscr{S}(\epsilon, T, \mathscr{K})| \;. \tag{2.47}$$

These numbers are interrelated by the following inequalities:

$$s(2\epsilon, T, \mathscr{K}) \;\leq\; r(\epsilon, T, \mathscr{K}) \;\leq\; s(\epsilon, T, \mathscr{K}) \;, \tag{2.48}$$

proved in Walters (1981), so that nets or separated subsets lead to similar counting.

2.4.2 Definition and properties

Since the trajectories separate at most exponentially with the time T in deterministic dynamical systems we expect that these numbers also increase exponentially with T because the points may become separated by smaller and smaller distances according to

$$d_T(\mathbf{X},\mathbf{Y}) \;\simeq\; \|\mathbf{X}-\mathbf{Y}\|\; \exp[+2\lambda^{(1)}(\mathbf{X})T] \;, \tag{2.49}$$

if $\mathbf{X}$ and $\mathbf{Y}$ are close enough. We are therefore interested in the maximum rate of growth of (2.46)–(2.47) which is the topological entropy per unit time with respect to the compact subset $\mathscr{K} \subseteq \mathscr{I}$:

$$h_{\text{top}}(\mathbf{\Phi}, \mathscr{K}) = \lim_{\epsilon\to 0} \lim \sup_{T\to\infty} \frac{1}{2T} \log r(\epsilon, T, \mathscr{K}) , \tag{2.50}$$

$$= \lim_{\epsilon\to 0} \lim \sup_{T\to\infty} \frac{1}{2T} \log s(\epsilon, T, \mathscr{K}) . \tag{2.51}$$

The *topological entropy per unit time* of the invariant subset $\mathscr{I}$ is then defined as

$$h_{\text{top}}(\mathbf{\Phi}, \mathscr{I}) = \text{Sup}_{\mathscr{K}\subseteq\mathscr{I}}\, h_{\text{top}}(\mathbf{\Phi}, \mathscr{K}) . \tag{2.52}$$

The topological entropy has the important property of being invariant under homeomorphisms, i.e., under continuous transformations of the dynamical system (Walters 1981).

When the topological entropy is positive the invariant subset $\mathscr{I}$ contains infinitely many trajectories in contrast to invariant subsets with a single or finitely many periodic orbits for which the topological entropy vanishes. Therefore, the motion may be expected to be nonperiodic on invariant subsets with positive topological entropy. We have therefore the definition of *topological chaos*:

An invariant subset $\mathscr{I}$ is said to be topologically chaotic if its topological entropy is positive.

Topological chaos is thus characterized by an exponential proliferation of orbits which are distinguishable over longer and longer time intervals. This exponential proliferation also manifests itself in the growth of the set of words of the symbolic dynamics with their symbolic length. All the words allowed in a symbolic dynamics form a tree graph, which gives all the possible paths followed by the dynamics discretized by a partition. In this regard, we remark that a system which is *deterministic* in a *real* phase space $\mathscr{M}$ may become, after partitioning, *nondeterministic* in the *discrete* state space formed by the cells of the partition (or by the alphabet). This is the main characteristic feature of topological chaos.

For a hyperbolic map homeomorphic to a Markov shift, the topological entropy per iteration is given by the logarithm of the largest eigenvalue of the transition matrix A, which gives the proliferation rate of the words in the symbolic dynamics (Walters 1981). Let us here mention that a Markov partition is said to be *generating* if it is fine enough for the symbolic dynamics to provide a faithful representation of the phase-space dynamics. In this case, the symbolic and the phase-space dynamics are homeomorphic and the topological entropy of the symbolic dynamics is equal to the topological entropy calculated in the phase space (see e.g. Grossmann and Thomae 1977).

Hereafter, we shall show that the exponential proliferation also concerns the periodic orbits of topologically chaotic systems.

2.5 The spectrum of periodic orbits

We propose to list the periodic orbits of an invariant subset in a flow. The previous sections have shown the importance of periodic orbits which are dense among the nonperiodic orbits in some hyperbolic systems, in particular. Listing and counting periodic orbits are thus important operations in the detailed analysis of dynamical systems because the periodic orbits characterize important aspects of the time evolution.

2.5.1 Periodic orbits and the topological zeta function

Countably many periodic orbits can be found in a topologically chaotic invariant set. We have already observed with (2.3) that a prime period T_p is associated with each periodic orbit and that repetitions of the period exist at multiples $T = rT_p$ of the prime period, where r is the repetition number. Let us denote by $\mathcal{P}$ the set of all the prime periodic orbits.

We introduce the dimensionless distribution of the periods according to

$$\rho(t) = \sum_{r=1}^{\infty} \sum_{p\in\mathcal{P}} T_p \; \delta(t - rT_p) \, , \tag{2.53}$$

where $\delta(t)$ denotes the Dirac distribution. The distribution (2.53) defines the spectrum of the periods and of their repetitions. We may also introduce the corresponding cumulative distribution function

$$R(T) = \int_0^T \rho(t)dt = \sum_{r=1}^{\infty} \sum_{p\in\mathcal{P}} T_p \; \theta(t - rT_p) \, , \tag{2.54}$$

which we expect to grow indefinitely in chaotic systems. In Eq. (2.54), $\theta(t)$ denotes the Heaviside step function such that $\delta(t) = (d/dt)\theta(t)$.

The Laplace transform of the distribution (2.53) can be written as

$$\begin{aligned} \int_0^{\infty} \rho(t) \; \exp(-st) \; dt &= \sum_{r=1}^{\infty} \sum_{p\in\mathcal{P}} T_p \; \exp(-srT_p) \\ &= -\frac{\partial}{\partial s} \sum_{p\in\mathcal{P}} \sum_{r=1}^{\infty} \frac{1}{r} \; \exp(-srT_p) \\ &= -\frac{\partial}{\partial s} \ln \zeta_{\rm top}(s) \\ &= -\frac{\zeta'_{\rm top}(s)}{\zeta_{\rm top}(s)} \, , \end{aligned} \tag{2.55}$$

in terms of the so-called *topological zeta function*

$$\zeta_{\text{top}}(s) = \prod_{p \in \mathcal{P}} \frac{1}{1 - \exp(-sT_p)} , \tag{2.56}$$

which is a product over all the prime periodic orbits of the invariant set $\mathcal{I}$ (Smale 1980b, Parry and Pollicott 1990). Consequently, the analytic structure of the zeta function determines the structure of the spectrum of periodic orbits and, in particular, of the rate of growth of the cumulative function (2.54).

The set of prime periodic orbits with periods $T_p < T$ form a separated subset inside $\mathcal{I}$ with respect to the distance (2.45). We may thus expect that the topological entropy per unit time is an upper bound on the growth of the cumulative function (2.54). Therefore, the periodic orbits proliferate at most exponentially in deterministic dynamics systems. Moreover, we may conclude that the zeta function (2.56) is absolutely convergent in the half complex plane (Ruelle 1978)

$$\text{Re}\, s > h_{\text{top}}(\mathbf{\Phi}, \mathcal{I}) . \tag{2.57}$$

In the following, we shall show that the zeta function of hyperbolic sets actually has a pole at the topological entropy.

2.5.2 Prime periodic orbits and fixed points

The study of the periodic orbits can be simplified by the introduction of a surface of section, which reduces the flow to a Poincaré map $\boldsymbol{\phi}$. The periodic orbits of the flow have several intersections with the surface of section, which are fixed points of some iterate $\boldsymbol{\phi}^n$ of the Poincaré map. The periodic orbits can thus be characterized by an integer n_p which is the number of intersections of the prime periodic orbit with the surface of section. A periodic orbit generates fixed points for all the iterates ϕ^n such that $n = rn_p$. Reciprocally, among the fixed points of $\boldsymbol{\phi}^n$, we must distinguish between those associated with a prime periodic orbit of period n and those associated with the rth repetition of a prime periodic orbit of smaller period $n_p = n/r$. In particular, a prime periodic orbit of period n contributes to n distinct fixed points of $\boldsymbol{\phi}^n$ although the rth repetition of a prime periodic orbit of period n_p contributes to only n_p distinct fixed points of $\boldsymbol{\phi}^{rn_p}$. Accordingly, we have that

$$|\text{Fix}\, \boldsymbol{\phi}^n| = \sum_{\substack{1 \leq m \leq n \\ m \text{ divides } n}} m\, |\mathcal{P}_m| , \tag{2.58}$$

where $\mathscr{P}_m$ denotes the set of all the prime periodic orbits of period m. The number of prime periodic orbits can be obtained by the Möbius inversion formula

$$|\mathscr{P}_n| = \frac{1}{n} \sum_{\substack{1 \leq l \leq n \\ l \text{ divides } n}} \mu(l) \, |\text{Fix } \phi^{n/l}| \,, \tag{2.59}$$

where the Möbius function is defined by $\mu(1) = 1$, $\mu(l) = (-)^k$ if $l = p_1 \dots p_k$ factorizes into distinct primes p_i, and $\mu(l) = 0$ if l contains a power of a prime (Smale 1980b, Artuso et al. 1990). This counting is illustrated in Table 2.1 for a Bernoulli shift.

Table 2.1 List of fixed points Fix σ^n and the corresponding prime periodic orbits $\mathscr{P}_n$ of the dyadic Bernoulli shift.

period n	Fix σ^n	$\lvert$Fix $\sigma^n\rvert$	$\mathscr{P}_n$	$\lvert\mathscr{P}_n\rvert$
1	0		0	
	1	2	1	2
2	00			
	01			
	10			
	11	4	01	1
3	000			
	001			
	010			
	011			
	100			
	101			
	110		001	
	111	8	011	2
4	0000			
	0001			
	0010			
	⋮		0001	
			0111	
	1111	16	0011	3
5	00000			
	00001		00001	
	00010		00011	
			00101	
	⋮		00111	
			01101	
	11111	32	01111	6
⋮	⋮	⋮	⋮	⋮

Let us remark that, for the Markov shift induced by a nonoverlapping partition, the number of fixed points of ϕ^n can be obtained directly from the transition matrix by

$$|\text{Fix}\ \phi^n| \ = \ |\text{Fix}\ \sigma^n| \ = \ \text{tr}\ \mathsf{A}^n\ . \tag{2.60}$$

2.5.3 How to order a sum over periodic orbits?

A strong analogy exists between the prime periodic orbits and the prime integer numbers (Ruelle 1992). A prime periodic orbit has a period which cannot be further divided into a smaller period in the same way as a prime integer cannot be divided by a smaller integer, as shown in Table 2.2.

Horizontally, the length n of the words of the symbolic dynamics increases. These words correspond to fixed points of the map ϕ^n. That n is divisible by an integer r means that there exist prime periodic orbits of smaller periods $n_p = n/r$. If we have all the divisors of n we have all the prime periods which exist in the set Fix ϕ^n. Table 2.2 gives all the possible factorizations of the lengths n into the product rn_p for the first few integers. The integer n_p is thus the prime period while r is the repetition number of the prime period n_p in the full period n.

According to this factorization, we can order a sum column after column (over n) and in each column from top to bottom (over n_p) in Table 2.2. On the other hand, the sum can be reordered by summing on the oblique lines with given repetition number r and, thereafter, on the prime period n_p which grows

Table 2.2 List of the periods n and of their factorization into the prime period n_p and the repetition number r in the full period.

$n =$	1	2	3	4	5	6	7	8	9	10	...
$(n_p, r) =$	$(1,1)$	(1,2)	(1,3)	(1,4)	(1,5)	(1,6)	(1,7)	(1,8)	(1,9)	(1,10)	...
		(2,1)		(2,2)		(2,3)		(2,4)		(2,5)	...
			(3,1)			(3,2)			(3,3)		...
				(4,1)				(4,2)			...
					(5,1)					(5,2)	...
						(6,1)					...
							(7,1)				...
								(8,1)			...
									(9,1)		...
										(10,1)	...
							⋮				⋱

without constraint on each oblique line. A sum over all the fixed points of ϕ can thus be decomposed as

$$\sum_{n=1}^{\infty} \sum_{\text{Fix } \phi^n} (\cdot) = \sum_{p \in \mathscr{P}} \sum_{r=1}^{\infty} n_p \, (\cdot) \, , \tag{2.61}$$

as a sum over the prime periodic orbits $p \in \mathscr{P}$ and their repetitions $r = 1, 2, 3, \ldots$ The factor n_p appears in the right-hand side because each periodic orbit contributes by n_p fixed points.

2.5.4 The topological entropy of a hyperbolic invariant set

In a hyperbolic system, the number of fixed points of ϕ^n is also the number of elements in a partition of phase space into cells corresponding to words $\omega_1 \cdots \omega_n$ of length n. Therefore, these fixed points form a minimal net for the invariant set. The growth rate of the number of cells in this minimal net is the topological entropy per unit time according to (2.50)–(2.52) (Walters 1981). The cumulative function $R(T)$ corresponds precisely to the number of fixed points of the Poincaré map because of Eq. (2.61) and the presence of the factor T_p in Eq. (2.54). Therefore, we have that

$$R(T) \sim \frac{\exp(T h_{\text{top}})}{h_{\text{top}}} \, . \tag{2.62}$$

The topological zeta function (2.56) has thus a pole at $s = h_{\text{top}}$ in the case of hyperbolic systems.

Since the number of prime periodic orbits with a prime period less than T is given approximately by dividing the number of fixed points by the period according to (2.59), we obtain the estimation

$$\text{Number} \left\{ p \in \mathscr{P} \, : \, T_p < T \right\} \sim \frac{\exp(T h_{\text{top}})}{T h_{\text{top}}} \, , \tag{2.63}$$

for the hyperbolic systems (Parry and Pollicott 1983, 1990).

2.6 The topological zeta function of hyperbolic systems

As aforementioned, we must distinguish between the case of nonoverlapping partition in which there is a one-to-one correspondence with a Markov shift σ and the case of overlapping partitions in which there is no such correspondence.

2.6.1 The topological zeta function for a nonoverlapping partition

In this case, a precise counting of the periodic orbits of the hyperbolic repeller can be carried out because the one-to-one correspondence (2.30) exists between the map $\boldsymbol{\phi}$ and the Markov shift $\boldsymbol{\sigma}$. Let us consider the distribution (2.53) of periods. If the first-return time function of (1.5) is known, the prime period can be written as

$$T_p \;=\; \tau_{\omega_.} \;+\; \tau_{\sigma\omega_.} \;+\; \ldots \;+\; \tau_{\sigma^{n_p}\omega_.} \;, \tag{2.64}$$

where $\tau_{\omega_.} = T[\mathbf{x} = \boldsymbol{\Theta}(\omega_.)]$ according to (2.30). Converting the sum over prime periodic orbits into a sum over the fixed points of the shift, we have

$$\begin{aligned}\rho(t) \;&=\; \sum_{r=1}^{\infty}\sum_{p\in\mathscr{P}}(\tau_{\omega_.} + \ldots + \tau_{\sigma^{n_p}\omega_.})\;\delta\Big[t - r(\tau_{\omega_.} + \ldots + \tau_{\sigma^{n_p}\omega_.})\Big]\\ &=\; \sum_{n=1}^{\infty}\frac{1}{n}\sum_{\omega_.\in\mathrm{Fix}\sigma^n}(\tau_{\omega_.} + \ldots + \tau_{\sigma^n\omega_.})\;\delta\Big(t - \tau_{\omega_.} - \ldots - \tau_{\sigma^n\omega_.}\Big)\;,\end{aligned} \tag{2.65}$$

where we used (2.61). Taking the Laplace transform of this distribution to get the corresponding topological zeta function, we obtain

$$\zeta_{\mathrm{top}}(s) \;=\; \exp\sum_{n=1}^{\infty}\frac{1}{n}\sum_{\omega_.\in\mathrm{Fix}\sigma^n}\exp\Big[-s(\tau_{\omega_.} + \ldots + \tau_{\sigma^n\omega_.})\Big]\;. \tag{2.66}$$

Rewriting this expression in terms of the Poincaré map (1.5) we find (Ruelle 1983)

$$\zeta_{\mathrm{top}}(s) \;=\; \exp\sum_{n=1}^{\infty}\frac{1}{n}\sum_{\mathbf{x}\in\mathrm{Fix}\boldsymbol{\phi}^n}\exp\Big[-s\sum_{j=1}^{n}T(\boldsymbol{\phi}^j\mathbf{x})\Big]\;. \tag{2.67}$$

The zeta function of the system $\boldsymbol{\Phi}$ can here be reduced to the zeta function of the Markov shift $\boldsymbol{\sigma}$

$$\zeta_{\mathrm{top}}(s;\boldsymbol{\Phi}) \;=\; \zeta_{\mathrm{top}}(s;\boldsymbol{\sigma})\;. \tag{2.68}$$

2.6.2 The topological zeta function for an overlapping partition

However, in bounded hyperbolic systems, the cells of the Markov partition overlap so that periodic orbits passing by the border of the partition may be counted several times and the one-to-one correspondence be lost.

Manning (1971) has shown that extra symbolic dynamics of subshifts $\boldsymbol{\sigma}_{\mathrm{i}}$ can be constructed on the basis of the Markov shift $\boldsymbol{\sigma}$. This construction can take into account the redundant counting. The extra subshifts $\boldsymbol{\sigma}_{\mathrm{i}}$ are labelled by

k-uples of integers $\mathbf{i} = (i_1, \ldots i_k)$. The number of fixed points of the map $\boldsymbol{\phi}$ can then be counted as

$$|\mathrm{Fix}\boldsymbol{\phi}^n| = \sum_{\mathbf{i}} (-)^{k+1} \, |\mathrm{Fix}\boldsymbol{\sigma}_{\mathbf{i}}^n| \, . \tag{2.69}$$

As a consequence, the zeta function of the system $\mathbf{\Phi}$ is a ratio of zeta functions for the different subshifts

$$\zeta_{\mathrm{top}}(s; \mathbf{\Phi}) = \frac{\prod_{\mathbf{i}:\, k \text{ odd}} \zeta_{\mathrm{top}}(s; \boldsymbol{\sigma}_{\mathbf{i}})}{\prod_{\mathbf{i}:\, k \text{ even}} \zeta_{\mathrm{top}}(s; \boldsymbol{\sigma}_{\mathbf{i}})} \, , \tag{2.70}$$

(see also Ruelle 1978).

In order to show the necessity of this counting, let us consider a simple example. For the r-adic map on the unit interval

$$\phi : \qquad x_{n+1} = (r x_n) \, , \quad \text{mod. } 1 \, , \tag{2.71}$$

the number of fixed points is simply $|\mathrm{Fix}\phi^n| = r^n$ and the zeta function is

$$\zeta_{\mathrm{top}}(s) = \frac{1}{1 - r e^{-s}} \, . \tag{2.72}$$

However, if the r-adic map is considered on the unit circle then ϕ has only one fixed point instead of two so that $|\mathrm{Fix}\phi^n| = r^n - 1$ and the zeta function is

$$\zeta_{\mathrm{top}}(s) = \frac{1 - e^{-s}}{1 - r e^{-s}} \, . \tag{2.73}$$

The same phenomenon happens in the Arnold cat map (MacKay 1993a).

Zeta functions similar to the topological zeta function also arise in the Liouvillian dynamics of such dynamical systems as we shall see in the next chapter.

Chapter 3

Liouvillian dynamics

3.1 Statistical ensembles

In the Introduction, we have argued that a statistical approach to chaotic dynamical systems is required for the study of long-time evolutions. With this aim, we consider ensembles of trajectories in phase space

$$\mathbf{X}_t^{(i)} = \mathbf{\Phi}^t \mathbf{X}_0^{(i)} \in \mathcal{M} \,, \tag{3.1}$$

where $i \in \mathbb{N}$. At each time t, these trajectories form a cloud of points in phase space. Let us remark that the statistical ensemble of initial conditions $\{\mathbf{X}_0^{(i)}\}$ is generated by launching repetitively the system into motion. The launching is often performed by a device external to the system itself. The precision achieved on the initial condition is generally limited by structural and mechanical irregularities of the launching device, which is at the origin of the need of statistical ensembles.

In a classical system, the physical observables are represented by functions $A(\mathbf{X})$ defined on the phase space. These functions are usually regular functions which may be analytic, or piecewise continuous or differentiable. Examples of observables are the momentum or the kinetic energy of the particles, or the indicator function of a domain of phase space.

The average of the observable A over the statistical ensemble (3.1) is defined by

$$\langle A \rangle = \lim_{N\to\infty} \frac{1}{N} \sum_{i=1}^{N} A(\mathbf{X}^{(i)}) . \tag{3.2}$$

The statistical ensemble may be represented by a density distribution which is here given by a weighted sum of Dirac distributions

$$f(\mathbf{X}) = \lim_{N\to\infty} \frac{1}{N} \sum_{i=1}^{N} \delta(\mathbf{X} - \mathbf{X}^{(i)}) , \tag{3.3}$$

which is normalized as

$$\int_{\mathcal{M}} f(\mathbf{X})\, d\mathbf{X} = 1 . \tag{3.4}$$

If the points $\{\mathbf{X}^{(i)}\}$ do not accumulate on a lower dimensional subset of phase space in the limit $N \to \infty$ the statistical ensemble can be represented by a regular probability density function f. We shall be concerned with such statistical ensembles which are statistically regular with respect to the system itself. The points $\{\mathbf{X}^{(i)}\}$ of statistically regular ensembles may be distributed for instance by using a pseudo-random generator which is known to emulate the probability density f (Knuth 1969, Press et al. 1986).

The average (3.2) may thus be expressed as

$$\langle A \rangle = \int_{\mathcal{M}} d\mathbf{X}\, A(\mathbf{X})\, f(\mathbf{X}) . \tag{3.5}$$

We remark that the convergence (3.3) to a regular density function f is an important question, especially, if we further consider the time evolution of the density f in the long-time limit.

3.2 Time evolution of statistical ensembles

3.2.1 Liouville equation

For flows, the time evolution of the probability density representing a statistical ensemble can be established by invoking the principle of conservation of probability according to which none of the members of the statistical ensemble disappears during the time evolution (1.1). In this respect, we obtain the continuity equation in phase space (Nicolis 1995)

$$\partial_t f + \mathrm{div}(\mathbf{F} f) = 0 . \tag{3.6}$$

Therefore, the time evolution of the probability density is ruled by a linear partial differential equation of first order in the time and in the phase-space variables, which is called the (generalized) Liouville equation

$$\partial_t f = \hat{L} f , \qquad \text{where} \quad \hat{L}(\cdot) = -\operatorname{div}(\mathbf{F} \cdot) , \tag{3.7}$$

is the so-called (generalized) Liouville operator.[1] The Liouville operator is the infinitesimal generator of the group of time translations induced in the space of the probability densities.

For Hamiltonian systems, the operator (3.7) reduces to the usual Liouville operator given by the Poisson bracket of the Hamiltonian (Balescu 1975)

$$\hat{L}(\cdot) = \{H, \cdot\} . \tag{3.8}$$

Because Liouville's equation is a linear equation the solution of the Liouville equation can be formally written as

$$f_t = e^{\hat{L}t} f_0 , \tag{3.9}$$

if the vector field $\mathbf{F}$ is time-independent. Otherwise, the solution is given by a Dyson series.

3.2.2 Frobenius–Perron and Koopman operators

Because Liouville's equation is a partial differential equation of first order in the variables $(t, \mathbf{X})$, its solutions are obtained by solving the underlying ordinary differential equation (1.1), according to the method of characteristics. Knowing the trajectories (1.2) of the system, the average (3.5) evolves in time as

$$\langle A \rangle_t = \int_{\mathcal{M}} d\mathbf{X}\, A(\mathbf{\Phi}^t \mathbf{X})\, f_0(\mathbf{X}) , \tag{3.10}$$

where f_0 is the initial density. This equation can be rewritten as

$$\langle A \rangle_t = \int_{\mathcal{M}^2} d\mathbf{X}\, d\mathbf{Y}\, A(\mathbf{X})\, \delta(\mathbf{X} - \mathbf{\Phi}^t \mathbf{Y})\, f_0(\mathbf{Y}) , \tag{3.11}$$

where $\delta(\mathbf{X} - \mathbf{\Phi}^t \mathbf{Y})$ may be considered as the conditional probability density for the trajectory to be located at the point $\mathbf{X}$ by the time t if the initial condition was the point $\mathbf{Y}$. This conditional probability density defines the kernel of the evolution operator. The propagator of the classical trajectories is thus given by

1. This equation has been named according to the French mathematician Liouville (1809–1882) who is at the origin of the famous Liouville's theorem of classical mechanics. This reference to Liouville for the case of his theorem has been discussed by J. Lützen (1990) who explained the role of L. Boltzmann in this citation. It turns out that the proof of the Liouville theorem in many textbooks since Tolman (1938) has more and more emphasized the importance of the equation itself which explains why the equation today bears his name (see for instance Prigogine 1962).

a Dirac distribution due to the deterministic, point-like character of classical time evolution.[2] The support of this Dirac distribution is punctual in the phase space of dimension $M = \dim \mathcal{M}$.

Equation (3.11) defines two operators which are adjoint of each other according to

$$\langle A \rangle_t = \langle \hat{P}^{\dagger t} A | f_0 \rangle = \langle A | \hat{P}^t f_0 \rangle , \tag{3.12}$$

with the notation $\langle A|f \rangle = \int A(\mathbf{X}) f(\mathbf{X}) d\mathbf{X}$. The time evolution of the probability density is ruled by the *Frobenius–Perron operator* defined by

$$f_t(\mathbf{X}) = \hat{P}^t f_0(\mathbf{X}) \equiv \int_{\mathcal{M}} \delta(\mathbf{X} - \mathbf{\Phi}^t \mathbf{Y}) \, f_0(\mathbf{Y}) \, d\mathbf{Y} , \tag{3.13}$$

while the time evolution of the observables is ruled by the *Koopman operator* which is its adjoint

$$A_t(\mathbf{Y}) = \hat{P}^{\dagger t} A(\mathbf{Y}) \equiv \int_{\mathcal{M}} \delta(\mathbf{X} - \mathbf{\Phi}^t \mathbf{Y}) \, A(\mathbf{X}) \, d\mathbf{X} . \tag{3.14}$$

For a system given by an invertible map $\mathbf{\Phi}^t$, the Frobenius–Perron operator becomes

$$\begin{aligned} f_t(\mathbf{X}) = \hat{P}^t f_0(\mathbf{X}) &= \frac{f_0(\mathbf{\Phi}^{-t}\mathbf{X})}{\left| \det \partial_{\mathbf{X}} \mathbf{\Phi}^t (\mathbf{\Phi}^{-t}\mathbf{X}) \right|} \\ &= f_0(\mathbf{\Phi}^{-t}\mathbf{X}) \exp\left[-\int_0^t \mathrm{div}\mathbf{F}(\mathbf{\Phi}^{\tau - t}\mathbf{X}) \, d\tau \right] , \end{aligned} \tag{3.15}$$

where we used Eq. (1.4). For an invertible conservative system, the Frobenius–Perron operator reduces to the substitution operator

$$f_t(\mathbf{X}) = \hat{P}^t f_0(\mathbf{X}) = f_0(\mathbf{\Phi}^{-t}\mathbf{X}) , \qquad \text{for} \quad \mathrm{div}\mathbf{F} = 0 . \tag{3.16}$$

3.2.3 Boundary conditions

The fact that Liouville's equation is a partial differential equation has the important consequence that boundary conditions are required for the resolution of Liouville's equation as for any partial differential equation. This is of particular importance in open systems with trajectories ingoing and outgoing the phase space $\mathcal{M}$. As an example, for the simulation of Rayleigh–Bénard convection under gravitation and a temperature gradient, the particles hitting the upper and lower walls are reinjected according to a Maxwellian velocity distribution with a temperature parameter fixed to the temperature of the corresponding

2. The point-like character of classical mechanics provides a separability between different points in phase space. As soon as $\mathbf{X}' \neq \mathbf{\Phi}^t \mathbf{X} \ \forall t \in \mathbb{R}$, the trajectories from the initial conditions $\mathbf{X}$ and $\mathbf{X}'$ may have completely independent long-time evolutions.

wall (Mareschal et al. 1988). In this case, the boundary condition is local in time and on the border $\partial\mathscr{M}$ of the phase space.

In general, we may separate the border of the phase space of an open system into a region where the trajectories enter and another where they exit

$$\partial\mathscr{M} = \partial_{\text{in}}\mathscr{M} \cup \partial_{\text{out}}\mathscr{M} , \qquad (3.17)$$

such that

$$\begin{aligned} \mathbf{n}(\mathbf{X}) \cdot \mathbf{F}(\mathbf{X}) &\leq 0 , \qquad \text{if} \quad \mathbf{X} \in \partial_{\text{in}}\mathscr{M} , \\ \mathbf{n}(\mathbf{X}) \cdot \mathbf{F}(\mathbf{X}) &> 0 , \qquad \text{if} \quad \mathbf{X} \in \partial_{\text{out}}\mathscr{M} , \end{aligned} \qquad (3.18)$$

where $\mathbf{n}(\mathbf{X})$ is a vector perpendicular to $\partial\mathscr{M}$ and pointing to the exterior of $\mathscr{M}$. Boundary conditions may only be imposed on $\partial_{\text{in}}\mathscr{M}$ because of causality.

In general, nonlocal boundary conditions may be considered in terms of the functional

$$f_t(\mathbf{X})\Big|_{\mathbf{X}\in\partial_{\text{in}}\mathscr{M}} = \mathscr{F}\Big\{\mathbf{X}, t \Big| f_\tau(\mathbf{Y}) \ : \quad \mathbf{Y} \in \partial\mathscr{M} , \ \tau \in]-\infty, t]\Big\} , \qquad (3.19)$$

to model different interaction processes with the surfaces. In this way, general nonequilibrium conditions may be imposed on the system. The following boundary conditions are often considered:

1. Periodic boundary conditions

The particles move for instance in a rectangular domain in which opposite walls are identified. In this way, particles exiting at one wall are reinjected at the opposite wall. Other domains of $\mathbb{R}^d$ may be considered under the condition that they tessellate $\mathbb{R}^d$ according to a lattice ℓ. At the level of the probability density, the periodic boundary conditions are expressed by

$$f_t(\mathbf{q}, \mathbf{p}) = f_t(\mathbf{q} + \mathbf{l}, \mathbf{p}) , \quad \text{with } \mathbf{l} \in \mathscr{L} = \ell^N , \qquad (3.20)$$

for a Hamiltonian like (1.7) with $\mathbf{X} = (\mathbf{q}, \mathbf{p})$.

2. Reflecting boundary conditions

Here, the particles are contained in a box on the walls of which they undergo elastic collisions. The box may have a general shape. If it is rectangular (or tessellate $\mathbb{R}^d$), it is equivalent to some periodic boundary condition. In both cases, the invariant measure is Liouville's one.

3. Quasiperiodic boundary conditions

Let us suppose that N particles move in a domain which tessellate $\mathbb{R}^d$ as for periodic boundary conditions. The vector field (1.1) or (1.9) is therefore symmetric under discrete position translations: $\mathbf{F}(\mathbf{q}, \mathbf{p}) = \mathbf{F}(\mathbf{q} + \mathbf{l}, \mathbf{p})$ with $\mathbf{l} \in$

$\mathcal{L} = \ell^N$. However, the probability density is allowed to extend nonperiodically over the whole lattice so that the periodic boundary condition (3.20) does not apply. Instead a Fourier transform must be carried out in position space to reduce the dynamics to the cell at the origin of the lattice ($\mathbf{l} = 0$). A wavenumber $\mathbf{k}$ is introduced which varies continuously in a Brillouin zone reciprocal to the lattice $\mathcal{L}$:

$$f_{t,\mathbf{k}}(\mathbf{q}, \mathbf{p}) = \sum_{\mathbf{l} \in \mathcal{L}} \exp(-i\mathbf{k} \cdot \mathbf{l}) \, f_t(\mathbf{q} + \mathbf{l}, \mathbf{p}) \,, \tag{3.21}$$

which is the Liouvillian analogue of Bloch's or Floquet's methods (Kittel 1976).

4. Absorbing boundary conditions

For this other case, the probability density is assumed to vanish at the border $\partial_{\mathrm{in}}\mathcal{M}$ of the phase-space domain $\mathcal{M}$

$$f_t(\mathbf{X})\Big|_{\mathbf{X} \in \partial_{\mathrm{in}}\mathcal{M}} = 0 \,. \tag{3.22}$$

As a consequence, the number N_t of trajectories inside the domain $\mathcal{M}$ decreases in time in a way that depends on the dynamics in $\mathcal{M}$, which can be viewed as the escape of trajectories out of $\mathcal{M}$. Absorbing boundary conditions thus lead to the escape-rate formalism (see below).

5. Flux boundary conditions

In order to describe a nonequilibrium system in which there is a nontrivial steady state, we assume that a flux of trajectories is continuously flowing across the phase space $\mathcal{M}$ between points of the boundary $\partial\mathcal{M}$. Such nonequilibrium situations are modelled by supposing that the density at the boundary is specified as

$$f_t(\mathbf{X})\Big|_{\mathbf{X} \in \partial_{\mathrm{in}}\mathcal{M}} = \mathcal{F}(\mathbf{X}) \,. \tag{3.23}$$

The different preceding boundary conditions may be combined together.

3.3 Invariant measures

3.3.1 Definition

An invariant measure is a measure which is stationary under the time evolution of the system, i.e., such that

$$\mu_{\mathrm{i}}(\mathbf{\Phi}^{-t} U) = \mu_{\mathrm{i}}(U) \,, \tag{3.24}$$

for any volume $U \subset \mathcal{M}$ in phase space. In Eq. (3.24), the subscript i stands for 'invariant'. The measure is a probability measure if it can be normalized to $\mu_{\rm i}(\mathcal{M}) = 1$.

The density which corresponds to the measure according to $f_{\rm i} = d\mu_{\rm i}/d\mathbf{X}$ is a stationary solution of Liouville's equation (3.6): $\hat{L}f_{\rm i} = 0$. Equivalently, it is an eigenstate of eigenvalue 1 for the Frobenius–Perron operator: $\hat{P}^t f_{\rm i} = f_{\rm i}$. Recent works on dissipative systems have emphasized the importance of singular invariant measures for which the density $f_{\rm i}$ is a distribution, as is the case for an attracting stationary point where $f_{\rm i}(\mathbf{X}) = \delta(\mathbf{X} - \mathbf{X}_s)$. Therefore, we may expect a large variety of different types of invariant measures, especially in nonequilibrium systems.

We notice that different boundary conditions can define dynamical systems with different phase spaces and different invariant measures, even if the dynamics $\mathbf{\Phi}^t$ is locally the same inside the boundaries.

The nature of the invariant measure can be investigated by considering the integral of Liouville's equation (3.6) over the phase space

$$\frac{d}{dt} \int_{\mathcal{M}} f \, d\mathbf{X} = - \int_{\partial_{\rm in}\mathcal{M}} f \, \mathbf{F} \cdot \mathbf{n} \, dS - \int_{\partial_{\rm out}\mathcal{M}} f \, \mathbf{F} \cdot \mathbf{n} \, dS \, , \tag{3.25}$$

where the first term in the right-hand member is positive according to (3.18) while the second is negative. A nontrivial invariant measure is supported by the plain phase space only if the right-hand member of (3.25) vanishes for $f_{\rm i}$. This is not the case for the absorbing boundary conditions (3.22) which should be treated separately in a following chapter on chaotic scattering.

3.3.2 Selection of an invariant measure

Even for specific boundary conditions, a system may have a lot of different invariant measures like those located on the stationary points, on the periodic orbits, or linear combinations of them. However, if the initial statistical ensemble is general and if the periodic orbits are unstable as in chaotic systems we may expect that the statistical ensemble will be distributed over a whole region of phase space after a long enough time. Due to the dynamical instability, the details of the initial ensemble can have disappeared on small scales in phase space such that regular observables evaluated at long times are statistically uncorrelated from the observables at the initial time. This is the property of mixing which we consider in the form of

$$\textit{asymptotic stationarity} \ : \qquad \langle A \rangle_t \ \to_{t \to \pm\infty} \int_{\mathcal{M}} \mu_{\rm i}^{\pm}(d\mathbf{X}) \, A(\mathbf{X}) \, . \tag{3.26}$$

This property requires that the average value of some observable A evolves in time toward the average evaluated with the invariant measure. We emphasize here that the forward time evolution may lead to an invariant measure $\mu_{\rm i}^{+}$ which is different from the invariant measure $\mu_{\rm i}^{-}$ reached by backward time evolution, even if the system is time-reversal symmetric. It is only if the system is time-reversal symmetric and volume preserving and if the boundary conditions are also time-reversal symmetric that $\mu_{\rm i}^{+} = \mu_{\rm i}^{-}$ as is known for Hamiltonian systems according to Liouville's theorem (Goldstein 1950). The above condition (3.26) requires the convergence in a weak sense, which is very important for volume-preserving systems where a strong convergence cannot hold.

For bounded systems in which a regular invariant measure exists, the above condition is equivalent to the mixing condition:

$$\textit{mixing} \ : \qquad \lim_{t\to\infty} \langle A_t B_0 \rangle_{\rm i} \ = \ \langle A \rangle_{\rm i} \, \langle B \rangle_{\rm i} \, , \tag{3.27}$$

where $A_t = A(\mathbf{\Phi}^t\mathbf{X})$, $B_0 = B(\mathbf{X}) = f_0(\mathbf{X})/f_{\rm i}(\mathbf{X})$, and $\langle \cdot \rangle_{\rm i} = \int \mu_{\rm i}(d\mathbf{X})(\cdot)$.

The mixing condition is known to imply ergodicity which assumes the equality between time averages and phase-space averages (Arnold and Avez 1968):

$$\begin{aligned} \textit{ergodicity} \ : \qquad & \lim_{T\to\pm\infty} \frac{1}{T} \int_0^T A(\mathbf{\Phi}^t \mathbf{X}_0) \, dt \ = \ \int_{\mathscr{M}} \mu_{\rm i}^{\pm}(d\mathbf{X}) \, A(\mathbf{X}) \\ & \forall \, \mathbf{X}_0 \in \tilde{\mathscr{M}} \subset \mathscr{M} \ : \ \mu_{\rm i}^{\pm}(\tilde{\mathscr{M}}) = 1 \, , \end{aligned} \tag{3.28}$$

i.e., for an initial condition $\mathbf{X}_0$ taken almost everywhere with respect to the measure $\mu_{\rm i}^{\pm}$. If this condition is satisfied the system is said to be ergodic. Otherwise, a decomposition into smaller ergodic components (or nonwandering invariant subsets) is needed as explained in the previous chapter.

3.3.3 The basic invariant measures of Hamiltonian systems

In Hamiltonian systems with f degrees of freedom, volumes are preserved in the phase space of the canonical variables $\mathbf{X} = (\mathbf{q}, \mathbf{p})$ so that an invariant measure is the Liouville measure $d\mathbf{X} = d^f q d^f p$. In systems where energy is conserved, an invariant measure can be constructed on each energy shell $H = E$ with the microcanonical measures

$$d\mu_{\rm i} \ = \ \mathscr{N} \, \delta(H - E) \, d^f q \, d^f p \ = \ \mathscr{N} \, \delta(\epsilon - E) \, d\epsilon \, \frac{d^{2f-1} S}{|{\rm grad} H|} \, , \tag{3.29}$$

where $d^{2f-1}S$ is an area element of the energy shell, ϵ is a supplementary coordinate perpendicular to the energy shell, and $\mathscr{N}$ is a normalization constant.

If the total linear momentum is moreover conserved then we must consider an invariant measure on the energy–momentum shell

$$d\mu_{\mathrm{i}} = \mathcal{N}\ \delta(H-E)\ \delta(\mathbf{P}_{\mathrm{tot}}-\mathbf{P})\ d^f q\ d^f p\ . \tag{3.30}$$

In systems with rotational symmetry, the total angular momentum is also conserved so that the microcanonical invariant measure has an energy–momenta shell

$$d\mu_{\mathrm{i}} = \mathcal{N}\ \delta(H-E)\ \delta(\mathbf{P}_{\mathrm{tot}}-\mathbf{P})\ \delta(\mathbf{L}_{\mathrm{tot}}-\mathbf{L})\ d^f q\ d^f p\ . \tag{3.31}$$

In *billiards*, we can introduce the Birkhoff coordinates which are the positions $\mathbf{q}_{\perp}$ at the times of collisions and the conjugate momenta $\mathbf{p}_{\perp}$ tangent to the border of the billiard: $\mathbf{X} = (\mathbf{q}, \mathbf{p}) \to (\mathbf{q}_{\perp}, \mathbf{p}_{\perp}, E, t) = (\mathbf{x}, E, t)$. The microcanonical invariant measure of the system is then equivalent to a uniform Lebesgue measure in the Birkhoff coordinates

$$d^{2f-2}\mathbf{x} = d^{f-1}q_{\perp}\ d^{f-1}p_{\perp}\ . \tag{3.32}$$

Whether the above measures are ergodic or not has to be investigated in each specific system by the topological methods of Chapter 2.

3.4 Correlation functions and spectral functions

In open systems with absorbing boundary conditions where $f_{\mathrm{i}} = 0$, the simple decay of the time average (3.26) to zero is already of great interest because it contains information on the time evolution of the statistical ensembles. More generally, we shall be interested in the following by the approach to the invariant measure, i.e., by the question of the asymptotic behaviour for $t \to \pm\infty$ of the time average (3.10) around its limit (3.26).

With this aim, we consider the effect of the dynamics on the *time-correlation functions* between observables

$$C_{AB}(t) = \langle A(\mathbf{\Phi}^t\mathbf{X})B(\mathbf{X})\rangle_{\mathrm{i}} - \langle A\rangle_{\mathrm{i}}\ \langle B\rangle_{\mathrm{i}}\ , \tag{3.33}$$

or more generally on the time averages of an observable $\langle A\rangle_t$. Such quantities decay to zero in systems with asymptotic stationarity, so that we may consider the time Fourier transform of the correlation functions which are the *spectral functions*:

$$S_{AB}(\omega) = \int_{-\infty}^{+\infty} e^{i\omega t}\ C_{AB}(t)\ dt\ . \tag{3.34}$$

These spectral functions contain information on the frequencies used by the system to perform its time evolution, which is given by the support of the spectral function for real or complex frequencies ω. Further information is contained in the spectral functions about the effect of the dynamics on the particular observables A and B considered.

Different theories have been developed to answer such questions. Spectral theories developed by Koopman (1931) and von Neumann (1932) in the thirties are concerned with the spectrum of real frequencies. Only recently, theories have been developed which use the spectrum of complex frequencies (Pollicott 1985, 1986; Ruelle 1986a, 1986b, 1987). Both aspects are certainly complementary in the sense that the spectral theory on real frequencies allows us to distinguish systems with a discrete spectrum of real frequencies from those with a continuous spectrum of real frequencies. Systems with a discrete real spectrum present almost-periodic oscillations while those with a continuous real spectrum present systematic decays. On the other hand, the spectral theory on complex frequencies treats essentially the systems with continuous real spectrum but allows us to distinguish different types of decays like exponential or algebraic.

3.5 Spectral theory on real frequencies

3.5.1 Spectral decomposition on a Hilbert space

The theory introduced by Koopman (1931) considers the time evolution operator $\hat{P}^t$ as a unitary operator acting on a Hilbert space defined with the scalar product $\langle f|g\rangle = \int f^* g d\mu_i$. As a consequence, the substitution operator $\hat{U}^t g(\mathbf{X}) = g(\mathbf{\Phi}^t\mathbf{X})$ is a unitary operator if $\mathbf{\Phi}^t$ is invertible (Arnold and Avez 1968, Cornfeld et al. 1982).

For continuous-time systems, the time evolution is a one-parameter Lie group whose generator is the self-adjoint operator $\hat{\mathscr{L}} = i\{H, \cdot\} = \hat{\mathscr{L}}^\dagger$ so that

$$\hat{U}^t = e^{i\hat{\mathscr{L}}t} . \tag{3.35}$$

For such a unitary operator, the spectrum is known to belong necessarily to the unit circle, which leads to the spectral decomposition (Stone's theorem)

$$\hat{U}^t = \sum_n \hat{E}_n \, e^{i\omega_n t} + \int_{\sigma_c} d\hat{E}(\omega) \, e^{i\omega t} . \tag{3.36}$$

The continuous spectrum is $\sigma_c \subset]-\pi, +\pi]$ for discrete-time systems with $t \in \mathbb{Z}$ and $\sigma_c \subset]-\infty, +\infty[$ for continuous-time flows. We assume here that there is no singular-continuous spectrum for which there is no known example in classical dynamical systems. The operators $\hat{E}_n$ and $\hat{E}(\omega)$ are self-adjoint projection operators in the Hilbert space. The operators $\hat{E}_n$ project onto the eigenspaces corresponding to the eigenvalues $\exp(i\omega_n t)$. On the other hand, the operators $\hat{E}(\omega)$ depend continuously on the real frequency ω and they project onto the subspaces corresponding to the continuous spectrum. The decomposition is complete on the Hilbert space in the sense that

$$\hat{I} = \sum_n \hat{E}_n + \int_{\sigma_c} d\hat{E}(\omega) , \tag{3.37}$$

as it should be since Eq. (3.36) reduces to Eq. (3.37) when $t = 0$. The decomposition (3.36) can be used to describe the time evolution of statistical averages $\langle A \rangle_t = \langle f_0 | \hat{U}^t A \rangle$.

The discrete spectrum $\{\exp(i\omega_n t)\}$ is composed of the eigenvalues of the Koopman operator $\hat{U}^t$. The real eigenfrequencies ω_n are the eigenvalues of the generator $\hat{\mathscr{L}}$, which correspond to eigenvalues of the Liouville operator $\hat{L} = -i\hat{\mathscr{L}}$ according to

$$\hat{L}\, \psi_n = s_n\, \psi_n , \qquad s_n = -i\omega_n . \tag{3.38}$$

The ergodic and mixing properties can be investigated at the level of the real frequency spectrum thanks to the following theorems:

A system is ergodic iff the eigenvalue 1 *of its Koopman operator is simply degenerate. A flow is ergodic iff the eigenvalue* $\omega = 0$ *of the Liouville operator is simply degenerate.*

The system is mixing iff the eigenvalue 1 *is simply degenerate and is the lone eigenvalue of the Koopman operator. The flow is mixing iff* $\omega = 0$ *is simply degenerate and is the lone eigenvalue of the Liouville operator.*

We notice that mixing systems have necessarily a continuous spectrum.

3.5.2 Purely discrete real spectrum

Completely integrable Hamiltonians are examples of systems with a purely discrete frequency spectrum. These Hamiltonians only depend on the action variables J_i which are conjugate to the angle variables $0 \leq \theta_i < 2\pi$:

$$H = H(J_1, J_2, \ldots, J_f) = H(\mathbf{J}) , \tag{3.39}$$

where f here denotes the number of degrees of freedom. The equations of motion of the action–angle variables are

$$\begin{aligned} \dot{\boldsymbol{\theta}} &= \partial_{\mathbf{J}} H = \boldsymbol{\Omega}(\mathbf{J}) , \\ \dot{\mathbf{J}} &= -\partial_{\boldsymbol{\theta}} H = 0 , \end{aligned} \tag{3.40}$$

so that the trajectories are $\boldsymbol{\theta}(t) = \boldsymbol{\Omega}(\mathbf{J})t + \boldsymbol{\theta}(0)$ (mod. 2π), which represent a quasiperiodic motion. In this case, the spectrum of the Koopman operator is purely discrete with the eigenvalues given by $\exp(i\omega_{\mathbf{n}}t)$ where

$$\omega_{\mathbf{n}} = \mathbf{n} \cdot \mathbf{\Omega}(\mathbf{J}) , \qquad \text{with} \quad \mathbf{n} \in \mathbb{Z}^f . \tag{3.41}$$

The corresponding eigenfunctions are $\psi_{\mathbf{n}}(\boldsymbol{\theta}) = \exp(i\mathbf{n}\cdot\boldsymbol{\theta})$ which form the Fourier basis of $\mathscr{L}^2(\mathbb{T}^f, \ell_{\rm i})$, where $\mathbb{T}^f$ denotes the torus $\mathbf{J} = \mathbf{J}_0$ with the invariant measure $d\ell_{\rm i} = d^f\theta/(2\pi)^f$. This invariant measure is represented by $d\mu_{\rm i} = \delta(\mathbf{J} - \mathbf{J}_0)d^fJd^f\theta/(2\pi)^f$ in the full phase space.

Clearly, these systems are never mixing because there are several eigenvalues in the discrete spectrum. The system is ergodic iff the eigenvalue $\omega_{\mathbf{n}} = 0$ is simply degenerate, i.e., iff

$$\omega_{\mathbf{n}} = \mathbf{n} \cdot \mathbf{\Omega} = 0 \quad \Rightarrow \quad \mathbf{n} = 0 , \tag{3.42}$$

which is precisely the condition of incommensurability between all the frequencies (Arnold and Avez 1968). Under this condition, there exists an orbit which is dense on the torus so that the system is ergodic iff it is transitive on the torus. Even if almost all the invariant tori are ergodic, we should nevertheless emphasize that an uncountable set of them are only partially ergodic and relatively periodic.[3] Besides, a countable set of tori are formed by continuous families of periodic orbits, which are not isolated.

3.5.3 Continuous real spectrum

We suppose here that the system is mixing so that the only eigenvalue of $\hat{U}^t$ is 1 with the corresponding projector $\hat{E}_0$. The interesting dynamics takes place on the orthocomplement of the invariant subspace of unit eigenvalue $\mathscr{H}_\perp = \mathscr{H} \ominus \mathscr{H}_0$ where $\mathscr{H}_0 = \hat{E}_0\mathscr{H}$. On the orthocomplement, the spectrum of real frequencies is continuous. It is furthermore assumed that the spectrum is absolutely continuous with respect to the Lebesgue measure on ω and, in particular, extends without rupture to the whole real axis, which is called a Lebesgue spectrum (Cornfeld et al. 1982).

For systems with a Lebesgue spectrum, a theorem by von Neumann[4] can be used to construct generalized eigenstates such as

$$\hat{U}^t \, \Psi_{l\omega} = e^{i\omega t} \, \Psi_{l\omega} , \tag{3.43}$$

3. A trajectory is relatively periodic if it is periodic with respect to a restricted set of coordinates. In this case, the trajectory is dense on a subset of dimension $d < f$ of the torus $\mathbb{T}^f$.
4. This theorem can be found in the Appendix 2 of the book by Cornfeld et al. (1982). The statements about the generalized eigenstates are due to the present author (see Fig. 3.1).

which satisfy the orthogonality relation

$$\langle \Psi_{l\omega} | \Psi_{l'\omega'} \rangle = 2\pi \, \delta_{ll'} \, \delta(\omega - \omega') \qquad \text{for} \quad \omega, \, \omega' \in \mathbb{R} \, , \tag{3.44}$$

such that the Koopman operator can be decomposed as

$$\hat{U}^t = \hat{E}_0 + \sum_{l=1}^{L} \int_{-\infty}^{+\infty} \frac{d\omega}{2\pi} \, |\Psi_{l\omega}\rangle \, e^{i\omega t} \, \langle \Psi_{l\omega}| \, , \tag{3.45}$$

where $l = 1, \ldots, L$ is the multiplicity of the Lebesgue spectrum. We emphasize that the generalized eigenstates do not belong to the Hilbert space and they are not even functions but linear functionals given by distributions with cumulative functions (see Fig. 3.1). This is in contrast with quantum mechanics where the generalized eigenstates of the continuous spectrum can be represented by functions. The difference is due to the fact that Schrödinger's equation is of 2nd order while Liouville's equation is of 1st order and corresponds to a dynamics of points following trajectories.

The comparison of (3.45) with Stone's decomposition (3.36) shows that

$$d\hat{E}(\omega) = \sum_{l=1}^{L} |\Psi_{l\omega}\rangle \, \langle \Psi_{l\omega}| \, \frac{d\omega}{2\pi} \, . \tag{3.46}$$

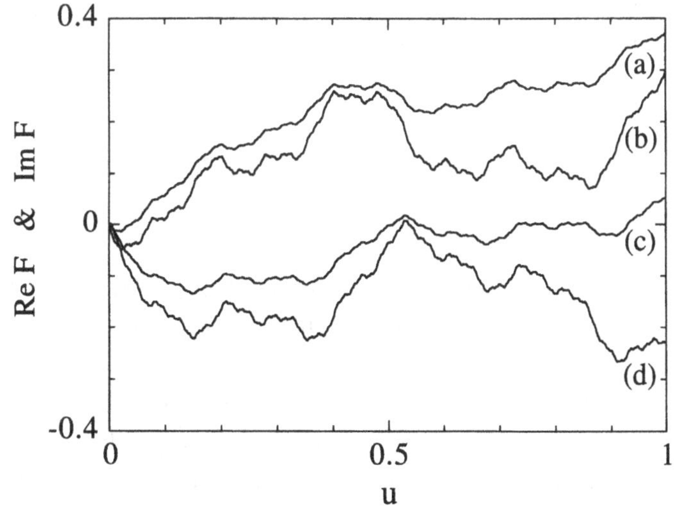

Figure 3.1 Cumulative function of the generalized eigenstate $\Psi_{l\omega} = \sum_{n=-\infty}^{+\infty} \exp(-i\omega n) \Upsilon_{\mathbf{l}n}$ defined with the shift states $\Upsilon_{\mathbf{l}n}(x, y) = \exp[2\pi i \mathbf{l} \cdot \boldsymbol{\phi}^n(x, y)]$ of the Arnold cat map $\boldsymbol{\phi}(x, y) = (x + y, x + 2y)$ (modulo 1). The cumulative function is defined by $F_{\mathbf{l}\omega}(s, u) = \int_0^s ds' \int_0^u du' \Psi_{\mathbf{l}\omega}(s'\mathbf{e}_s + u'\mathbf{e}_u)$ where $\mathbf{e}_s$ and $\mathbf{e}_u$ are the stable and unstable directions of this map (Arnold and Avez 1968). The figure depicts the cumulative function versus u for $\omega = 1$: Re $F_{\mathbf{l}\omega}$ at (a) $s = 0.5$, (b) $s = 1$; Im $F_{\mathbf{l}\omega}$ at (c) $s = 0.5$, (d) $s = 1$.

The Fourier transform of the generalized eigenstates defines another basis of states which are defined along the time axis

$$\Upsilon_{lt} = \int_{-\infty}^{+\infty} \frac{d\omega}{2\pi}\, e^{i\omega t}\, \Psi_{l\omega} \,. \tag{3.47}$$

These states correspond to the shift states in discrete-time systems because they obey

$$\hat{U}^{\tau}\, \Upsilon_{lt} = \Upsilon_{l\,t+\tau}\,, \quad t,\, \tau \in \mathbb{R}\,, \quad l = 1,2,3,\ldots,L\,, \tag{3.48}$$

$$\langle \Upsilon_{lt} | \Upsilon_{l't'} \rangle = \delta_{ll'}\, \delta(t-t')\,, \tag{3.49}$$

and form a complete basis. Moreover, they are eigenstates of a time operator

$$\hat{\mathscr{T}} = \sum_{l=1}^{L} \int_{-\infty}^{+\infty} dt\, |\Upsilon_{lt}\rangle\, t\, \langle \Upsilon_{lt}|\,, \tag{3.50}$$

which is conjugate to the self-adjoint Liouvillian operator

$$[\hat{\mathscr{L}}, \hat{\mathscr{T}}] = i\,\hat{I}\,, \tag{3.51}$$

on $\mathscr{H}_{\perp}$. For the time operator (3.50) to exist, it is important that ω may vary without restriction over $-\infty < \omega < +\infty$, i.e., that the spectral measures $\langle f | d\hat{E}(\omega) g\rangle$ are absolutely continuous with respect to the Lebesgue measure for every $\omega \in \mathbb{R}$.[5] The Liouvillian and the time operators are Lie generators of conjugated families of unitary operators

$$\hat{U}^{t} = e^{i\hat{\mathscr{L}}t} \quad \text{and} \quad \hat{V}^{\omega} = e^{i\omega\hat{\mathscr{T}}}\,, \tag{3.52}$$

which obey the Weyl-type relation (Cornfeld et al. 1982, App. 2)

$$\hat{U}^{t}\, \hat{V}^{\omega} = e^{-i\omega t}\, \hat{V}^{\omega}\, \hat{U}^{t}\,. \tag{3.53}$$

The above structure was invented in the sixties by Sinai (1961, 1966; Lax and Phillips 1967, Misra 1978). The operator (3.50) has been called ‘time operator’ by Prigogine (1980).

The condition that the system has a Lebesgue spectrum with moreover a countable multiplicity has been shown by Sinai (1961, 1966) to be equivalent to the so-called *Kolmogorov property* defined in ergodic theory (Arnold and Avez 1968). The Kolmogorov property is therefore much stronger than the property of mixing because an absolutely continuous spectrum of infinite multiplicity is assumed.

5. This condition excludes a continuous spectrum over a semi-infinite range like $-\infty < \omega \leq 0$, as is the case for the time evolution of a free particle in quantum mechanics where we would take $\omega = -E/\hbar$, E being the energy and $\hbar$ the Planck constant.

3.5.4 Continuous real spectrum and the Wiener–Khinchin theorem

The decomposition in terms of the generalized eigenstates is very convenient because it leads to the Wiener–Khinchin theorem concerning time-correlation functions

$$C_{ij}(t) = \langle A_i(\Phi^t \mathbf{X}) A_j(\mathbf{X})\rangle - \langle A_i\rangle\langle A_j\rangle . \qquad (3.54)$$

between several observables $\{A_i\}_{i=1}^n$.

The correlation functions form a matrix such that

$$\mathsf{C}(t) = \mathsf{C}^{\mathrm{T}}(-t) . \qquad (3.55)$$

A matrix of spectral densities can be defined as

$$\mathsf{S}(\omega) = \int_{-\infty}^{+\infty} dt\, e^{i\omega t}\, \mathsf{C}(t) . \qquad (3.56)$$

This matrix is nonnegative according to the following version of the

Wiener–Khinchin theorem. If the dynamical system is mixing and has a Lebesgue continuous spectrum, the matrix of spectral densities is non-negative definite for all frequencies:

$$\mathsf{S}(\omega) \geq 0 , \quad or \quad \sum_{i,j=1}^{n} S_{ij}(\omega) z_i^* z_j \geq 0 , \qquad (3.57)$$

for all $z_i \in \mathbb{C}$. In particular, $S_{ii}(\omega) \geq 0$.

This result is proved by considering the spectral decomposition of the observables onto the generalized eigenvectors

$$A_i(\mathbf{X}) = \langle A_i\rangle_\mu + \sum_{l=1}^{L} \int_{-\infty}^{+\infty} \frac{d\omega}{2\pi}\, a_{il}(\omega)\, \Psi_{l\omega}(\mathbf{X}) , \qquad (3.58)$$

with $a_{il}(\omega) = \langle \Psi_{l\omega} | A_i\rangle_\mu$. Using the orthonormality and the generalized eigenvalue property, we obtain that

$$S_{ij}(\omega) = \sum_{l=1}^{L} a_{il}^*(\omega)\, a_{jl}(\omega) , \qquad (3.59)$$

from which we infer (3.57). *Q.E.D.*

We notice that the matrix of spectral densities is Hermitian which implies the Onsager reciprocity relations in the context of nonequilibrium thermodynamics (Onsager 1931, deGroot and Mazur 1962).

3.5.5 Continuous real spectrum and Gaussian fluctuations

In systems where the correlation functions decay fast enough, the convergence to the ergodic limit (3.28) may present Gaussian dynamical fluctuations in the sense of the central limit theorem:

$$\lim_{T\to\infty} \mu\left\{ \mathbf{X} : \frac{\int_0^T A(\mathbf{\Phi}^t\mathbf{X})dt - T\langle A\rangle_\mu}{\sqrt{2DT}} < y \right\} = \frac{1}{\sqrt{2\pi}} \int_{-\infty}^{y} e^{-z^2/2}\, dz \, , \quad (3.60)$$

where D is the generalized diffusion coefficient defined by the following variance

$$D = \lim_{T\to\infty} \frac{1}{2T} \left\langle \left[\int_0^T A(\mathbf{\Phi}^t\mathbf{X})dt - T\langle A\rangle_\mu \right]^2 \right\rangle_\mu . \quad (3.61)$$

The diffusion coefficient can here be reduced as follows to the integral of the autocorrelation function $C(t) = C_{AA}(t)$. Performing the change of time variables

$$\begin{cases} \tau = t - t' \, , \\ \tau' = (t + t')/2 \, , \end{cases} \quad (3.62)$$

the double time integral in (3.61) can be transformed to get

$$\begin{aligned} \int_0^T dt \int_0^T dt'\, C(t - t') &= \int_{-T}^{+T} d\tau \int_{|\tau|/2}^{T-|\tau|/2} d\tau'\, C(\tau) \\ &= \int_{-T}^{+T} d\tau\, (T - |\tau|)\, C(\tau) \, . \end{aligned} \quad (3.63)$$

Therefore, the generalized diffusion coefficient is given by

$$D = \frac{1}{2} \int_{-\infty}^{+\infty} C(\tau)\, d\tau \, , \quad (3.64)$$

under the condition that

$$\lim_{T\to\infty} \frac{1}{T} \int_{-T}^{+T} d\tau\, |\tau|\, C(\tau) = 0 \, . \quad (3.65)$$

This last condition implies that the autocorrelation function should decay faster than $1/|\tau|$.

Equation (3.64) is known as the Green–Kubo formula which is fundamental for nonequilibrium statistical mechanics (Green 1951–60, Kubo 1957, Résibois and De Leener 1977). This formula expresses the different transport coefficients in terms of the correlation functions of microscopic observables. Here, we have shown that this formula already appears in dynamical systems theory as a consequence of the statistical properties of the dynamics itself. In classical mechanics, this formula is thus already of application to systems with a compact phase space under the necessary condition that the mixing property holds, i.e. that the spectrum is continuous. This is one of the results of nonequilibrium statistical mechanics which can be directly constructed in terms of the intrinsic dynamical properties.

There is a nondegeneracy condition which should be satisfied in order for the diffusion coefficient to be positive. This condition is that the equation

$$\mathbf{F}(\mathbf{X}) \cdot \text{grad } w(\mathbf{X}) = A(\mathbf{X}) - \langle A \rangle_\mu , \tag{3.66}$$

has no bounded solution $w(\mathbf{X})$. If this was the case, then we would have $D = 0$ because

$$\frac{d}{dt} w(\mathbf{\Phi}^t\mathbf{X}) = \mathbf{F}(\mathbf{\Phi}^t\mathbf{X}) \cdot \text{grad } w(\mathbf{\Phi}^t\mathbf{X}) , \tag{3.67}$$

so that the variance in (3.61) would be bounded rather than linearly increasing.

A further remark concerns the status of the central limit theorem under time reversal. Equation (3.60) can also be expressed with $A(\mathbf{\Phi}^{-t}\mathbf{X})$ instead of $A(\mathbf{\Phi}^t\mathbf{X})$. Then, the time average would have to be carried out over the past half-trajectory. In this respect, the central limit theorem may have to be formulated with a different invariant set than for the future half-trajectory if $\mu_{\rm i}^- \neq \mu_{\rm i}^+$. The central limit theorem is time-reversal invariant only when $\mu_{\rm i}^- = \mu_{\rm i}^+$, i.e., only for systems which are time-reversal symmetric *and* volume-preserving.

The diffusion coefficient can also be expressed in term of the spectral function associated with the correlation function C_{AA} as

$$D = \frac{1}{2} S_{AA}(0) , \tag{3.68}$$

which is non-negative for systems with a Lebesgue continuous spectrum according to the Wiener–Khinchin theorem and even positive if the nondegeneracy condition holds. If the spectral function is not continuous at $\omega = 0$, the dynamical fluctuations cannot be Gaussian and the transport coefficient does not exist. In such circumstances, the fluctuations may no longer be Gaussian and anomalous transport properties may exist which can be studied from the behaviour of $S(\omega)$ near $\omega = 0$ (Geisel et al. 1985, Wang and Hu 1993).

3.6 Spectral theory on complex frequencies or resonance theory

3.6.1 Analytic continuation to complex frequencies

For systems with a continuous spectrum, the spectral theory gives relatively little information on the way the relaxation proceeds toward the long-time limit. If the observables belong to a Hilbert space many different decays are possible from fast decays to very slow decays. The possible transient behaviour of the system would be much better characterized if the intrinsic relaxation rates observed in natural phenomena could be theoretically defined. With this purpose, we must consider the analytic continuation of the spectral functions toward complex frequencies. If the correlation function presents damped oscillations we may expect the existence of an intrinsic complex frequency whose real part gives the period of oscillations and imaginary part the relaxation rate. Such complex frequencies are known as resonances.

The concept of resonances has been introduced in the theory of probabilistic random processes where they are defined as the poles of the spectral functions $S(\omega)$ (Doob 1953, Yaglom 1962).[6] Here, we would like to develop the theory of classical resonances. Since classical dynamical systems with an invariant probability measure have been shown to be rigorously equivalent to probabilistic random processes (Ornstein 1974) and to be characterized by the very same quantities as in the theory of random processes, like the spectral functions $S(\omega)$, it is natural to expect that resonances at complex frequencies also exist for classical deterministic systems (Isola 1988, Eckmann 1989, Baladi et al. 1989). Such resonances would characterize transient behaviour in the time evolution of statistical ensembles of trajectories of the classical system. In this context, they have been studied in particular by Pollicott (1985, 1986) and Ruelle (1986a, 1986b, 1987, 1989a, 1989b) and we shall refer to them as the Pollicott–Ruelle resonances.

We would like to present here a method of construction which is as general as possible and which concerns any possible dynamical system whether it is bounded or open.

Let us consider the spectral function $S(\omega)$ of a correlation function $C(t)$. Its analytic continuation toward complex frequencies $z = \text{Re}z + i\text{Im}z$ with $\omega = \text{Re}z$ may reveal several types of complex singularities such as

$$\text{simple or multiple poles :} \quad S(z) = \sum_{l=1}^{\infty} \frac{a_l^{(j)}}{(z - z_j)^{m_j - l + 1}} , \tag{3.69}$$

$$\text{branch points :} \quad S(z) = \sum_{l=1}^{\infty} b_l^{(c)} \, (z - z_c)^{\alpha_c + l - 1} , \tag{3.70}$$

6. There are two important differences with respect to quantum-mechanical scattering resonances (Joachain 1975). First, the present transient behaviour concerns the time evolution of a statistical ensemble of realizations of the process. However it concerns the time evolution of the Schrödinger wavefunction in quantum systems. The second difference is that here the frequency is analytically continued instead of the energy. In quantum systems, the frequency is given by Bohr's relation $\omega = (E - E')/\hbar$.

as well as natural boundaries. In Eq. (3.69), the integer m_j is the multiplicity of the pole.

The correlation function can be recovered by an inverse Fourier transform if the spectral function is known at real frequencies

$$C(t) = \int_{-\infty}^{+\infty} e^{-i\omega t} S(\omega) \frac{d\omega}{2\pi} = \int_{\mathbb{R}} e^{-izt} S(z) \frac{dz}{2\pi} , \tag{3.71}$$

which holds for both positive and negative times. Here, the integral is performed along the real frequency axis in the plane of the complex frequency z. The Koopman operator of the system is assumed to have a continuous spectrum which is located in z along the real axis. Viewed as a spectral decomposition, Eq. (3.71) is thus reminiscent of the unitary character of the Koopman operator defined over a Hilbert space $\mathscr{L}^2(\mathscr{M}, \mu)$. As a consequence of Koopman's unitarity, the time evolution can be continued to the whole real time axis (and even at complex times).

The knowledge of the complex singularities of $S(z)$ may be exploited to simplify the integral by deforming the integration contour in the lower or the upper half complex plane z. Let us first remark that we have removed the trivial simple pole at zero frequency $z = 0$ which corresponds to the unique invariant measure in ergodic systems, when we assume that $\lim_{t\to\infty} C(t) = 0$. If the system is not ergodic there might exist other poles at $z = 0$ or even a branch cut to be removed.

We consider the deformation of the integration contour toward a new contour $\mathscr{C}_R$ at which $z = \omega + iR$ except for the segments of the contour which are required to circumvent the different singularities as shown in Fig. 3.2. A pole is circumvented along a small circle of radius ε followed clockwise if $\mathrm{Im} z_j < 0$ or anticlockwise if $\mathrm{Im} z_j > 0$. This amounts to carry out the change of variable $z = z_j + \varepsilon e^{i\theta}$ so that $dz = \varepsilon e^{i\theta} d\theta$. Each branch point is the starting point of a branch cut which may be chosen parallel to the imaginary axis. In this case, the branch cut must be circumvented by two lines parallel to the branch cut with $z = z_c + i\eta \pm 0$ and $dz = i d\eta$, where 0 stands for an infinitesimally small positive number and $0 \leq |\eta| < |R - \mathrm{Im} z_c|$.

In the limit $R \to -\infty$ where $\mathscr{C}_R \to \mathscr{C}^+$, the correlation function is then given by a sum of contributions coming from the different complex singularities in the lower half plane ($\mathrm{Im} z < 0$), under the condition that the time is strictly positive ($t > 0$). On the other hand, for $R \to +\infty$ where $\mathscr{C}_R \to \mathscr{C}^-$, the correlation function is given by a sum of contributions from the complex singularities of the upper half plane ($\mathrm{Im} z > 0$) under the condition that the time is strictly negative ($t < 0$). Therefore, the two contours $\mathscr{C}^\pm$ define two distinct representations of the correlation function which are valid for either forward or backward time evolutions

$$C(t) = \int_{\mathscr{C}\pm} e^{-izt}\, S(z)\, \frac{dz}{2\pi}\,, \qquad \text{if } t>0 \quad \text{or} \quad t<0\,. \tag{3.72}$$

Let us now calculate the time dependences of the correlation function due to the different complex singularities. For positive times, we get

$$C(t) = \sum_{j:\ \mathrm{Im} z_j<0} C_j(t) + \sum_{c:\ \mathrm{Im} z_c<0} C_c(t) \qquad (t>0)\,, \tag{3.73}$$

with the contribution from each pole

$$C_j(t) = \sum_{l=1}^{m_j} \frac{a_l^{(j)} e^{i\pi(l-m_j-1)/2}}{(m_j-l)!}\, t^{m_j-l}\, e^{-iz_jt}\,, \qquad (t>0)\,, \tag{3.74}$$

and the contribution from each branch cut

$$C_c(t) = \sum_{l=1}^{\infty} \frac{b_l^{(c)} e^{i\pi(l+\alpha_c-1)/2}}{\Gamma(1-\alpha_c-l)}\, \frac{e^{-iz_ct}}{t^{\alpha_c+l}}\,, \qquad (t>0)\,. \tag{3.75}$$

Therefore, we find that each singularity contributes as follows to the correlation function

$$\text{simple poles}: \quad C(t) \sim \exp(-iz_jt)\,, \tag{3.76}$$

$$\text{multiple poles}: \quad C(t) \sim t^{m_j-1} \exp(-iz_jt)\,, \tag{3.77}$$

$$\text{branch points}: \quad C(t) \sim \frac{\exp(-iz_ct)}{t^{\alpha_c+1}}\,. \tag{3.78}$$

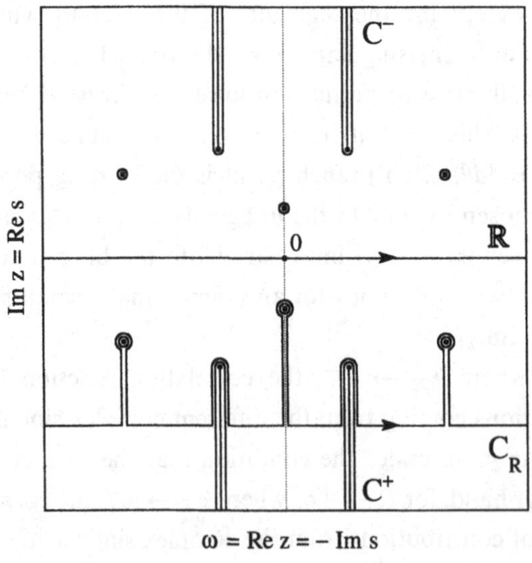

Figure 3.2 Four contours of integration in the plane of complex frequency z: $\mathbb{R}$, $\mathscr{C}_R$ with $R<0$, $\mathscr{C}^+$, and $\mathscr{C}^-$.

A simple pole leads to a purely exponential decay. In general, the decay due to a multiple pole is slower than exponential because it is an exponential multiplied by a positive power of time. A similar power of time is combined with an exponential in the case of a branch point. The power factor increases the decay if $\alpha_c > -1$ but decreases the decay if $\alpha_c < -1$. If the branch point is located at some real frequency ($\text{Im}z_c = 0$ and $S(\omega) \sim |\omega - \omega_c|^{\alpha_c}$), we must have that $\alpha_c > 0$ (otherwise the spectral function is not continuous but has a singularity in contradiction with previous assumptions) and the decay is algebraic with possible oscillatory modulations. If there is no branch cut the decay is therefore exponential as determined by the set of poles. If there is no pole and no branch cut the decay is entire as is the case if the correlation function is a Gaussian.

Expressions similar to (3.73)–(3.78) hold for the correlation function at negative times.

3.6.2 Singularities as a spectrum of generalized eigenvalues

The location of the complex singularities may turn out to be independent of the observables[7] considered in the correlation function and thus intrinsic to the dynamical system itself. We may thus suppose that the complex singularities characterize the properties of the Frobenius–Perron operator $\hat{P}^t$ rather than of the specific observables considered in the correlation function. This idea leads us to consider in particular the poles associated with the resonances as generalized eigenvalues of the Frobenius–Perron operator.

We should however distinguish the complex singularities in the lower half plane from those of the upper half plane of complex frequencies because they define two different representations of the time evolution which hold either for positive or for negative times. These two new representations form two distinct semigroups defined by two new evolution operators $\hat{P}^{\pm t} = \exp(\hat{L}^{\pm} t)$ for either $t > 0$ or $t < 0$. The forward semigroup has its spectrum in the lower half plane of complex frequencies and the backward semigroup has its spectrum in the upper half plane. The three operators are *a priori* distinct but they are related by the conditions

$$\begin{aligned} \langle g|\hat{P}^{+t}|f\rangle &= \langle g|\hat{P}^{t}|f\rangle \,, \qquad \text{for} \quad t>0 \,, \\ \langle g|\hat{P}^{-t}|f\rangle &= \langle g|\hat{P}^{t}|f\rangle \,, \qquad \text{for} \quad t<0 \,, \end{aligned} \tag{3.79}$$

for smooth enough observables f and g. In this sense, the operators $\hat{P}^{\pm}$ are extensions of the Frobenius–Perron operator $\hat{P}^t$ and they always carry the same

7. Nevertheless, the observables may have to be taken in a suitable functional space which depends on the system.

physical information on the time evolution but expressed in different functional spaces.

The generators $\hat{L}^{\pm}$ of the semigroup evolution operators $\hat{P}^{\pm t} = \exp(\hat{L}^{\pm} t)$ are linear operators which are no longer anti-self-adjoint because of the extension. We therefore consider generalized eigenvalue problems on spaces which are larger than the Hilbert space since the generalized eigenstates are expected to be linear functionals or Schwartz distributions in general (Dunford and Schwartz 1958–71, Gelfand and Shilov 1968). We drop the subscript $\pm$ for convenience but we should keep in mind that the generalized eigenstates of $\hat{L}^+$ differ from those of $\hat{L}^-$. The eigenvalues and the right eigenvectors are defined by

$$\hat{L}\, \psi_j \;=\; s_j\, \psi_j\,, \tag{3.80}$$

while the left eigenvectors are defined as the eigenvectors of the adjoint operator

$$\hat{L}^{\dagger}\, \tilde{\psi}_j \;=\; s_j^{*}\, \tilde{\psi}_j\,. \tag{3.81}$$

With Dirac's notations, we have

$$\begin{aligned} \hat{L}\, |\psi_j\rangle &= s_j\, |\psi_j\rangle\,, \\ \langle\tilde{\psi}_j|\, \hat{L} &= s_j\, \langle\tilde{\psi}_j|\,. \end{aligned} \tag{3.82}$$

Because the operator $\hat{L}$ is not expected to be anti-self-adjoint the eigenvectors do not necessarily form a complete basis of the functional space when eigenvalues are degenerate. This may lead to Jordan-block structures in the decomposition of such an operator (Gelfand 1961, Gohberg and Klein 1971). A Jordan-block structure arises if the operator $\hat{L}$ admits vectors ψ satisfying

$$(\hat{L} - s_j\hat{I})^{m-1}\, \psi \;\neq\; 0\,, \tag{3.83}$$

while

$$(\hat{L} - s_j\hat{I})^{m}\, \psi \;=\; 0\,, \tag{3.84}$$

for some integer $m > 1$ ($\hat{I}$ being the identity operator). Such vectors are called root vectors of rank m associated with the degenerate eigenvalue s_j. The eigenvectors are the root vectors of rank one. The linear subspace formed by all the root vectors associated with the eigenvalue s_j is an invariant subspace under successive applications of $\hat{L}$, which is called the root subspace associated with s_j. The dimension of this root subspace is equal to the multiplicity m_j of the eigenvalue s_j (Friedman 1956). In the case where not all root vectors are of rank one, the set of all the eigenvectors is not sufficient to span the whole functional space and other root vectors of superior ranks need to be introduced in the vector basis. A possible basis can be formed by the root vectors $\{\psi_{j,i}\}$ satisfying

$$\hat{L}\ \psi_{j,i} = s_j\ \psi_{j,i} + \psi_{j,i-1}\ , \quad \text{with } i = 1, 2, \ldots, m_j\ , \tag{3.85}$$

where m_j is the dimension of the root subspace of s_j, $\psi_{j,0} = 0$, and $\psi_{j,1} = \psi_j$ is the right eigenvector. The operator $\hat{L}$ admits therefore the following Jordan-type decomposition for the contribution of the poles $\{s_j = -iz_j\}$

$$\hat{L}\Big|_{\text{poles}} = \sum_j \begin{pmatrix} |\psi_{j,1}\rangle & |\psi_{j,2}\rangle & \cdots & |\psi_{j,m_j}\rangle \end{pmatrix} \times \begin{pmatrix} s_j & 1 & 0 & \ldots & 0 \\ 0 & s_j & 1 & \ldots & 0 \\ 0 & 0 & s_j & \ldots & 0 \\ \vdots & \vdots & \vdots & \ddots & \vdots \\ 0 & 0 & 0 & \ldots & s_j \end{pmatrix} \begin{pmatrix} \langle\tilde{\psi}_{j,1}| \\ \langle\tilde{\psi}_{j,2}| \\ \vdots \\ \langle\tilde{\psi}_{j,m_j}| \end{pmatrix} , \tag{3.86}$$

where the right and left root vectors $\{\psi_{j,i}\}$ and $\{\tilde{\psi}_{j,i}\}$ form a biorthonormal complete basis such that

$$\langle\tilde{\psi}_{j,i}|\psi_{j',i'}\rangle = \delta_{jj'}\ \delta_{ii'}\ , \quad \text{for} \quad i = 1, 2, \ldots, m_j\ , \quad i' = 1, 2, \ldots, m_{j'}\ , \tag{3.87}$$

and

$$\sum_j \sum_{i=1}^{m_j} |\psi_{j,i}\rangle\ \langle\tilde{\psi}_{j,i}| = \hat{I}\Big|_{\text{poles}} . \tag{3.88}$$

The previous relations require that the antilinear functionals $|\psi_{j,i}\rangle$ may act on the linear functionals $\langle\tilde{\psi}_{j,i}|$.

The Jordan-type decomposition (3.86) can be expanded to show that the right root vectors satisfy (3.85) while the left root vectors satisfy

$$\hat{L}^\dagger\ \tilde{\psi}_{j,i} = s_j^*\ \tilde{\psi}_{j,i} + \tilde{\psi}_{j,i+1}\ , \quad \text{with } i = 1, 2, \ldots, m_j\ , \tag{3.89}$$

$\tilde{\psi}_{j,m_j+1} = 0$ and $\tilde{\psi}_{j,m_j} = \tilde{\psi}_j$ is the left eigenvector.

On the other hand, the branch cuts if they exist are expected to correspond to continuous spectra of complex frequencies which contribute to the operator according to

$$\hat{L}\Big|_{\text{cuts}} = \sum_c \int_{-\infty}^{0} d\eta\ |\psi_\eta^{(c)}\rangle\ (s_c + \eta)\ \langle\tilde{\psi}_\eta^{(c)}|\ , \tag{3.90}$$

where the generalized eigenvectors of the branch cuts are supposed to be biorthogonal to those of the resonances.

The generalized spectral decomposition of the semigroup (for instance the forward one) can then be obtained by taking the exponential of the operator $\hat{L}^+$. A Jordan-block matrix J_j as in (3.86) can be decomposed into

$$\mathsf{J}_j = s_j \, 1 + \Delta_j \,, \tag{3.91}$$

where 1 is the $m_j \times m_j$ identity matrix while Δ_j is a $m_j \times m_j$ matrix such as

$$\Delta_j = \begin{pmatrix} 0 & 1 & 0 & 0 & \dots & 0 & 0 \\ 0 & 0 & 1 & 0 & \dots & 0 & 0 \\ 0 & 0 & 0 & 1 & \dots & 0 & 0 \\ 0 & 0 & 0 & 0 & \dots & 0 & 0 \\ \vdots & \vdots & \vdots & \vdots & \ddots & \vdots & \vdots \\ 0 & 0 & 0 & 0 & \dots & 0 & 1 \\ 0 & 0 & 0 & 0 & \dots & 0 & 0 \end{pmatrix} , \tag{3.92}$$

and which satisfies

$$\Delta_j^n = 0 \qquad \text{for} \qquad n \geq m_j \,. \tag{3.93}$$

Accordingly, we have that

$$\exp(\mathsf{J}_j t) = \exp(s_j t) \sum_{n=0}^{m_j - 1} \frac{t^n}{n!} \Delta_j^n \,, \tag{3.94}$$

so that the same product of an exponential with a power law is recovered as in the case of multiple poles.

Therefore, the forward semigroup admits the following generalized spectral decomposition

$$\begin{aligned} C_{fg}(t) &= \langle g | \hat{P}^{+t} | f \rangle \\ &= \sum_j e^{s_j t} \sum_{l=1}^{m_j} \frac{t^{m_j - l}}{(m_j - l)!} \sum_{i=1}^{l} \langle g | \psi_{j,i} \rangle \, \langle \tilde{\psi}_{j,i+m_j-l} | f \rangle \\ &\quad + \sum_c e^{s_c t} \int_{-\infty}^{0} d\eta \; e^{\eta t} \, \langle g | \psi_\eta^{(c)} \rangle \, \langle \tilde{\psi}_\eta^{(c)} | f \rangle \,, \end{aligned} \tag{3.95}$$

where $\mathrm{Re}\, s_j < 0$, $\mathrm{Re}\, s_c < 0$, and $t > 0$. If we integrate the contributions of the continuous spectra we would obtain contributions of the form $\exp(s_c t)/t^{\alpha_c + 1}$ as in (3.78) since $s_c = -iz_c$.

In this way, we have obtained a decomposition which is comparable with the decomposition (3.73)–(3.78) of the correlation function in terms of the singularities of the spectral function. We can then identify the coefficients. The advantage of the form (3.95) is that the coefficients are expressed in terms of linear (and antilinear) functionals which can be interpreted as vectors determined by some generalized eigenvalue problem. The obtained expressions are no longer restricted to a specific correlation function but can be applied to every correlation function $C_{fg}(t)$ as long as the functions f and g belong to the functional spaces on which the decomposition is valid. The eigenvectors $\psi_{j,1}$ define states which approach exponentially the invariant state and which are intrinsically defined by the equations of motion. As we previously noticed, such generalized spectral decomposition cannot hold on the Hilbert space but only on smaller subspaces of test functions f and g because the eigenvectors (and the root vectors) are no longer functions but linear functionals or distributions (Dunford and Schwartz 1958–71, Gelfand and Shilov 1968). This is a general conclusion which appeared for a long time as a barrier to the construction of such decompositions. In the following, we shall present several methods for the explicit construction of such generalized decompositions.

3.6.3 Definition of a trace

In order to calculate the resonance spectrum of the semigroups, a method has been proposed which is based on trace formulas (Artuso et al. 1990, Cvitanović and Eckhardt 1991). Such trace formulas have been developed only recently and proved to be extremely powerful since the resonances can be calculated in terms of the critical elements of the dynamical system which are the stationary points and the periodic orbits.

A formal definition of a trace can be introduced in the spaces of distributions on which the semigroup evolution operators act. If we suppose that the spectrum of complex singularities consists only of poles, i.e., is discrete, a trace can formally be defined as

$$\mathrm{Tr}\ \hat{P}^{+t} = \sum_{\substack{j,i \\ \mathrm{Re}s_j<0}} \langle \tilde{\psi}_{j,i} | \hat{P}^{+t} | \psi_{j,i} \rangle = \sum_{\substack{j \\ \mathrm{Re}s_j<0}} m_j\, e^{s_j t} \ , \tag{3.96}$$

where the last equality is obtained from the decomposition (3.95) by assuming the absence of branch cuts and by using the biorthonormality (3.87). The eigenvalues can then be obtained as poles in the Laplace transform of (3.96)

$$\int_0^{+\infty} \mathrm{Tr}\ \hat{P}^{+t}\, e^{-st}\, dt = \sum_{\substack{j \\ \mathrm{Re}s_j<0}} \frac{m_j}{s - s_j} = \mathrm{Tr}\ \frac{1}{s - \hat{L}^{+}} \ , \tag{3.97}$$

which is equivalent to the trace of the resolvent of the generator of the semigroup. A similar relation holds for the backward semigroup.

However, such an equation is only formal and is not useful because it requires that the eigenvalue problem is already solved.

Useful and efficient trace formulas are based on the expression (3.13) of the evolution operator in terms of its kernel which is a Dirac distribution. By analogy with integral operators, a trace can be defined by equating the final and the initial variables, $\mathbf{X} = \mathbf{Y}$. Accordingly, we introduce

$$\mathrm{Tr}_{\mathscr{M}}\ \hat{P}^{\pm t} \equiv \int_{\mathscr{M}} \delta(\mathbf{X} - \mathbf{\Phi}^t\mathbf{X})\ d\mathbf{X}\ . \tag{3.98}$$

We notice that the trace (3.98) is invariant under smooth nonlinear changes of variables $\mathbf{X}$ so that this trace is intrinsic to the operator itself and does not depend on the representation. The contributions to the trace are thus given by all the solutions of

$$\mathbf{X} = \mathbf{\Phi}^t\mathbf{X}\ , \tag{3.99}$$

for some time t. Two types of such solutions exist:

1. The stationary points where the condition (3.99) is satisfied for all times t;
2. The periodic orbits where the condition (3.99) is satisfied for the discrete values of the time given by the repetitions of the prime periods: $t = rT_p$.

We have the very important:

Validity condition. The critical elements, i.e., the stationary points and the periodic orbits must be isolated in the phase space $\mathscr{M}$.

A critical element is isolated if the solution of (3.99) is locally unique so that there is no continuous family of solutions. Indeed, if there is a continuous family of solutions to (3.99) then there is at least a direction of integration $d\mathbf{X}$ along which the subintegral of (3.98) is infinite, which is not allowed in the use of the Dirac distribution.

Continuous families of critical elements may exist in systems with continuous symmetries or with constants of motion as in Hamiltonian systems. In such cases, the phase space $\mathscr{M}$ must be reduced to a smaller one in which the critical elements are isolated.

In autonomous Hamiltonian systems defined on the full phase space of the canonical variables $(\mathbf{q}, \mathbf{p})$, it turns out that the stationary points are typically isolated. Indeed, they appear only at some critical energies $E_s = H(\mathbf{q}_s, \mathbf{p}_s)$ and are isolated within the corresponding energy shell. In the case where the only critical element is a stationary point, the trace can be defined over the whole $2f$-dimensional phase space

$$\mathrm{Tr}_{\mathbb{R}^{2f}}\ e^{\hat{L}\pm t} \equiv \int_{\mathbb{R}^{2f}} \delta[(\mathbf{q}, \mathbf{p}) - \mathbf{\Phi}^t(\mathbf{q}, \mathbf{p})]\ d^{2f}q\ d^{2f}p\ , \tag{3.100}$$

where $\mathbf{\Phi}^t$ is the Hamiltonian flow given by Hamilton's equations.

However, periodic orbits are not typically isolated but form continuous families as energy changes. These continuous families may undergo abrupt transformations at some bifurcations. Therefore, Eq. (3.100) cannot apply in the presence of typical periodic orbits. The phase space must be reduced by using energy conservation because periodic orbits are typically isolated in each energy shell $H(\mathbf{q},\mathbf{p}) = E$. The variables may be taken as $\mathbf{X} = (E,\mathbf{x})$ where $\mathbf{x}$ are $2f-1$ variables which are for instance the time and the canonical variables transverse to a given periodic orbit $\mathbf{x} = (t,\mathbf{q}_\perp,\mathbf{p}_\perp)$. In these variables, the flow is

$$\mathbf{\Phi}^t\mathbf{X} = \mathbf{\Phi}^t(E,\mathbf{x}) = (E,\boldsymbol{\phi}_E^t\mathbf{x}) , \tag{3.101}$$

where $\boldsymbol{\phi}_E^t$ is the flow within the energy shell $H = E$. The Frobenius–Perron operator becomes

$$\begin{aligned} \hat{P}^t f(\mathbf{X}) &= \hat{P}^t f(E,\mathbf{x}) \\ &= \int_{\mathbb{R}^{2f}} \delta(E-\epsilon)\,\delta(\mathbf{x}-\boldsymbol{\phi}_\epsilon^t\mathbf{y})\,f(\epsilon,\mathbf{y})\,d\epsilon\,d^{2f-1}y \\ &= \int_{H=E} \delta(\mathbf{x}-\boldsymbol{\phi}_E^t\mathbf{y})\,f(E,\mathbf{y})\,d^{2f-1}y . \end{aligned} \tag{3.102}$$

Therefore, the trace must here be defined by

$$\mathrm{Tr}_{H=E}\, e^{\hat{L}\pm t} \equiv \int_{H=E} \delta(\mathbf{x}-\boldsymbol{\phi}_E^t\mathbf{x})\,d^{2f-1}x . \tag{3.103}$$

Further reductions must be carried out in the presence of continuous symmetries like translational or rotational symmetries, in which cases linear or angular momenta are also conserved.

It should be noticed that a continuous family of periodic orbits may still exist within the energy shell in systems without continuous symmetry as in the stadium billiard or in other billiards with parallel walls (Bunimovich 1979). In such a system, the continuous family of bouncing ball orbits is known to be responsible for algebraic decay in the correlation functions (Artuso et al. 1996). Here, another method than a phase-space reduction must be used. In a sense, the bouncing ball orbits responsible for algebraic behaviour must be removed by considering observables which vanish on these special orbits. Such a procedure can be defined in particular within the large-deviation formalism exposed in the next chapter.

In the following, we shall calculate the trace formula for stationary points in order to obtain the resonances and we shall then construct the corresponding generalized spectral decomposition. Afterwards, the same construction will be developed for periodic orbits.

3.7 Resonances of stationary points

3.7.1 Trace formula and eigenvalues

Let us consider a flow with a single isolated stationary point

$$\dot{\mathbf{X}} = \mathbf{F}(\mathbf{X}) \, , \quad \mathbf{F}(\mathbf{X}_s) = 0 \, , \tag{3.104}$$

($\mathbf{X} \in \mathbb{R}^M$), for which all the stability eigenvalues

$$\det\left[\frac{\partial \mathbf{F}}{\partial \mathbf{X}}(\mathbf{X}_s) - \xi \, \mathsf{1}\right] = 0 \, , \tag{3.105}$$

have no vanishing real part $\mathrm{Re}\xi_j \neq 0$, i.e., the stationary point is assumed to be hyperbolic. The stable directions correspond to the eigenvalues such that $\mathrm{Re}\xi_j < 0$ and the unstable directions to those with $\mathrm{Re}\xi_j > 0$. Since the flow is real the eigenvalues are either real or form pairs of complex conjugate eigenvalues $\xi_j = \rho_j \pm i\omega_j$, in which case the stationary point has the character of a stable or unstable focus in some direction whether $\rho_j < 0$ or $\rho_j > 0$ respectively. Let us recall that the Lyapunov exponents are given by the real parts of the stability eigenvalues: $\lambda_j = \mathrm{Re}\xi_j$. We furthermore assume for simplicity that none of the eigenvalues is degenerate, which is the generic situation. Therefore, the matrix of the linear part of the vector field can be diagonalized.

Using a well-known property of Dirac's distribution, the kernel of the Frobenius–Perron operator which appears in the trace (3.98) becomes

$$\delta(\mathbf{X} - \mathbf{\Phi}^t \mathbf{X}) = \frac{\delta(\mathbf{X} - \mathbf{X}_s)}{\left|\det\left[\mathsf{1} - \frac{\partial \mathbf{\Phi}^t}{\partial \mathbf{X}}(\mathbf{X}_s)\right]\right|} \, , \tag{3.106}$$

where we used the assumption that there is a single stationary point, i.e., a single solution of Eqs. (3.99) or (3.104) which exists for all times t.

Now, the Jacobian matrix of the flow evaluated at the stationary point is given in terms of the matrix of the linearized vector field as

$$\frac{\partial \mathbf{\Phi}^t}{\partial \mathbf{X}}(\mathbf{X}_s) = \exp\left[\frac{\partial \mathbf{F}}{\partial \mathbf{X}}(\mathbf{X}_s) \, t\right] = \mathsf{R}^{-1} \cdot \exp(\Xi t) \cdot \mathsf{R} \, , \tag{3.107}$$

where Ξ is the diagonal matrix containing the stability eigenvalues $\{\xi_j\}_{j=1}^M$. Since the determinant can be calculated on any basis, we obtain

$$\mathrm{Tr} \, \hat{P}^{\pm t} = \frac{1}{\left|\det\left[\mathsf{1} - \frac{\partial \mathbf{\Phi}^t}{\partial \mathbf{X}}(\mathbf{X}_s)\right]\right|} = \prod_{j=1}^{M} \frac{1}{|1 - \exp(\xi_j t)|} \, . \tag{3.108}$$

We notice that the trace function decays always for $t \to \pm\infty$ independently of the stability of the stationary point. In order to display this decay (here for $t \to +\infty$), we separate the stable and unstable directions

$$\mathrm{Tr}\ \hat{P}^{\pm t} = \prod_{\mathrm{Re}\xi_j>0} \frac{\exp(-\xi_j t)}{1-\exp(-\xi_j t)} \prod_{\mathrm{Re}\xi_j<0} \frac{1}{1-\exp(\xi_j t)} . \tag{3.109}$$

We observe that the factors associated with the stable directions tend to one as $t \to +\infty$, while the factors associated with the unstable directions decrease to zero like $\exp(-\xi_j t)$ as $t \to +\infty$. The reverse property holds for $t \to -\infty$ because the stabilities are exchanged.

Since we have assumed that none of the directions is neutral we find that all the denominators in $1/(1-x)$ satisfy $|x| < 1$ so that we are allowed to expand it in Taylor series to obtain

$$\mathrm{Tr}\ \hat{P}^{\pm t} = \sum_{\mathbf{l},\mathbf{m}=0}^{\infty} \exp\left[- \sum_{\mathrm{Re}\xi_j>0} (m_j+1)\ \xi_j + \sum_{\mathrm{Re}\xi_j<0} l_j\ \xi_j \right] , \tag{3.110}$$

where $\mathbf{l} = (l_1, \ldots, l_{M_s})$ and $\mathbf{m} = (m_1, \ldots, m_{M_u})$ with $M_s + M_u = M$. Therefore, we obtain a sum of exponential decays so that the spectrum of resonances is given by

forward semigroup :

$$s_{\mathbf{lm}} = - \sum_{\mathrm{Re}\xi_j>0} (m_j+1)\ \xi_j + \sum_{\mathrm{Re}\xi_j<0} l_j\ \xi_j ,$$

$$\text{with} \quad l_j, m_j = 0, 1, 2, 3, \ldots . \tag{3.111}$$

We observe that the Liouvillian resonances are determined by the linear stability properties of the stationary point. We obtain a fully explicit expression in terms of the stability eigenvalues. We have the remarkable result that all the resonances belong to the lower half plane, which guarantees the existence of the forward semigroup. The remarkable fact is that the stationary point may be stable or unstable with negative or positive Lyapunov exponents, the statistical ensembles of the Liouvillian dynamics are always decaying toward an asymptotic invariant measure. The Liouvillian dynamics of statistical ensembles has thus a property of stability so that the unstable stationary points become allowable.

Among the different decay rates or resonances (3.111) the most important role in the ensemble dynamics is played by those which have the smallest real part in absolute value. Indeed, $\mathrm{Re}s_{\mathbf{lm}}$ controls the exponential decay of the corresponding eigenstate. The transient component of the ensemble which survives in the long-time limit is the one corresponding to the slowest decay rate, i.e., to the resonance $\mathbf{l} = \mathbf{m} = 0$. When the stationary point is stable, this resonance vanishes, which corresponds to the invariant measure which is centered on the attracting fixed point.

On the other hand, if the point is unstable the slowest decay rate is $s_0 = -\gamma^+$ with $\mathbf{l} = \mathbf{m} = 0$, which defines the so-called *escape rate* γ^+. The escape rate characterizes the long-time exponential decay of a statistical ensemble of trajectories escaping from the vicinity of the unstable stationary point. Such a situation is commonly encountered in open systems as discussed in the following chapters (see in particular Section 4.4). Because the Lyapunov exponents are directly related to the stability eigenvalues of the stationary point according to $\lambda_j = \mathrm{Re}\xi_j$, we have the general result:

The escape rate of an unstable stationary point (with respect to the forward semigroup) is given by the sum of positive Lyapunov exponents:

$$\text{forward semigroup :} \qquad \gamma^+ = \sum_{\lambda_j > 0} \lambda_j \, . \tag{3.112}$$

All the other resonances are found in the complex plane s below the resonance corresponding to the escape rate: $\mathrm{Re}s_{\mathbf{lm}} \leq -\gamma^+$.

By reversing time, the antiresonances, i.e., the resonances of the backward semigroup, are given by

$$\begin{aligned} &\text{backward semigroup :} \\ &s_{\mathbf{lm}} = \sum_{\mathrm{Re}\xi_j > 0} l_j \, \xi_j - \sum_{\mathrm{Re}\xi_j < 0} (m_j + 1) \, \xi_j \, , \\ &\text{with} \quad l_j, m_j = 0, 1, 2, 3, \ldots \, . \end{aligned} \tag{3.113}$$

Thanks to the above formulas, the resonances of several types of stationary points can be calculated systematically.

3.7.2 Eigenstates and spectral decompositions

We have obtained the general expression for the Liouvillian resonances at stationary points. The derivation is based on the trace formula which does not provide the generalized eigenvectors associated with the resonances. In order to construct these eigenstates and the complete spectral decomposition, we need to go back to the general expression for the time evolution of a correlation function or of the average of an observable. In this way, we obtain an alternative calculation of the resonances and, furthermore, the corresponding eigenstates.

Different methods are possible. In general, the method proceeds by first solving the equations of motion to have an explicit form for the flow $\mathbf{\Phi}^t\mathbf{X}$. This explicit form is then expanded in Taylor series together with the observable and the initial density around the stationary point which we assume to be isolated and of hyperbolic character. Instead of a brute force Taylor expansion, successive terms may be obtained by first looking after the slowest decay modes in the long-time limit. Thereafter, the asymptotic terms associated with the slowest decay modes are subtracted from the general expression of the time evolution and the next-to-slowest decay modes are obtained. The procedure is developed recursively to obtain all the decay modes from the slowest to the most rapid (see Gaspard et al. 1995a).

A very convenient method to solve the equations of motion proceeds by a linearization of the vector field (Arnold 1982). However, a linearization requires strong conditions on the vector field which is restrictive although the spectral decomposition (with possible Jordan-block structures) may still exist under more general conditions.

In the general case of a stationary point of saddle type, the real parts of the stability eigenvalues may be positive or negative. The system is said to belong to the Siegel domain if the convex envelope of the eigenvalues $(\xi_1, \ldots, \xi_M)$ contains the origin of the complex plane ξ (Arnold 1982). We may therefore expect that a slight perturbation of the vector field will create commensurabilities of the type

$$\xi_i = \sum_j n_j \, \xi_j \,, \qquad n_j \geq 0 \,, \quad \sum_j n_j \geq 2 \,. \tag{3.114}$$

In this sense, the commensurabilities are dense in the Siegel domain. When there is no commensurability the vector field can be formally linearized in the vicinity of the stationary point (Arnold 1982). When the stationary point shows commensurabilities the vector field can no longer be formally linearized in its vicinity, which may lead to Jordan-block structure in the spectral decomposition of the Liouvillian dynamics. We shall analyze such a Jordan-block structure in the case of Hamiltonian systems where commensurabilities are generic. However, as we argued above, commensurabilities are not generic for general dynamics systems which can be linearized in most cases. This observation has been made precise by Siegel in the following terms (Arnold 1982).

Definition. The stationary point is said to be of type (C, ν) if its stability eigenvalues $\boldsymbol{\xi} = (\xi_1, \xi_2, \ldots, \xi_M) \in \mathbb{C}^M$ satisfy

$$\left| \xi_i - \sum_{j=1}^{M} n_j \, \xi_j \right| \geq \frac{C}{|\mathbf{n}|^\nu} \,, \tag{3.115}$$

for all the integer vectors $\mathbf{n}$ *with nonnegative components such that* $|\mathbf{n}| = \sum_{j=1}^M n_j \geq 2$.

This definition can be used to formulate the condition that the stability eigenvalues are sufficiently far from commensurability. Such a condition guarantees the linearization of the vector field with a holomorphic transformation according to the (Arnold 1982)

Siegel theorem. If the stability eigenvalues ξ *of a holomorphic vector field are of type* (C, ν) *then the vector field is biholomorphically equivalent to its linear part near the stationary point.*

The holomorphic transformation can be represented by the so-called Poincaré series which are therefore convergent for almost all linear parts. The linearizing transformation can be expressed in new coordinates which separate the stable and the unstable directions, $\mathbf{Y} = (\mathbf{Y}_s, \mathbf{Y}_u)$,

$$\begin{cases} \mathbf{Y}_s = \mathbf{G}_s(\mathbf{X}) , \\ \mathbf{Y}_u = \mathbf{G}_u(\mathbf{X}) . \end{cases} \tag{3.116}$$

The linearized vector field is

$$\begin{cases} \dot{\mathbf{Y}}_s = \Xi_s \cdot \mathbf{Y}_s , & (\Xi_s < 0) , \\ \dot{\mathbf{Y}}_u = \Xi_u \cdot \mathbf{Y}_u , & (\Xi_u > 0) . \end{cases} \tag{3.117}$$

Now, we decompose the time evolution of the average $\langle A \rangle_t$ of an observable $A(\mathbf{X})$ under the assumption of linearization. We first perform the above linearizing change of variables. Thereafter, we make an extra change of variables in order to let appear, in the average $\langle A \rangle_t$, the quantities $\exp(\Xi_s t)$ and $\exp(-\Xi_u t)$ which both vanish in the limit $t \to +\infty$ defining the forward semigroup. This will allow us to perform a Taylor expansion with respect to these quantities, leading to a series representation in terms of exponential relaxations.

Using the linearization, the time evolution of an average can be written as

$$\begin{aligned} \langle A \rangle_t &= \int d\mathbf{X}\, f_0(\mathbf{X})\, A(\mathbf{\Phi}^t\mathbf{X}) \\ &= \int d\mathbf{Y}_s\, d\mathbf{Y}_u \frac{f_0[\mathbf{G}^{-1}(\mathbf{Y}_s, \mathbf{Y}_u)]}{|\det \partial_{\mathbf{X}}\mathbf{G}[\mathbf{G}^{-1}(\mathbf{Y}_s, \mathbf{Y}_u)]|} \\ &\quad \times A\Big[\mathbf{G}^{-1}\big(e^{\Xi_s t}\mathbf{Y}_s, e^{\Xi_u t}\mathbf{Y}_u\big)\Big] . \end{aligned} \tag{3.118}$$

We introduce the variable $\mathbf{Y}'_u = \exp(\Xi_u t)\mathbf{Y}_u$ so that the average becomes

$$\begin{aligned} \langle A \rangle_t &= \det\big(e^{-\Xi_u t}\big) \int d\mathbf{Y}_s\, d\mathbf{Y}'_u \frac{f_0[\mathbf{G}^{-1}(\mathbf{Y}_s, e^{-\Xi_u t}\mathbf{Y}'_u)]}{|\det \partial_{\mathbf{X}}\mathbf{G}[\mathbf{G}^{-1}(\mathbf{Y}_s, e^{-\Xi_u t}\mathbf{Y}'_u)]|} \\ &\quad \times A\Big[\mathbf{G}^{-1}\big(e^{\Xi_s t}\mathbf{Y}_s, \mathbf{Y}'_u\big)\Big] . \end{aligned} \tag{3.119}$$

In this formula, the trajectory is expressed in terms of the stable coordinates of the initial condition at $t = 0$ and of the unstable coordinates of the final point at time t. The expression is now expanded in multiple Taylor series of the variables $[\exp(+\xi_s t), \exp(-\xi_u t)]$ where $\xi = (\xi_s, \xi_u)$ are the stability eigenvalues. We obtain

$$\langle A\rangle_t = \sum_{\mathbf{l},\mathbf{m}=0}^{\infty} \exp\Big\{[\mathbf{l}\cdot\xi_s - (\mathbf{m}+\mathbf{1})\cdot\xi_u]t\Big\} \langle A|\psi_{\mathbf{lm}}\rangle \langle\tilde{\psi}_{\mathbf{lm}}|f_0\rangle , \tag{3.120}$$

with

$$\begin{aligned}\langle A|\psi_{\mathbf{lm}}\rangle &= \int d\mathbf{Y}'_u \frac{\mathbf{Y}'^{\mathbf{m}}_u}{\mathbf{m}!} \partial^{\mathbf{l}}_{\mathbf{Y}'_s} A\big[\mathbf{G}^{-1}(\mathbf{Y}'_s, \mathbf{Y}'_u)\big]\Big|_{\mathbf{Y}'_s=0} , \\ \langle\tilde{\psi}_{\mathbf{lm}}|f_0\rangle &= \int d\mathbf{Y}_s \frac{\mathbf{Y}^{\mathbf{l}}_s}{\mathbf{l}!} \partial^{\mathbf{m}}_{\mathbf{Y}_u} \frac{f_0[\mathbf{G}^{-1}(\mathbf{Y}_s,\mathbf{Y}_u)]}{|\det\partial_{\mathbf{X}}\mathbf{G}[\mathbf{G}^{-1}(\mathbf{Y}_s,\mathbf{Y}_u)]|}\Big|_{\mathbf{Y}_u=0} .\end{aligned} \tag{3.121}$$

Therefore, the right and left eigenstates of the forward semigroup are given by the distributions

$$\begin{aligned}\psi_{\mathbf{lm}}(\mathbf{X}) &= \partial^{\mathbf{l}}_{\mathbf{Y}'_s}\left\{ |\det\partial_{\mathbf{X}}\mathbf{G}(\mathbf{X})| \frac{\mathbf{G}_u(\mathbf{X})^{\mathbf{m}}}{\mathbf{m}!} \delta^{M_s}\big[\mathbf{Y}'_s - \mathbf{G}_s(\mathbf{X})\big]\right\}\Big|_{\mathbf{Y}'_s=0} , \\ \tilde{\psi}_{\mathbf{lm}}(\mathbf{X}) &= \partial^{\mathbf{m}}_{\mathbf{Y}_u}\left\{ \frac{\mathbf{G}^*_s(\mathbf{X})^{\mathbf{l}}}{\mathbf{l}!} \delta^{M_u}\big[\mathbf{Y}_u - \mathbf{G}^*_u(\mathbf{X})\big]\right\}\Big|_{\mathbf{Y}_u=0} ,\end{aligned} \tag{3.122}$$

where M_s and M_u are the dimensions of the stable and unstable manifolds: $M_s + M_u = M$. We remark that the right and left eigenstates are distributions as long as there exist stable or unstable directions. If there is no unstable directions as for an attracting point ($M_u = 0$) the left eigenstates are given by functions while the right eigenstates are given by distributions and vice versa for anti-attracting points. It should be noticed that the support of the right eigenstates of the forward semigroup is the unstable manifold while it is the stable manifold for the left eigenstates. The supports are exchanged in the case of the backward semigroup. The eigenstates (3.122) are biorthonormal and form a complete basis if the observable A and the initial density f_0 are regular enough to admit derivatives and moments of arbitrarily high orders along the stable and unstable manifolds, and for the series over $\mathbf{l}$ and $\mathbf{m}$ to converge. These conditions can be made explicit in specific systems (Gaspard et al. 1995a).

We recover the result of the trace formula that the Liouvillian eigenvalues are the linear combinations (3.111) of the stability eigenvalues. The leading eigenvalue with the smallest real part is given in absolute value by the forward escape rate (3.112). The corresponding left and right eigenstates are

$$\begin{aligned}\psi_{00}(\mathbf{X}) &= |\det\partial_{\mathbf{X}}\mathbf{G}(\mathbf{X})| \,\delta^{M_s}\big[\mathbf{G}_s(\mathbf{X})\big] , \\ \tilde{\psi}_{00}(\mathbf{X}) &= \delta^{M_u}\big[\mathbf{G}^*_u(\mathbf{X})\big] .\end{aligned} \tag{3.123}$$

We notice that these leading eigenstates are positive and they define conditionally invariant measures supported by the invariant manifolds. This property is a consequence of the Frobenius–Perron theorem which says that the eigenvectors associated with the leading eigenvalue of a nonnegative operator are always positive (Gantmacher 1959). Indeed, the present operator is the Frobenius–Perron operator (3.13) which has a nonnegative kernel giving the transition probabilities in phase space. Another consequence of the Frobenius–Perron theorem is that the leading eigenvalue of the Frobenius–Perron operator is always simple if the system is transitive (Gantmacher 1959).

3.7.3 Hamiltonian stationary point of saddle type

The case of a saddle point is very common in scattering Hamiltonian systems. A typical example with only one degree of freedom is given by the Eckart Hamiltonian $H = p^2/(2) + v_0 \cosh^{-2}(q/a)$.

We suppose that a single saddle point exists in the two-dimensional phase space (q,p). The stability eigenvalues are $\xi = (+\lambda, -\lambda)$ with the Lyapunov exponent $\lambda = \sqrt{(\partial^2_{qp}H)^2 - \partial^2_q H \partial^2_p H}$ where the second derivatives of the Hamiltonian $H(q,p)$ are evaluated at the saddle point translated to the origin: $(q,p) = (0,0)$. According to (3.111), the Liouvillian eigenvalues are given by

$$s_m = -(m+1)\,\lambda\,, \qquad m = 0,1,2,3,\ldots \tag{3.124}$$

with the multiplicity $m+1$. This is a consequence of commensurabilities between the stability eigenvalues $+\lambda$ and $-\lambda$, which is itself a consequence of volume conservation in phase space. As a corollary, the Hamiltonian cannot be linearized. The minimal form of the vector field still contains nonlinear terms called resonant terms (Arnold 1978, 1982, 1988). There exists a canonical transformation, $(x,y) = \mathbf{G}(q,p)$, which brings the Hamiltonian to the Birkhoff normal form

$$H(x,y) = h(I) = \lambda\, I + \frac{\alpha}{2!}\, I^2 + \frac{\beta}{3!}\, I^3 + \cdots\,, \quad \text{with} \quad I = x\, y\,, \tag{3.125}$$

where α, β,... are coefficients determined by the reduction to the normal form and which characterizes the intrinsic nonlinear behaviour near the saddle point.

In the new coordinates, the equations of motion can be integrated and the flow is obtained as

$$\mathbf{\Phi}^t(x,y) = \left(e^{+h't}\, x\,,\, e^{-h't}\, y\right), \tag{3.126}$$

where $h' = \partial_I h(xy)$. We notice that $I = xy$ is a constant of motion so that $x_0 y_0 = x_t y_t$.

Let us now suppose that the transformation $(x, y) = \mathbf{G}(q, p)$ from the original variables (q, p) to the new coordinates (x, y) has been carried out on the observable and on the initial density which are transformed into

$$\begin{aligned} B(x,y) &= A[\mathbf{G}^{-1}(x,y)] \,, \\ g_0(x,y) &= f_0[\mathbf{G}^{-1}(x,y)] \,. \end{aligned} \tag{3.127}$$

After this change of variables, the time evolution of the average of the observable becomes

$$\begin{aligned} \langle A \rangle_t &= \int dx\, dy\, B[\mathbf{\Phi}^t(x,y)]\, g_0(x,y) \\ &= \int dx\, dy\, B\left[x e^{+\lambda t} e^{\eta(xy)t}, y e^{-\lambda t} e^{-\eta(xy)t}\right] g_0(x,y) \,, \end{aligned} \tag{3.128}$$

where

$$\eta(I) = h'(I) - \lambda = \alpha\, I + \frac{\beta}{2!}\, I^2 + \cdots \,. \tag{3.129}$$

We perform the change of variable $x' = e^{\lambda t} x$ in order for the average to depend on the time through the decaying exponential function $e^{-\lambda t}$. Thereafter, we expand in Taylor series of the small quantity $\varepsilon = e^{-\lambda t}$. In the present case, the Taylor expansion generates terms which are in powers of the time t, which yields a Jordan-block structure. Expanding up to the term corresponding to $s_m = -2\lambda$ with $m = 1$, we get

$$\begin{aligned} \langle A \rangle_t = \; & e^{-\lambda t} \int dx\, B(x,0) \int dy\, g_0(0,y) \\ & + e^{-2\lambda t} \int dx\, x\, B(x,0) \int dy\, \partial_x g_0(0,y) \\ & + e^{-2\lambda t} \int dx\, \partial_y B(x,0) \int dy\, y\, g_0(0,y) \\ & + \alpha\, t\, e^{-2\lambda t} \int dx\, x^2\, \partial_x B(x,0) \int dy\, y\, g_0(0,y) \\ & + \mathcal{O}\left(t^2 e^{-3\lambda t}\right) \,, \end{aligned} \tag{3.130}$$

where we replaced x' by x for simplicity and where $\alpha = \partial_I^2 h(0)$. In the first factor of the last term, an integration by part is carried out to reduce a power of x with a derivative ∂_x, assuming that the observable $B(x,0)$ vanishes at large values of x. A comparison with the Jordan-type expansion (3.86) up to the same order, i.e.,

$$\begin{aligned}\langle A\rangle_t &= e^{-\lambda t}\,\langle B|\psi_{00}\rangle\,\langle\tilde{\psi}_{00}|g_0\rangle \\ &+ e^{-2\lambda t}\,\big(\langle B|\psi_{10}\rangle\;\langle B|\psi_{11}\rangle\big)\begin{pmatrix}1 & t\\ 0 & 1\end{pmatrix}\begin{pmatrix}\langle\tilde{\psi}_{10}|g_0\rangle\\ \langle\tilde{\psi}_{11}|g_0\rangle\end{pmatrix} \\ &+ \mathcal{O}\left(t^2 e^{-3\lambda t}\right), \end{aligned} \tag{3.131}$$

allows us to identify the first eigenstates and other root states as follows

$$\begin{aligned}\psi_{00}(x,y) = \delta(y)\,, &\quad \tilde{\psi}_{00}(x,y) = \delta(x)\,,\\ \psi_{10}(x,y) = x\,\delta(y)\,, &\quad \tilde{\psi}_{10}(x,y) = -\delta'(x)\,,\\ \psi_{11}(x,y) = +\frac{1}{2\partial_I^2 h(0)}\,\delta'(y)\,, &\quad \tilde{\psi}_{11}(x,y) = -2\partial_I^2 h(0)\; y\;\delta(x)\,, \end{aligned} \tag{3.132}$$

where the indices of ψ_{mj} are $m = 0, 1, 2, \ldots$ and $j = 0, 1, \ldots, m$. Biorthonormality and the completeness relations hold up to this order. Higher root states can be obtained by a systematic Taylor expansion and by a proper identification of the root states through a comparison with the formal Jordan-type decomposition (3.86) by using Eq. (3.95). Now, the spectral decomposition in the original canonical variables can be obtained by inverting the canonical transformation.

We remark that the Jordan-block structure disappears and the time evolution operator can be diagonalized for the quadratic Hamiltonian $H = (p^2/2) - v_0(q/a)^2$ with an inverted harmonic potential.

For interpretation, it is interesting to have a construction of the eigenstates which avoids the use of the canonical transformation to the normal form. This is possible if we observe that the eigenstates of the forward semigroup are localized on the unstable manifolds which can be represented in the parametric form

$$W_u \;:\quad (q,p) = \left[q_{u\pm}(t),\; p_{u\pm}(t)\right], \tag{3.133}$$

where the parameter is the time $-\infty < t < +\infty$ and both signs correspond to both branches of the manifold. The unstable manifold approaches the saddle point as $t \to -\infty$. The eigenstates are concentrated along the branches of W_u according to a Dirac distribution transverse to W_u, which can be expressed as

$$\psi_{m0}(q,p) = \sum_{\epsilon=\pm}\int_{-\infty}^{+\infty} dt\;\exp(-s_m t)\;\delta\left[q - q_{u\epsilon}(t)\right]\,\delta\left[p - p_{u\epsilon}(t)\right]. \tag{3.134}$$

The exponential function is obtained in order for the eigenstates to satisfy $\hat{P}^{+t}\psi_{m0}(q,p) = \psi_{m0}\left[\mathbf{\Phi}^{-t}(q,p)\right] = \exp(s_m t)\psi_{m0}(q,p)$. Similar reasonings can in principle be used to obtain the other root states.

A general conclusion is that the spectral decomposition of the Liouvillian dynamics near a Hamiltonian saddle point will in general involve a Jordan-block structure in spite of the fact that the saddle point is strictly hyperbolic. The Jordan-block structure appears because of the commensurabilities between the stability eigenvalues which are always present in generic Hamiltonian systems. There is thus a general relationship between the presence of resonant terms in the Birkhoff normal form of a vector field and the presence of Jordan-block structures in the spectral decomposition of its Liouvillian dynamics. This relationship should not surprise us because the Liouvillian dynamics plays a central role in obtaining the constants of motion as well as the normal form, for instance, with the Birkhoff–Gustavson method (Birkhoff 1927; Gustavson 1966; Arnold 1982, 1988; Lichtenberg and Lieberman 1983). Here, we have simply the extension of the same considerations to the time evolution of statistical ensembles.

3.8 Resonances for hyperbolic sets with periodic orbits

3.8.1 Trace formula

As we noted before, the trace (3.98) involves the critical elements of a flow. We have already treated the invariant sets composed of a stationary point. We shall here treat the invariant sets with periodic orbits. We assume that the set is hyperbolic so that each periodic orbit p has M_s stable directions and M_u unstable directions ($M_s + M_u = M - 1$) and so many stability eigenvalues

$$|\Lambda^{(u)}_{p,1}| ,\ldots, |\Lambda^{(u)}_{p,M_u}| \; > \; 1 \; > \; |\Lambda^{(s)}_{p,1}| ,\ldots, |\Lambda^{(s)}_{p,M_s}| \; . \tag{3.135}$$

These stability eigenvalues are obtained by diagonalization of the linearized Poincaré map (1.5) evaluated at a fixed point $\mathbf{x}_p$ of the periodic orbit

$$\mathsf{m}_p \;=\; \partial_{\mathbf{x}} \phi^{n_p}(\mathbf{x}_p) \; , \tag{3.136}$$

which is the monodromy matrix of the periodic orbit. The periodic orbits are supposed to be isolated and their periods are

$$T_p \;=\; \sum_{j=1}^{n_p} \; T(\phi^j \mathbf{x}_p) \; , \tag{3.137}$$

in terms of the first-return time function of Eq. (1.5). We should thus expect that a trace like (3.98) will be zero for almost all times except for the repetition times $t = rT_p$ where the trace displays delta distributions. The trace (3.98) is evaluated by introducing coordinates which are parallel and transverse to the periodic orbits. The coordinates $(\mathbf{x}, \tau)$ of the suspended flow (1.6) are precisely such coordinates. The time coordinate τ is longitudinal to the flow while the coordinate $\mathbf{x}$ is transverse to the flow. Since the trace formula (3.98) is invariant under a nonlinear change of coordinates like $\mathbf{X} = \mathbf{G}(\mathbf{x}, \tau)$, we get

$$\mathrm{Tr}\, \hat{P}^{\pm t} = \int d\mathbf{x} \int_0^{T(\mathbf{x})} d\tau\, \delta(\mathbf{x} - \boldsymbol{\phi}^n \mathbf{x})\, \delta\left[t - \sum_{j=0}^{n-1} T(\boldsymbol{\phi}^j \mathbf{x})\right], \tag{3.138}$$

where the integer $n = n(\mathbf{x}, \tau)$ depends on the current phase-space point.

We proceed with a decomposition of the domain of integration into tubular neighbourhoods of each periodic orbit, which is possible because they are isolated. Each of these tubes intersects the Poincaré surface of section into n_p small neighbourhoods of the intersections $\{\boldsymbol{\phi}^j \mathbf{x}_p\}_{j=1}^{n_p}$ of the periodic orbit with the surface of section. Accordingly, we have the decomposition

$$\delta(\mathbf{x} - \boldsymbol{\phi}^n \mathbf{x}) = \sum_{p,r:n=rn_p} \sum_{j=1}^{n_p} \frac{\delta(\mathbf{x} - \boldsymbol{\phi}^j \mathbf{x}_p)}{|\det(1 - \mathsf{m}_p^r)|}, \tag{3.139}$$

where the sum extends over all the repetitions r of the prime periods such that $n = rn_p$ (cf. Section 2.5). Noting that $T_p = \sum_{j=1}^{n_p} \int_0^{T(\boldsymbol{\phi}^j \mathbf{x}_p)} d\tau$ because of (3.137), Eq. (3.138) becomes

$$\mathrm{Tr}\, \hat{P}^{\pm t} = \sum_{p \in \mathscr{P}} \sum_{r=1}^{\infty} T_p \frac{\delta(t - rT_p)}{|\det(1 - \mathsf{m}_p^r)|}, \tag{3.140}$$

which was discovered by Cvitanović and Eckhardt (1991). This trace formula and its time-discrete version play a crucial role in many applications as discussed in the following (Christiansen et al. 1990a, 1990b; Cvitanović 1991; Gaspard 1991a; Eckhardt 1993). We shall show below that the trace formula is related to the so-called Zeta functions, which provides us with an analytic and direct method to calculate the Liouvillian resonances, i.e., the decay or relaxation rates and, in particular, the escape rate of the dynamical system under study.

In relation to the trace formula (3.140), we should here mention a sum rule over periodic orbits obtained by Hannay and Ozorio de Almeida (1984). If the system is *bounded* and ergodic, Eq. (3.28) is equivalent to

$$\lim_{T \to +\infty} \frac{1}{T} \int_\tau^T dt\, \delta(\mathbf{X} - \boldsymbol{\Phi}^t \mathbf{X}_0) = f_{\mathrm{i}}(\mathbf{X}), \tag{3.141}$$

where $f_{\mathrm{i}}(\mathbf{X})$ is the density of the invariant measure and τ is a small and positive time. The identity holds for almost all the initial conditions $\mathbf{X}_0$ and we may suppose that it also holds for the current point $\mathbf{X}_0 = \mathbf{X}$. Integrating both members of Eq. (3.141) with this identification, we obtain an identity for the trace (3.98)

$$\lim_{T\to+\infty} \frac{1}{T} \int_{\tau}^{T} dt \ \mathrm{Tr}\ \hat{P}^{+t} = \int f_{\mathrm{i}}(\mathbf{X})\, d\mathbf{X} = 1 . \tag{3.142}$$

Substituting the trace formula (3.140) in Eq. (3.142), we infer the sum rule

$$\lim_{T\to+\infty} \frac{1}{T} \sum_{0<rT_p<T} \frac{T_p}{|\det(1-\mathsf{m}_p^r)|} = 1 , \tag{3.143}$$

by Hannay and Ozorio de Almeida (1984) (see also Ozorio de Almeida 1988). This sum rule expresses the fact that the exponential proliferation (2.63) of periodic orbits is counterbalanced by their Lyapunov instability in *bounded* chaotic systems. The trace formula (3.140) goes beyond this sum rule and allows the study of the relaxation mechanism toward the invariant measure in bounded and unbounded systems.

3.8.2 The Selberg–Smale Zeta function

The Laplace transform of the trace (3.140) of the Frobenius–Perron operator defines formally the trace of the resolvent of the Liouville operator

$$\mathrm{Tr}\ \frac{1}{s-\hat{L}^{+}} = \int_{0}^{+\infty} e^{-st}\ \mathrm{Tr}\ \hat{P}^{+t}\ dt = \frac{\partial}{\partial s} \ln Z(s) , \tag{3.144}$$

where we introduced the Selberg–Smale Zeta function (Smale 1980b)

$$Z(s) \equiv \exp - \sum_{p\in\mathscr{P}} \sum_{r=1}^{\infty} \frac{1}{r} \frac{\exp(-srT_p)}{|\det(1-\mathsf{m}_p^r)|} . \tag{3.145}$$

This Zeta function can be transformed into a product over all the prime periodic orbits as follows. After diagonalization of the monodromy matrix (3.136) in terms of its eigenvalues (3.135), the determinant in the denominator of (3.145) becomes

$$\det(1-\mathsf{m}_p^r) = \prod_{i=1}^{M_s} [1-\Lambda_{p,i}^{(s)r}] \prod_{j=1}^{M_u} [1-\Lambda_{p,j}^{(u)r}] , \tag{3.146}$$

Now, we consider the inverse of the determinant taken in absolute value. Since the stable and unstable stability eigenvalues are less or greater than one in absolute value they lead to the different Taylor expansions:

$$\frac{1}{|1-\Lambda|} = \frac{1}{1-\Lambda} = \sum_{l=0}^{\infty} \Lambda^l , \quad \text{if } |\Lambda|<1 ,$$

$$\frac{1}{|1-\Lambda|} = \frac{1}{|\Lambda|(1-\Lambda^{-1})} = \sum_{m=0}^{\infty} \frac{1}{|\Lambda|\Lambda^m} , \quad \text{if } |\Lambda|>1 . \tag{3.147}$$

Consequently, we get

$$\frac{1}{|\det(1-\mathsf{m}_p^r)|} = \prod_{i=1}^{M_s}\sum_{l_i=0}^{\infty}(\Lambda_{p,i}^{(s)l_i})^r \prod_{j=1}^{M_u}\sum_{m_j=0}^{\infty}\left(\frac{1}{|\Lambda_{p,j}^{(u)}|\Lambda_{p,j}^{(u)m_j}}\right)^r . \tag{3.148}$$

We now replace the results we obtained into the Zeta function (3.145). The sums over the prime periodic orbits and over the integers $\{l_i, m_j\}$ are moved outside the exponential where they become products. The sum over the repetition number r is performed with

$$\sum_{r=1}^{\infty} \frac{1}{r}\, x^r = -\ln(1-x) . \tag{3.149}$$

The Zeta function finally becomes

$$Z(s) = \prod_p \prod_{l_1\cdots l_{M_s} m_1\cdots m_{M_u}=0}^{\infty}\left(1 - \frac{e^{-sT_p}\prod_{i=1}^{M_s}\Lambda_{p,i}^{(s)l_i}}{\prod_{j=1}^{M_u}|\Lambda_{p,j}^{(u)}|\Lambda_{p,j}^{(u)m_j}}\right) \tag{3.150}$$

which is a product over all the prime periodic orbits of factors involving the period and the stability eigenvalues of each prime periodic orbit.

Because of Eq. (3.144), the poles of the Liouvillian resolvent are now given by the complex zeros of the Zeta function, $Z(s) = 0$. Indeed, if s_j is a simple zero of the Zeta function, we have that $Z(s) = A_j(s)(s - s_j)$ where $A_j(s) \neq 0$ in the vicinity of $s = s_j$. Accordingly, the logarithmic derivative of the Zeta function has a simple pole at $s = s_j$ because $\partial_s \ln Z(s) = \partial_s \ln A_j(s) + (s - s_j)^{-1}$. Equation (3.144) implies that $s = s_j$ is also a pole of the Liouvillian resolvent. The zeros of the Zeta function give therefore the generalized eigenvalues of the Liouvillian operator. This fundamental result shows that the resonances are intrinsic to the system itself because they are given in terms of the periodic orbits and their characteristic quantities.

For *symplectic systems* where the stability eigenvalues form pairs (Λ, Λ^{-1}), the stable and unstable eigenvalues are related by $\Lambda_{p,i}^{(s)} = 1/\Lambda_{p,i}^{(u)} = 1/\Lambda_{p,i}$. As a consequence, many factors in the Zeta function (3.150) are identical and we obtain

$$Z(s) = \prod_p \prod_{m_1\cdots m_{M_u}=0}^{\infty}\left(1 - \frac{e^{-sT_p}}{\prod_{j=1}^{M_u}|\Lambda_{p,j}|\Lambda_{p,j}^{m_j}}\right)^{(m_1+1)\cdots(m_{M_u}+1)} . \tag{3.151}$$

In the case of a *periodic hyperbolic set* with a single periodic orbit, the Zeta function is only a product over the integers $\{l_i, m_j\}$ which is absolutely convergent because of hyperbolicity. Therefore, the resonances are given by the zeros of the individual factors in the infinite product as

forward semigroup :

$$s = \frac{1}{T_p}\left[\sum_{i=1}^{M_s} l_i \ln\Lambda_{p,i}^{(s)} - \sum_{j=1}^{M_u}\left(\ln|\Lambda_{p,j}^{(u)}| + m_j \ln\Lambda_{p,j}^{(u)}\right) + 2\pi i n\right] . \tag{3.152}$$

These resonances belong to the lower half plane of the complex plane s. The resonance which is the closest to the imaginary axis controls the evolution on the longest time scale, which defines the escape rate of the system. This leading resonance is obtained by setting $\mathbf{l} = \mathbf{m} = 0$. Using (1.29), we obtain the result that the escape rate is given as the sum of the positive Lyapunov exponents of the periodic orbit

forward semigroup :

$$s = -\gamma^+ = -\sum_{j=1}^{M_u} \lambda_{p,j}^{(u)} , \tag{3.153}$$

which is identical to (3.112). All the other resonances of the forward semigroup are found below this leading resonance which thus defines a gap in the complex plane s. If the system is bounded, there is no escape so that the escape rate vanishes: $\gamma^+ = 0$. The leading resonance is then found at the origin $s = 0$ where it corresponds to the eigenvalue of the Koopman operator at zero frequency $\omega = 0$. This eigenvalue is thus associated with the invariant measure of the system. For bounded systems, we thus obtain the relation $Z(0) = 0$ with the Zeta functions (3.150) or (3.151), which gives a product rule obeyed by the stability eigenvalues of the periodic orbits.

We note that the above results apply equally well to periodic attractors like limit cycles where there is no unstable direction ($M_u = 0$).

For a *chaotic hyperbolic set*, the periodic orbits form a countable set so that the product over the periodic orbits is an infinite product. Moreover, the periodic orbits proliferate exponentially as shown in Chapter 2. In such systems, the infinite product over periodic orbits is nontrivial (Rugh 1992; Eckhardt et al. 1993b, 1995).

In order to investigate the convergence properties of these infinite products, we factorize the Zeta function as

$$Z(s) = \prod_{l_1\cdots l_{M_s} m_1\cdots m_{M_u}=0}^{\infty} \frac{1}{\zeta_{l_1\cdots l_{M_s} m_1\cdots m_{M_u}}(s)} , \tag{3.154}$$

into inverses of the so-called Ruelle zeta functions (Ruelle 1976, 1978, 1994)

$$\zeta_{l_1\cdots l_{M_s} m_1 \cdots m_{M_u}}(s) \equiv \exp \sum_p \sum_{r=1}^{\infty} \frac{1}{r} \left(\frac{e^{-sT_p} \prod_{i=1}^{M_s} \Lambda_{p,i}^{(s)l_i}}{\prod_{j=1}^{M_u} |\Lambda_{p,j}^{(u)}| \Lambda_{p,j}^{(u)m_j}} \right)^r . \tag{3.155}$$

The Ruelle zeta functions are bounded according to

$$|\zeta_{l_1\cdots l_{M_s} m_1 \cdots m_{M_u}}(s)| \leq \exp \sum_p \sum_{r=1}^{\infty} \frac{1}{r} \left(\frac{e^{-\mathrm{Re}sT_p} \prod_{i=1}^{M_s} |\Lambda_{p,i}^{(s)}|^{l_i}}{\prod_{j=1}^{M_u} |\Lambda_{p,j}^{(u)}|^{m_j+1}} \right)^r , \tag{3.156}$$

which is a series like $\exp(\sum_{n=1}^{\infty} c_n/n)$ with $n = rn_p$. This series converges if c_n vanishes for arbitrarily large integers n, so that the domain of convergence can be obtained from the decay rate of c_n at large values of n. Using Eq. (2.61), we obtain the result that the Ruelle zeta functions are absolutely convergent for

$$\mathrm{Re}\, s > -\Gamma^+_{l_1\cdots l_{M_s} m_1 \cdots m_{M_u}} , \tag{3.157}$$

where

$$\Gamma^+_{l_1\cdots l_{M_s} m_1 \cdots m_{M_u}} = \lim_{T\to\infty} -\frac{1}{T} \ln \sum_{r,p:rT_p<T} n_p \left(\frac{\prod_{i=1}^{M_s} |\Lambda_{p,i}^{(s)}|^{l_i}}{\prod_{j=1}^{M_u} |\Lambda_{p,j}^{(u)}|^{m_j+1}} \right)^r , \tag{3.158}$$

which are non-negative. In the domains (3.157), the Ruelle zeta function may neither be infinite nor vanish because the series only appears inside an exponential in Eq. (3.155).

Each of the domains (3.157) forms an upper half plane in complex plane s, which is limited by a line located below the imaginary axis because the Γ's are non-negative (see Fig. 3.3). Below the limit $\mathrm{Re}s = -\Gamma^+_{l_1\cdots l_{M_s} m_1 \cdots m_{M_u}}$, we may expect poles for the corresponding Ruelle zeta function. The larger are the integers $l_1, \ldots, l_{M_s}, m_1, \ldots, m_{M_u}$, the larger is the corresponding Γ and the deeper is the limit of convergence in the lower half plane. Hence, the limit which is above all the others is obtained when all the integers vanish, $l_1 = \cdots = l_{M_s} = m_1 = \cdots = m_{M_u} = 0$, which defines the leading Ruelle zeta function $\zeta_{0\cdots 0}(s)$. Moreover, we notice that, when the stability eigenvalues are real and the integers $l_1, \ldots, l_{M_s}, m_1, \ldots, m_{M_u}$ are even, the absolute values are not required in the bound (3.156). As a consequence, a pole exists on the limit of convergence in such cases. This is the case in particular when all the integers vanish so that the leading eigenvalue is given by

$$s = -\gamma^+ = -\Gamma^+_{0\cdots 0} , \tag{3.159}$$

which defines the escape rate of the forward semigroup for a chaotic repeller. We notice that the escape rate (3.159) vanishes for bounded systems. All the zeros of the Selberg–Smale Zeta function and, therefore, all the generalized eigenvalues are below the escape rate. Above this limit, the Liouville operator of the forward semigroup has no generalized eigenvalues.

The zeros of the Zeta function between the limit $\text{Re}s = -\gamma^+$ and the next-to-leading limit of absolute convergence $\Gamma^+_{(1)} = \text{Min}_{l_1+\cdots+m_{M_u}>0}\Gamma^+_{l_1\cdots m_{M_u}}$ may be obtained by analytic continuation of the leading Ruelle zeta function $\zeta_{0\cdots 0}(s)$. In this domain, the leading Ruelle zeta function is not absolutely convergent but the other Ruelle zeta functions are absolutely convergent so that they do not produce any zero or pole in the Zeta function (3.150). We shall see below that the Zeta function is expected to be an entire function so that all its zeros should be given by the zeros of the inverse of the leading Ruelle zeta function $1/\zeta_{0\cdots 0}(s)$ for $-\Gamma^+_{(1)} < \text{Re}s \leq -\gamma^+$.

However, below the limit of $-\Gamma^+_{(1)}$, the Zeta function is the product of at least two non absolutely convergent Ruelle zeta functions which can therefore no longer be factorized. As a consequence, the Zeta function should in general be considered as a global infinite product over the prime periodic orbits and over the integers l_i and m_j.[8] We should thus emphasize that for values of s below the limit $-\Gamma^+_{(1)}$, the factorization may no longer be justified. The reason is that the infinite product over prime periodic orbits may also develop poles in the individual factors of the Zeta function due to negative signs of the stability eigenvalues Λ, even if the full infinite product of the Zeta function is an entire

8. This delicate convergence of the infinite product over the periodic orbits is in contrast with the more robust convergence with respect to the sole product over the integers l_i and m_j. This difference may be attributed to the unavoidable dependency on the various period T_p through $\exp(-sT_p)$, which imposes a dependency of the radius of absolute convergence on the value of s.

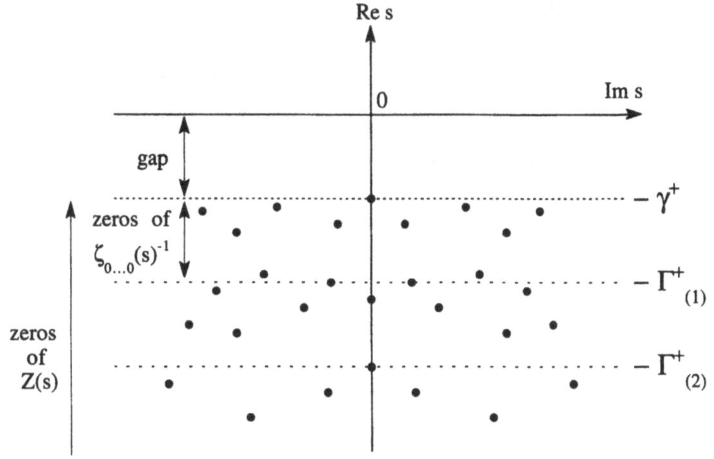

Figure 3.3 Location of the eigenvalues of the Liouville operator $\hat{L}^+$ of the forward semigroup in the complex plane of s.

function (Rugh 1992; Eckhardt et al. 1993b, 1995). Because of the property that the Zeta function is entire, if a pole appears in one of the inverse Ruelle zeta functions it will turn out to be cancelled by a corresponding zero of another inverse Ruelle zeta function. In such circumstances, the full infinite product should be considered for calculations.

Let us add that the consequences on the Zeta functions of the invariance of the dynamical system under a symmetry group have also been studied (Robbins 1989, Lauritzen 1990, Cvitanović and Eckhardt 1993).

3.8.3 Formulation of the eigenvalue problem

The trace formula is extremely useful to obtain the resonances of the Liouvillian dynamics, which are called the Pollicott–Ruelle resonances (Pollicott 1985, 1986; Ruelle 1986a, 1986b, 1987, 1989a, 1989b). In order to obtain the associated eigenstates, the formalism must be developed by going back to the spectral functions (3.34) of a correlation function $\langle A|\hat{P}^{+t}|f\rangle$ which is assumed to vanish as $t \to +\infty$. The spectral function can be expressed in terms of the Laplace transforms of the correlation function at $s = -i\omega$. These Laplace transforms are nothing other than the resolvent of the Liouvillian operator

$$\begin{aligned} \int_0^{+\infty} e^{-st}\, \langle A|\hat{P}^{+t}|f\rangle\, dt &= \int_0^{+\infty} e^{-st}\, \langle A|\exp(\hat{L}^{+}t)|f\rangle\, dt \\ &= \langle A|\frac{1}{s-\hat{L}^{+}}|f\rangle\, . \end{aligned} \tag{3.160}$$

Such time integrals can be evaluated if we introduce the Poincaré map (1.5) and expresses the flow as the suspended flow (1.6) with the change of coordinates $\mathbf{X} = \mathbf{G}(\mathbf{x}, \tau)$. For the simplicity of the calculation, we assume that the flow is volume-preserving.

Because the time evolution is piecewise defined in terms of the suspended flow (1.6), the integral over the time can be transformed into a sum of integrals over the evolution between the successive intersections with the hypersurface of section as (Pollicott 1985)

$$\begin{aligned} \int_0^{\infty} dt\, e^{-st}\, \hat{P}^{+t} f(\mathbf{X}) &= \int_0^{\infty} dt\, e^{-st}\, f\left[\tilde{\mathbf{\Phi}}^{-t}(\mathbf{x},\tau)\right] \\ &= \int_0^{\tau} dt\, e^{-st}\, f(\mathbf{x}, \tau - t) \\ &\quad + \sum_{n=1}^{\infty} \int_{\tau+T(\phi^{-1}\mathbf{x})+\cdots+T(\phi^{-n+1}\mathbf{x})}^{\tau+T(\phi^{-1}\mathbf{x})+\cdots+T(\phi^{-n}\mathbf{x})} dt \\ &\quad \times\, e^{-st}\, f\left[\phi^{-n}\mathbf{x}, \tau - t + \sum_{j=1}^{n} T(\phi^{-j}\mathbf{x})\right] \end{aligned}$$

$$
\begin{aligned}
&= e^{-s\tau} \int_0^\tau d\tau' \, e^{s\tau'} \, f(\mathbf{x},\tau') \\
&\quad + \sum_{n=1}^{\infty} e^{-s\tau} \exp\left[-s\sum_{j=1}^{n} T(\boldsymbol{\phi}^{-j}\mathbf{x})\right] \int_0^{T(\boldsymbol{\phi}^{-n}\mathbf{x})} d\tau' \\
&\quad \times e^{s\tau'} \, f(\boldsymbol{\phi}^{-n}\mathbf{x},\tau') \\
&= e^{-s\tau} \left[\int_0^\tau d\tau' \, e^{s\tau'} \, f(\mathbf{x},\tau') + \sum_{n=1}^{\infty} \hat{K}_s^n v_s(\mathbf{x})\right],
\end{aligned}
\tag{3.161}
$$

where we performed changes of integration variables and we introduced the function

$$
v_s(\mathbf{x}) = \int_0^{T(\mathbf{x})} d\tau' \, e^{s\tau'} \, f(\mathbf{x},\tau') \,, \tag{3.162}
$$

and the following Frobenius–Perron operator of the mapping $\boldsymbol{\phi}$

$$
\hat{K}_s v(\mathbf{x}) = \int d\mathbf{y} \; \delta(\mathbf{x} - \boldsymbol{\phi}\mathbf{y}) \; e^{-sT(\mathbf{y})} \; v(\mathbf{y}). \tag{3.163}
$$

This new Frobenius–Perron operator depends on the complex variable s which should give the relaxation rate of the system. The variable s is accompanied by the first-return time function which contains information on the time evolution in the flow. It is a general result that the Frobenius–Perron operator of a flow can be reduced to the one for the Poincaré map in this way (Pollicott 1985).

For a nonconservative invertible map, the Frobenius–Perron operator (3.163) becomes

$$
\hat{K}_s v(\mathbf{x}) = e^{-sT(\boldsymbol{\phi}^{-1}\mathbf{x})} \frac{v(\boldsymbol{\phi}^{-1}\mathbf{x})}{|\det \partial_{\mathbf{x}} \boldsymbol{\phi}(\boldsymbol{\phi}^{-1}\mathbf{x})|} \,. \tag{3.164}
$$

The eigenvalue problem can be formulated in terms of this reduced Frobenius–Perron operator as

$$
\begin{aligned}
\hat{K}_{s_j} \, \psi_j(\mathbf{x}) &= \psi_j(\mathbf{x}) \,, \\
\hat{K}_{s_j}^{\dagger} \, \tilde{\psi}_j(\mathbf{x}) &= \tilde{\psi}_j(\mathbf{x}) \,,
\end{aligned}
\tag{3.165}
$$

in analogy with (3.82). The left and right eigenvectors are required to be biorthonormal: $\langle \tilde{\psi}_j | \psi_j \rangle = 1$. Here also, the operator $\hat{K}_s$ and its adjoint do not coincide in general so that nontrivial root vectors may be expected.

When the system has a single isolated periodic orbit, the eigenvalue problem and the root vectors can be constructed by reducing the equations of the Poincaré map to its normal form near its fixed point and, possibly, by linearizing the Poincaré map. This procedure would be the analogue of the calculation carried out near stationary points with Eqs. (3.114)–(3.122) and would give eigenstates similar to (3.122). We shall not carry out this calculation again which we leave as an exercise to the reader. We proceed with the general case with countably many periodic orbits.

Relation between the eigenstates of the map and of the flow

The eigenvalue problem for the map is connected as follows to the one for the flow,

$$\hat{P}^{+t}\Psi_j(\mathbf{x},\tau) \;=\; e^{s_j t}\;\Psi_j(\mathbf{x},\tau)\;. \tag{3.166}$$

If we take the Laplace transform of both members of Eq. (3.166) and if we use Eq. (3.161) introducing the transform (3.162) of the eigenstate as

$$\Upsilon_{j,s}(\mathbf{x}) \;=\; \int_0^{T(\mathbf{x})} d\tau'\; e^{s\tau'}\;\Psi_j(\mathbf{x},\tau')\;, \tag{3.167}$$

the eigenvalue equation becomes

$$\int_0^{\tau} d\tau'\; e^{s\tau'}\;\Psi_j(\mathbf{x},\tau') + \sum_{n=1}^{\infty} \hat{K}_s^n\;\Upsilon_{j,s}(\mathbf{x}) \;=\; \frac{e^{s\tau}}{s-s_j}\;\Psi_j(\mathbf{x},\tau)\;, \tag{3.168}$$

for $0 \leq \tau < T(\mathbf{x})$.

Using the relation

$$\sum_{n=1}^{\infty} \hat{K}_s^n \;=\; \frac{\hat{K}_s}{\hat{I}-\hat{K}_s}\;, \tag{3.169}$$

and setting $\tau = 0$, Eq. (3.168) reduces to

$$(\hat{I}-\hat{K}_s)\;\Psi_j(\mathbf{x},0) \;=\; (s-s_j)\;\hat{K}_s\;\Upsilon_{j,s}(\mathbf{x})\;. \tag{3.170}$$

The equality $s = s_j$ leads to

$$(\hat{I}-\hat{K}_{s_j})\;\Psi_j(\mathbf{x},0) \;=\; 0\;, \tag{3.171}$$

which shows that the eigenstate of the flow at $\tau = 0$ can be identified with the eigenstate of the mapping defined by (3.165)

$$\Psi_j(\mathbf{x},0) \;=\; \psi_j(\mathbf{x})\;. \tag{3.172}$$

To determine the eigenstate of the flow for the other values of τ, we differentiate Eq. (3.168) with respect to τ to get

$$\partial_\tau\;\Psi_j(\mathbf{x},\tau) \;=\; -\,s_j\;\Psi_j(\mathbf{x},\tau)\;, \tag{3.173}$$

whose solution is

$$\Psi_j(\mathbf{x},\tau) \;=\; \exp(-s_j\tau)\;\psi_j(\mathbf{x})\;, \qquad \text{for} \qquad 0 \leq \tau < T(\mathbf{x})\;. \tag{3.174}$$

We emphasize that Eq. (3.174) holds for $\tau < T(\mathbf{x})$. We also notice that, since $\text{Re}s_j < 0$, the eigenstate (3.174) presents an exponential growth which only lasts for a finite time interval after which the exponential damping excepted from Eq. (3.166) prevails. In this way, the eigenvalue problem formulated at the level of the flow and of the mapping are interconnected.

3.8.4 Fredholm determinant

The Zeta function (3.145) itself can be expressed in terms of the trace of the Frobenius–Perron operator (3.163) of the Poincaré map if we observe that

$$\text{Tr}\, \hat{K}_s^n = \sum_{\mathbf{x}\in\text{Fix}\phi^n} \frac{\exp\left[-s\sum_{j=1}^n T(\phi^j\mathbf{x})\right]}{|\det[1-\partial_{\mathbf{x}}\phi^n(\mathbf{x})]|} \,. \tag{3.175}$$

Using the counting relation (2.61), the Zeta function (3.145) can then be transformed into

$$\begin{aligned} Z(s) &= \exp - \sum_{n=1}^{\infty} \frac{1}{n}\, \text{Tr}\hat{K}_s^n \\ &= \exp\, \text{Tr}\, \ln(\hat{I}-\hat{K}_s) \\ &= \text{Det}(\hat{I}-\hat{K}_s)\,, \end{aligned} \tag{3.176}$$

which is the Fredholm determinant of the operator $\hat{K}_s$. Therefore, the Pollicott–Ruelle resonances are given by the zeros of the Fredholm determinant

$$Z(s) = \text{Det}(\hat{I}-\hat{K}_s) = 0\,, \tag{3.177}$$

in the complex plane of the variable s. Therefore, the infinite product (3.150) is equal to the Fredholm determinant of a Frobenius–Perron operator.

The form (3.177) of the characteristic equation of the eigenvalue problem is suitable for approximations. The Frobenius–Perron operator can be discretized into a large transfer matrix built onto phase-space cells corresponding to the words $\omega_1\omega_2\cdots\omega_n$ of the symbolic dynamics of a Markov partition:

$$\begin{aligned} &(\mathsf{K}_s^{(n)})_{\omega_1\cdots\omega_n,\omega_1'\cdots\omega_n'} = \\ &\frac{\exp(-sT^{(n)}_{\omega_1\cdots\omega_n})}{|\Lambda^{(n)}_{\omega_1\cdots\omega_n}|}\, A_{\omega_1\omega_2}\cdots A_{\omega_{n-1}\omega_n}\, A_{\omega_n\omega_n'}\, \delta_{\omega_1'\omega_2}\cdots \delta_{\omega_{n-1}'\omega_n}\,, \end{aligned} \tag{3.178}$$

where the quantities $T^{(n)}_{\omega_1\cdots\omega_n}$ and $\Lambda^{(n)}_{\omega_1\cdots\omega_n}$ are approximate first-return times and stretching factors for the evolution between the cell $C_{\omega_1\omega_2\cdots\omega_n}$ and its image by ϕ. The transfer matrix is nonvanishing only for allowed transitions according to the topological transition matrix (2.37) (Sinai 1972, Gaspard and Rice 1989a).

3.8.5 Fredholm theory for the eigenstates

According to the previous observation, we should expect that the eigenstates are given in general by the Fredholm–Grothendieck theory which generalizes the Fredholm theory of linear integral operators (Vernon-Lovitt 1950; Grothendieck 1955; Ruelle 1976, 1978; Mayer 1991; Rugh 1992).

The operator (3.163) is taken at a fixed value of the variable s considered as a parameter: $\hat{K} = \hat{K}_s$. This operator is supposed to act on a certain Banach space $\mathscr{B}$, which is a vector space equiped with a norm and which is complete with respect to this norm. The dual space $\mathscr{B}^\dagger$ is the space of all the bounded linear functionals on $\mathscr{B}$. We shall here be essentially interested in the formal aspect of the theory.

The solution of the inhomogeneous equation,

$$v - z\hat{K}v = w , \tag{3.179}$$

where z is a complex number, is given by

$$v = w + z\hat{R}(z)\,w , \tag{3.180}$$

in terms of the resolvent operator

$$\hat{R}(z) = \frac{\hat{K}}{\hat{I} - z\hat{K}} . \tag{3.181}$$

The idea of the Fredholm theory is to write the resolvent as the ratio

$$\hat{R}(z) = \frac{\hat{N}(z)}{D(z)} , \tag{3.182}$$

where both $D(z)$ and $\hat{N}(z)$ are entire functions of z (Vernon-Lovitt 1950).

In order to achieve this goal, the inversion of the equation (3.179) is considered as for a matrix equation so that the denominator is taken as the corresponding determinant which is thus the Fredholm determinant

$$D(z) = \mathrm{Det}(\hat{I} - z\hat{K}) . \tag{3.183}$$

Combining Eqs. (3.181) and (3.182), the numerator is obtained as

$$\hat{N}(z) = D(z)\frac{\hat{K}}{\hat{I} - z\hat{K}} = -D(z)\,\partial_z \ln(\hat{I} - z\hat{K}) , \tag{3.184}$$

so that we get the relation

$$\mathrm{Tr}\,\hat{N}(z) = -\partial_z D(z) . \tag{3.185}$$

By analogy with (3.176), the numerator and the denominator are given by exponentials of Taylor series in z as

$$D(z) = \exp - \sum_{n=1}^{\infty} \frac{z^n}{n} \operatorname{Tr} \hat{K}^n ,$$

$$\hat{N}(z) = \hat{K} \exp \sum_{n=1}^{\infty} \frac{z^n}{n} (\hat{K}^n - \operatorname{Tr} \hat{K}^n) . \tag{3.186}$$

In order to represent the numerator and the denominator of (3.182) as direct Taylor series, the algebraic structure of the exterior product needs to be introduced. The nth exterior product $\wedge^n \mathscr{B}$ of the Banach space is defined by the space of the vectors of n variables which are

$$\left(v_1 \wedge v_2 \wedge \cdots \wedge v_n\right)(\mathbf{x}_1, \ldots, \mathbf{x}_n) = \det\left[v_i(\mathbf{x}_j)\right]_{i,j=1}^{n} , \tag{3.187}$$

where $v_i(\mathbf{x}) \in \mathscr{B}$. The nth exterior product of the operator $\hat{K}$ is defined by

$$\left(\bigwedge^n \hat{K}\right)(v_1 \wedge \cdots \wedge v_n) = \hat{K} v_1 \wedge \cdots \wedge \hat{K} v_n . \tag{3.188}$$

The kernel of this operator is

$$K\begin{pmatrix} \mathbf{x}_1 & \cdots & \mathbf{x}_n \\ \mathbf{y}_1 & \cdots & \mathbf{y}_n \end{pmatrix} = \det \begin{pmatrix} K(\mathbf{x}_1, \mathbf{y}_1) & \ldots & K(\mathbf{x}_n, \mathbf{y}_1) \\ \vdots & \ddots & \vdots \\ K(\mathbf{x}_1, \mathbf{y}_n) & \ldots & K(\mathbf{x}_n, \mathbf{y}_n) \end{pmatrix} . \tag{3.189}$$

This algebraic structure allows us to expand the Fredholm determinant and the numerator as

$$D(z) = 1 + \sum_{n=1}^{\infty} \frac{(-z)^n}{n!} \operatorname{Tr}_{\wedge^n \mathscr{B}} \bigwedge^n \hat{K} ,$$

$$\hat{N}(z) = \hat{K} + \sum_{n=1}^{\infty} \frac{(-z)^n}{n!} \operatorname{Tr}_{\wedge^n \mathscr{B}} \bigwedge^{n+1} \hat{K} , \tag{3.190}$$

in terms of the traces

$$\operatorname{Tr}_{\wedge^n \mathscr{B}} \bigwedge^n \hat{K} = \int \cdots \int K\begin{pmatrix} \mathbf{x}_1 & \cdots & \mathbf{x}_n \\ \mathbf{x}_1 & \cdots & \mathbf{x}_n \end{pmatrix} d\mathbf{x}_1 \cdots d\mathbf{x}_n ,$$

$$\left(\operatorname{Tr}_{\wedge^n \mathscr{B}} \bigwedge^{n+1} \hat{K}\right)(\mathbf{x}, \mathbf{y}) = \int \cdots \int K\begin{pmatrix} \mathbf{x} & \mathbf{x}_1 & \cdots & \mathbf{x}_n \\ \mathbf{y} & \mathbf{x}_1 & \cdots & \mathbf{x}_n \end{pmatrix} d\mathbf{x}_1 \cdots d\mathbf{x}_n . \tag{3.191}$$

If the operator $\hat{K}$ was a finite matrix, the expansions (3.190) would be polynomial, which suggests their great smoothness. For standard integral operators, Fredholm has shown that $D(z)$ and $\hat{N}(z)$ are entire functions of z (Vernon-Lovitt 1950). Grothendieck (1955) has generalized this result to distribution-type operators.

The eigenvalue problem for z at fixed value of s may be transformed into the eigenvalue problem (3.165) for s with z fixed to unity ($z = 1$). If the dependency on the parameter s is regular enough, the quantities (3.190) may also be entire functions of the parameter s.

The Fredholm theory also provides expressions for the projectors onto the eigenspaces of the operator. If the operator $\hat{K}_s$ has a simple eigenvalue at $s = s_j$, this eigenvalue is a simple zero of the Fredholm determinant: $D(s_j; 1) = 0$ but $\partial_z D(s_j; 1) \neq 0$. Because of (3.185), the numerator does not vanish near the eigenvalue where the operator acts like a projection operator $\hat{K}_s \leftrightarrow |\psi_j\rangle\langle\tilde{\psi}_j|$. Using Eq. (3.184), the eigenprojector is thus given by

$$|\psi_j\rangle \, \langle\tilde{\psi}_j| \;=\; - \left. \frac{\hat{N}(s;z)}{\partial_z D(s;z)} \right|_{z=1, s=s_j} . \qquad (3.192)$$

Similar expressions can be obtained for the root vectors.

3.8.6 Periodic-orbit averages of observables

Averages of observables over eigenstates like (3.192) can also be obtained in terms of the periodic orbits of the hyperbolic set.

Before coming to this point, let us remark that the support of the right eigenstates of the forward semigroup is formed by the unstable manifolds of a hyperbolic repeller. This is because the trajectories are attracted toward the unstable manifolds due to contraction in the transverse stable directions during the forward time evolution. On the other hand, the left eigenstates – which may be considered as the right eigenstates of the backward time evolution – have the stable manifolds for support. Furthermore, Eq. (3.192) shows that an average is evaluated with the product of the left and right eigenstates. Therefore, the average is evaluated for a distribution concentrated at the intersection of the stable and unstable manifolds of the repeller, which is the repeller itself according to (2.15). This result may be summarized by

$$\begin{aligned} \text{support of } \; |\Psi_j\rangle \;&=\; W_u(\mathscr{I}) \, , \\ \text{support of } \; \langle\tilde{\Psi}_j| \;&=\; W_s(\mathscr{I}) \, , \\ \text{support of } \; |\Psi_j\rangle \, \langle\tilde{\Psi}_j| \;&=\; W_u(\mathscr{I}) \cap W_s(\mathscr{I}) \;=\; \mathscr{I} \, . \end{aligned} \qquad (3.193)$$

For Axiom-A repellers in which the periodic orbits are dense, we may therefore hope to express averages in terms of periodic orbits. This goal is reached by considering a Frobenius–Perron operator which involves the observable $A(\mathbf{X})$ to be averaged

$$\hat{P}^t_\alpha \, f(\mathbf{X}) \;=\; \int \delta(\mathbf{X} - \mathbf{\Phi}^t\mathbf{Y}) \, \exp\left[\alpha \int_0^t A(\mathbf{\Phi}^\tau\mathbf{Y}) \, d\tau\right] f(\mathbf{Y}) \, d\mathbf{Y} \, . \qquad (3.194)$$

An eigenvalue problem is set up for this operator

$$\hat{P}^t_\alpha \, |\Psi_{j,\alpha}\rangle \;=\; \exp(s_{j,\alpha}t) \, |\Psi_{j,\alpha}\rangle \;, \qquad \text{with} \quad \langle\tilde{\Psi}_{j,\alpha}|\Psi_{j,\alpha}\rangle = 1 \;. \tag{3.195}$$

Taking derivatives of Eqs. (3.195) with respect to the parameter α, we obtain that

$$\langle\tilde{\Psi}_{j,\alpha}| \int_0^t A(\mathbf{\Phi}^\tau \mathbf{X}) d\tau |\Psi_{j,\alpha}\rangle \;=\; t \, \partial_\alpha \, s_{j,\alpha} \;. \tag{3.196}$$

Differentiating with respect to time and noticing that the right-hand member is then time-independent, we can set $t = 0$ and $\alpha = 0$ to get

$$\langle\tilde{\Psi}_j|A|\Psi_j\rangle \;=\; \partial_\alpha \, s_{j,0} \;, \tag{3.197}$$

where $\Psi_j \;=\; \Psi_{j,0}$ are the eigenstates of the unperturbed problem. Therefore, the average of an observable is given by the derivative of the corresponding eigenvalue with respect to a parameter.

The Frobenius–Perron operator of the Poincaré map which corresponds to (3.194) is

$$\hat{K}_{s,\alpha} \, v(\mathbf{x}) \;=\; \int \delta(\mathbf{x} - \phi\mathbf{y}) \, \exp\left[\alpha a(\mathbf{y}) - sT(\mathbf{y})\right] \, v(\mathbf{y}) \, d\mathbf{y} \;,$$

$$\text{with} \quad a(\mathbf{x}) \;=\; \int_0^{T(\mathbf{x})} A[\mathbf{\Phi}^\tau \circ \mathbf{G}(\mathbf{x}, 0)] \, d\tau \;, \tag{3.198}$$

in terms of the suspended flow (1.6). For a Hamiltonian system with two degrees of freedom the corresponding Fredholm determinant is given by the Zeta function (3.151)

$$\begin{aligned} Z(s;\alpha) \;&=\; \mathrm{Det}(\hat{I} \;-\; \hat{K}_{s,\alpha}) \\ &=\; \prod_p \prod_{m=0}^{\infty} \left(1 \;-\; \frac{e^{-sT_p} \, e^{\alpha A_p}}{|\Lambda_p|\Lambda_p^m}\right)^{m+1} \\ &=\; \prod_p \prod_{m=0}^{\infty} (1 \;-\; t_{pm})^{m+1} \;, \end{aligned} \tag{3.199}$$

where

$$A_p \;=\; \sum_{i=1}^{n_p} a(\phi^i \mathbf{x}_p) \;=\; \int_0^{T_p} A(\mathbf{\Phi}^t \mathbf{X}_p) \, dt \;. \tag{3.200}$$

The eigenvalue is given by the zero:

$$Z(s_{j,\alpha};\alpha) \;=\; 0 \;. \tag{3.201}$$

Differentiating (3.201) with respect to α, we get the average (3.197) in the form

$$\langle\tilde{\Psi}_j|A|\Psi_j\rangle \;=\; -\left.\frac{\partial_\alpha Z}{\partial_s Z}\right|_{s=s_j,\alpha=0} \;. \tag{3.202}$$

Moreover, the Zeta function (3.199) can be expanded into a sum over pseudo-cycles, i.e., over all distinct nonrepeating combinations of prime cycles, which is the so-called cycle expansion (Cvitanović 1988, Artuso et al. 1990)

$$Z(s;\alpha) = \sum_{l=0}^{\infty} \sum_{p_1 m_1 \cdots p_l m_l}{}' \ \imath_{m_1 \cdots m_l} \, t_{p_1 m_1} \cdots t_{p_l m_l} \,, \tag{3.203}$$

where the coefficients $\imath_{m_1 \cdots m_l}$ are integers in $\mathbb{Z}$. Replacing this cycle expansion in Eq. (3.202), we obtain the periodic-orbit expansion of the average

$$\langle \tilde{\Psi}_j | A | \Psi_j \rangle = \frac{\sum_{l=0}^{\infty} \sum'_{p_1 m_1 \cdots p_l m_l} \imath_{m_1 \cdots m_l} \frac{(A_{p_1} + \cdots + A_{p_l}) \exp\left[-s(T_{p_1} + \cdots + T_{p_l})\right]}{|\Lambda_{p_1}| \cdots |\Lambda_{p_l}| \Lambda_{p_1}^{m_1} \cdots \Lambda_{p_l}^{m_l}}}{\sum_{l=0}^{\infty} \sum'_{p_1 m_1 \cdots p_l m_l} \imath_{m_1 \cdots m_l} \frac{(T_{p_1} + \cdots + T_{p_l}) \exp\left[-s(T_{p_1} + \cdots + T_{p_l})\right]}{|\Lambda_{p_1}| \cdots |\Lambda_{p_l}| \Lambda_{p_1}^{m_1} \cdots \Lambda_{p_l}^{m_l}}} \,, \tag{3.204}$$

with $s = s_j$. The average is here evaluated in terms of the characteristic quantities of the periodic orbits: their period T_p, their stability eigenvalue Λ_p, and the time average (3.200) of the observable along the periodic orbit. This is the result announced earlier that averages can be evaluated in terms of the periodic orbits which are dense on the repeller.[9] In the case where the eigenvalue $s_j = 0$ corresponds to the invariant measure, Eq. (3.204) provides therefore an expression for this invariant measure in terms of the dense set of periodic orbits (Artuso et al. 1990). For the other eigenvalues $s_j \neq 0$, Eq. (3.204) also defines invariant measures corresponding to decaying states.

3.9 Resonance spectrum at bifurcations

The previous theory allows us to study the time evolution of statistical ensembles near bifurcations as well (Gaspard et al. 1995a). At the bifurcation, the critical elements of a vector field undergo a change in their stability so that we should expect a drastic change in the resonance spectrum. Just at the bifurcation, it is known that the time evolution displays critical slowing down (Nicolis 1995). The stability of the critical elements is neutral which leads to algebraic time evolutions. Since power-law decays can be expressed as integrals of exponential decays we should expect branch-cut singularities in the resonance spectrum of the critical flow.

Let us here summarize the situation for several of the simplest bifurcations.

9. We notice that the property (3.193) prevents us from obtaining the eigenstates themselves in terms of the periodic orbits because their support extends over the invariant manifolds $W_{u,s}(p)$ as well.

3.9.1 Pitchfork bifurcation

The normal form describing the pitchfork bifurcation is the one-dimensional vector field (Nicolis 1995)

$$\dot{x} = \mu x - x^3 + \mathcal{O}(x^5) . \tag{3.205}$$

The pitchfork bifurcation occurs when the bifurcation parameter takes the value $\mu = 0$.

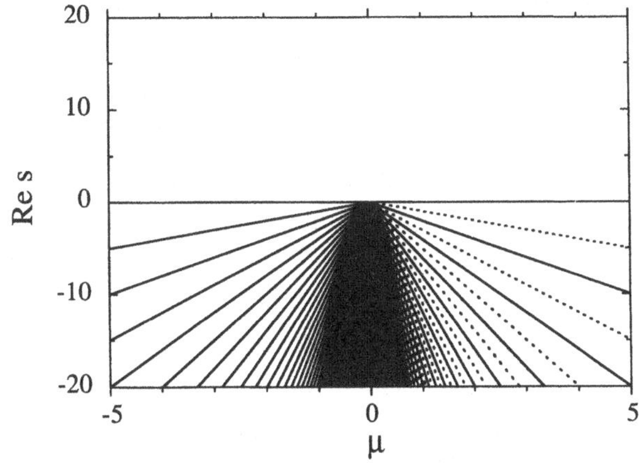

Figure 3.4 Real parts of the eigenvalues (3.206)–(3.207) as a function of the bifurcation parameter μ of the pitchfork bifurcation. The imaginary parts are zero in this case (Nicolis and Gaspard 1994, Gaspard et al. 1995a).

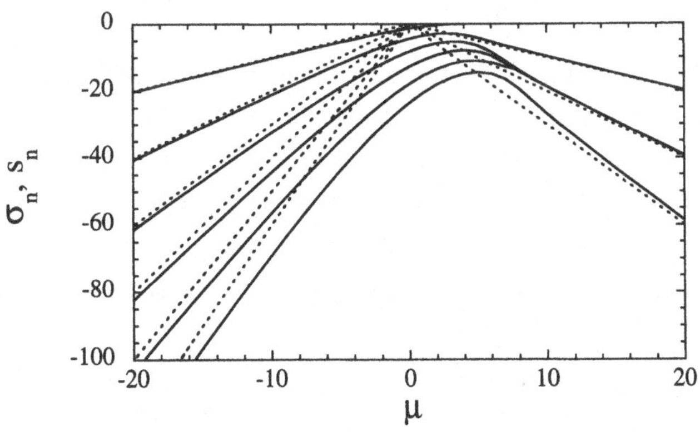

Figure 3.5 Eigenvalues σ_n of the Fokker–Planck equation (3.208) of the pitchfork bifurcation versus the bifurcation parameter μ. They are compared with the eigenvalues s_n of the deterministic Liouville equation (Gaspard et al. 1995a).

For $\mu < 0$, the vector field has a single stationary point which is the origin: $x = 0$. The stability eigenvalue is $\xi = -|\mu|$ so that the point is attracting. According to (3.111), the resonances are

$$\mu < 0 \; : \qquad x = 0 \; : \qquad s_l \; = \; -\,l\,|\mu| \, , \qquad l = 0, 1, 2, 3, \ldots \, . \tag{3.206}$$

The Liouville operator is diagonalizable in this case.

For $\mu > 0$, the vector field has three stationary points: the origin $x = 0$ and two others at $x = \pm\sqrt{\mu} + \mathcal{O}(\mu^{3/2})$. The origin has become unstable with stability eigenvalue $\xi_0 = \mu$, while the two new points have picked up the lost stability of the origin and their stability eigenvalues are $\xi_\pm = -2\mu + \mathcal{O}(\mu^2)$. According to (3.111), the Liouvillian resonances are

$$\begin{aligned} \mu > 0 \; : \qquad\qquad x = 0 \; : & \qquad s_m \; = \; -(m+1)\,\mu \, , \\ x = \pm\sqrt{\mu} + \mathcal{O}(\mu^{3/2}) \; : & \qquad s_l \; = \; l\,\left[-2\mu + \mathcal{O}(\mu^2)\right] \, , \end{aligned} \tag{3.207}$$

with $l, m = 0, 1, 2, 3, \ldots$. Here, the eigenvalues s_l are doubly degenerate because of the symmetry of the vector field (3.205) under $x \to -x$. The eigenstates corresponding to the resonances (3.206) and (3.207) have been constructed elsewhere (Gaspard et al. 1995a).

At $\mu = 0$, the resonances accumulate on the negative half of the real axis as shown in Fig. 3.4. Therefore, the accumulation of the discrete eigenvalues generates a continuous spectrum at criticality. A branch cut is present starting at $s = 0$, which leads to algebraic decays like $1/\sqrt{t}$.

The spectrum of Liouvillian resonances can be compared with the spectrum of the Fokker–Planck equation corresponding to the addition of a Langevin noise in (3.205) (Kramers 1940, van Kampen 1981, Risken 1984)

$$\partial_t f \; + \; \mathrm{div}(\mathbf{F}f) \; = \; D\,\Delta f \, , \tag{3.208}$$

where $\Delta = \nabla^2$ is the Laplacian operator. The Fokker–Planck equation has a discrete eigenvalue spectrum even at criticality as shown in Fig. 3.5 (van Kampen 1977, 1978; Dekker and van Kampen 1979). A remarkable result is that the eigenvalues of the Fokker–Planck equation approach the eigenvalues of the Liouville equation as the amplitude of the noise decreases $D \to 0$. This establishes a nice relationship between those equations (Gaspard et al. 1995a).

3.9.2 Hopf bifurcation

The normal form describing the supercritical Hopf bifurcation is the two-dimensional vector field (Nicolis 1995)

$$\dot{z} \; = \; (\mu + i\omega)\,z \; - \; (1 + i\alpha)\,|z|^2\,z \; + \; \mathcal{O}(|z|^4 z) \, , \tag{3.209}$$

with $z = x + iy$. The Hopf bifurcation occurs at the parameter value $\mu = 0$ where the stable focus which existed before criticality becomes repelling and generates an attracting limit cycle. The focus is located at the origin $z = 0$ while the periodic orbit is at $|z| = \sqrt{\mu} + \mathcal{O}(\mu^{3/2})$ and its period is $T_p = 2\pi/\omega + \mathcal{O}(\mu)$.

According to (3.111), the resonances of the stable focus are

$$\mu < 0 \ : \qquad \text{stable focus} \ : \qquad s = (l_+ + l_-)\mu + i(l_+ - l_-)\omega \ , \tag{3.210}$$

with $l_\pm = 0, 1, 2, 3, \ldots$.

On the other hand, above the bifurcation, we need Eq. (3.111) to determine the resonances of the unstable focus and also Eq. (3.152) for the resonances of the limit cycle:

$$\begin{aligned} &\mu > 0 \ : \\ &\text{unstable focus} \ : \qquad s = -(m_+ + m_- + 2)\mu + i(m_- - m_+)\omega \ , \\ &\quad \text{stable cycle} \ : \qquad s = -[2\mu + \mathcal{O}(\mu^2)]l + in[\omega - \alpha\mu + \mathcal{O}(\mu^2)] \ , \end{aligned} \tag{3.211}$$

with $m_\pm, l = 0, 1, 2, 3, \ldots$ and $n = 0, \pm 1, \pm 2, \pm 3, \ldots$. The situation is depicted in Fig. 3.6. As the Hopf bifurcation is approached, the pyramidal array of the

(a)

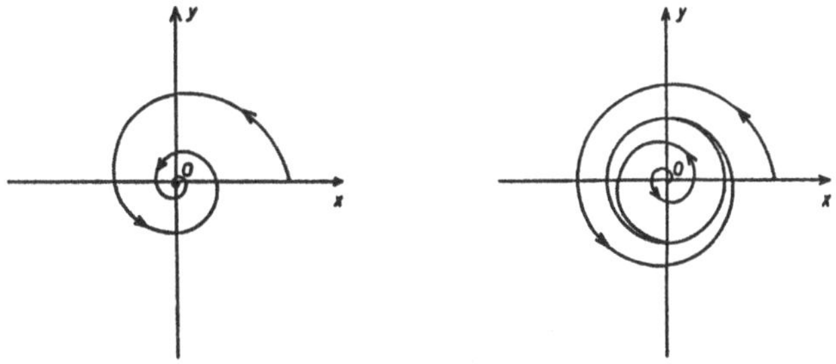

(b)

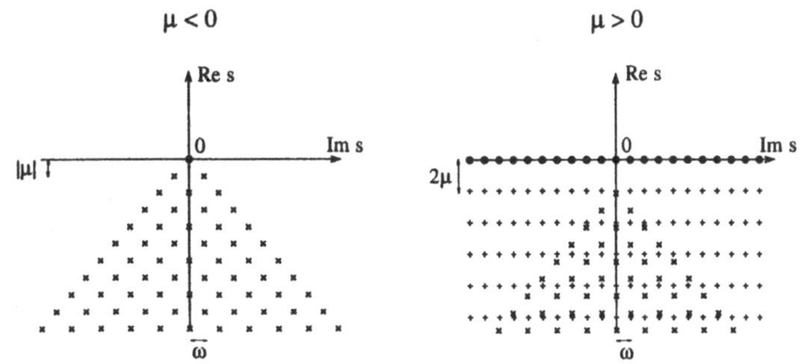

Figure 3.6 (a) The vector field (3.209) before and after the Hopf bifurcation. (b) Spectrum of Liouvillian resonances (3.210)–(3.211) before and after the Hopf bifurcation.

resonances of the stable focus tends to the imaginary axis. After the bifurcation, the stable cycle contributes by a half lattice of resonances while the unstable focus contributes by another pyramidal array but with origin shifted inside the lower half plane due to its instability.

3.9.3 Homoclinic bifurcation in a two-dimensional flow

In two-dimensional vector fields of dissipative systems, it may happen that the period of a limit cycle becomes infinite because the cycle approaches a saddle point in phase space (Arnold 1982). A bifurcation occurs when the cycle collides with the saddle point and creates, at criticality, a homoclinic loop to the saddle as depicted in Fig. 3.7a. The analysis of the geometry of trajectories near the homoclinic loop shows that the flow can be locally reduced to the one-dimensional Poincaré map

$$\begin{cases} y_{n+1} = a\,\mu + b\,y_n^{|\lambda_s/\lambda_u|}\,, \\ t_{n+1} = t_n + (1/\lambda_u)\,\ln(c/y_n)\,, \end{cases} \tag{3.212}$$

where y_n is the coordinate of the intersection of the trajectory with a line perpendicular to the unstable manifold in the vicinity of the saddle point (Gaspard 1990). λ_s and λ_u are the stable and unstable Lyapunov exponents of the saddle point at criticality. If $|\lambda_s| > \lambda_u$, a stable periodic orbit bifurcates from

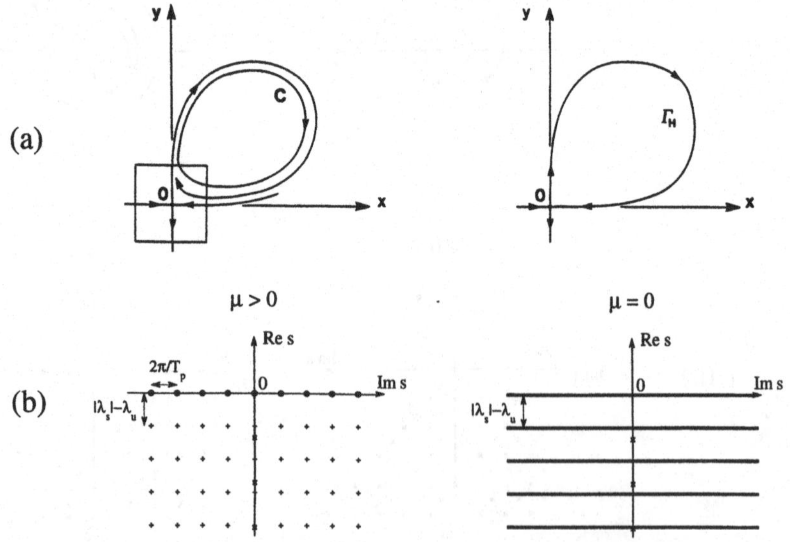

Figure 3.7 (a) Two-dimensional vector field at a saddle-loop homoclinic bifurcation. (b) Spectrum of Liouvillian resonances (3.214) before and at the homoclinic bifurcation for $|\lambda_s| > \lambda_u$.

the homoclinic orbit for $\mu > 0$ with a period and a stability eigenvalue given by

$$T_p \simeq \frac{1}{\lambda_u} \ln \frac{c}{a\mu} ,$$
$$\Lambda_p \sim \mu^{|\lambda_s/\lambda_u|-1} . \qquad (3.213)$$

As a consequence of Eq. (3.152), the spectrum of resonances of the stable periodic orbit is therefore

$$s = \frac{2\pi i n}{T_p} + \frac{m}{T_p} \ln \Lambda_p \simeq \frac{2\pi i n \lambda_u}{\ln(1/\mu)} - m \left(|\lambda_s| - \lambda_u\right) , \qquad (3.214)$$

as $\mu \to 0$, with $m = 0, 1, 2, 3, \ldots$ and $n = 0, \pm 1, \pm 2, \pm 3, \ldots$. In this case which is depicted in Fig. 3.7b, there remains a gap between the eigenvalues on the imaginary axis and the next-to-leading resonances, which is the constant $\Delta \mathrm{Re} s = |\lambda_s| - \lambda_u$ characteristic of the saddle point. On the other hand, the resonances accumulate to form horizontal lines because the period becomes infinite so that the imaginary separation of the resonances, which is $(2\pi/T_p)$, goes to zero.

The extension of the probabilistic approach to other types of homoclinicities is a most challenging problem. Birkhoff (1927) proved the existence of a chaotic repeller near robust homoclinic orbits to saddle-type periodic orbits. Similarly, a theorem by Shilnikov (1970) shows the existence of a chaotic repeller near nonrobust homoclinic orbits to a saddle-focus in three- and higher-dimensional flows. The methods of the present chapter can in principle be applied to obtain Liouvillian resonances of Birkhoff's and Shilnikov's repellers. Such results could provide a better understanding of chaos on these invariant subsets which are commonly observed in models of physico-chemical processes.

3.10 Liouvillian dynamics of one-dimensional maps

For noninvertible one-dimensional maps, the Frobenius–Perron operator is

$$\hat{P} f(x) = \sum_i \frac{f(\phi_i^{-1} x)}{|\phi'(\phi_i^{-1} x)|} , \qquad (3.215)$$

where the sum extends over all the inverse branches of ϕ. Recently, such operators have been actively investigated (Lasota and Mackey 1985; Dörfle 1985; Roepstorff 1987; Mayer and Roepstorff 1987, 1988; Gaspard 1992b; Hasegawa and Saphir 1992a, 1992b; Driebe and Ordóñez 1996; Antoniou and Qiao 1996). The previous periodic-orbit theory applies to them if they are everywhere expanding, i.e., hyperbolic. In particular, the Fredholm determinant of the Frobenius–Perron operator (3.215) is then given by the Zeta function (3.150) with $M_s = 0$, $M_u = 1$, and $T_p = n_p$:

$$Z(s) = \prod_{p\in\mathscr{P}} \prod_{m=0}^{\infty} \left[1 - \frac{\exp(-sn_p)}{|\Lambda_p|\Lambda_p^m}\right], \tag{3.216}$$

where $\Lambda_p = \prod_{j=1}^{n_p} \phi'(\phi^j x_p)$. The generalized eigenvalues of the Frobenius–Perron operator are obtained as the zeros of this Zeta function.[10]

As an example, we consider the particular case of the r-adic maps of the unit interval:

$$x_{n+1} = (rx_n), \qquad (\text{mod. } 1), \tag{3.217}$$

in which $\Lambda_p = r^{n_p}$ uniformly for each periodic orbit. The Frobenius–Perron operator of the r-adic map is

$$\hat{P}\, f(x) = \frac{1}{r} \sum_{i=0}^{r-1} f\left(\frac{x+i}{r}\right). \tag{3.218}$$

Using Eqs. (3.175)–(3.176), the Zeta function can be calculated as

$$Z(s) = \prod_{m=0}^{\infty} \left[1 - \frac{\exp(-s)}{r^m}\right], \tag{3.219}$$

because $\text{Tr}\hat{P}^n = |\text{Fix}\phi^n|/(r^n - 1)$ with $|\text{Fix}\phi^n| = r^n$. Consequently, the resonances are given by

$$\chi_m = \exp s_m = \frac{1}{r^m}, \qquad m = 0, 1, 2, \ldots . \tag{3.220}$$

This result can more simply be obtained by applying the Frobenius–Perron operator (3.218) to the monomials $\{x^m\}$. In this basis, the operator is represented by an infinite triangular matrix on the diagonal of which the eigenvalues (3.220) are found.

The comparison of the operator (3.218) with the Euler–Maclaurin formula [Eq. (23.1.32) in Abramowitz and Stegun 1972] shows that the iterates of the Frobenius–Perron operator have the spectral decomposition:

$$\hat{P}^n = \sum_{m=0}^{\infty} |\Psi_m\rangle \frac{1}{r^{mn}} \langle\tilde{\Psi}_m|, \tag{3.221}$$

with the right eigenfunctions given by the Bernoulli polynomials

$$\Psi_m(x) = \frac{B_m(x)}{m!}, \tag{3.222}$$

10. If the eigenfunctions $f(x)$ are allowed to have discontinuities on the periodic orbits of the map, extra eigenvalues of the Frobenius–Perron operator should be expected as shown by MacKernan and Nicolis (1994).

while the left eigenstates

$$\tilde{\Psi}_0 = \theta(x)\,\theta(1-x)\,,$$
$$\tilde{\Psi}_m(x) = (-)^{m-1}\left[\delta^{(m-1)}(x-1) - \delta^{(m-1)}(x)\right]\,, \qquad (3.223)$$

are distributions given in terms of derivatives of the Dirac delta distribution: $\delta^{(n)}(x) = (d/dx)^n\delta(x)$. The spectral decomposition (3.221), which is known as a polynomial expansion (Boas and Buck 1958), holds for densities $f(x)$ which are analytic entire functions of exponential type $\tau < 2\pi$, i.e., such that $\forall\epsilon > 0$, $\exists\kappa_\epsilon(f) > 0$,

$$|f(z)| \leq \kappa_\epsilon \exp[(\tau+\epsilon)|z|]\,, \qquad \forall\, z = x+iy \in \mathbb{C}\,. \qquad (3.224)$$

We remark that, if the r-adic map is defined on the unit circle instead of the unit interval as considered in Section 2.7, the Zeta function is simply

$$Z(s) = 1 - \exp(-s)\,, \qquad (3.225)$$

because $|\mathrm{Fix}\phi^n| = r^n - 1$. We have already observed a similar simplification in the case of the topological zeta function (2.73). The result (3.225) shows that the constant function is now the only nontrivial eigenstate of the Frobenius–Perron operator. On the other hand, the trigonometric functions form nested null spaces $\mathcal{N}^{(n)}$ because

$$\hat{P}\,\exp(i2\pi r^n x) = \exp(i2\pi r^{n-1}x)\,, \qquad n \geq 1\,,$$
$$\hat{P}\,\exp(i2\pi x) = 0\,. \qquad (3.226)$$

A similar structure was reported by Dörfle (1985) for general piecewise linear maps.

Chapter 4

Probabilistic chaos

4.1 Dynamical randomness and the entropy per unit time

If dynamical instability is quantitatively measured by the Lyapunov exponents, on the other hand, dynamical randomness is characterized by the entropy per unit time. The entropy per unit time is a transposition of the concept of thermodynamic entropy per unit volume from space translations to time translations. As Boltzmann showed, the entropy is the logarithm of the number of complexions, i.e., the number of microscopic states which are possible in a certain volume and under certain constraints. In the time domain, the number of complexions becomes the number of possible trajectories in a given time interval. The entropy per unit time is therefore an estimation of the rate at which the number of possible trajectories grows with the length of the time interval.

This scheme is not in contradiction with the famous Cauchy theorem which asserts the uniqueness of the trajectory issued from given initial conditions. Indeed, as in statistical mechanics, the counting proceeds with the constraint that the trajectories belong to cells of phase space. Since each cell is a continuum, the counting becomes nontrivial. Indeed, an initial cell may be stretched into a long and thin cell which will overlap several other cells at the next time step. In this way, the stretching and folding mechanism in phase space implies that the tree of possible trajectories has a number of branches which grows exponentially with a positive branching rate.

The counting may be purely topological, which yields the definition of the topological entropy per unit time of Chapter 2. On the other hand, the tree may be weighted by transition probabilities, which leads to the definition to the Kolmogorov–Sinai (KS) entropy per unit time (Eckmann and Ruelle 1985).

The concept of entropy per unit time was introduced by Shannon in his famous information theory (Shannon and Weaver 1949). The purpose was to characterize random processes which produce symbols per unit time. Physico-chemical systems are therefore considered as sources of information. Later, Kolmogorov (1958, 1959) and Sinai (1959) proposed a rigorous definition in the context of dynamical systems theory. More recently, the works by Chaitin (1987) and others have shown that the entropy per unit time plays a fundamental role in our logical understanding of what is randomness by introducing the concept of algorithmic complexity.

In the following, we shall first introduce a model of observation of a dynamical system and proceed to the definitions of the algorithmic complexity and of the KS entropy.

4.1.1 A model of observation

The observation of natural systems and, in particular, of their time evolution may be summarized with the following model.

A measuring device of finite resolution records the time evolution of certain observables of the system under study in a laboratory. The measuring device encodes the observations in binary or decimal form. In general, the amount of data never grows faster than linearly with time. Data accumulate at a certain rate C which characterizes the measuring device.[1] We suppose that the observable $A(\mathbf{X}_t)$ is measured at equal times $t_n = n\Delta t$ where Δt is the time interval between the observations. The observable $A_n = A(\mathbf{X}_{t_n})$ is digitalized, i.e., an integer or symbol ω_n is assigned if A_n belongs to a certain cell C_{ω_n} of its domain of variation. Several observables are possibly recorded in this way and the full set of variables $\mathbf{X}_t$ are hopefully recorded through the observation of the given observables. The recorded data appear in the form of a sequence

$$\boldsymbol{\omega}_N = \omega_0\, \omega_1\, \omega_2 \cdots \omega_n \cdots \omega_N \,, \tag{4.1}$$

of integers or symbols in some set or alphabet $\omega_n \in \mathscr{A}$. The total time of observation is $T = N\Delta t$. In an ideal measurement, the symbols ω_n correspond to a partition $\mathscr{C}$ into cells which cover the whole phase space without leaving holes and without overlap: $\mathscr{M} = \cup_{\omega\in\mathscr{A}} C_\omega$ and $C_\omega \cap C_{\omega'} = \emptyset$ if $\omega \neq \omega'$. In

1. This rate C is called the capacity of the channel or here of the measuring device.

this model, the flow is discretized into the map $\mathbf{\Phi}^{\Delta t}$, which should not be confused with the Poincaré map (1.5) because this latter is not isochronic while observation is supposed to be stroboscopic in the present model.[2] In the ideal case, the sequence (4.1) gives the successive cells which are visited by the trajectory during its time evolution: $\mathbf{X}_{t_n} \in C_{\omega_n}$.

4.1.2 Information redundancy and algorithmic complexity

It turns out that the system itself may generate information at a rate which is much smaller than the recording rate C. Indeed, regularities (or redundancies) may appear in a sequence like (4.1) which can be exploited to perform a compression of information. This is the case for periodic time behaviour where it is enough to record the signal during a period to know the signal at any time. The compression of information can be studied with the so-called *algorithmic complexity*, which is the length of the smallest program Π_N able to reproduce the trajectory (4.1) on a universal Turing machine $\mathbb{U}$

$$K(\omega_N) = \mathrm{Min}_{\Pi_N \,:\, \omega_N = \mathbb{U}(\Pi_N)} \, |\Pi_N| \,, \tag{4.2}$$

where $|\Pi|$ denotes the binary length of the code or program Π (Kolmogorov 1983, Chaitin 1987). The smallest program realizes the compression of information because it is sufficient to record Π_N with $N = T/\Delta t$ instead of the whole sequence ω_N. The whole sequence is of length CT_N although the program is of length $K(\omega_N)$ which is always smaller than the original sequence. If no regularity is observed in the sequence, we can always keep the sequence itself. The increase of the algorithmic complexity with time can therefore characterize the intrinsic production of information in the system, and thus its dynamical randomness.

If the trajectory is periodic $\omega_N = p^r$ for $N = rn_p$, the program only needs to memorize the pattern p of the period and to repeat this pattern over the total length N. The length of this program is thus equal to the finite length $n_p = |p|$ of the periodic pattern p, plus the length of the binary expansion of the repetition number r of the period in the total time T, plus a finite and constant number of instructions to the Turing machine $\mathbb{U}$, i.e.,

$$K(\omega_T) \sim \log T \qquad \text{(periodicity)}\,, \tag{4.3}$$

where $\omega_T = \omega_N$ for $T = N\Delta t$. Let us remark that, in fact, Eq. (4.3) applies also to quasiperiodic and many other aperiodic patterns. As examples, we can mention the binary or decimal expansions of the number π as well as time

2. For this reason, the symbols ω_n in Eq. (4.1) have a different meaning than the symbols assigned by a Markov partition in the Poincaré surface of section when the first-return time function $T(\mathbf{x})$ is not constant in Eq. (1.5).

series generated by the Feigenbaum attractor at the threshold of chaos in the period-doubling scenario (Grassberger 1986a).

There may exist random processes with strong correlation in time where the signal shows long periods of order interspaced by bursts of randomness in such a way that the algorithmic complexity is not extensive with the time T. This *sporadicity or sporadic randomness* is characterized by the condition that the algorithmic complexity increases as

$$K(\omega_T) \sim T^\alpha \; (0 < \alpha < 1) \quad \text{or}$$
$$\frac{T}{(\log T)^\beta} \; (\beta > 0) \qquad \text{(sporadic randomness)}\,, \tag{4.4}$$

(Gaspard and Wang 1988). Sporadicity manifests itself, in particular, in the Manneville–Pomeau intermittent maps (Manneville 1980, Pomeau and Manneville 1980). In sporadic processes, the information contained in a trajectory of length T can be compressed to a number $K(\omega_T)$ of bits by taking advantage of the large redundancy in the trajectory. However, this compression cannot be as complete as in the case of periodic processes. Consequently, a certain degree of randomness remains in the sporadic signals.

On the other hand, if the trajectory is 'random' there is no way to reproduce it other than to memorize the whole trajectory or, at least, a sequence of length increasing linearly with time so that

$$K(\omega_T) \sim T \qquad \text{(regular randomness)}\,. \tag{4.5}$$

Since we expect that the regular randomness (4.5) is the generic behaviour it is thus natural to characterize the intrinsic production of information by the rate of linear increase

$$h_{\text{KS}} = \lim_{T\to\infty} \frac{K(\omega_T)}{T}\,, \tag{4.6}$$

for typical trajectories ω_T. Equation (4.6) can be used as a theoretical definition of the KS entropy per unit time.[3] The actual rate of production of information in the system is thus a quantity which is in principle intrinsic to the dynamical system itself and independent of the measuring device.

In order to evaluate the efficiency of the measurement, the KS entropy has to be compared with the accumulation rate C in order to know if the measuring device is suitable or not to faithfully track the trajectories of the system. If $h_{\text{KS}} < C$, we remark that the recorded data are redundant and that some compression is possible. On the other hand, if $C < h_{\text{KS}}$, the measuring device records too little information to determine the trajectory. We may thus conclude that:

3. However, a more practical and systematic definition will be given below.

The KS entropy per unit time is the minimal data accumulation rate which is sufficient to reconstruct the system trajectory from the recorded data.

4.1.3 Kolmogorov–Sinai entropy per unit time

After the previous theoretical motivations, we proceed with the probabilistic definition of the KS entropy per unit time (Kolmogorov 1958, 1959; Sinai 1959; Arnold and Avez 1968; Cornfeld et al. 1982).

The dynamical system $\mathbf{\Phi}^t$ under study is supposed to be a stationary random process of invariant probability measure μ taking place in a bounded phase space $\mathcal{M}$. The invariant measure gives a probability weight to each cell in phase space and, thus, to each classical path defined by the set of all trajectories which visit successively the cells $C_{\omega_0}, C_{\omega_1}, \ldots, C_{\omega_{n-1}}$ at times $t = 0, \Delta t, \ldots, (n-1)\Delta t$. The corresponding probabilities are actually the n-time correlation functions

$$\begin{aligned} \mu(\omega_0\omega_1\ldots\omega_{n-1}) &= \int_{\mathcal{M}} \mu(d\mathbf{X})\, I_{\omega_0}(\mathbf{X})\, I_{\omega_1}(\mathbf{\Phi}^{\Delta t}\mathbf{X}) \;\ldots\; I_{\omega_{n-1}}(\mathbf{\Phi}^{(n-1)\Delta t}\mathbf{X}) \\ &= \langle \hat{U}^{t_0} I_{\omega_0}\, \hat{U}^{t_1} I_{\omega_1} \ldots \hat{U}^{t_{n-1}} I_{\omega_{n-1}} \rangle_\mu \,, \quad \text{with } t_j = j\,\Delta t \,, \end{aligned} \tag{4.7}$$

where $I_{\omega_j}(\mathbf{X})$ are the indicator functions of the cells C_{ω_j} and where $\hat{U}^t A(\mathbf{X}) = A(\mathbf{\Phi}^t\mathbf{X})$ is the Koopman evolution operator. This remark is important in the context of nonequilibrium statistical mechanics where the hierarchy of the n-time correlation functions is at the basis of our knowledge of the thermodynamic system. At the top of the hierarchy, we find the thermodynamic averages like $\langle A \rangle_\mu$ which are time independent and, next, the two-time correlation functions which define the transport properties by the Green–Kubo formulas. Since the KS entropy per unit time is based on the n-time correlation functions (4.7), we see that it characterizes extremely fine correlations between the observables at different successive times.

The entropy per unit time of the partition $\mathscr{C}$ is then defined by

$$h(\mathscr{C}) = \lim_{n\to\infty} -\frac{1}{n\Delta t} \sum_{\omega_0\ldots\omega_{n-1}} \mu(\omega_0\omega_1\ldots\omega_{n-1}) \ln \mu(\omega_0\omega_1\ldots\omega_{n-1}) \,. \tag{4.8}$$

The KS entropy per unit time of the system is defined by

$$h_{\text{KS}} = \text{Sup}_{\mathscr{C}}\, h(\mathscr{C}) \,. \tag{4.9}$$

Taking the supremum over all the partitions $\mathscr{C}$ is a way to define a quantity which no longer refers to a particular partition but which is intrinsic to the dynamical system $\mathbf{\Phi}^t$ and the invariant measure μ.

The interpretation of $h(\mathscr{C})$ is provided by the Shannon–McMillan–Breiman theorem which states that, if the system is ergodic and if a trajectory successively visits the cells C_{ω_j} at times $t_j = j\Delta t$, we have

$$\mu(\omega_0\omega_1\ldots\omega_{n-1}) \sim \exp\big[-n\Delta t\; h(\mathscr{C})\big] , \tag{4.10}$$

for almost all trajectories of the system (Billingsley 1965). In this sense, the entropy per unit time is the decay rate of the n-time correlation functions (4.7). We emphasize that this rate is taken by increasing the number of time intervals but not by increasing the time intervals between the observables $I_\omega(\mathbf{X})$, as done in defining the property of mixing.

Let us moreover add that the connection to the algorithmic complexity is established with Eq. (4.6) which is known to hold for almost all trajectories (Zvonkin and Levin 1970). Therefore, we can deduce the interpretation of the KS entropy in terms of information compression as previously discussed.

4.2 The large-deviation formalism

The KS entropy per unit time is the dynamical analogue of the entropy per unit volume of equilibrium statistical mechanics. This analogy suggests that the whole formalism of equilibrium statistical mechanics can be extended to dynamical systems theory. This formalism is known in probability theory as the large-deviation formalism because it is concerned with the large fluctuations as opposed to the majority of moderate fluctuations which are ruled by the central limit theorem. Therefore, it will be important to keep in mind both the methods and results of equilibrium statistical mechanics of classical systems, since this subject motivates a large part of the discussion to follow, as was first demonstrated by Sinai (1972), Bowen (1975), and Ruelle (1978). In this context, the large-deviation formalism is considered in time instead of space in contrast with equilibrium statistical mechanics.

The central quantity of interest in the large-deviation, or thermodynamic, formalism is an invariant probability measure on the set of phase-space trajectories. Here, we outline the construction of the probability measure using a fundamental quantity which is the topological pressure (Ruelle 1978). The topological pressure plays a role in dynamical systems very similar to that of the free energy for statistical-mechanical systems. In addition, it is also essential for establishing an important connection between the invariant probability measure on trajectories and the microcanonical measure on the constant energy surface. This connection lies at the heart of the large-deviation formalism.

In this chapter, we consider the dynamics on a hyperbolic repeller (or attractor) of Axiom-A type. We suppose that the periodic orbits proliferate exponentially on the repeller which is thus topologically chaotic. The idea is to start from the topological properties introducing a counting device based on the distance (2.45) introduced by Bowen (1975) and Walters (1981), as explained in Chapter 2. The topological properties are used to construct a family of probability invariant measures on the repeller (Gaspard and Dorfman 1995).

4.2.1 Separated subsets

With this aim, we introduce separated subsets of the repeller. The use of separated subsets allows us to construct a very general procedure for analyzing chaotic properties of both closed and open hyperbolic systems.

These subsets are composed of points which are separated by a distance d_T larger than ϵ over a time interval $[-T, +T]$ [cf. (2.45)]. That is, we construct a set of points $\mathcal{S} = \{\mathbf{Y}_1, \ldots, \mathbf{Y}_S\}$ of the invariant set $\mathcal{I}$ such that $d_T(\mathbf{Y}_i, \mathbf{Y}_j) > \epsilon$ for $i, j = 1, \ldots, S$. If the invariant set is compact, i.e., bounded, then one can always find a subset $\mathcal{S}$ with a finite number of points. This set is called a *(ϵ, T)-separated subset* of the invariant set.

Since (ϵ, T)-separated subsets exist with a finite number of points, there exists at least one set, $\mathcal{S}_{\max}(\epsilon, T)$, with the maximum number of points, say $s(\epsilon, T, \mathcal{I})$ points. Since the trajectories separate exponentially with T we expect that $s(\epsilon, T, \mathcal{I})$ will increase exponentially with T also because the points may become separated by smaller and smaller distances. This rate of growth is the topological entropy (2.51)–(2.52) defined in Chapter 2.

4.2.2 Topological pressure and the dynamical invariant measures

The idea of the large-deviation formalism is to introduce a functional of physical observables which is the generating functional of the average and of the multitime correlation functions of the given observable $A(\mathbf{X})$. The functional is called the topological pressure (Ruelle 1978) and is defined as

$$\mathcal{P}(A) = \lim_{\epsilon \to 0} \limsup_{T \to \infty} \frac{1}{2T} \ln \mathcal{Z}(\epsilon, T, A) , \tag{4.11}$$

with the partition functional

$$\mathcal{Z}(\epsilon, T, A) = \text{Sup}_{\mathcal{S}} \sum_{\mathbf{Y} \in \mathcal{S}} \exp \int_{-T}^{+T} A(\mathbf{\Phi}^t \mathbf{Y}) dt , \tag{4.12}$$

where $\mathcal{S}$ is a (ϵ, T)-separated subset of the invariant set $\mathcal{I}$.

The topological pressure has remarkable properties (Walters 1981). In particular, it is a convex functional of the observable, i.e.,

$$\mathscr{P}[\nu A + (1-\nu)B] \;\leq\; \nu\, \mathscr{P}(A) \;+\; (1-\nu)\, \mathscr{P}(B)\,, \tag{4.13}$$

for $0 \leq \nu \leq 1$ and any two observables A and B. The convexity results from the Hölder inequality $\sum a^\nu b^{1-\nu} \leq (\sum a)^\nu (\sum b)^{1-\nu}$ for the positive quantities $a = \exp(\sum A)$ and $b = \exp(\sum B)$ (Marcus and Minc 1964). Another property of the pressure function is that

$$A \;\leq\; B \;\Rightarrow\; \mathscr{P}(A) \;\leq\; \mathscr{P}(B)\,. \tag{4.14}$$

Indeed, the left-hand member of (4.14) implies $\sum \exp(\sum A) \leq \sum \exp(\sum B)$. The right-hand member is inferred from the definition (4.11) and by considering integrals as limits of sums. The following inequality is also important:

$$\mathscr{P}(A+B) \;\leq\; \mathscr{P}(A) \;+\; \mathscr{P}(B)\,, \tag{4.15}$$

which is due to the inequality $\sum ab \leq (\sum a)(\sum b)$ for positive quantities like $a = \exp(\sum A)$ and $b = \exp(\sum B)$, and to the definition (4.11).

When the observable is everywhere vanishing, $A = 0$, the topological pressure reduces to the topological entropy because

$$s(\epsilon, T, \mathscr{I}) = \mathscr{Z}(\epsilon, T, A=0)\,, \qquad \text{so that} \qquad h_{\text{top}}(\mathbf{\Phi}, \mathscr{I}) = \mathscr{P}(0)\,. \tag{4.16}$$

Let us now consider another observable B of the dynamical system. The average of this observable is defined by

$$\langle B \rangle_{\mu_A} \;\equiv\; \mu_A(B) \;=\; \frac{d}{d\nu}\mathscr{P}(A+\nu B)\Big|_{\nu=0}\,. \tag{4.17}$$

Introducing the definition of the pressure, we get that $\mu_A(B) = \int B(\mathbf{X})\mu_A(d\mathbf{X})$ with the measure

$$\mu_A(d\mathbf{X}) \;=\; \lim_{\epsilon\to 0} \limsup_{T\to\infty} \text{Sup}_{\mathscr{S}} \sum_{\mathbf{Y}\in\mathscr{S}} \frac{\exp \int_{-T}^{+T} A(\mathbf{\Phi}^t\mathbf{Y})dt}{\mathscr{Z}(\epsilon, T, A)}$$
$$\times \frac{1}{2T}\int_{-T}^{+T} \delta(\mathbf{X} - \mathbf{\Phi}^t\mathbf{Y})dt\; d\mathbf{X}\,, \tag{4.18}$$

which is referred to as a dynamical measure. We observe that each trajectory of the (ϵ, T)-separated subset is weighted by a Boltzmann-type probability given by

$$\pi_A(\epsilon, T, \mathbf{Y}) = \frac{\exp \int_{-T}^{+T} A(\mathbf{\Phi}^t\mathbf{Y})dt}{\mathscr{Z}(\epsilon, T, A)}\,, \tag{4.19}$$

where $\int_{-T}^{+T} A(\mathbf{\Phi}^t\mathbf{X})dt$ plays the role of $-\beta E$. Accordingly, such measures have been called Gibbs canonical measure. Here, the role of the energy is played by the average of the observable A over the time interval $[-T, +T]$. We emphasize that the Gibbs measure, Eq. (4.18) is determined by a time interval $[-T, +T]$ rather than by a number of particles, interaction range, etc., as in equilibrium statistical mechanics. In the limit $T \to \infty$ the probability measures, Eq. (4.18), tend to measures which are invariant under time evolution, known as 'equilibrium states' as proved by Bowen and Ruelle (1975) for Axiom-A systems. In this limit, a connection can be established between the measures Eq. (4.18) and the invariant probability measure on the invariant set, a connection which will be important for us subsequently.

In this formalism, correlation functions between two observables B_1 and B_2 can be obtained as second derivatives

$$\frac{\partial^2}{\partial v_1 \partial v_2} \mathscr{P}(A + v_1 B_1 + v_2 B_2)\Big|_{v_1=v_2=0} = \int_{-\infty}^{+\infty} \Big[\mu_A(B_1 B_2 \circ \Phi^t) - \mu_A(B_1)\mu_A(B_2)\Big] dt\,, \tag{4.20}$$

in a straightforward notation, provided the various limits exist (Ruelle 1978). Moreover, the Kolmogorov–Sinai entropy per unit time of the dynamical system with respect to the invariant measure μ_A may here be defined in analogy with Eq. (4.8)–(4.9) by

$$h_{\text{KS}}(\mu_A) = \lim_{\epsilon \to 0} \limsup_{T \to \infty} \left(-\frac{1}{2T}\right) \text{Sup}_{\mathscr{S}} \sum_{\mathbf{Y} \in \mathscr{S}} \pi_A(\epsilon, T, \mathbf{Y}) \, \ln \pi_A(\epsilon, T, \mathbf{Y})\,. \tag{4.21}$$

The limit $\epsilon \to 0$ is taken which corresponds to the supremum over finer and finer partitions. Using Eq. (4.19), we can now combine Eqs. (4.17) and (4.21) with $B = A$ to obtain the fundamental identity of importance to us,

$$h_{\text{KS}}\,(\mu_A) = -\mu_A(A) + \mathscr{P}(A)\,. \tag{4.22}$$

In equilibrium statistical mechanics, this identity is the relation between the entropy, the internal energy, and the free energy, respectively. We remark that the identity (4.22) is at the basis of a variational principle proposed by Ruelle (1978) according to which

$$\mathscr{P}(A) \;=\; \text{Sup}_{\mu} \left[h_{\text{KS}}(\mu) \;+\; \mu(A)\right]\,. \tag{4.23}$$

The supremum is reached for the invariant measure $\mu = \mu_A$, where the identity (4.22) holds.

4.2.3 Pressure functions based on the Lyapunov exponents

By varying the observable A, we can obtain many invariant measures, and it is not yet clear which among these many possible measures is the one appropriate to a specific numerical simulation of the dynamical system starting from a given statistical ensemble. The answer to this question will be delayed till the next two sections where the question is answered for closed and open hyperbolic systems. Nevertheless, one thing that is common to the closed and open cases is the special role played by the sum over the local stretching rates multiplied by a parameter $-\beta$ in order to emphasize the formal analogy with the Gibbs states

$$A(\mathbf{X}) = -\beta\, u(\mathbf{X}) = -\beta \sum_{\lambda_i>0} \chi_i(\mathbf{X}) \,, \tag{4.24}$$

in terms of the local dispersion rates (1.48), (1.74), or (1.128). This observable measures the local dispersion of trajectories emanating from the vicinity of the point $\mathbf{X}$. The larger the dynamical instability on the trajectory, the smaller the probability (4.19) of visiting the neighbourhood of this trajectory. This reasoning is at the basis of the choice (4.24). In this case, the pressure functional becomes the following pressure function

$$P(\beta) = \mathscr{P}\left[-\beta \sum_{\lambda_i>0} \chi_i(\mathbf{X})\right] , \tag{4.25}$$

which defines an invariant probability measure μ_β depending on the parameter β. The sum of averaged stretching rates is then given by

$$-\frac{dP}{d\beta}(\beta) = \mu_\beta\Big(\sum_{\lambda_i>0} \chi_i\Big) \,. \tag{4.26}$$

Using the time invariance of the dynamical measure μ_β applied to Eq. (1.43), we obtain that the averages of the local stretching rates are identical with the averages of the corresponding local positive Lyapunov exponent

$$\mu_\beta(\chi_i) = \mu_\beta(\lambda_i) \,. \tag{4.27}$$

Therefore, the fundamental identity (4.22) becomes here

$$h_{\mathrm{KS}}(\mu_\beta) = \beta \sum_{\lambda_i>0} \mu_\beta(\lambda_i) + P(\beta) = \beta\, \mu_\beta(u) + P(\beta) \,. \tag{4.28}$$

We notice that the parameter β has nothing directly to do with the inverse of a thermodynamic temperature. It is worth noting, however, that the large-deviation formalism does make possible deep connections between the calculation of thermodynamic functions and dynamical properties.

The individual average Lyapunov exponents can be obtained by defining a multivariate pressure function $P(\beta_1,\ldots,\beta_L)$ from the observable

$$A(\mathbf{X}) \;=\; -\sum_{\lambda_i>0} \beta_i\; \chi_i(\mathbf{X}) \,, \tag{4.29}$$

depending on the L parameters $\boldsymbol{\beta} = (\beta_1,\ldots,\beta_L)$.[4] The average Lyapunov exponents are now given by

$$-\frac{\partial P}{\partial \beta_i}(\boldsymbol{\beta}) \;=\; \mu_{\boldsymbol{\beta}}(\lambda_i) \,. \tag{4.30}$$

When all the parameters are equal $\beta = \beta_1 = \ldots = \beta_L$, the pressure function (4.25) is recovered: $P(\beta) = P(\beta,\ldots,\beta)$ (see Fig. 4.1).

4.2.4 Entropy function and Legendre transform

It is convenient to introduce also an entropy function by considering the number of points of the (ϵ, T)-separated subset such that the time average of their associated local stretching rates, calculated over a time interval $[-T,+T]$, take values in the interval $[\varphi_i, \varphi_i + d\varphi_i[$ according to

$$\text{Number}\left\{ \mathbf{Y} \in \mathscr{S} : \; \frac{1}{2T}\int_{-T}^{+T} \chi_i(\mathbf{\Phi}^t\mathbf{Y})dt \in [\varphi_i, \varphi_i + d\varphi_i[, \; i = 1,\ldots,L \right\} =$$

$$C(T,\mathscr{S},\boldsymbol{\varphi}) \; \exp\big[2TS(\boldsymbol{\varphi})\big] \; d^L\varphi \,, \tag{4.31}$$

in the limit where $\epsilon \to 0$ and $T \to \infty$, where $C(T,\mathscr{S},\boldsymbol{\varphi})$ is a slowly varying function of the time T. The entropy function $S(\boldsymbol{\varphi})$ is known to be a concave function (Lanford 1973, Bohr and Rand 1987, Rand 1989).

4. $L = f - 1$ for dynamically unstable Hamiltonian systems with f degrees of freedom.

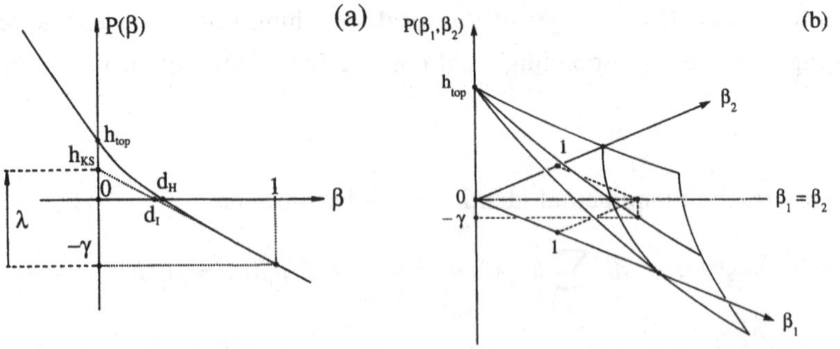

Figure 4.1 (a) Schematic shape of the topological pressure function $P(\beta)$ in a two-degree-of-freedom system and geometric construction of the different characteristic quantities of the repeller. (b) Schematic behaviour of the multivariate pressure function $P(\beta_1, \beta_2)$ for a three-degree-of-freedom system.

The entropy function is related to the pressure function $P(\boldsymbol{\beta})$ by a Legendre transform. Indeed, the sum over all the points of the (ϵ, T)-separated subset $\mathscr{S}$ in the definition (4.11)–(4.12) with the observable (4.29) can be replaced by an integral over $\boldsymbol{\varphi}$

$$P(\boldsymbol{\beta}) \;=\; \lim_{T\to\infty} \frac{1}{2T} \ln \int d^L\varphi \; C(T,\mathscr{S},\boldsymbol{\varphi}) \; \exp[2TS(\boldsymbol{\varphi})] \; \exp(-2T\boldsymbol{\varphi}\cdot\boldsymbol{\beta}) \, . \tag{4.32}$$

In the limit $T \to \infty$, the integral can be evaluated by the steepest-descent method which selects the maximum $\boldsymbol{\varphi}_{\boldsymbol{\beta}}$ of the function in the argument of the exponential as solution of

$$\frac{\partial \boldsymbol{S}}{\partial \boldsymbol{\varphi}}(\boldsymbol{\varphi}_{\boldsymbol{\beta}}) = \boldsymbol{\beta} \, . \tag{4.33}$$

Therefore, the pressure function is given by

$$\boldsymbol{P}(\boldsymbol{\beta}) = \boldsymbol{S}(\boldsymbol{\varphi}_{\boldsymbol{\beta}}) - \boldsymbol{\varphi}_{\boldsymbol{\beta}} \cdot \boldsymbol{\beta} \, . \tag{4.34}$$

Reciprocally, once the pressure function is known, the entropy function is obtained as

$$\boldsymbol{S}(\boldsymbol{\varphi}) = \boldsymbol{P}(\boldsymbol{\beta}_{\boldsymbol{\varphi}}) + \boldsymbol{\varphi} \cdot \boldsymbol{\beta}_{\boldsymbol{\varphi}} \qquad \text{with} \qquad \boldsymbol{\varphi} = -\frac{\partial \boldsymbol{P}}{\partial \boldsymbol{\beta}}(\boldsymbol{\beta}_{\boldsymbol{\varphi}}) \, . \tag{4.35}$$

In particular, the topological entropy is given by

$$h_{\text{top}}(\boldsymbol{\Phi}) = P(0) = S(\boldsymbol{\varphi}_0) \quad \text{with} \quad \boldsymbol{\varphi}_0 = -\frac{\partial P}{\partial \boldsymbol{\beta}}(0) \quad \text{or} \quad \frac{\partial S}{\partial \boldsymbol{\varphi}}(\boldsymbol{\varphi}_0) = 0 \, , \tag{4.36}$$

and the KS entropy by

$$\boldsymbol{h}_{\text{KS}}(\boldsymbol{\mu}_{\boldsymbol{\beta}}) = \boldsymbol{S}(\boldsymbol{\varphi}_{\boldsymbol{\beta}}) \qquad \text{since} \qquad \boldsymbol{\varphi}_{\boldsymbol{\beta}} = \boldsymbol{\mu}_{\boldsymbol{\beta}}(\boldsymbol{\lambda}) \, , \tag{4.37}$$

which justifies the name. Here also, we can define a univariate entropy function $S(\varphi) \equiv S(\varphi, \ldots, \varphi)$, which is related to the univariate pressure function (4.25) by a Legendre transform.

At this stage, we need a proper interpretation of the different terms appearing in the fundamental identity (4.28), especially of the pressure $P(\beta)$ and we need to fix the value of the parameter β. We address these questions in the next two sections.

4.3 Closed hyperbolic systems

4.3.1 The microcanonical measure as a Sinai–Ruelle–Bowen dynamical measure

We consider a hyperbolic system which is closed, i.e., from which no trajectory may escape. If the system is Hamiltonian and time-independent, the appropriate invariant measure is, of course, the microcanonical measure $\mu_e(d\mathbf{X})$ given by (3.29).

Our purpose is to identify the microcanonical measure with one of the dynamical measures introduced in Section 4.2. With this purpose, we consider one of the small regions surrounding a point $\mathbf{Y}$ in a finite (ϵ, T)-separated subset $\mathscr{S}$ and the probability of this region can be computed using the microcanonical measure once the region is identified. This same region is a domain in which the trajectories of all points will remain separated by a distance less than ϵ over a time interval $[-T, +T]$, which is known as a ball $\mathscr{B}(\epsilon, T, \mathbf{Y})$. The microcanonical probability of such a ball can be estimated using the results of Chapter 1 on dynamical instability as follows

$$
\begin{aligned}
&\mu_e\Big[\mathscr{B}(\epsilon, T, \mathbf{Y})\Big] \\
&= \mu_e\Big\{\mathbf{X} \in \mathscr{M} : \quad \|\mathbf{\Phi}^t\mathbf{X} - \mathbf{\Phi}^t\mathbf{Y}\| \leq \epsilon, \ \forall t \in [-T, +T]\Big\} \\
&\simeq \mu_e\Big\{\mathbf{X} \in \mathscr{M} : \quad \|\mathsf{M}(t, \mathbf{Y}) \cdot (\mathbf{X} - \mathbf{Y})\| \leq \epsilon, \ \forall t \in [-T, +T]\Big\} \\
&= \mu_e\Big\{\mathbf{X} \in \mathscr{M} : \ \sum_j \sigma_j(t, \mathbf{Y}) |\mathbf{u}_j(t, \mathbf{Y}) \cdot (\mathbf{X} - \mathbf{Y})|^2 \leq \epsilon^2, \ \forall t \in [-T, +T]\Big\} \\
&\sim \prod_{\sigma_j(+T,\mathbf{Y})>1} \frac{1}{\sqrt{\sigma_j(+T, \mathbf{Y})}} \prod_{\sigma_j(-T,\mathbf{Y})>1} \frac{1}{\sqrt{\sigma_j(-T, \mathbf{Y})}} \\
&\sim \exp - \int_{-T}^{+T} \sum_{\lambda_i>0} \chi_i(\mathbf{\Phi}^t\mathbf{Y}) dt \\
&\sim \exp - \int_{-T}^{+T} u(\mathbf{\Phi}^t\mathbf{Y}) dt \\
&\sim \pi_u(\epsilon, T, \mathbf{Y}) \, .
\end{aligned}
\tag{4.38}
$$

At the first line, the definition (2.45) of the distance has been used. We supposed that ϵ is small enough and we used Eq. (1.23) to get the second line. The third line results from the spectral decomposition (1.24). The fourth line is based on the fact that the quadratic form defines a small ellipsoid with axes determined by the quantities $\sigma_j(t, \mathbf{Y})$. Half of these quantities increase exponentially for $t > 0$ and the other half increase exponentially for $t < 0$, while the vectors $\mathbf{u}_j(t, \mathbf{Y})$ are slowly varying functions of time. The fifth line is a consequence of the relations (1.25) and (1.43) of the eigenvalues $\sigma_j(t, \mathbf{Y})$ to the local stretching rates and of the pairing rule (1.52) that, to every stretching rate, there corresponds a contracting rate with the same absolute value by time reversibility. Finally, the last two lines result from the definition (4.24) of the observable $u(\mathbf{X})$ and of the probability (4.19).

To get this last result, we used the fact that the normalization factor $\mathscr{Z}(\epsilon, T, u)$ is a slowly varying (subexponential) function of time in the case of *closed* hyperbolic systems. Indeed, the sum of all the probabilities (4.38) is approximately constant. On the other hand, all the equalities in Eq. (4.38) hold up to factors which are slowly varying with time. According to the fifth line, the sum of the exponential factors involving the local stretching rates is slowly varying. With the last line, this implies that the dynamical partition function $\mathscr{Z}(\epsilon, T, u)$ is also slowly varying with time.[5] Therefore, all of the exponential dependence should be completely taken into account by the dispersion factor given on the right-hand side of Eq. (4.38). This observation suggests that the member of the family of measures Eq. (4.18) based on the observable $A = -u$ which corresponds to the value $\beta = 1$ is the natural invariant measure corresponding to the microcanonical measure

$$\mu_e = \mu_{\beta=1} \ . \tag{4.39}$$

This result is general and extends to open hyperbolic systems (Gaspard and Dorfman 1995).

We now develop some consequences of this identification. Since the system is closed and the equilibrium measure of the constant energy surface is finite, both the microcanonical and the dynamic measures must be normalizable to unity, say. Since we have already concluded that the normalization factor $\mathscr{Z}(\epsilon, T, u)$ does not depend on time in an exponential way,[6] but increases at most in a subexponential way with T when $\beta = 1$, it follows from Eq. (4.11) that the pressure at this value of β must be zero, i.e.,

$$P(1) = 0 \ , \qquad \text{for closed systems} \ . \tag{4.40}$$

Inserting this result in the fundamental identity, Eq. (4.28) with $\beta = 1$ we recover Pesin's identity

$$h_{KS}(\mu_e) = \sum_{\lambda_i > 0} \mu_e(\lambda_i) \tag{4.41}$$

so that the KS entropy of the invariant natural measure is the sum of the

5. We notice that this result is connected to the periodic-orbit sum rule (3.143).
6. As it would if particles were escaping at an exponential rate from an open system (see Section 4.4).

positive Lyapunov exponents averaged over the same measure (Pesin 1976, 1977). We remark that the dynamical invariant measure which is the microcanonical measure is absolutely continuous with respect to the Lebesgue measure along the unstable manifolds. When this property holds which is a corollary of the Pesin identity, the dynamical measure is referred to as a Sinai–Ruelle–Bowen (SRB) measure.

4.3.2 The pressure function for closed systems

Another remarkable and useful identity for the pressure $P(\beta)$ can be obtained in an alternative way as an average over the microcanonical measure according to

$$P(\beta) = \limsup_{T\to\infty} \frac{1}{2T} \ln \int_{\mathscr{M}} \mu_e(d\mathbf{X}) \exp\left[(1-\beta)\int_{-T}^{+T} \sum_{\lambda_i>0} \chi_i(\mathbf{\Phi}^t\mathbf{X})dt\right] . \tag{4.42}$$

The original definition of the pressure can be recovered as follows. Considering an (ϵ, T)-separated subset $\mathscr{S}$ of the phase space $\mathscr{M}$, the integral can be discretized into a sum over the points $\{\mathbf{Y}\}$ of $\mathscr{S}$ replacing the volume elements $d\mathbf{X}$ by small balls $\mathscr{B}(\epsilon, T, \mathbf{Y})$. This sum would be equal to the integral after the appropriate limits are taken. Therefore, the right-hand side of Eq. (4.42) is given by

$$\lim_{\epsilon\to 0}\limsup_{T\to\infty} \frac{1}{2T} \ln \text{ Sup}_{\mathscr{S}} \sum_{\mathbf{Y}\in\mathscr{S}} \mu_e[\mathscr{B}(\epsilon, T, \mathbf{Y})] \exp\left[(1-\beta)\int_{-T}^{+T} u(\mathbf{\Phi}^t\mathbf{Y})dt\right]. \tag{4.43}$$

Now, according to Eq. (4.38), the probabilities of the small balls are exponentially decreasing as $\exp - \int_{-T}^{+T} u(\mathbf{\Phi}^t\mathbf{Y})dt$, which introduces an extra inverse power of the dispersing factor and explains the power $(1-\beta)$ in Eq. (4.42). Finally, we obtain the definition of the pressure, namely

$$P(\beta) = \lim_{\epsilon\to 0}\limsup_{T\to\infty} \frac{1}{2T} \ln \text{ Sup}_{\mathscr{S}} \sum_{\mathbf{Y}\in\mathscr{S}} \exp\left[-\beta\int_{-T}^{+T} u(\mathbf{\Phi}^t\mathbf{Y})dt\right]. \tag{4.44}$$

We note that for the purpose of numerical calculations, we can substitute the stretching factors $\sigma_j(t, \mathbf{X})$ for the exponentials of the integrated local stretching factors and use

$$P(\beta) = \lim_{T\to\infty} \frac{1}{T} \ln \int_{\mathscr{M}} \mu_e(d\mathbf{X}) \prod_{\sigma_j>1} \sqrt{\sigma_j(T, \mathbf{X})}^{1-\beta} \tag{4.45}$$

instead of Eq. (4.42). Here, there is no factor of two in the denominator because the stretching factors correspond to the time interval $[0, +T]$ rather than $[-T, +T]$ as before.

Equations (4.42) or (4.45) provide a convenient way to numerically compute the pressure function as a phase-space average of an expression involving the local stretching factors. A typical pressure function is illustrated in Fig. 4.2a for a closed hyperbolic system. Note that it is a convex, monotonic function of β and vanishes at $\beta = 1$. This is not true of open systems. The KS entropy is obtained by finding the slope of the pressure at $\beta = 1$, and carrying out the linear extrapolation illustrated in Fig. 4.2a. Finally, we note that the value of $P(\beta)$ at $\beta = 0$, is the topological pressure $h_{\text{top}}(\mathbf{\Phi})$ of the dynamical system, as can be seen from Eqs. (4.36). The fact that $h_{\text{KS}} \leq h_{\text{top}}$ puts a bound on the slope of $P(\beta)$ at $\beta = 1$.

4.3.3 Generating functions of observables and transport coefficients

Within the large-deviation formalism, it is possible to characterize in detail the random time evolution of some observable by introducing the generating function

$$Q(\alpha, \beta) = \lim_{T\to\infty} \frac{1}{2T} \ln \mu_\beta \left\{ \exp\left[\alpha \int_{-T}^{+T} A(\mathbf{\Phi}^t \mathbf{X})\, dt\right] \right\} . \qquad (4.46)$$

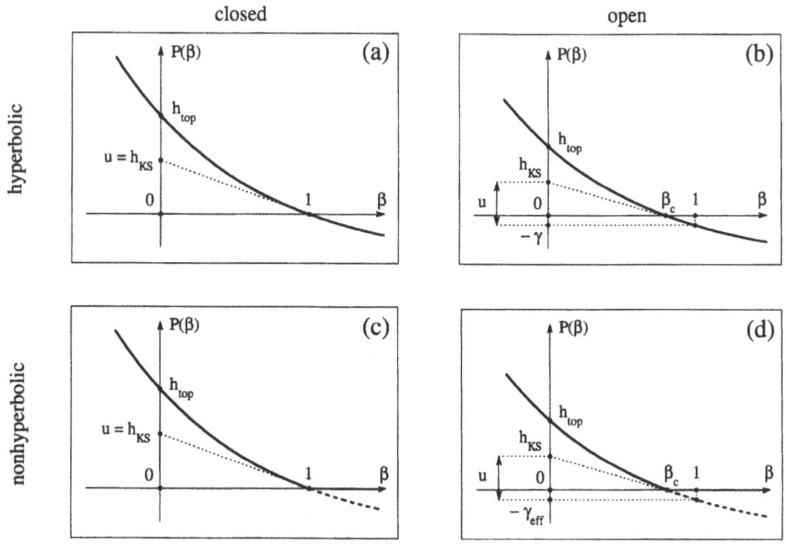

Figure 4.2 Schematic behaviour of the pressure function in the cases of: (a) a closed hyperbolic system; (b) an open hyperbolic system; (c) a closed nonhyperbolic system; (d) an open nonhyperbolic system. γ denotes the escape rate; γ_{eff} the effective escape rate; u the sum of the mean positive Lyapunov exponents; h_{KS} the KS entropy per unit time; h_{top} the topological entropy per unit time.

The first derivative with respect to the parameter α gives the average over the invariant measure: $\mu_\beta(A) = \partial_\alpha Q\big|_{\alpha=0}$. The second derivative gives the time integral of the autocorrelation of the observable A [cf. Eq. (4.20)] and, therefore, the generalized diffusion coefficient of the central limit theorem (3.60). In this regard, the transport coefficient with respect to the SRB measure μ_β can be obtained in the present large-deviation formalism. The standard transport coefficient is given by the value at $\beta = 1$ at which we recover the microcanonical measure. The higher derivatives give the higher-order transport coefficients such as the Burnett coefficients (van Beijeren 1982, Gaspard 1992d).

We can also define a large-deviation function according to

$$\mu_\beta\left\{\mathbf{Y} \in \mathscr{S} : \frac{1}{2T}\int_{-T}^{+T} A(\mathbf{\Phi}^t\mathbf{Y})dt \in [\eta, \eta + d\eta[\right\} = C(T, \mathscr{S}, \eta, \beta) \exp\big[2TH(\eta, \beta)\big]\, d\eta \ , \qquad (4.47)$$

in the limit $T \to \infty$. In Eq. (4.47), C is a slowly varying function and the observable is integrated over the trajectory $\mathbf{Y}$ of an (ϵ, T)-separated subset of the phase space $\mathscr{M}$. The large-deviation function $H(\eta, \beta)$ can be considered as a generalization to dynamical systems theory of the Onsager–Machlup action function (Oono 1993). The function H is related to the generating function by a Legendre transform according to

$$Q(\alpha, \beta) = H(\eta, \beta) + \alpha\,\eta \ , \qquad \text{for} \qquad \alpha = -\frac{\partial H}{\partial \eta} \quad \text{and} \quad \eta = \frac{\partial Q}{\partial \alpha} \ . \qquad (4.48)$$

These considerations are also useful in the study of anomalous transport (Wang and Hu 1993).

4.4 Open hyperbolic systems

In this section, the main focus of our analysis is on the chaotic behaviour of open systems, i.e., systems with trajectories which escape from a bounded region of phase space in a finite amount of time. The behaviour of such systems can be easily visualized by considering an escape-time function as explained in Chapter 2. The connection between transport coefficients and dynamical quantities is based on the escape-rate formula (see Chapter 6) and here we describe the derivation of this formula as a special case of the fundamental identity, Eq. (4.22).

4.4.1 The nonequilibrium invariant measure of the repeller

We now define the invariant probability measure of the repeller. We first consider a probability measure $\nu_0(d\mathbf{X})$ on the region U corresponding to a particular statistical ensemble of N_0 initial conditions at phase-space points $\{\mathbf{X}^{(i)}\}$ where these points are distributed uniformly in U with respect to the microcanonical ensemble, say. This measure ν_0 is taken to be of the form

$$\nu_0(d\mathbf{X}) = \lim_{N_0\to\infty} \frac{1}{N_0} \sum_{i=1}^{N_0} \delta\left(\mathbf{X} - \mathbf{X}^{(i)}\right) d\mathbf{X} \tag{4.49}$$

Of these N_0 initial points, the number N_T still contained in U over the time interval $[0, +T]$ decays according to

$$\lim_{N_0\to\infty} \frac{N_T}{N_0} = \nu_0[\Upsilon_U^{(+)}(T)] = \int_{\Upsilon_U^{(+)}(T)} \nu_0(d\mathbf{X}). \tag{4.50}$$

We note that the limit $N_0 \to \infty$ is essential to define a smooth function of time T, since N_T typically shows large statistical fluctuations when $N_T < 10$, and drops to zero after a finite time. Similar expressions hold for the time intervals $[-T, 0]$ and $[-T, +T]$.

Assuming that almost all the trajectories escape, i.e., that $\lim_{t\to\infty} \nu_0[\Upsilon_U^{(+)}(T)] = 0$, the decay curve, Eq. (4.50), may be exponential or slower than exponential in general systems. However, in hyperbolic systems (where all orbits are of saddle type) the decay is exponential. Thus we can define an *escape rate* according to

$$\gamma = -\lim_{T\to\infty} \frac{1}{2T} \ln \nu_0 \left[\Upsilon_U(T)\right] \tag{4.51}$$

$$= -\lim_{T\to\infty} \frac{1}{T} \ln \nu_0 \left[\Upsilon_U^{(+)}(T)\right] = \gamma^+ \tag{4.52}$$

$$= -\lim_{T\to\infty} \frac{1}{T} \ln \nu_0 \left[\Upsilon_U^{(-)}(T)\right] = \gamma^- , \tag{4.53}$$

where $\Upsilon_U(T) = \Upsilon_U^{(+)}(T) \cap \Upsilon_U^{(-)}(T)$ (Kadanoff and Tang 1984). All these definitions are equivalent because of time reversibility. For noninvertible systems like 1D maps, only Eq. (4.52) is applicable. The above definitions are in agreement with the escape rate defined as a Pollicott–Ruelle resonance in Chapter 3 because $\nu_0[\Upsilon_U^{(+)}(T)] = \langle I_U|\hat{P}^t|f_0\rangle$ where I_U is the indicator function of the neighbourhood U and $f_0 = d\nu_0/d\mathbf{X}$. This relation holds in the case of absorbing boundary conditions at ∂U so that there is no reentry of escaped trajectories.

In the long-time limit, the trajectories remaining in the domain U are distributed according to a probability measure which is invariant for the dynamics on the repeller and which is given in Chapter 3 by the periodic-orbit measure (3.204). In order to construct such an invariant probability measure on the repeller in a way that only uses the time evolution of running trajectories instead of periodic orbits, we use ideas familiar from ergodic theory. In the usual arguments of ergodic theory, one considers the time average of some dynamical quantity for a system whose phase-space point is confined to a closed energy shell.

If the system is ergodic then the long-time average of any dynamical quantity is equal to its ensemble average taken with respect to the microcanonical measure. Consider now a system whose phase-space trajectory is confined to the repeller. If the trajectories on the repeller are ergodic with respect to a natural measure on the repeller, then the long-time average of any dynamical quantity on the repeller should be equal to the ensemble average of this quantity with respect to the natural measure. To construct the natural nonequilibrium measure on the repeller, then, we begin with the definition of the time average of some observable $A(\mathbf{X})$ on the repeller as

$$\mu_{\rm ne}(A) = \lim_{T\to\infty} \lim_{\tilde{N}_T\to\infty} \frac{1}{\tilde{N}_T} \sum_{i=1}^{\tilde{N}_T} \frac{1}{2T} \int_{-T}^{+T} A(\mathbf{\Phi}^t \mathbf{X}^{(i)})\, dt \,. \tag{4.54}$$

In Eq. (4.54), the sum extends over the $\tilde{N}_T$ phase-space points whose trajectories remain in region U over the time interval $-T < t < +T$. This is required in order to balance the escape of typical trajectories out of the repeller. This is a procedure which is specific to open systems, in contrast to the case of closed systems where the trajectories remain on the hyperbolic set.

The time average can be rewritten in terms of the initial measure as

$$\langle A\rangle_{\rm ne} \;=\; \mu_{\rm ne}(A) = \int A(\mathbf{X})\mu_{\rm ne}(d\mathbf{X}) \,, \tag{4.55}$$

with

$$\begin{aligned}\mu_{\rm ne}(d\mathbf{X}) \;=\; & \lim_{T\to\infty} \frac{1}{\nu_0[\Upsilon_U(T)]} \int \nu_0(d\mathbf{Y})\, I_{\Upsilon_U(T)}(\mathbf{Y}) \\ & \times \frac{1}{2T} \int_{-T}^{+T} \delta(\mathbf{X} - \mathbf{\Phi}^t \mathbf{Y}) dt d\mathbf{X} \,. \end{aligned} \tag{4.56}$$

Here $I_\Upsilon(\mathbf{X})$ is the characteristic function of the set Υ in phase space, i.e., $I_\Upsilon(\mathbf{X}) = 1$ if $\mathbf{X} \in \Upsilon$ and zero otherwise. In this way, we have defined a normalized probability measure on the repeller: $\mu_{\rm ne}$. This is the desired invariant measure with the fractal repeller as its support. We also note that the subscript 'ne' of $\mu_{\rm ne}$ refers to the nonequilibrium character of the natural invariant measure. This measure is the natural generalization of the microcanonical canonical ensemble measure to open systems where the dynamics takes place on the fractal repeller, $\mathscr{I}$. Using this measure, we can consider the long-time limit and define the average of an observable according to Eq. (4.55). This procedure amounts to performing the statistics on the set of $\tilde{N}_T$ initial conditions which are still in the domain U at times $\pm T$. As $T \to \infty$ these trajectories approach more and more closely to the trajectories on the repeller. As a consequence, $\mu_{\rm ne}$ is an invariant probability measure on the repeller. We will return to this invariant measure presently.

4.4.2 Connection with the dynamical invariant measure and the pressure function

Now, we wish to construct a dynamical measure on the repeller similar to that used for closed systems in Section 4.3. This will enable us to continue the development of the previous sections and to obtain the fundamental identity, Eq. (4.22), for open systems. The large-deviation formalism suggests considering the family of dynamical invariant measures defined in analogy with Eq. (4.18)

$$\begin{aligned}\mu_\beta(d\mathbf{X}) &= \lim_{\epsilon\to 0} \limsup_{T\to\infty} \mathrm{Sup}_{\mathcal{S}} \sum_{\mathbf{Y}\in\mathcal{S}} \frac{\exp -\beta \int_{-T}^{+T} u(\mathbf{\Phi}^t\mathbf{Y})dt}{z(\epsilon, T, \beta)} \\ &\quad \times \frac{1}{2T} \int_{-T}^{+T} \delta(\mathbf{X} - \mathbf{\Phi}^t\mathbf{Y})dt \; d\mathbf{X} \,, \end{aligned} \tag{4.57}$$

with the dynamical partition function

$$z(\epsilon, T, \beta) = \mathrm{Sup}_{\mathcal{S}} \sum_{\mathbf{Y}\in\mathcal{S}} \exp -\beta \int_{-T}^{+T} u(\mathbf{\Phi}^t\mathbf{Y})dt \,, \tag{4.58}$$

in terms of the dispersion rate $u = \sum_{\lambda_i>0} \chi_i$. Let us emphasize that the local stretching rates as well as the local Lyapunov exponents are well defined for the trajectories of any subset $\mathcal{S}$ of the repeller $\mathcal{I}$ since those trajectories remain in the compact domain U forever. A disadvantage of this definition is that the repeller $\mathcal{I}$ needs to be known since a subset $\mathcal{S}$ of it is considered. Physically and numerically, the repeller of systems of scattering type appears out of the dynamics after a very long time as described above. The previously defined invariant measure avoids this *a priori* knowledge of the repeller since the invariant measure (4.56) is automatically constructed by the time evolution.

We now show that both measures (4.56) and (4.57) are equivalent for open hyperbolic systems when $\beta = 1$. We do this by using the neighbourhoods of the points of a (ϵ, T)-separated subset, $\mathcal{S}$, of the repeller $\mathcal{I}$. Note that we take the time T to be the same in both sets so that we can closely approximate the repeller by means of the separated subsets as we take the limit $T \to \infty$. In this situation, we can transform the expression (4.56) into

$$\mu_{\rm ne}(d\mathbf{X}) = \lim_{\epsilon\to 0}\limsup_{T\to\infty} \mathrm{Sup}_{\mathscr{S}} \sum_{\mathbf{Y}\in\mathscr{S}} \frac{\nu_0[\mathscr{B}(\epsilon,T,\mathbf{Y})\cap \Upsilon_U(T)]}{\nu_0[\Upsilon_U(T)]} \times \frac{1}{2T}\int_{-T}^{+T} \delta(\mathbf{X}-\mathbf{\Phi}^t\mathbf{Y})dtd\mathbf{X}\,. \tag{4.59}$$

It is clear that this measure is normalized. Since the points $\mathbf{Y}$ belongs to a (ϵ, T)-separated subset of the repeller $\mathscr{I}$, they belong to the repeller itself. If the points $\mathbf{Y}$ are not close to the boundary ∂U and if ϵ is small enough, the balls $\mathscr{B}(\epsilon, T, \mathbf{Y})$ are contained inside $\Upsilon_U(T)$ so that

$$\nu_0[\mathscr{B}(\epsilon,T,\mathbf{Y})\cap\Upsilon_U(T)] \simeq \nu_0[\mathscr{B}(\epsilon,T,\mathbf{Y})] \sim \exp - \int_{-T}^{+T} u(\mathbf{\Phi}^t\mathbf{Y})dt\,, \tag{4.60}$$

where considerations similar to the ones of Eq. (4.38) have been applied here to ν_0. As a consequence, Eq. (4.59) becomes

$$\begin{aligned}\mu_{\rm ne}(d\mathbf{X}) &= \lim_{\epsilon\to 0}\limsup_{T\to\infty} \mathrm{Sup}_{\mathscr{S}} \sum_{\mathbf{Y}\in\mathscr{S}} \frac{\exp - \int_{-T}^{+T} u(\mathbf{\Phi}^t\mathbf{Y})dt}{z(\epsilon,T,1)} \\ &\quad\times \frac{1}{2T}\int_{-T}^{+T} \delta(\mathbf{X}-\mathbf{\Phi}^t\mathbf{Y})dtd\mathbf{X} \\ &= \mu_1(d\mathbf{X})\,.\end{aligned} \tag{4.61}$$

Accordingly, we recover the dynamical invariant measure corresponding to the value $\beta = 1$ as in the case of closed hyperbolic systems [see Eq.(4.39)].

We note that the normalization factor is required here for the following reason. In order to estimate the normalization factor, we decompose the set $\Upsilon_U(T)$ into small balls $\mathscr{B}(\epsilon, T, \mathbf{Y})$ centered on the points $\mathbf{Y}$ of an (ϵ, T)-separated subset $\mathscr{S}$ as

$$\begin{aligned}\nu_0[\Upsilon_U(T)] &\sim \mathrm{Sup}_{\mathscr{S}} \sum_{\mathbf{Y}\in\mathscr{S}} \nu_0[\mathscr{B}(\epsilon,T,\mathbf{Y})\cap\Upsilon_U(T)] \\ &\sim \mathrm{Sup}_{\mathscr{S}} \sum_{\mathbf{Y}\in\mathscr{S}} \exp - \int_{-T}^{+T} u(\mathbf{\Phi}^t\mathbf{Y})dt = z(\epsilon,T,1) \\ &\sim \exp 2TP(1)\,.\end{aligned} \tag{4.62}$$

Comparing with the definition Eq. (4.51) of the escape rate γ, this result shows that the normalization factor decays exponentially like $z(\epsilon, T, 1) \sim \exp(-2\gamma T)$ and, moreover, that the pressure function $P(\beta)$ has, for an open system of hyperbolic type, the value

$$P(1) = -\gamma\,. \tag{4.63}$$

Furthermore, we can now define the important quantities of the large-deviation formalism of Section 4.2 and the fundamental identity (4.22) follows once again.

We can identify all of the quantities appearing in this equation at $\beta = 1$. For this case we find

$$\gamma = \sum_{\lambda_i>0} \mu_{\rm ne}(\lambda_i) - h_{\rm KS}(\mu_{\rm ne}) , \tag{4.64}$$

(Kantz and Grassberger 1985, Eckmann and Ruelle 1985, Chernov and Markarian 1997). Equation (4.64) is the escape-rate formula for an open system, giving the relation between the escape rate, γ, and the sum of the positive Lyapunov exponents and Kolmogorov–Sinai entropy for trajectories on the fractal repeller, $\mathscr{I}$, using the natural measure $\mu_1 = \mu_{\rm ne}$.[7] This result generalizes Pesin's formula to open hyperbolic systems. Indeed, when the system is closed, the escape rate vanishes, $\gamma = 0$, and Pesin's formula is recovered as well as Eq. (4.40). The average of the sum of positive Lyapunov exponents over the nonequilibrium invariant measure can be calculated as before as a derivative of the pressure function

$$\sum_{\lambda_i>0} \mu_{\rm ne}(\lambda_i) = -\left.\frac{dP(\beta)}{d\beta}\right|_{\beta=1} , \tag{4.65}$$

according to Eqs. (4.26)–(4.27).

The escape-rate formula (4.64) may be interpreted by saying that the dynamical instability characterized by the sum of positive Lyapunov exponents induces two joint but distinct effects: (1) an escape γ of trajectories out of the neighbourhood U of the repeller because the system is open; (2) a dynamical randomness $h_{\rm KS}$ because of chaotic transients near the repeller. For a given dynamical instability, the larger the chaos, the smaller the escape, and vice versa.

We now describe a practical way to calculate the pressure function in open systems. As we have seen with Eq. (4.25), the pressure function requires the knowledge of the local Lyapunov exponents. In practice, we only know the stretching factors $\sigma_j(T, \mathbf{X}^{(i)})$ of the trajectories of the initial ensemble (4.49). In analogy with Eq. (4.45), we here propose

$$P(\beta) = \lim_{T\to\infty} \frac{1}{T} \ln \int_{\Upsilon_U^{(+)}(T)} \nu_0(d\mathbf{X}) \prod_{\sigma_j>1} \sqrt{\sigma_j(T, \mathbf{X})}^{\,1-\beta} . \tag{4.66}$$

This definition is equivalent to the previous one for the following reasons. By time reversibility, we can convert the average over the forward set $\Upsilon_U^{(+)}(T)$ into an average over the set $\Upsilon_U(T)$ by considering the time interval $[-T, +T]$ rather than $[0, +T]$. Using an (ϵ, T)-separated subset $\mathscr{S}$ to discretize the integral, the

7. We shall also use the notations $\bar{\lambda}_i(\mathscr{I}) = \mu_{\rm ne}(\lambda_i) = \mu_1(\lambda_i)$ and $h_{\rm KS}(\mathscr{I}) = h_{\rm KS}(\mu_{\rm ne}) = h_{\rm KS}(\mu_1)$ in the following.

right-hand member of (4.66) becomes

$$\lim_{\epsilon\to 0}\limsup_{T\to\infty}\frac{1}{2T}\ln \mathrm{Sup}_{\mathscr{S}}\sum_{\mathbf{Y}\in\mathscr{S}}\nu_0[\mathscr{B}(\epsilon,T,\mathbf{Y})\cap \Upsilon_U(T)]$$
$$\times\exp\left[(1-\beta)\int_{-T}^{+T}u(\mathbf{\Phi}^t\mathbf{Y})dt\right] . \tag{4.67}$$

According to Eq. (4.60), we recover the original definition of the pressure

$$P(\beta)=\lim_{\epsilon\to 0}\limsup_{T\to\infty}\frac{1}{2T}\ln z(\epsilon,T,\beta) , \tag{4.68}$$

with the partition function (4.58).

We emphasize that the nonequilibrium invariant measure, given by (4.56) or (4.61), is not absolutely continuous with respect to the Lebesgue measure along the unstable manifolds but singular with a fractal for support. In this way, the nonequilibrium invariant measure differs from the equilibrium one (4.39). We shall now characterize this fundamental difference in terms of the concept of fractal dimension.

4.4.3 Fractal repellers in two-degrees-of-freedom systems and their dimensions

We have already mentioned that a chaotic hyperbolic repeller is a fractal object in phase space. This conclusion is based on both facts (1) that a chaotic set contains uncountably many trapped trajectories and (2) that the repeller is of zero Liouville measure in phase space. As a consequence, the repeller is a Cantor set and thus a fractal (Mandelbrot 1982). Such fractals are invariant under the dynamics and can be characterized in terms of fractal generalized dimensions which can be calculated within the large-deviation formalism and, especially, in terms of the pressure function. In systems with many degrees of freedom there exist many directions along which *partial* fractal dimensions can be defined (Eckmann and Ruelle 1985). These partial dimensions can be evaluated in the case of attractors thanks to the Kaplan–Yorke formula (Kaplan and Yorke 1979). However, few results are known on the numerical algorithms appropriate to calculate the partial dimensions in high-dimensional systems. In view of this limitation, we shall here restrict ourselves to symplectic systems with two degrees of freedom. The time-reversal symmetry maps the stable manifolds onto the unstable manifolds so that their partial dimensions are equal, which simplifies their determination. We mention that many works have been devoted to the characterization of fractal repellers (Szepfalusy and Tél 1986; Tél 1986, 1987, 1990; Bohr and Tél 1988; Grassberger et al. 1988; Grebogi et al. 1988; Kovács and Tél 1990; Ott and Tél 1993; Ott 1993).

The fractal character of the repeller is a phase-space geometric consequence of the dynamical randomness of the time evolution. The connection between both aspects is due to the dynamical instability in phase space. As we discussed in Chapter 1, the dynamical instability is characterized by the local stretching factors (1.37). In symplectic systems with two degrees of freedom, there is at most one unstable direction, one stretching factor $|\Lambda(t,\mathbf{X})| > 1$, and one corresponding positive Lyapunov exponent. For billiards, the stretching factor can be estimated by using Eq. (1.141).

The pressure function can be defined with Eq. (4.66) where we notice that the single eigenvalue $\sigma_j(T,\mathbf{X})$ estimates the local stretching factor in absolute value $|\Lambda(T,\mathbf{X})|$ [cf. Eqs. (1.25) and (1.38)]. Therefore, the pressure function can be calculated by

$$\begin{aligned} P(\beta) &= \lim_{T\to+\infty}\lim_{N_0\to\infty}\frac{1}{T}\ln\frac{1}{N_0}\sum_{i=1}^{N_T}|\Lambda(T,\mathbf{X}_0^{(i)})|^{1-\beta} \\ &= -\gamma + \lim_{T\to+\infty}\lim_{N_0\to\infty}\frac{1}{T}\ln\langle\,|\Lambda(T,\mathbf{X})|^{1-\beta}\rangle_{\rm ne}\,. \end{aligned} \tag{4.69}$$

where we used the property (4.50) and the nonequilibrium average (4.54)–(4.55) (Gaspard and Baras 1995).

In Hamiltonian systems with two degrees of freedom, the fundamental identity (4.64) reduces to

$$\gamma = \bar{\lambda} - h_{\rm KS}\,, \tag{4.70}$$

where both the Lyapunov exponent and the KS entropy are evaluated on the natural invariant measure $\mu_{\beta=1}$ on the repeller. In particular, the average Lyapunov exponent is given by

$$\bar{\lambda} = -P'(1)\,, \tag{4.71}$$

where the prime denotes a derivative with respect to the argument β of the pressure function. On the other hand, the escape rate is given by (4.63), so that the KS entropy is obtained as

$$h_{\rm KS} = P(1) - P'(1)\,. \tag{4.72}$$

This result can be interpreted geometrically in Fig. 4.1a. The escape rate is given by the value of the pressure function at $\beta = 1$. The KS entropy is the ordinate of the intersection of a straight line tangent to the pressure function at $\beta = 1$ with the vertical axis. The Lyapunov exponent is the total distance from the ordinate of the pressure function at $\beta = 1$ up to the ordinate corresponding to the KS entropy.

The partial generalized fractal dimensions

Our aim here is to characterize the fractal properties of the repeller. We consider a Poincaré section which is transverse to the flow on the repeller. In the surface of section, the stable and unstable manifolds of each trajectory form lines. Therefore, the intersections of an arbitrary line $\mathscr{L}$ with the stable manifolds of the fractal repeller form another fractal with generalized dimensions taking their values in the interval $0 \leq d_q \leq 1$, which are known as the partial dimensions of the repeller (Eckmann and Ruelle 1985). In contrast, the full dimensions of the repeller itself in the three-dimensional phase space belong to the interval $0 \leq D_q \leq 3$ because they are related to the partial dimensions by $D_q = 2d_q + 1$.

The generalized fractal dimensions of the repeller can be estimated as follows (Hasley et al. 1986, Bessis et al. 1988). The line $\mathscr{L}$ can be taken as the support of the ensemble ν_0 of initial conditions (4.49). The forward escape-time function (2.17) can be calculated on this line. Since the escape-time function has singularities on the stable manifolds of the repeller, the set of initial conditions for which the escape time is larger than T is composed of many small intervals which cover the fractal set $\mathscr{L} \cap W_s(\mathscr{I})$. The widths of the intervals of the covering are approximately given by the inverse of the absolute values of the stretching factors up to time T:

$$\ell_i \sim \frac{1}{|\Lambda(T, \mathbf{X}_0^{(i)})|} , \tag{4.73}$$

for trajectories of initial conditions $\{\mathbf{X}_0^{(i)}\}$ in the initial ensemble ν_0. On the other hand, the nonequilibrium measure (4.61) weights small balls $\mathscr{B}(\epsilon, T, \mathbf{X}_0^{(i)})$ centered on the intervals (4.73) with probabilities

$$p_i \sim \frac{\exp(\gamma T)}{|\Lambda(T, \mathbf{X}_0^{(i)})|} , \tag{4.74}$$

which are normalized according to $\sum_{i=1}^{N_T} p_i = 1$. These probabilities are actually given by (4.19) with the denominator (4.62).

In the limit $T \to \infty$, the partial generalized dimensions, d_q, associated with the invariant measure μ_{ne} are determined by the condition that the following quantity must remain of the order of one

$$\sum_{i=1}^{N_T} \frac{p_i^q}{\ell_i^{(q-1)d_q}} \sim 1 , \qquad \text{for} \quad N_T \to \infty , \tag{4.75}$$

where the number of intervals in the covering of the fractal may here be taken as the number N_T of trajectories still in the region U at time T (Hasley et al. 1986). The left-hand side of (4.75) can be rewritten as a statistical average (4.55) over the measure $\mu_{\rm ne}$

$$\sum_{i=1}^{N_T} p_i \, \frac{p_i^{q-1}}{\ell_i^{(q-1)d_q}} \;=\; \left\langle \frac{p_i^{q-1}}{\ell_i^{(q-1)d_q}} \right\rangle_{\rm ne} . \tag{4.76}$$

Introducing the results (4.73)–(4.74) in Eqs. (4.75)–(4.76), the generalized dimensions must satisfy

$$\exp\Big[(q-1)\gamma T\Big] \, \Big\langle |\Lambda(T,\mathbf{X})|^{(q-1)(d_q-1)} \Big\rangle_{\rm ne} \sim 1 \,, \quad \text{for } T \to +\infty \,. \tag{4.77}$$

Using the relation (4.69) for the pressure, we obtain the result that the generalized dimensions are given by the root of the following equation (Bessis et al. 1988)

$$P\Big[q \,+\, (1-q)d_q\Big] \;=\; -q\,\gamma \,. \tag{4.78}$$

When $q = 0$, the generalized dimension is equal to the Hausdorff dimension, $d_0 = d_{\rm H}$, which is given as the root of the pressure itself

$$P(d_{\rm H}) \;=\; 0 \,. \tag{4.79}$$

Since the escape rate is nonvanishing in open systems, the Hausdorff dimension is smaller than unity so that the repeller does not completely fill the phase space as already anticipated. When the escape rate vanishes the Hausdorff dimension approaches the unit value and the repeller fills the full phase space. These results can be geometrically obtained from Fig. 4.1a where the partial Hausdorff dimension is given by the intersection of the pressure function with the horizontal axis.

Another important dimension is the so-called information dimension which is given when $q = 1$, $d_{\rm I} = d_1$. Differentiating (4.78) with respect to q and setting $q = 1$, Young's formula is obtained

$$d_{\rm I} \;=\; 1 \,-\, \frac{P(1)}{P'(1)} \;=\; 1 \,-\, \frac{\gamma}{\bar{\lambda}} \;=\; \frac{h_{\rm KS}}{\bar{\lambda}} \,, \tag{4.80}$$

where we used (4.71)–(4.72) (Young 1982). Accordingly, the information dimension is directly related to the KS entropy and the average Lyapunov exponent. In Fig. 4.1a, the information dimension is given as the intersection of the straight line tangent to the pressure at $\beta = 1$ with the horizontal axis. This geometric construction proves the inequality

$$d_{\rm I} \;\leq\; d_{\rm H} \,, \tag{4.81}$$

coming from the convexity of the pressure function. The $f(\alpha)$-spectrum can also be derived by a Legendre transform of the generalized dimension d_q (Hasley et al. 1986).

It is convenient to introduce the codimensions which are the defects of the partial dimensions with respect to one

$$c_q = 1 - d_q \, . \tag{4.82}$$

In particular, we have the Hausdorff and the information codimensions

$$c_{\rm H} = 1 - d_{\rm H} \, , \tag{4.83}$$

$$c_{\rm I} = 1 - d_{\rm I} \, . \tag{4.84}$$

The codimensions also belong to the interval, $0 \leq c_q \leq 1$ (Gaspard and Baras 1992, 1995).

The Hausdorff codimension can be obtained using the following numerical algorithm developed by the group of Maryland (McDonald et al. 1985, Ott 1993). An ensemble of pairs of initial conditions separated by ϵ is considered along the line $\mathscr{L}$ in the neighbourhood U of the repeller. The pair is said to be uncertain if there is a singularity of the escape-time function between both initial conditions. If we associate to each trajectory a symbolic sequence, $\omega_0\omega_1\ldots\omega_n$, which gives the labels of successive phase-space cells visited by the trajectory, the pair is certain if the symbolic sequences corresponding to both initial conditions are identical. When the pair is certain, both initial conditions of the pair belong to an interval of continuity of the escape-time function since they are not separated by a singularity. On the other hand, the symbolic sequences are different when the pair is uncertain. The fraction $f(\epsilon)$ of uncertain pairs in the initial ensemble ν_0 is known to behave like

$$f(\epsilon) \sim \epsilon^{c_{\rm H}} \, . \tag{4.85}$$

A derivation of this result can be found elsewhere (McDonald et al. 1985, Ott 1993). Equation (4.85) provides therefore a direct numerical method to calculate the Hausdorff dimension.

4.5 Generalized zeta functions

4.5.1 Frobenius–Perron operators in the large-deviation formalism

The large-deviation formalism is closely related to the periodic-orbit theory exposed in Chapter 3. The aim of the generalized eigenvalue problem of the Frobenius–Perron operator is to obtain the eigenstates of the time evolution of statistical ensembles. On the other hand, the large-deviation formalism is only concerned with the leading eigenstate corresponding to the slowest decay when

$\beta = 1$. Both formalisms coincide at the value $\beta = 1$ (except in nonhyperbolic systems).

Actually, it is possible to define some generalized Frobenius–Perron operators in order to compare both formalisms for each value of β. This is done by defining operators like (3.194) by taking an observable which is proportional to the local dispersion rate $u(\mathbf{X})$:

$$\hat{P}^t_\beta\, f(\mathbf{X}) \;=\; \int \delta(\mathbf{X} - \mathbf{\Phi}^t\mathbf{Y})\, \exp\left[(1-\beta)\int_0^t u(\mathbf{\Phi}^\tau\mathbf{Y})\, d\tau\right] f(\mathbf{Y})\, d\mathbf{Y}\,. \tag{4.86}$$

This operator is a particular case of the more general Frobenius–Perron operator

$$\hat{P}^t_A\, f(\mathbf{X}) \;=\; \int \delta(\mathbf{X} - \mathbf{\Phi}^t\mathbf{Y})\, \exp\left\{\int_0^t [u(\mathbf{\Phi}^\tau\mathbf{Y}) + A(\mathbf{\Phi}^\tau\mathbf{Y})]\, d\tau\right\} f(\mathbf{Y})\, d\mathbf{Y}\,. \tag{4.87}$$

An eigenvalue problem can be set up for such Frobenius–Perron operators. For two-degree-of-freedom Hamiltonian systems, the local dispersion rate is related to the stability eigenvalue of each periodic orbit by

$$\exp \int_0^{T_p} u(\mathbf{\Phi}^t\mathbf{X}_p)\, dt \;=\; |\Lambda_p|\,. \tag{4.88}$$

Accordingly, the generalized eigenvalues of the operator (4.87) are given by the zeros of the Zeta function (3.199), which here becomes

$$Z(s;A) \;=\; \prod_p \prod_{m=0}^{\infty} \left(1 \;-\; \frac{e^{A_p - sT_p}}{\Lambda_p^m}\right)^{m+1}, \tag{4.89}$$

in terms of the integral (3.200) of the observable A along the periodic orbit. According to the discussion of Section 3.8, the leading eigenvalues are given by the zeros of the leading inverse zeta function introduced by Ruelle (1976, 1978, 1994)

$$\zeta(s;A) \;=\; \prod_{p\in\mathscr{P}} \left\{1 \;-\; \exp\int_0^{T_p} [A(\mathbf{\Phi}^t\mathbf{X}_p) - s]dt\right\}^{-1}, \tag{4.90}$$

where $\mathbf{X}_p$ is a point of the periodic orbit p. The Ruelle zeta function generalizes the topological zeta function (2.56) which is recovered when the observable A vanishes

$$\zeta(s;A=0) \;=\; \zeta_{\text{top}}(s)\,. \tag{4.91}$$

On the other hand, the Ruelle zeta function of the Liouvillian dynamics, which is given by the factorization of the Zeta function (3.151) for $M_u = 1$

as $Z(s) = \prod_{m=0}^{\infty} \zeta_m(s)^{-m-1}$, is recovered for the local dispersion rate taken as the observable A

$$\zeta(s; A = -u) \;=\; \zeta_0(s) \, . \tag{4.92}$$

4.5.2 The topological pressure as an eigenvalue

The analytic properties of the Ruelle zeta function (4.90) can be obtained by considering the Zeta function (4.89) as the Fredholm determinant (3.176)

$$Z(s; A) \;=\; \mathrm{Det}(\hat{I} - \hat{K}_{s,A}) \;=\; \exp - \sum_{n=1}^{\infty} \frac{1}{n} \,\mathrm{Tr}\, \hat{K}_{s,A}^{n} \, , \tag{4.93}$$

of the Frobenius–Perron operator of the Poincaré map $\boldsymbol{\phi}$

$$\hat{K}_{s,A}\, v(\mathbf{x}) \;=\; \int \delta(\mathbf{x} - \boldsymbol{\phi}\mathbf{y}) \, \exp\big[U(\mathbf{y}) + a(\mathbf{y}) - sT(\mathbf{y})\big] \, v(\mathbf{y}) \, d\mathbf{y} \, , \tag{4.94}$$

defined like (3.198) with

$$U(\mathbf{x}) \;=\; \int_0^{T(\mathbf{x})} u[\boldsymbol{\Phi}^{\tau} \circ \mathbf{G}(\mathbf{x}, 0)] \, d\tau \, . \tag{4.95}$$

The trace of the nth iterate of the operator (4.94) can be evaluated with the same method as Eq. (3.175). The denominator can be approximated in terms of the dispersion rate by

$$|\det[1 - \partial_{\mathbf{x}} \boldsymbol{\phi}^n(\mathbf{x})]| \;\simeq\; \exp \sum_{j=0}^{n-1} U(\boldsymbol{\phi}^j \mathbf{x}) \, . \tag{4.96}$$

A comparison with the partition function (4.12) shows that this trace has the following leading exponential behaviour

$$\begin{aligned} \mathrm{Tr}\, \hat{K}_{s,A}^{n} \;&=\; \sum_{\mathbf{x} \in \mathrm{Fix}\boldsymbol{\phi}^n} \frac{\exp\Big\{\sum_{j=1}^{n} \big[U(\boldsymbol{\phi}^j\mathbf{x}) + a(\boldsymbol{\phi}^j\mathbf{x}) - sT(\boldsymbol{\phi}^j\mathbf{x})\big]\Big\}}{|\det[1 - \partial_{\mathbf{x}} \boldsymbol{\phi}^n(\mathbf{x})]|} \\ &\sim\; \exp(-sT_n) \sum_{\mathbf{X} \in \mathrm{Fix}\boldsymbol{\phi}^n} \exp \int_0^{T_n} A(\boldsymbol{\Phi}^{\tau}\mathbf{X}) \, d\tau \, , \\ &\sim\; \exp(-sT_n) \, \mathscr{Z}(\epsilon, T_n/2, A) \, , \end{aligned} \tag{4.97}$$

where T_n is an estimation of the mean time taken to perform n intersections with the Poincaré surface of section. According to the definition (4.11), we may conclude that the leading pole $s_1(A)$ of the Ruelle zeta function (4.90) for a *real* function A is the topological pressure:

$$s_1(A) \;=\; \mathscr{P}(A) \, . \tag{4.98}$$

For general functions A, the zeta function (4.90) is holomorphic for

$$\mathrm{Re}\, s \;>\; \mathscr{P}(\mathrm{Re}A)\,. \tag{4.99}$$

We notice that, if the function A is complex, a pole is not necessarily located at the real value $\mathscr{P}(\mathrm{Re}A)$.

When $A = 0$, Eq. (4.16) confirms that the leading pole of the topological zeta function is the topological entropy

$$s_1(A=0) \;=\; \mathscr{P}(A=0) \;=\; P(\beta=0) \;=\; h_{\mathrm{top}}(\mathbf{\Phi}, \mathscr{I})\,, \tag{4.100}$$

[cf. Eq. (2.62)].

The eigenvalues of the particular operator (4.86) are given by the zeros of the Zeta function

$$Z(s;\beta) \;=\; \prod_p \prod_{m=0}^{\infty} \left(1 \;-\; \frac{e^{-sT_p}}{|\Lambda_p|^{\beta}\Lambda_p^m}\right)^{m+1} \;=\; \prod_{m=0}^{\infty} \frac{1}{\zeta_m(s;\beta)^{m+1}}\,. \tag{4.101}$$

The invariant measures μ_β can thus be calculated as a sum over periodic orbits using (3.204) with the Zeta function (4.101). The leading factor of (4.101) which determines the leading eigenstate studied in the large-deviation formalism is the inverse of the Ruelle zeta function

$$\zeta_0(s;\beta) \;=\; \zeta(s;A=-\beta u) \;=\; \prod_p \left(1 \;-\; \frac{e^{-sT_p}}{|\Lambda_p|^{\beta}}\right)^{-1}\,. \tag{4.102}$$

It is remarkable that this zeta function interpolates between the topological zeta function (4.91) at $\beta = 0$ and the Ruelle zeta function (4.92) of the Liouvillian dynamics at $\beta = 1$. According to (4.98), the leading eigenvalue of (4.86), i.e., the leading pole of the Ruelle zeta function (4.102) can be identified with the topological pressure function (4.25)

$$s_1(\beta) \;=\; P(\beta)\,, \tag{4.103}$$

which provides a powerful method to evaluate the pressure function in terms of the periodic orbits of the repeller, for instance, with a transfer matrix like (3.178) or a cycle expansion like (3.203).

We should here point out that the pressure function can also be obtained using the entire Fredholm determinants introduced by Cvitanović and Vattay (1993).

4.6 Probabilistic Markov chains and lattice gas automata

The large-deviation formalism can also be applied to exact or approximate Markov chain models, in order to characterize the chaotic properties of these dynamical systems with the quantities defined above. The large-deviation formalism can be expressed in terms of the probabilistic Markov chain itself, as we present below.

4.6.1 Markov chain models

Many dynamical systems can be modelled in terms of Markov chains. Indeed, the equivalence of deterministic systems with random processes like coin tossing suggests that some dynamical systems may be isomorphic to Markov chains. This is the case for model systems like the r-adic maps of the interval, the baker map, or the Arnold cat map, where finite Markov partitions can be constructed so that the dynamics becomes strictly equivalent not only to a topological Markov chain but also to a probabilistic Markov chain (Seidel 1933, Adler and Weiss 1970, Ornstein 1974). Accordingly, the random process equivalent to the deterministic map shows a loss of correlation after two iterations, which is characteristic of Markov chains. However, we should emphasize that, in general, hyperbolic systems can only be approximated by probabilistic Markov chains constructed *on finer and finer* Markov partitions.

On the other hand, in view of the generality of hyperbolic behaviour, the above results suggest modelling systems of statistical mechanics with probabilistic Markov chains, which is the purpose of the lattice gas automata (Dab et al. 1991, Lawniczak et al. 1991, Kapral et al. 1992, Ernst 1992). Lattice gas automata are probabilistic Markov chains over system states defined by the set of discrete states taken by individual particles. The single particle state for particle i, ξ_i takes one among m integers which determine its position and velocity. The collection of all the single-particle states $(\xi_1,\xi_2,\xi_3,\ldots,\xi_N) = \omega$ defines a state of the whole system. As a consequence the total number of possible system states is m^N. This set of possible system states plays the role of an alphabet $\mathscr{A} = \{1,2,3,\ldots,m^N\}$ of a symbolic dynamics $\mathscr{M} = \Sigma_{\mathscr{A}}$ in which the trajectories of the system correspond to bi-infinite sequences of states

$$\boldsymbol{\omega}_{.} = \ldots\omega_{-2}\omega_{-1}\cdot\omega_0\omega_1\omega_2\ldots \,. \tag{4.104}$$

The successive symbols ω_n give the state of the system at successive times $t_n = n\Delta t$. The set of all bi-infinite sequences defines the phase space $\Sigma_{\mathscr{A}}$ of the lattice gas automaton. This so-defined phase space is a continuum, as it should be. For example a simple automaton with an alphabet of two symbols, 0 and 1, and a set of trajectories consisting of all bi-infinite sequences of these two symbols can be mapped onto a unit square, as is done in the baker's transformation. We remark that a symbolic dynamics establishes a one-to-one correspondence between the system trajectories and the bi-infinite sequences.

A Markov chain is fully characterized by the transition matrix P of the random process on the system states. The elements $P_{\omega\omega'}$ of the matrix P give the probabilities of transition between two successive states ω and ω'. The Markov chain admits an invariant vector $\{p_\omega\}$ giving the probabilities of finding the system in the states $\omega \in \{1, 2, \ldots, L\}$. Since probability is conserved we infer that

$$\sum_{\omega'} P_{\omega\omega'} = 1, \qquad \text{and} \qquad \sum_{\omega} p_\omega P_{\omega\omega'} = p_{\omega'} \, , \tag{4.105}$$

with $\sum_\omega p_\omega = 1$ (Billingsley 1965, Feller 1968). We may assume that the Markov chain is ergodic so that the invariant vector is unique. The invariant probability measure is then defined as

$$\mu_{\mathscr{A}}(\omega_0\omega_1 \ldots \omega_{n-1}) = p_{\omega_0} P_{\omega_0\omega_1} P_{\omega_1\omega_2} \ldots P_{\omega_{n-2}\omega_{n-1}} \, . \tag{4.106}$$

Bernoulli chains are particular Markov chains for which $P_{\omega\omega'} = p_{\omega'}$.

4.6.2 Isomorphism between Markov chains and area-preserving maps

If a deterministic dynamical system can be modelled by a Markov chain, on the other hand, an area-preserving map can be constructed which is isomorphic to a given Markov chain (Gaspard and Wang 1993).

Let us first prove the simpler statement that a Bernoulli chain is equivalent to an area-preserving map of the unit square. We suppose that the unit square is divided into L vertical rectangles $\mathscr{R}_\omega$ of area p_ω and corresponding to the L states of the chain (see Fig. 4.3). The map acts on these rectangles by horizontal stretching into horizontal rectangles $\phi(\mathscr{R}_\omega)$ of unit width in such a way that the

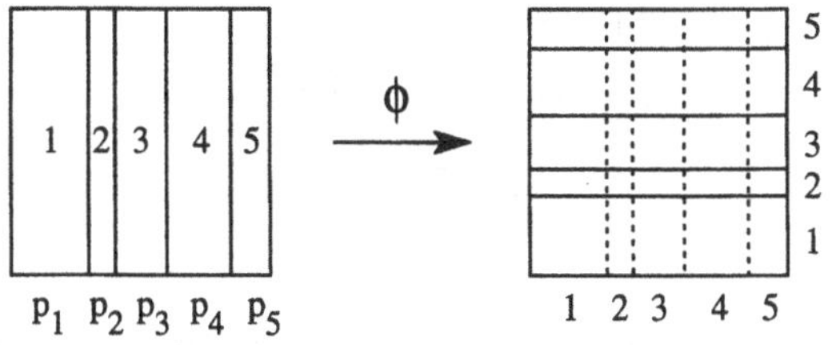

Figure 4.3 Two-dimensional map of the unit square which is isomorphic to the Bernoulli process of probabilities $\{p_\omega\}$ (Gaspard and Wang 1993).

area of each of them is preserved. The height of the horizontal rectangle $\phi(\mathcal{R}_\omega)$ is therefore p_ω. The first horizontal rectangle is placed at the bottom of the square. The next one on top of it and so on. Let us denote by $\mathcal{R}_{\omega'}$ the vertical rectangles on the right-hand square which correspond to the states ω' after one iteration ϕ. The joint probability of visiting successively the states ω and ω' is the area $a_{\omega\omega'}$ of the rectangle $(\omega, \omega') = \phi(\mathcal{R}_\omega) \cap \mathcal{R}_{\omega'}$. This rectangle (ω, ω') has a width p_ω and a height $p_{\omega'}$. Whereupon, the transition probability from the state ω to the state ω' is

$$P_{\omega\omega'} = \frac{a_{\omega\omega'}}{p_\omega} = \frac{p_\omega \, p_{\omega'}}{p_\omega} = p_{\omega'} , \tag{4.107}$$

which agrees with the definition of the Bernoulli chain.

The construction is similar for the Markov chains (see Fig. 4.4). Vertical rectangles $\mathcal{R}_\omega$ of unit height are associated with the different states ω. The areas of these rectangles are given by the invariant probabilities $\{p_\omega\}$, which are thus equal to their widths. The vertical division lines of the rectangles are reported on the right-hand square where they now represent the states ω' after the transition. Let us consider the transition from the state $\omega = 1$ to the state ω'. Since the transition probabilities $P_{\omega\omega'}$ are now depending on both ω and ω', the stretching can no longer be uniform within the rectangle $\mathcal{R}_1$. This vertical rectangle $\mathcal{R}_1$ is therefore mapped on a series of L small rectangles, each one of them occupying the bottom of each vertical rectangle $\mathcal{R}_{\omega'}$ but with different heights that we must now determine. The area of the rectangle $(\omega, \omega') = \phi(\mathcal{R}_\omega) \cap \mathcal{R}_{\omega'}$ divided by the area of the rectangle $\mathcal{R}_\omega$ should be equal to the transition probability $P_{\omega\omega'}$ as for the Bernoulli chains

$$P_{\omega\omega'} = \frac{a_{\omega\omega'}}{p_\omega} . \tag{4.108}$$

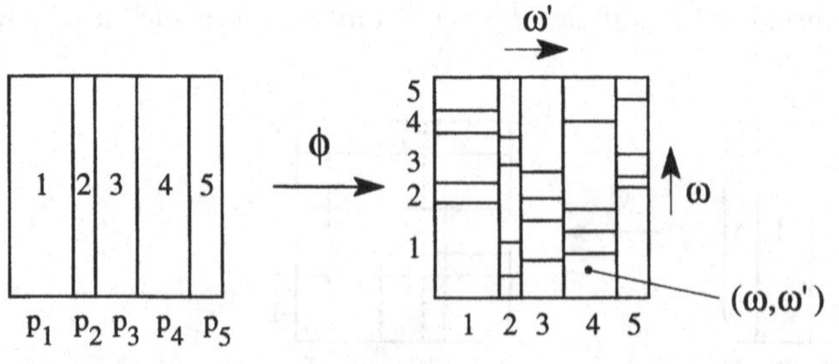

Figure 4.4 Two-dimensional map of the unit square which is isomorphic to a Markov chain of probabilities $\{p_\omega\}$ and of transition matrix $P_{\omega\omega'}$ (Gaspard and Wang 1993).

Since the rectangle (ω, ω') has a width equal to $p_{\omega'}$ by our previous construction, its height should be equal to

$$\frac{a_{\omega\omega'}}{p_{\omega'}} = \frac{p_\omega}{p_{\omega'}} P_{\omega\omega'} , \tag{4.109}$$

which ends the construction of the equivalent area-preserving map. Summing over the heights of all the rectangles (ω, ω') in the same vertical rectangle ω', we must recover the unity

$$\sum_\omega \frac{p_\omega}{p_{\omega'}} P_{\omega\omega'} = 1 , \tag{4.110}$$

which is the case according to the definition (4.105) of the invariant probability vector.

The preceding construction shows that every Markov chain on a countable number of states is isomorphic to a deterministic area-preserving map of the unit square. This result is the converse of a series of results obtained by Sinai, Ornstein, and others who showed that certain uniformly hyperbolic mappings like the baker transformation or the Arnold cat are isomorphic to a Markov chain (Cornfeld et al. 1982, Ornstein 1974, Adler and Weiss 1970).

Thanks to the isomorphism, we can establish a connection between some properties of the Markov chain and those of a deterministic dynamical system. However, the dimensionality is lost in this connection. As a consequence, geometric properties of dimension which are typical of differentiable dynamical systems cannot be recovered. Nevertheless, we are interested here in the quantities like Eq. (4.64) which only contains the sum of positive Lyapunov exponents, a global quantity which does not require the knowledge of individual Lyapunov exponents. Such quantities can be defined for Markov chains like lattice gas automata (Dorfman et al. 1995, Gaspard and Dorfman 1995).

4.6.3 Repellers of Markov chains

Although the dynamics of the Markov chain (4.106) takes place within the full set of states, $\mathscr{A}$, there may exist a subset of states, $\mathscr{B} \subset \mathscr{A}$, which are visited in a transient way (Gaspard 1992a, 1993b). This might happen, for instance, if we impose an absorbing boundary condition (3.22) on the system. If the initial state belongs to this subset, the state will escape from this subset after a finite time for almost all of the trajectories Eq. (4.104). Nevertheless there exist trajectories which remain forever on the subset $\mathscr{B}$. These trapped trajectories are exceptional in the sense that their measure, Eq. (4.106), vanishes and consequently they form a repeller $\mathscr{I}_{\mathscr{B}} = \Sigma_{\mathscr{B}}$. The first time at which a given trajectory escapes from the subset $\mathscr{B}$ defines a problem of first passage. The dynamics of the full system with respect to the repeller is similar to a scattering process where the trajectory visits, for a while, a vicinity of the repeller. In this image, where a trajectory is going in and out of the subset $\mathscr{B}$, the repeller may be considered as a predefined nonequilibrium fluctuation.

Since the dynamics is transient on the repeller there exists an escape rate γ. We can calculate this escape rate by constructing the Markov subchain between states of the subset $\mathscr{B}$. The values of the transition probabilities are given by a submatrix which is contained in the matrix P of the full Markov chain Eq. (4.105). This submatrix Q is composed of the elements of P between the states of the subset $\mathscr{B}$

$$\mathsf{P} = \begin{matrix} & \mathscr{A}\setminus\mathscr{B} & \mathscr{B} \\ \mathscr{A}\setminus\mathscr{B} & * & * \\ \mathscr{B} & * & \mathsf{Q} \end{matrix} , \tag{4.111}$$

where $*$ denotes other submatrices.

Because it is obtained by truncating the full matrix P, the submatrix Q is not a stochastic matrix obeying Eq. (4.105). In particular, the leading eigenvalue of Q is no longer 1 but is smaller than 1. Actually the eigenvalues of the submatrix Q give the decay rates out of the subset $\mathscr{B}$ so that the leading eigenvalue gives the escape rate γ according to

$$\mathsf{Q} \cdot \mathbf{v} = \exp(-\gamma)\, \mathbf{v} , \qquad \text{and} \qquad \mathbf{u}^{\mathrm{T}} \cdot \mathsf{Q} = \exp(-\gamma)\, \mathbf{u}^{\mathrm{T}} , \tag{4.112}$$

in terms of the leading right and left eigenvectors of Q. The escape rate is here in units of the time steps, Δt, of the automaton. If Q is a nonnegative, irreducible, aperiodic matrix then according to the Frobenius–Perron theorem all of the other eigenvalues are less in absolute value than the largest eigenvalue, $\exp(-\gamma)$ (Gantmacher 1959).

Using the right and left eigenvectors we can construct the Markov subchain of the repeller dynamics in terms of the following matrix and its associated invariant vector

$$\Pi_{\omega\omega'} = \exp(\gamma) Q_{\omega\omega'} \frac{v_{\omega'}}{v_\omega} , \qquad \text{and} \qquad \pi_\omega = \frac{u_\omega v_\omega}{\mathbf{u}\cdot\mathbf{v}} , \tag{4.113}$$

which is now stochastic since

$$\sum_{\omega'\in\mathscr{B}} \Pi_{\omega\omega'} = 1, \qquad \sum_{\omega\in\mathscr{B}} \pi_\omega \Pi_{\omega\omega'} = \pi_{\omega'}, \qquad \text{and} \qquad \sum_{\omega\in\mathscr{B}} \pi_\omega = 1 \, . \tag{4.114}$$

This stochastic matrix defines an invariant measure on the repeller which is given by

$$\mu_{\mathscr{B}}(\omega_0\omega_1 \ldots \omega_{n-1}) = \pi_{\omega_0} \Pi_{\omega_0\omega_1} \Pi_{\omega_1\omega_2} \ldots \Pi_{\omega_{n-2}\omega_{n-1}} , \tag{4.115}$$

where $\omega_j \in \mathscr{B}$. This new invariant measure $\mu_{\mathscr{B}}$ gives nonvanishing probabilities only for trajectories staying on the repeller.

4.6.4 Large-deviation formalism of Markov chains

We now have the necessary elements to proceed with a derivation of the formula (4.64). The escape rate γ was already defined in terms of the leading eigenvalue of the submatrix Q. On the other hand, the KS entropy of the dynamics on the repeller is defined as the KS entropy per time step Δt of the Markov chain Eqs. (4.114) and (4.115) (Cornfeld et al. 1982)

$$h_{\text{KS}}(\mu_{\mathscr{B}}) = - \sum_{\omega,\omega' \in \mathscr{B}} \pi_\omega \Pi_{\omega\omega'} \ln \Pi_{\omega\omega'} \ . \tag{4.116}$$

We need to find the quantity which plays the role of the sum of the positive Lyapunov exponents, $\sum_{\lambda_i>0} \lambda_i$. It is not obvious at first sight how to do this. The solution can be found by mapping a Markov chain onto a deterministic map as explained above, and then computing the sum of the positive Lyapunov exponents for the deterministic map (Dorfman et al. 1995, Gaspard and Dorfman 1995). One finds in this way that the inverses of the probabilities, $Q_{\omega\omega'}$, for the separate steps of the Markov chain play the role of the stretching factors by which trajectories separate in the map or by which the probability is dispersed in the Markov chain. The inverse of the dispersion factor for the trajectories visiting successively the states $\omega_0\omega_1\omega_2\ldots\omega_{n-1}$ is therefore given by

$$\exp -U(\omega_0\omega_1\ldots\omega_{n-1}) = Q_{\omega_0\omega_1} Q_{\omega_1\omega_2} \cdots Q_{\omega_{n-2}\omega_{n-1}} \ . \tag{4.117}$$

The role of the sum of the Lyapunov exponents is then played by the average quantity

$$\mu_{\mathscr{B}}(u) = \lim_{n\to\infty} \frac{1}{n} \sum_{\omega_0,\ldots,\omega_{n-1} \in \mathscr{B}} \mu_{\mathscr{B}}(\omega_0\ldots\omega_{n-1}) \ \ln \exp U(\omega_0\ldots\omega_{n-1}) \ , \tag{4.118}$$

with the definition Eq. (4.117). Equations (4.117)–(4.118) define for Markov chains – and in particular for lattice gas automata – the analogue of the dispersion rates given by Eq. (1.48) for systems with smooth potentials and by Eq. (1.128) for hard spheres.

Using the factorization property of the dispersion factors, Eq. (4.117), and the Markov property of the measure Eqs. (4.114) and (4.115), we obtain

$$\mu_{\mathscr{B}}(u) = \sum_{\omega,\omega' \in \mathscr{B}} \pi_\omega \Pi_{\omega\omega'} \ln \frac{1}{Q_{\omega\omega'}} \ , \tag{4.119}$$

which is positive since the matrix elements are smaller than one: $Q_{\omega\omega'} < 1$. The dispersion rate u has the same units, Δt^{-1}, as the escape rate γ and the

KS entropy, Eq. (4.116). The isomorphism between the Markov chain and a deterministic, area-preserving map shows that the sum of the positive Lyapunov exponents for the deterministic systems is identical to $\mu_{\mathscr{B}}(u)$ given above,

$$\mu_{\mathscr{B}}(u) = \sum_{\lambda_i > 0} \mu_{\mathscr{B}}(\lambda_i) , \tag{4.120}$$

both calculated on the repeller $\Sigma_{\mathscr{B}}$. This last property guarantees that the averaged quantity u is the unique analogue of the sum of the positive mean Lyapunov exponents for lattice gas automata.

Replacing the definitions Eq. (4.113) of the matrix $\mathbf{\Pi}$ and of the invariant vector $\boldsymbol{\pi}$ in the expression Eq. (4.116) for the KS entropy, we obtain the identity

$$\gamma = \mu_{\mathscr{B}}(u) - h_{\mathrm{KS}}(\mu_{\mathscr{B}}) , \tag{4.121}$$

which is the escape-rate formula (4.64) for Markov chains and lattice gas automata. This formula gives the escape rate from the repeller $\Sigma_{\mathscr{B}}$ as the difference between an average dispersion rate playing the role of the sum of the positive Lyapunov exponents and the KS entropy.

Similarly, we have the analogue of the Pesin theorem. When we relax the constraint on the subset of allowed states, then $\mathscr{A} = \mathscr{B}$, the repeller becomes the full phase space, and $\mathsf{Q} = \mathsf{P}$. As a consequence, the escape rate vanishes and

$$\mu_{\mathscr{A}}(u) = h_{\mathrm{KS}}(\mu_{\mathscr{A}}) . \tag{4.122}$$

We can also define the dynamical pressure function according to

$$P(\beta) = \lim_{n\to\infty} \frac{1}{n} \ln \sum_{\omega_0,\ldots,\omega_{n-1}\in\mathscr{B}} (Q_{\omega_0\omega_1} Q_{\omega_1\omega_2} \cdots Q_{\omega_{n-2}\omega_{n-1}})^{\beta} . \tag{4.123}$$

This pressure function has all of the properties described in Section 4.2, which relate it to the preceding quantities like the escape rate, the dispersion rate u and the KS entropy

$$\gamma = -P(1) , \qquad \mu_{\mathscr{B}}(u) = -P'(1) , \qquad \text{and} \qquad h_{\mathrm{KS}}(\mu_{\mathscr{B}}) = P(1) - P'(1) . \tag{4.124}$$

It is a straightforward calculation to show that the quantities γ, $\mu_{\mathscr{B}}(u)$, and $h_{\mathrm{KS}}(\mu_{\mathscr{B}})$ derived from the pressure function are identical with those given by Eqs. (4.112), (4.116), and (4.119), respectively. This is accomplished by using the spectral decomposition of the matrix Q, in terms of its eigenvalues and eigenvectors. If this matrix is irreducible and aperiodic the dominant behaviour of Q^n, for large n is determined by the largest eigenvalue and the associated left and right eigenvectors (Gantmacher 1959, Feller 1968). The correspondence of the two kinds of expressions for the dynamical quantities then follows immediately.

Moreover, the analogue of the topological entropy per time step Δt is given by

$$h_{\text{top}}(\mathscr{B}) = P(0) \,. \tag{4.125}$$

The topological entropy is the largest eigenvalue of the topological transition matrix associated with the Markov subchain $\mathbf{\Pi}$, i.e., the matrix with elements 0 or 1 depending upon whether $\Pi_{\omega\omega'}$ is zero or positive. If the subchain matrix Q is a strictly positive $L_{\mathscr{B}} \times L_{\mathscr{B}}$ matrix with $L_{\mathscr{B}} \leq m^N$, then the topological entropy is equal to $h_{\text{top}}(\mathscr{B}) = \ln L_{\mathscr{B}} \leq N \ln m$.

4.7 Special fractals generated by uniformly hyperbolic maps

The process of escape out of a finite region in phase space generates a fractal repeller as we have seen above. The escape process can be considered as a problem of first passage by assuming that the boundary of the region is absorbing. However, there are other conditions which may be imposed on the way trajectories are selected (Gaspard 1992a). We shall here consider two such conditions in closed systems. We suppose that the system is uniformly hyperbolic and has only one positive Lyapunov exponent λ which is the same for all the trajectories. This is the case for uniformly expanding one-dimensional maps of the interval and for uniformly hyperbolic area-preserving maps. We moreover suppose that the map is isomorphic to a probabilistic Markov chain with a topological transition matrix $A_{\omega\omega'}$.

4.7.1 Condition on the mean sojourn time in a domain

A nonequilibrium constraint can be considered where the particle is allowed to leave a domain U for a while and to come back repetitively later so that the mean sojourn time in U is a given fraction $0 \leq \alpha \leq 1$ of the total time of the experiment.

Let U be an arbitrary domain in the phase space $\mathscr{M}$ and $I_U(X)$ be its indicator function. A fractal repeller is defined as

$$\mathscr{I}(\alpha) = \left\{ \mathbf{x} \in \mathscr{M} : \frac{1}{n} \sum_{j=0}^{n-1} I_U(\boldsymbol{\phi}^j \mathbf{x}) \to \alpha \,, \quad \text{for } n \to \infty \right\} . \tag{4.126}$$

We introduce the set

$$\Upsilon_n(\alpha) = \left\{ \mathbf{x} \in \mathcal{M} : \frac{1}{n} \sum_{j=0}^{n-1} I_U(\boldsymbol{\phi}^j \mathbf{x}) \in [\alpha , \alpha + d\alpha[\right\} , \tag{4.127}$$

and we define a nonpositive function $F(\alpha)$ by

$$\mu\big[\Upsilon_n(\alpha)\big] = C(\alpha;n) \exp\big[nF(\alpha)\big] \, d\alpha , \tag{4.128}$$

where μ is the invariant probability measure of the system which is the Lebesgue measure in the present case of uniformly hyperbolic maps. $C(\alpha;n)$ is the prefactor which is a slowly varying function of the time n and which plays no essential role in the following considerations. From the definition (4.128), $F(\alpha)$ appears to be minus the escape rate from the set $\mathcal{I}(\alpha)$ of trajectories

$$\gamma = - F(\alpha) . \tag{4.129}$$

A generating function $G(\eta)$ is defined by

$$G(\eta) \equiv \lim_{n\to\infty} \frac{1}{n} \ln \langle \exp \eta \sum_{j=0}^{n-1} I_U(\boldsymbol{\phi}^j \mathbf{x}) \rangle , \tag{4.130}$$

where $\langle\cdot\rangle$ denotes the average over the equilibrium measure μ. The generating function is related to $F(\alpha)$ by a Legendre transform according to

$$G(\eta) = F(\alpha) + \eta \, \alpha , \tag{4.131}$$

where

$$\eta = - \frac{dF}{d\alpha} , \quad \text{or} \quad \alpha = \frac{dG}{d\eta} . \tag{4.132}$$

The topological pressure of the set $\mathcal{I}(\alpha)$ of trajectories is defined by Eq. (4.123) where the transition matrix is here given by $Q_{\omega\omega'} = \exp(-\lambda) A_{\omega\omega'}$ so that

$$P(\beta) = \lim_{n\to\infty} \frac{1}{n} \ln \sum_{\omega_0 \cdots \omega_{n-1} \in \Upsilon_n(\alpha)} \exp(-\lambda\beta n) \, A_{\omega_0\omega_1} \, A_{\omega_1\omega_2} \cdots A_{\omega_{n-2}\omega_{n-1}} . \tag{4.133}$$

Since the pressure at $\beta = 1$ is equal to the escape rate which here takes the value (4.129) we obtain the relation

$$P(\beta) = F(\alpha) + (1 - \beta) \, \lambda . \tag{4.134}$$

We emphasize that the parameter α in (4.134) characterizes the fractal set (4.126). According to the formula (4.72), the KS entropy is given by

$$h_{\text{KS}} = \lambda + F(\alpha) , \tag{4.135}$$

and the partial Hausdorff dimension by

$$d_{\mathrm{H}} = 1 + \frac{F(\alpha)}{\lambda} . \tag{4.136}$$

The previous problem was studied by Eggleston (1949) and Billingsley (1965) for the dyadic map of real numbers of the unit interval: $x_{n+1} = (2x_n)$ modulo 1. This system has a Lyapunov exponent $\lambda = \ln 2$ and is known to be isomorphic to a Bernoulli process of probabilities $(1/2, 1/2)$. A Markov partition is constructed by dividing the unit interval in two equal intervals with corresponding symbols $\{0, 1\}$. This partition induces the binary expansion of the initial conditions of the points of the interval

$$x_0 = \sum_{n=0}^{\infty} \frac{\omega_n}{2^{n+1}} . \tag{4.137}$$

The system is thus a probabilistic Markov chain (4.105) with

$$\mathsf{P} = \begin{pmatrix} 1/2 & 1/2 \\ 1/2 & 1/2 \end{pmatrix} , \quad \text{and} \quad \mathbf{p} = \begin{pmatrix} 1/2 \\ 1/2 \end{pmatrix} . \tag{4.138}$$

Eggleston and Billingsley have then considered the fractal of initial conditions which have a binary expansion with a fraction α of symbol $\omega_n = 1$. This fractal is defined by (4.126) with the indicator function of the interval $U = [1/2, 1]$. According to the Markov property (4.106), the generating function (4.130) can be calculated as

$$\begin{aligned} G(\eta) &= \lim_{n\to\infty} \frac{1}{n} \ln \sum_{\omega_0\omega_1\cdots\omega_{n-1}} \mu(\omega_0\omega_1\cdots\omega_{n-1}) \exp\left[\eta \sum_{j=0}^{n-1} I_U(\omega_j)\right] \\ &= \lim_{n\to\infty} \frac{1}{n} \ln \operatorname{tr} \mathsf{r}\, \mathsf{R}^{n-1} , \end{aligned} \tag{4.139}$$

with

$$\mathsf{R}_{\omega\omega'} = P_{\omega\omega'} \exp\left[\eta\, I_U(\omega')\right] = A_{\omega\omega'} \exp\left[\eta\, I_U(\omega') - \lambda\right] , \tag{4.140}$$

$$\mathsf{r}_{\omega\omega'} = p_{\omega'} \exp\left[\eta\, I_U(\omega')\right] , \tag{4.141}$$

hence

$$G(\eta) = \ln \chi(\eta) , \tag{4.142}$$

where $\chi(\eta)$ is the largest eigenvalue of the matrix R.

For the Eggleston–Billingsley problem, the matrix (4.140) is given by

$$\mathsf{R} = \begin{matrix} & \begin{matrix} 0 & \quad 1 \end{matrix} \\ \begin{matrix} 0 \\ 1 \end{matrix} & \begin{pmatrix} 1/2 & e^{\eta}/2 \\ 1/2 & e^{\eta}/2 \end{pmatrix} \end{matrix} , \tag{4.143}$$

which has the eigenvalues 0 and $(e^\eta + 1)/2$. The generating function (4.142) is thus

$$G(\eta) = \ln \frac{1+e^\eta}{2} , \tag{4.144}$$

and its Legendre transform (4.131)–(4.132) gives the escape rate as

$$\gamma = - F(\alpha) = \alpha \ln \alpha + (1-\alpha) \ln (1-\alpha) + \ln 2 . \tag{4.145}$$

These functions are drawn in Fig. 4.5. According to (4.136), the Hausdorff dimension of the Eggleston–Billingsley fractal $\mathscr{I}(\alpha)$ is then

$$d_{\mathrm{H}} = \frac{1}{\ln 2} \left[-\alpha \ln \alpha - (1-\alpha) \ln(1-\alpha)\right] . \tag{4.146}$$

The KS entropy of $\mathscr{I}(\alpha)$ is then

$$h_{\mathrm{KS}} = - \alpha \ln \alpha - (1-\alpha) \ln (1-\alpha) . \tag{4.147}$$

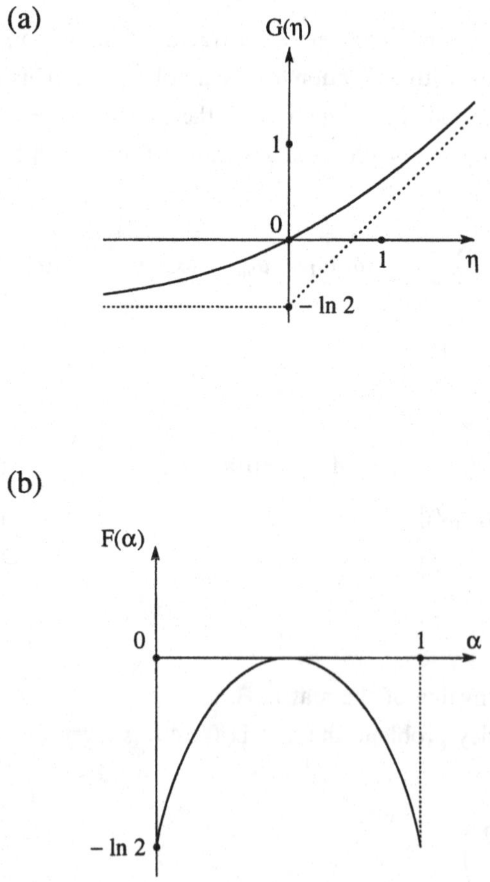

Figure 4.5 (a) Generating function $G(\eta)$ of the residence time in the interval $1/2 \le x < 1$ for the map $x \to 2x$ (mod. 1) of the unit interval; (b) Legendre transform $F(\alpha)$ of $G(\eta)$ giving the escape rate out of the fractal set of trajectories spending a fraction α of time in the interval $1/2 \le x < 1$.

4.7.2 Fractal repeller generated in trajectory reconstruction

We consider a further example of a fractal repeller generated by nonequilibrium constraints on the dynamics of a uniformly hyperbolic map of Lyapunov exponent λ (Gaspard 1992a). We suppose that a measuring device observes the time evolution of the system as described in Section 4.1. The measuring device has a finite resolution and resolves the system with a partition $\mathscr{C}$ of phase space $\mathscr{M}$ into cells labelled by integers $\omega_n \in \mathscr{A}$. The cells are here arbitrary. The trajectory $\boldsymbol{\phi}^n\mathbf{x}$ from the initial condition $\mathbf{x}$ will produce an arbitrarily long sequence (4.1) which appears as the recorded data from the measuring device. Once the observation has been made we would like to reconstruct the trajectory of the system from the observed data (4.1). We expect that this reconstruction will be possible only if the observation is carried out with a high enough resolution, i.e., if the partition is sufficiently fine. Otherwise, some ambiguity will arise in the reconstruction of the sequence $\boldsymbol{\omega}$ and we shall obtain a set of trajectories rather than a single trajectory.

The question is thus the following. Given the partition $\mathscr{C}$, what is the Hausdorff dimension of the set of trajectories $\mathscr{I}_{\boldsymbol{\omega}}$ which correspond to the recorded sequence $\boldsymbol{\omega}$? If the partition $\mathscr{C}$ is generating, we know that a unique trajectory corresponds to $\boldsymbol{\omega}$ and the Hausdorff dimension is then zero.[8]

According to the Shannon–McMillan–Breiman theorem (4.10), we observe that the entropy per unit time $h(\mathscr{C})$ of the partition $\mathscr{C}$ with respect to the dynamical system is identical with the escape rate from the set $\mathscr{I}_{\boldsymbol{\omega}}$

$$\gamma(\mathscr{I}_{\boldsymbol{\omega}}) = h(\mathscr{C}) . \tag{4.148}$$

Because the system is uniformly hyperbolic, the pressure function is a linear function of the parameter β which can be fixed by the Lyapunov exponent λ and by Eq. (4.148) as

$$P(\beta) = \lambda\,(1-\beta) - h(\mathscr{C}) , \tag{4.149}$$

so that the KS entropy per unit time of the set $\mathscr{I}_{\boldsymbol{\omega}}$ is

$$h_{\mathrm{KS}}(\mathscr{I}_{\boldsymbol{\omega}}) = \lambda - \gamma(\mathscr{I}_{\boldsymbol{\omega}}) = \lambda - h(\mathscr{C}) . \tag{4.150}$$

Because the system is uniformly hyperbolic, all the generalized dimensions are equal. Assuming that the map is a symplectic area-preserving map, the dimensions of the fractal set selected by the observation of one trajectory is

8. Eventually, a finite number of trajectories could correspond to $\boldsymbol{\omega}$ if the partition has overlapping cells.

equal to twice the partial dimensions obtained as the zero of the pressure function (4.149) so that

$$\dim_{\rm H}(\mathscr{I}_\omega) = \dim_{\rm I}(\mathscr{I}_\omega) = 2 - 2\,\frac{h(\mathscr{C})}{\lambda}\,. \tag{4.151}$$

The set $\mathscr{I}_\omega$ is thus a fractal repeller. The Hausdorff dimension is depicted in Fig. 4.6 as a function of the entropy of the partition.

When the partition $\mathscr{C}$ is generating we have that $h(\mathscr{C}) = \lambda$, $h_{\rm KS}(\mathscr{I}_\omega) = 0$, and $\dim_{\rm H,I}(\mathscr{I}_\omega) = 0$ since $\mathscr{I}_\omega$ contains at most a finite number of trajectories and is thus non-chaotic. In this case, the escape rate reaches its maximum, $\gamma(\mathscr{I}_\omega) = \lambda$.

For a measuring device with a bad resolution, $h(\mathscr{C})$ is close to zero and the device can resolve the dynamics only into fractal sets with a dimension close to the phase-space dimension. On the other hand, we see that the measuring device needs to have a data accumulation rate close to the KS entropy of the system, namely $h_{\rm KS} = \lambda$, to be able to resolve individual trajectories. There is a transition at $h_{\rm KS}$ where the Hausdorff dimension of the selected fractal set drops to zero. In this way, a fractal dimension characterizes the resolving power of the measuring device observing a given dynamical system. This result justifies our interpretation of the KS entropy per unit time given in Section 4.1 where we conclude that the KS entropy is the minimal data accumulation rate allowing the reconstruction of the trajectory from the recorded data without ambiguity. If the partition was coarser and the entropy thus lower the set reconstructed from the recorded data would be a fractal with a nonvanishing dimension so that there would be an ambiguity among an uncountable set of trajectories in

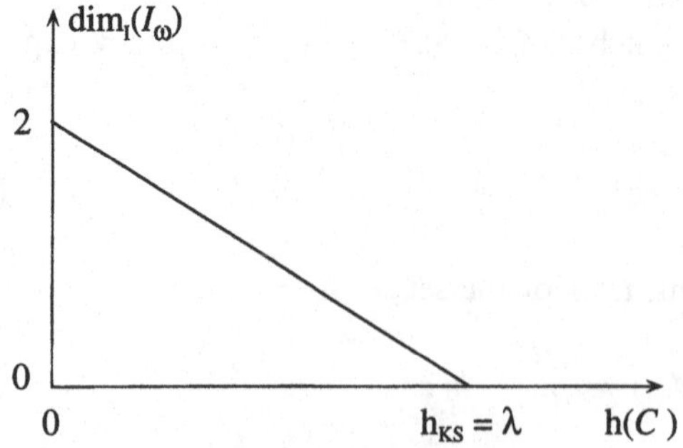

Figure 4.6 Information or Hausdorff dimension of the fractal repeller $\mathscr{I}_\omega$ formed by all the trajectories which visit the same sequence ω of cells of an arbitrary partition $\mathscr{C}$ of phase space, versus the entropy per unit time $h(\mathscr{C})$ of this partition $\mathscr{C}$. The information dimension vanishes when the partition $\mathscr{C}$ is the generating partition in which case $h(\mathscr{C})$ is the KS entropy $h_{\rm KS} = \lambda$ of the mapping.

the answer to the question of which trajectory was actually followed by the system during the record of the sequence (4.1).

4.8 Nonhyperbolic systems

In many cases, the system possesses periodic orbits with vanishing Lyapunov exponents corresponding to nontrivial directions. This is the case, in particular, for the stadium billiard, the Sinai billiard, and the hard-sphere gas with periodic boundary conditions or placed in a rectangular box. The periodic orbits that we have in mind are those special orbits where the particles move without collisions or bounce between parallel walls. These special orbits form sets of zero Lebesgue measure so that they do not prevent the system from being ergodic, or mixing, or from having positive average Lyapunov exponents. However, all the Lyapunov exponents of these special periodic orbits vanish because perturbed trajectories may separate from the periodic orbit in an algebraic way: $\|\mathbf{\Phi}^t(\mathbf{X}_p + \delta\mathbf{X}) - \mathbf{\Phi}^t(\mathbf{X}_p)\| \sim t$. As a consequence, if the system is open and the domain U contains such a marginally unstable periodic orbit, the decay in Eq. (4.50) is nonexponential as

$$\nu_0[\Upsilon_U^{(+)}(t)] \sim \frac{1}{t^{f-1}} , \tag{4.152}$$

where f is the total number of degrees of freedom of the system, and the escape rate vanishes.[9]

The large-deviation formalism is very useful in this context because the pressure function is still nontrivial in this case (Wang 1989). If the Lyapunov exponents vanish in some regions of phase space, some of the dispersion rates $u(\mathbf{Y})$ may be vanishing for a few values of $\mathbf{Y}$ in the (ϵ, T)-separated subsets. For large and positive values of β, the partition function (4.58) is dominated by the few terms with $u(\mathbf{Y}) = 0$ because all the other terms are exponentially vanishing in the limit $T \to \infty$. As a result, the pressure function is equal to zero for large values of β. On the other hand, for negative values of β, the terms with nonvanishing dispersion rates $u(\mathbf{Y})$ dominate the sum in the partition function so that the pressure is then positive and nontrivial. There exists a critical, lowest value of β above which the pressure vanishes

$$P(\beta) = 0 \qquad \text{for} \qquad \beta_c \leq \beta . \tag{4.153}$$

9. We note that this algebraic decay results from purely geometric effects – such as appear when particles travel down long corridors without hitting anything – and thus has no direct connection to the long-time tails which appear in the time correlation function expressions for transport coefficients (Dorfman and Cohen 1970, 1972, 1975; Ernst et al. 1970; Pomeau and Résibois 1975).

Schematic pressure functions are depicted in Figs. 4.2c, 4.2d for nonhyperbolic closed and open systems. For closed systems, we observe that the pressure is zero above $\beta = 1$ which is in agreement with the vanishing of the escape rate due to Eq. (4.152), so that the critical value is $\beta_c = 1$. Moreover, the sum of positive Lyapunov exponents must be defined as the left-sided derivative of the pressure at $\beta = 1$ in the case of a closed system.

The discontinuity in the shape of the pressure function is the evidence of a phenomenon of dynamical phase transitions – so-called in analogy with statistical mechanics (Artuso et al. 1989, Beck and Schlögl 1993). We have the following interpretation. The system is described by the continuous family of invariant measures, μ_β. The parameter β acts like a filtering parameter. When $\beta > \beta_c$, the measure is concentrated on the regular trajectories which have vanishing Lyapunov exponents. For these measures, we can talk about an ordered or regular phase since the corresponding invariant states are ordered. On the other hand, when $\beta < \beta_c$, the measure μ_β gives dominant probability weights to the nonperiodic trajectories which are uncountable. In this case, we can speak of a chaotic or a disordered phase. Tuning the parameter β therefore reveals the chaotic features of the dynamics. This is currently done when we refer to the topological pressure per unit time, i.e., $P(\beta = 0)$, as an indicator of chaos.

However, if the pressure function is analytic away from critical points, it has the further advantage that it can be extrapolated from below criticality up to the value at $\beta = 1$ so as to define a supercritical measure, for instance, in the case of open nonhyperbolic systems. In this way, it is possible to define an effective escape rate, γ_{eff}, as well as an effective value for the sum of positive Lyapunov exponents at $\beta = 1$. The effective rate can be evidenced in numerical simulations from the transient behaviour of the decay function Eq. (4.50). In many nonhyperbolic systems with a large number of degrees of freedom, the power law decay Eq. (4.152) may remain a very small effect which is visible only after extremely long times because marginally unstable periodic orbits are very rare and the decay could appear exponential within statistical errors. In nonhyperbolic systems, although we may know from theoretical arguments that the escape rate defined by Eq. (4.51) actually vanishes, the concept of an effective escape rate is useful to characterize in a rigorous way a numerically observed exponential decay.

Chapter 5

Chaotic scattering

5.1 Classical scattering theory

5.1.1 Motivations

Matter is often studied by scattering with beams of particles such as photons, electrons, neutrons, or others. The quantities of interest are the cross-sections which give the effective surface offered by the target for the realization of a certain scattering event. Scattering processes are usually conceived in a statistical approach. For instance, a cross-section cannot be determined by a single collision but by a statistical ensemble of collisions with a uniform distribution of the incoming impact parameters. In this regard, a natural relation appears between scattering theory and the Liouvillian dynamics.

Many different processes may be considered in scattering theory, for instance elastic or inelastic collisions (Joachain 1975). Among the latter, the reaction processes between molecules or nuclei are of particular importance because they play a crucial role in the transformation of matter. Beside the cross-sections, other important quantities are the reaction rates which characterize the time evolution of statistical ensembles during reactions. The rates have the inverse of a time as unit. We may thus expect that reaction rates belong to the same class of properties as the relaxation rates of Liouvillian dynamics. This is the case, in particular, for unimolecular reactions which are dissociation processes (Gaspard and Rice 1989a, 1989b). The reaction rates can here be assimilated with the lifetimes of the metastable states of the transition complex, i.e., of the transient states formed when the fragments of the reactions are still in interaction. Here also, these lifetimes are essentially statistical properties of the time evolution instead of properties of individual trajectories of the system.

The scattering processes are fundamentally quantum mechanical but in many circumstances the wavelength is so small that the collisions may be considered as classical. If the most famous collision processes such as the Coulomb scattering are essentially integrable motion, on the other hand, the generality of chaotic behaviour in classical mechanics suggests that the majority of collision processes are classically chaotic. We may expect chaotic behaviour as soon as the number of degrees of freedom increases as in inelastic collisions but also if the object which undergoes collisions is composed of several obstacles as for the collisions of particles with molecules or solids. When the scattering becomes chaotic the differential cross-sections may be expected to be very complex functions of the incoming trajectories.

In this chapter, we shall study such classical chaotic scattering processes with simple examples in order to describe their different properties. In this way, we shall show that the chaotic scattering process is classically controlled by sets of trajectories forming hyperbolic repellers in phase space. Such hyperbolic repellers defined in Chapter 2 turn out to be the classical analogue of the transition complex of a reaction. Indeed, we have seen in Chapters 3 and 4 that hyperbolic repellers are characterized by the Pollicott–Ruelle resonances. These classical resonances of the Liouvillian dynamics define the classical lifetimes of the dissociation process which takes place when the particles are still in interaction. A remarkable feature is that strictly hyperbolic sets are very common in scattering systems, especially in direct reactions taking place at high energies well above the barriers. In this way, the unimolecular reaction rate can be defined classically as the escape rate of the hyperbolic repeller. At lower energies, the presence of barriers reduces the escape rate and, possibly, generates nonhyperbolic behaviour. We shall discuss these different aspects and characterize them in terms of the quantities introduced in Chapters 2 to 4.

5.1.2 Classical scattering function and time delay

A typical scattering process is described by a Hamiltonian like

$$H(\mathbf{q},\mathbf{p}) = H_0(\mathbf{q},\mathbf{p}) + V(\mathbf{q}) , \tag{5.1}$$

which becomes a free Hamiltonian H_0 at large distances, i.e., such that the potential vanishes fast enough when the particles are separated:

$$V(\mathbf{q}) \to 0 , \qquad \text{for} \quad \mathbf{q} \to \infty . \tag{5.2}$$

As a consequence, the particles of the collision fly in free motion before and after the collisional events (with possible rotations or vibrations around their centre of mass). These free trajectories are described by the Hamiltonian H_0.[1]

The Hamiltonian H is supposed to conserve energy because we assume that there is no time-dependent external field in the collision region. When the scattering process involves two (or more) beams of particles, linear and angular momenta are also conserved. When scattering occurs between a beam and an infinitely heavy solid, only energy is conserved.

Each of the Hamiltonians H and H_0 defines a flow in the phase space $\mathbf{X} = (\mathbf{q}, \mathbf{p})$, which are denoted by $\mathbf{\Phi}^t$ and $\mathbf{\Phi}_0^t$, respectively. They leave invariant the energy shells $H = E$ and $H_0 = E$, respectively.

A *scattering function* $\mathbf{\Sigma}$ may be defined which maps ingoing trajectories of H_0 onto outgoing trajectories. In analogy with quantum scattering theory this function may be defined by

$$\mathbf{X}_{\text{out}} = \mathbf{\Sigma}(\mathbf{X}_{\text{in}}) = \lim_{t\to\infty} \mathbf{\Phi}_0^{-t/2} \circ \mathbf{\Phi}^t \circ \mathbf{\Phi}_0^{-t/2} (\mathbf{X}_{\text{in}}) , \tag{5.3}$$

(Narnhofer 1980, Narnhofer and Thirring 1981). This function is interpreted as follows. The point $\mathbf{X}_{\text{in}}$ is taken in the collision region of phase space and is supposed to belong to an incoming trajectory of the free Hamiltonian. The purpose of the scattering function (5.3) is to emulate the collision process undergone by this incoming trajectory and hence to determine the corresponding outgoing trajectory. With this aim, the current point $\mathbf{X}_{\text{in}}$ is first evolved backward under the free Hamiltonian during a time $t/2$ corresponding to half a collision. If t is large enough, the point $\mathbf{\Phi}_0^{-t/2}(\mathbf{X}_{\text{in}})$ can be taken as the initial condition of the actual trajectory of the collision under the full Hamiltonian. The final point after the collision will thus be obtained by evolving the initial condition under the full Hamiltonian H during the time t: $\mathbf{\Phi}^t \circ \mathbf{\Phi}_0^{-t/2}(\mathbf{X}_{\text{in}})$. This final point is supposed to be nearby the outgoing trajectory but it does not characterize the outgoing trajectory in the same way as $\mathbf{X}_{\text{in}}$ does for the ingoing trajectory because it is at the end of this trajectory and not in the same region as $\mathbf{X}_{\text{in}}$. To have a comparable point to characterize the outgoing trajectory we have thus to evolve the final point backward under the free Hamiltonian to obtain the point (5.3) which defines the outgoing trajectory (Naisse 1980).

Different coordinates may be considered to define the scattering function. This function defines a map for the collision process. Since the evolution is Hamiltonian, this map is symplectic, conserves energy, and preserves volumes in

1. For further information on cross-sections of classical and quantum collisions, see Joachain (1975).

phase space. We note that a generating function for this map can be taken as the reduced action

$$A = -\int \mathbf{q} \cdot d\mathbf{p} , \tag{5.4}$$

which is well defined for an open trajectory since the difference between the ingoing and the outgoing momenta **p** is finite (Jung 1986, 1987; Smilansky 1991; Blümel et al. 1992). Another quantity of particular interest is the time delay due to the collision. This time delay may be taken as the derivative of the action (5.4) with respect to the energy

$$T = \partial_E A , \tag{5.5}$$

which is a function of the scattering orbit (Narnhofer 1980, Narnhofer and Thirring 1981). This function gives an estimation of the time spent by the trajectory in the collision as compared for instance with a free flight. This function is the classical analogue of the Wigner time delay defined in quantum mechanics, for instance, as the energy derivative of the phase shift (Gaspard 1993b).

This function has similarities with the escape-time function defined in Chapter 2, except that the escape-time function is defined inside the region of collision U while the time-delay function is defined outside this region on the ingoing trajectories. This discussion brings us to the problem of the set of trajectories on which the above functions are defined.

5.1.3 The different types of trajectories

A distinction must be established between nonhyperbolic and hyperbolic regimes of scattering.

In *nonhyperbolic regimes* of two-degrees-of-freedom systems, there exist whole regions of phase space which do not play any role in the scattering process beside a phase-space exclusion. These regions are the elliptic islands of the invariant set $\mathcal{I}$ where trajectories are indefinitely trapped (MacKay and Meiss 1987, MacKay 1993b). The time to escape from these regions is infinite. The escape-time function of Chapter 2 is not defined over these regions of positive area. As a corollary, these regions do not belong to the domain of definition of the scattering function (5.3) and of the time-delay function (5.5). Indeed, these functions are only defined for ingoing trajectories.

In systems with more than two degrees of freedom, there is no topological constraint preventing trajectories escaping from 'inside' the invariant tori. This property is at the origin of the so-called *Arnold diffusion* which is a very slow transport taking place in phase-space regions where quasiperiodic motions dominate (Arnold 1978, Lichtenberg and Lieberman 1983). The time to escape from these regions may often be extremely long without being infinite (Gaspard and Rice 1989b). As long as the escape time exists, the scattering function (5.3) also exists.

Trajectories which spend a finite time in the collision region U and then exit U are called the *scattering orbits.* For such a trajectory, both the functions (5.3) and (5.5) are well defined because the outgoing trajectory exists as the image of the ingoing trajectory under the map (5.3) and the time delay (5.5) is finite.

Moreover, there also exist trajectories which are asymptotic to a trapped trajectory and which remain trapped in U for all future times. This is the case if the ingoing trajectory belongs to the stable manifold of the invariant set $\mathcal{I}$. In this case, the outgoing trajectory does not exist and the time delay is infinite. This singular character of the time delay on the stable manifolds of the invariant set is of course similar to that which concerns the escape-time function of Chapter 2.

By reversing time, we can also discuss the existence of the inverse of the scattering function Σ^{-1}, which is defined on the scattering orbits but not on the unstable manifolds of the invariant set where the time delay of the reversed process is infinite.

In the *hyperbolic regime,* the situation not very different except that the invariant set is hyperbolic and of zero Lebesgue measure. The scattering function and the time delay function are here also defined only on the scattering orbits and on the stable (or unstable) manifolds of the invariant set at the exclusion of the invariant set itself which is not reached from outside except after an infinite time.

The invariant set plays an active role in half-collisions as in dissociation processes where the initial conditions are taken inside the domain U. In experiments observing dissociations, it is the escape-time function of Chapter 2 which is more relevant than the scattering function and the time delay. Such experiments are repeated over an ensemble of initial conditions, which allows us to define correlation function or decay curves which give the number of trajectories inside the domain U at each time. The behaviour of such functions of time is controlled by the invariant set and its stable and unstable manifolds.

Using the resonance spectral theory of Chapter 3, the time evolution of such functions can be decomposed in terms of the Pollicott–Ruelle resonances like the escape rate. These resonances provide the intrinsic lifetimes of the invariant set. On the other hand, we may also use the Lyapunov exponents and the fractal dimensions to characterize the repeller in its phase space. These quantities give us information on the fine scale properties of the repeller and on the behaviour of trajectories in the vicinity of the repeller.

5.1.4 Scattering operator for statistical ensembles

We mention here the contribution of Lax and Phillips (1967) to the classical scattering theory. These authors have considered a scattering operator in analogy with the quantum scattering theory. However, in contrast with quantum scattering theory, their work concerns the scattering of statistical ensembles of classical trajectories instead of individual quantum wavefunctions. The theory is developed from the spectral properties of the Koopman operator based on the assumption of a Lebesgue continuous spectrum. Such conditions are satisfied for systems like the disk scatterers which have a hyperbolic repeller of Axiom-A type.

Koopman operators are associated with the Hamiltonian flows used in Eq. (5.3) as

$$\hat{U}^t \, g(\mathbf{X}) \;=\; \exp(-\hat{L}t)\, g(\mathbf{X}) \;=\; g(\mathbf{\Phi}^t\mathbf{X})\,. \tag{5.6}$$

Møller-type operators are defined as

$$\hat{W}_{\pm} \;=\; \lim_{t\to+\infty} \hat{U}^{\pm t}\, \hat{U}_0^{\mp t}\,. \tag{5.7}$$

A *scattering operator* is defined as

$$\hat{S} \;=\; \hat{W}_+^{-1}\, \hat{W}_-\,. \tag{5.8}$$

This operator maps the ingoing probability density onto the outgoing one in terms of the scattering function (5.3)

$$f_{\text{out}}(\mathbf{X}) \;=\; \hat{S}\, f_{\text{in}}(\mathbf{X}) \;=\; f_{\text{in}}(\mathbf{\Sigma}^{-1}\mathbf{X})\,. \tag{5.9}$$

5.2 Hard-disk scatterers

5.2.1 Generalities

In order to illustrate the previous discussion, we consider the scattering of a point particle in elastic collisions on hard disks fixed in the plane. These billiards have been much studied because they provide simple models of chaotic dynamics (Eckhardt 1987, Jung and Scholz 1987, Gaspard and Rice 1989a, Kovács and Tél 1990, Korsch and Wagner 1991). Thanks to these models, it becomes possible to study how kinetic and transport properties can appear in mechanical systems.

In these billiards, the defocusing character of the collisions on disks induces a dynamical instability so that all the trapped orbits are unstable and these systems are hyperbolic, as discussed in Chapters 1 and 2.

We may consider billiards formed by any number of disks as shown in Fig. 5.1, from the one-disk scatterer up to the Lorentz gas composed of an infinite number of disks filling the whole plane. In this progression, the system transits from finite scatterers to spatially extended systems which may sustain a transport process like diffusion. In the following, we shall successively investigate the properties of the disk scatterers as their size increases.

The disk scatterers are two-dimensional billiards of free Hamiltonian, $H = (p_x^2+p_y^2)/2m$, with a domain $\mathscr{Q}$ in position plane $\mathbf{q} = (x, y)$ defined by $\|\mathbf{q}-\mathbf{q}_c\| \geq a$, where $\{\mathbf{q}_c\}_{c=1}^N$ are the centres of disks of radius a. The coordinates of the centres fix the geometry of the disk billiards. The disks may overlap or not. The disk scatterers are therefore Hamiltonian-type flows with two degrees of freedom.

5.2.2 The dynamics

The dynamics is governed by the rule of elastic collisions (1.11). The collision events are determined as follows.

An efficient algorithm of integration of the trajectories of the billiard is the following one which uses the Cartesian coordinates (Gaspard and Baras 1995). The particle velocities before and after the nth collision, which we denote by $\mathbf{v}_n^{(-)}$

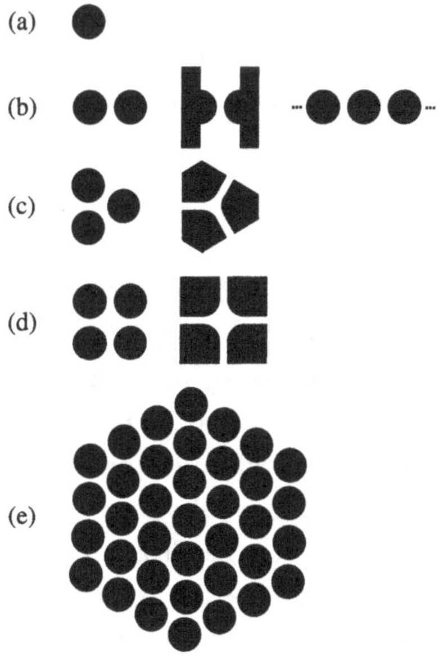

Figure 5.1 Geometry of the disk scatterers: (a) the one-disk scatterer; (b) the two-disk scatterer and variants; (c) the three-disk scatterer and the crossroad billiard with three connected wires (Eckhardt 1993); (d) the four-disk scatterer and the crossroad billiard with four connected wires (Gaspard et al. 1994); (e) the Lorentz-type scatterer.

and $\mathbf{v}_n^{(+)}$, are related by the law of geometric optics

$$\mathbf{v}_n^{(+)} = \mathbf{v}_n^{(-)} - 2\left(\mathbf{n}_n \cdot \mathbf{v}_n^{(-)}\right)\mathbf{n}_n . \tag{5.10}$$

Starting from the initial conditions $(\mathbf{q}_n, \mathbf{v}_n^{(+)})$ issued from a previous collision, the trajectory between two successive collisions at times t_n and t_{n+1} is given by the free flight

$$\mathbf{q}(t) = \mathbf{q}_n + (t - t_n)\,\mathbf{v}_n^{(+)} . \tag{5.11}$$

If the next collision occurs on the disk centered at $\mathbf{q}_c$ the time of the collision is determined by the condition

$$[\mathbf{q}(t) - \mathbf{q}_c]^2 = a^2 , \tag{5.12}$$

which amounts to solving the quadratic equation

$$A_n\,(t - t_n)^2 + 2\,B_n\,(t - t_n) + C_n = 0 , \tag{5.13}$$

with

$$\begin{aligned} A_n &= \left(\mathbf{v}_n^{(+)}\right)^2 , \\ B_n &= \mathbf{v}_n^{(+)} \cdot (\mathbf{q}_n - \mathbf{q}_c) , \\ C_n &= (\mathbf{q}_n - \mathbf{q}_c)^2 - a^2 . \end{aligned} \tag{5.14}$$

The time of the next collision is therefore given by

$$t_{n+1} = t_n - \frac{B_n}{A_n} - \sqrt{\left(\frac{B_n}{A_n}\right)^2 - \frac{C_n}{A_n}} , \tag{5.15}$$

where the minus sign is chosen between both determinations of the square root because the collision occurs at the first intersection of the free trajectory (5.11) with the disk. The new position on the disk c is given by $\mathbf{q}_{n+1} = \mathbf{q}(t_{n+1})$, while the normal at this point of impact is

$$\mathbf{n}_{n+1} = \frac{\mathbf{q}_{n+1} - \mathbf{q}_c}{\|\mathbf{q}_{n+1} - \mathbf{q}_c\|} = \frac{1}{a}\,(\mathbf{q}_{n+1} - \mathbf{q}_c) , \tag{5.16}$$

and the incident velocity at $\mathbf{q}_{n+1}$ is $\mathbf{v}_{n+1}^{(-)} = \mathbf{v}_n^{(+)}$. The outgoing velocity and the free flight after the collision are calculated identically starting again from Eq. (5.10). In this way, the complete trajectory can be calculated by recurrence.

The collision times for all the disks need to be compared in order to determine the shortest positive collision time for each free flight. In the presence of walls or of periodic boundary conditions, extra propagation rules should be added to the collision rules on the disks. If the lengths of the free flight are collected as well as the angles of collision it is possible to determine the Lyapunov exponent of the running orbit during the integration according to Chapter 1.

Energy is conserved as well as the magnitude of the velocity. Consequently, every quantity per unit time is proportional to a quantity per unit length up to a velocity factor. In the following, we shall suppose that the velocity is fixed to unity $v = 1$.

5.2.3 The Birkhoff mapping

Although it is not an efficient algorithm for numerical purposes, it is interesting to obtain the analytic expression of the Birkhoff (or Poincaré) mapping between two disks as defined by Eq. (1.5). The disk radius is unity $a = 1$. The geometry of the collisions is depicted in Fig. 5.2.

We denote by θ_n the angle of the impact point taken anticlockwise with respect to the x-axis on the nth disk. ϕ_n is the angle between the incident ray and the normal at this collision. In order to form a pair of conjugate canonical variables, we take the component of the velocity which is tangent to the disks and which is positive as θ_n increases: $\varpi_n = \sin \phi_n$. In the Birkhoff coordinates $\mathbf{x}_n = (\theta_n, \varpi_n)$, the mapping $\mathbf{x}_{n+1} = \boldsymbol{\phi}(\mathbf{x}_n)$ is area-preserving (Birkhoff 1927, MacKay 1993b).

$r_{n,n+1}$ denotes the distance between the centres C_n and C_{n+1} of the disks n and $n+1$ and $\alpha_{n,n+1}$ the angle between the vector joining the disk n to the disk $n+1$ and the x-axis.

By observation of the angles, we obtain that

$$\theta_n + \phi_n = \theta_{n+1} - \phi_{n+1} - \pi \, , \qquad (\text{modulo } 2\pi) \, . \tag{5.17}$$

On the other hand, we take the projection Π of the collision geometry on a perpendicular to the trajectory between the impact points I_n and I_{n+1}. On this

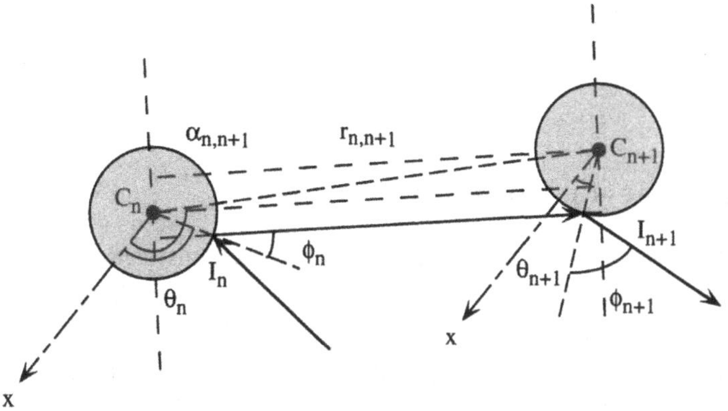

Figure 5.2 Geometry of a flight between two disks to determine the Birkhoff mapping.

perpendicular, we find the four following segments:

$$\begin{aligned}\Pi(C_nI_n) &= \sin\phi_n \ ,\\ \Pi(C_{n+1}I_{n+1}) &= \sin\phi_{n+1} \ ,\\ \Pi(C_nC_{n+1}) &= r_{n,n+1}\ \sin(\phi_n+\theta_n-\alpha_{n,n+1}) \ ,\\ \Pi(I_nI_{n+1}) &= 0 \ .\end{aligned} \tag{5.18}$$

According to the identity

$$\Pi(C_nI_n) = \Pi(C_nC_{n+1})+\Pi(C_{n+1}I_{n+1})+\Pi(I_{n+1}I_n) \ , \tag{5.19}$$

we obtain a second relation

$$\sin\phi_{n+1} = \sin\phi_n - r_{n,n+1}\ \sin(\phi_n+\theta_n-\alpha_{n,n+1}) \ . \tag{5.20}$$

Thanks to Eqs. (5.17) and (5.20), the expression of the Birkhoff mapping can be obtained explicitly as

$$\begin{aligned}\varpi_{n+1} &= \varpi_n - r_{n,n+1}\left[\varpi_n\cos(\theta_n-\alpha_{n,n+1})+\sqrt{1-\varpi_n^2}\,\sin(\theta_n-\alpha_{n,n+1})\right],\\ \theta_{n+1} &= \theta_n+\pi+\arcsin\varpi_n+\arcsin\varpi_{n+1} \ , \quad (\text{mod. } 2\pi) \ .\end{aligned} \tag{5.21}$$

On the other hand, the time of the free flight is given by the length from I_n to I_{n+1}, which is the first-return time function

$$\begin{aligned}T(\theta_n,\varpi_n) = \Big\{ & r_{n,n+1}^2+2r_{n,n+1}\left[\cos(\theta_{n+1}-\alpha_{n,n+1})-\cos(\theta_n-\alpha_{n,n+1})\right]\\ & +2-2\cos(\theta_{n+1}-\theta_n)\Big\}^{1/2} \ ,\end{aligned} \tag{5.22}$$

for $v=1$. The knowledge of the configuration of the disks allows us to fix the values of the parameters $r_{n,n+1}$ and $\alpha_{n,n+1}$ in this expression of the mapping, which is therefore completely explicit.

After these generalities, let us consider the disk scatterers by increasing order of complexity.

5.2.4 The one-disk scatterer

This billiard is one of the simplest examples of scattering systems (see Fig. 5.1a). Every trajectory is of scattering type: coming from infinity, colliding on the disk, and returning to infinity. This scatterer has rotational symmetry so that angular momentum is conserved. Its domain of definition $\mathscr{Q}$ is complementary to the domain of the circle billiard.

The total cross-section is given by the diameter of the disk: $\sigma = 2a$. For particles arriving on the disks parallel to the x-axis from the left-hand side ($x < 0$) with an impact parameter y, the deflection angle θ is given by $y = a \sin \phi = a \cos \frac{\theta}{2}$, because the deflection angle θ is related to the velocity angle ϕ by $\theta = \pi - 2\phi$. Assuming a uniform distribution of the impact parameter y in the incident beam ($d\sigma = |dy|$), the differential cross-section is obtained as

$$\frac{d\sigma}{d\theta} = \frac{a}{2} \left| \sin \frac{\theta}{2} \right| . \tag{5.23}$$

We notice that there is no orbit which is trapped at finite distance so that there is no repeller in this scatterer.

5.2.5 The two-disk scatterer

Adding a disk in the scatterer has an important effect on the properties of the scatterer. Let us denote by r the distance between the centres of the disks (see Fig. 5.1b).

First of all, the rotational symmetry is broken so that angular momentum is no longer conserved. The total cross-section varies between $2a$ and $4a$ according to the direction of the incident beam.

Almost all the trajectories are of scattering type since they only have a finite number of collisions on the disks. However, there is now the possibility of trajectories undergoing an infinite number of collisions on the scatterer when $r > 2a$ and to be indefinitely trapped between the two disks. Indeed, there exists a periodic orbit bouncing between the two disks along the line joining the two centres. It is the unique orbit which remains trapped at finite distance for $t \to +\infty$ and $t \to -\infty$. This unique periodic orbit forms therefore a repeller because it is unstable. Its Lyapunov exponent can be calculated as above with Eq. (1.141). The length between the impacts here is $L = r - 2a$ so that

$$\lambda = \frac{1}{r - 2a} \ln \frac{r - a + \sqrt{r(r - 2a)}}{a} , \tag{5.24}$$

in the energy shell corresponding to the unit velocity $v = 1$. For $r \to 2a$, the Lyapunov exponent diverges according to

$$\lambda \simeq \sqrt{\frac{2}{a(r - 2a)}} , \quad \text{for } r \simeq 2a . \tag{5.25}$$

On the other hand, it vanishes for large interdisk distances like

$$\lambda \simeq \frac{1}{r} \ln \frac{2r}{a} , \quad \text{for } r \gg a . \tag{5.26}$$

The periodic orbit controls the scattering process in the system. Since it is unstable and the system is symplectic there exist a stable and an unstable manifold associated with the periodic orbit. These invariant manifolds constitute the separatrix between the trajectories which cross the bottleneck formed by the periodic orbit and those which are back scattered. The stable manifold W_s is composed of all the trajectories which are coming from infinity and which are trapped in the scatterer as $t \to +\infty$. The stable manifold is responsible for the divergences in the escape-time function which occur when the impact parameter y of an incoming trajectory meets the stable manifold. On the other hand, the unstable manifold W_u is composed of trajectories which are trapped for $t \to -\infty$ but escape to infinity as $t \to +\infty$. Accordingly, the repeller and its invariant manifolds control the scattering process. The stable and unstable manifolds are mapped onto each other by time-reversal symmetry so that we can understand how the dynamical instability is compatible with time reversibility.

We notice that other variants of the two-disk scatterer may be considered in which the ingoing and outgoing dynamics is confined to a wire (see Fig. 5.1b). Another variant defines a grating. In each variant, the repeller is composed of the same unstable periodic orbit and, eventually, of its images by reflection or translation. Such billiards are models of electronic mesoscopic circuits in the form of a constriction separating two reservoirs (Roukes and Alerhand 1989, Jalabert et al. 1990).

The Liouvillian resonances of the two-disk scatterer are obtained as the zeros of the Selberg–Smale Zeta function (4.101) (with $\beta = 1$), expressed here for a single unstable periodic orbit in the form

$$Z(s) = \prod_{m=0}^{\infty} \left\{1 - \exp[-s2L - 2L\lambda(1+m)]\right\}^{m+1} , \tag{5.27}$$

because the stability eigenvalue is positive and $L = r - 2a$. We recall that the velocity is fixed to unity $v = 1$. The resonances are thus given in terms of the Lyapunov exponent (5.24) as

$$s = i \frac{\pi n}{r - 2a} - \lambda (m+1) , \tag{5.28}$$

with the multiplicity $m+1$. The escape rate is the leading resonance for $m = n = 0$ and it is here equal to the Lyapunov exponent because the repeller is periodic: $\gamma = \lambda$.

The pressure function per unit length is obtained as the leading pole of the Ruelle zeta function (4.102) as

$$P(\beta) = \beta \lambda . \tag{5.29}$$

so that the entropies and the partial dimensions vanish as expected for a periodic repeller.

5.2.6 The three-disk scatterer

A dramatic effect happens when a further disk is added to the scatterer. We consider here the three-disk scatterer in which the disk centres form an equilateral triangle of side r (see Figs. 5.1c and 5.3). Figure 5.4 depicts the escape-time function of the three-disk scatterer, constructed from initial conditions with varying impact parameter. Here, we observe that the function has divergences on a set with self-similar properties. The number of divergence appears to multiply by two on successive scales. This is evidence for the fractal character of the underlying repeller. This billiard is one of the simplest example of chaos in a scattering system. Gaspard and Rice (1989a) have considered this billiard as a vehicle for the study of unimolecular reactions.

The symbolic dynamics

The trapped trajectories can be labelled by bi-infinite sequences of symbols of the ternary alphabet composed of the labels a, b, and c of the three disks as

$$\omega = \cdots \omega_{-2}\omega_{-1}\omega_0\omega_1\omega_2 \cdots \quad \text{with} \quad \omega_n \in \mathscr{A} = \{a, b, c\} . \tag{5.30}$$

Since the same disk cannot be hit twice successively, $\omega_{n+1} \neq \omega_n$ so that the topological transition matrix is

$$\mathsf{A} = \begin{array}{c} \\ a \\ b \\ c \end{array} \begin{array}{c} \begin{array}{ccc} a & b & c \end{array} \\ \begin{pmatrix} 0 & 1 & 1 \\ 1 & 0 & 1 \\ 1 & 1 & 0 \end{pmatrix} \end{array} . \tag{5.31}$$

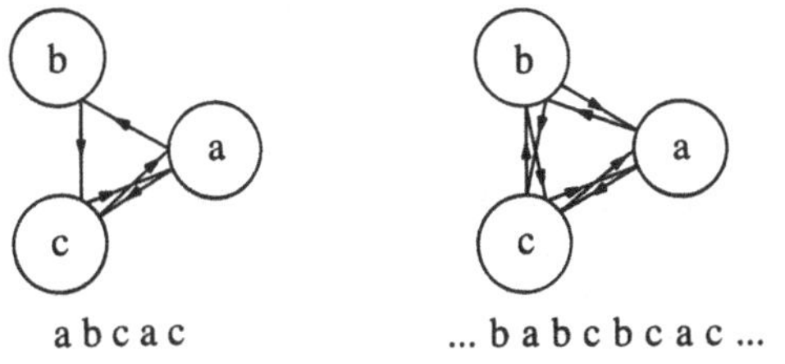

Figure 5.3 Geometry of the three-disk scatterer and examples of periodic and nonperiodic trajectories of the repeller.

This constitutes a natural symbolic dynamics for the three-disk scatterer. We have still to verify if this symbolic dynamics is in one-to-one correspondence with the trajectories of the system, or not. If (5.30)–(5.31) is a faithful symbolic dynamics, the trajectory has the choice between two possible disks at each successive collision. Hence, the number of possible trajectories multiplies by two at each collision so that the topological entropy is here equal to

$$\tilde{h}_{\rm top} = \ln 2 \quad \text{per symbol} . \tag{5.32}$$

This qualitative reasoning is confirmed by counting the number of fixed points in the $n^{\rm th}$ iterate of the mapping

$$|{\rm Fix}\, \boldsymbol{\phi}^n| = |{\rm Fix}\, \boldsymbol{\sigma}^n| = {\rm tr}\, {\sf A}^n = 2^n + 2\,(-)^n , \tag{5.33}$$

as well as by the fact that the topological entropy is the logarithm of the largest positive eigenvalue of the transition matrix (5.31).

The set of trajectories in this symbolic dynamics is uncountable while the periodic orbits are countable. On the other hand, because of the defocusing character of the collisions on disks, all these orbits are unstable and almost all the trajectories escape to infinity after a finite number of collisions. Therefore, the repeller is of zero Lebesgue measure in the phase space. Accordingly, the

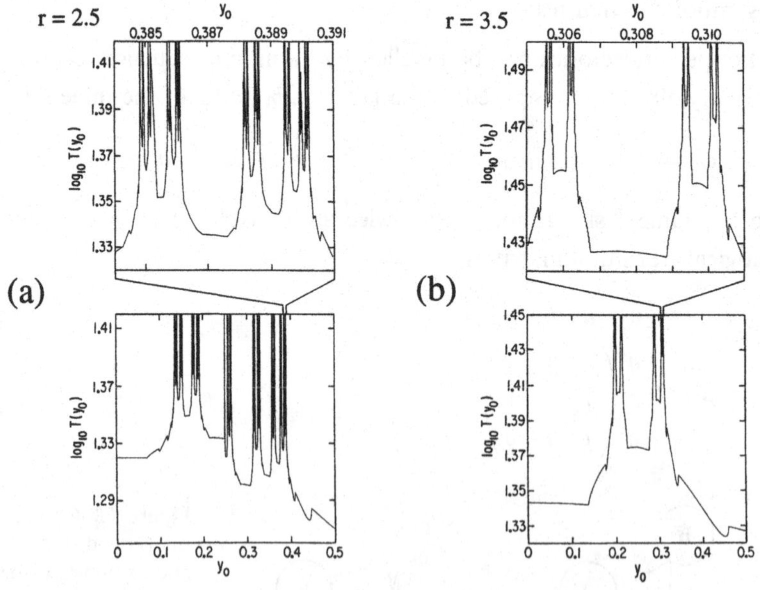

Figure 5.4 The escape-time function of the three-disk scatterer for (a) $r = 2.5$ and (b) $r = 3.5$ (with $a = 1$). y_0 is the impact parameter of the ingoing trajectory of initial velocity perpendicular to the line joining the centres of the disks b and c (Gaspard and Rice 1989a).

repeller is now a fractal object in contrast with the two-disk scatterer. This fractal can be visualized by light scattering on a set of spherical mirrors, which generates by reflection a nested structure of images (Walker 1988, Korsch and Wagner 1991).

In order to construct the fractal repeller, we shall analyze in detail the dynamics of escape to delimit as well as possible the region of phase space where trajectories remain trapped under forward and backward time evolution. This analysis will allow us to show the reverse statement that a unique trajectory corresponds to each symbolic sequence (5.30) satisfying the transition rule (5.31). This reverse statement is valid under certain geometric conditions to be determined by the analysis. We may anticipate by noting that the reverse correspondence is in danger when the disks become too close to each other. Indeed, trajectories may then become tangent to a disk which leads to a pruning of the corresponding trajectories. In this case, there may exist symbolic sequences (5.30) without corresponding trajectories. There exists therefore a value of the distance between the centres of the disks below which the correspondence no longer holds. In this respect, we have the theorem (Gaspard and Rice 1989a):

The three-disk scatterer of triangular and equilateral geometry such that $r > r_c$ *has a continuously hyperbolic repeller which is in one-to-one correspondence with the Markov topological shift* $\Sigma(\mathscr{A}, \mathsf{A}, \mathbb{Z})$. *This repeller is transitive and of Axiom-A type.*

Geometric construction of the fractal repeller

Let us turn to the proof of the above result. In order to establish the reverse correspondence, we shall construct a Markov partition of the phase space into cells corresponding to the transitions between the different disks. The cells must contain all the trajectories which remain trapped between the disks. The impact points are represented by the coordinates θ_a, θ_b, or θ_c, which are the angles along the perimeter of the disks counted anticlockwise from the intersection of the disks with the equilateral triangle joining their centres. First, we observe that the trajectories which are trapped must necessarily hit the disks within the intervals

$$0 \leq \theta_a, \theta_b, \theta_c \leq \frac{\pi}{3}, \tag{5.34}$$

otherwise the trajectory escapes to infinity either backward or forward in time. Therefore, the trapped trajectories are found in the cells $\{b{\cdot}a, c{\cdot}a, a{\cdot}b, c{\cdot}b, a{\cdot}c, b{\cdot}c\}$. For instance, the cell $b \cdot a$ contains all the trajectories currently on disk a and coming from disk b. Figure 5.5 shows the trajectories which delimit the borders of the cell $b \cdot a$.

On the one hand, we have the trajectories starting from the corner $\theta_b = \pi/3$ and reaching the current point θ_a. If ℓ_{ba} denotes the length of the path and if we resolve the triangle formed by the centre of a and the two impact points on a and b, we get

$$\begin{aligned} \ell_{ab} \sin(\tilde{\phi}_{a,1} - \theta_a) &= \sin\theta_a , \\ \ell_{ab} \cos(\tilde{\phi}_{a,1} - \theta_a) &= r - 1 - \cos\theta_a , \end{aligned} \tag{5.35}$$

so that the velocity angle of this border of the cell is obtained as

$$\tilde{\phi}_{a,1} = \theta_a + \arctan\frac{\sin\theta_a}{r-1-\cos\theta_a} . \tag{5.36}$$

On the other hand, we have the trajectories starting from the corner $\theta_b = 0$ and reaching the current point θ_a. The calculation is similar and we obtain the second border of the cell

$$\tilde{\phi}_{a,2} = \theta_a - \arctan\frac{(\sqrt{3}/2) - \sin\theta_a}{r-1-\cos\theta_a} . \tag{5.37}$$

The cell is therefore obtained as the domain

$$\mathscr{C}_{b\cdot a} = \{(\theta_a, \varpi_a)|0 \le \theta_a \le \pi/3 \ \& \ \sin\tilde{\phi}_{a,2} \le \varpi_a \le \sin\tilde{\phi}_{a,1}\} . \tag{5.38}$$

We note that, for $r \gg 2a$, the velocity angles (5.36) and (5.37) become

$$\begin{aligned} \tilde{\phi}_{a,1} &= \theta_a + \frac{a}{r}\sin\theta_a + \mathcal{O}(a^2/r^2) , \\ \tilde{\phi}_{a,2} &= \theta_a + \frac{a}{r}\sin\theta_a - \frac{\sqrt{3}a}{2r} + \mathcal{O}(a^2/r^2) , \end{aligned} \tag{5.39}$$

so that the width of the cell is determined by

$$\tilde{\phi}_{a,1} - \tilde{\phi}_{a,2} = \frac{\sqrt{3}a}{2r} + \mathcal{O}(a^2/r^2) . \tag{5.40}$$

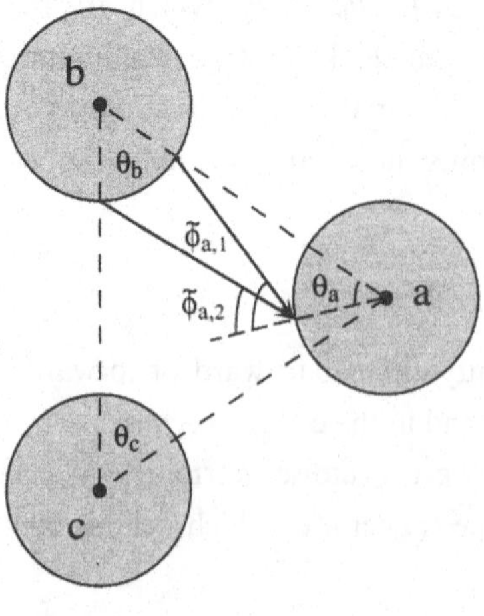

Figure 5.5 Construction of the partition of the three-disk scatterer.

Therefore, the cell becomes thinner and thinner as the distance r between the disks increases, which has important consequences on the fractal properties of the repeller as discussed below.

We remark that the scatterer is symmetric under the group C_{3v} so that it is enough to construct the partition on one disk, for instance a, in order to obtain the full partition by extension using the symmetry group. We also remark that the time-reversal operation $\boldsymbol{\iota}(\theta_\omega, \varpi) = (\theta_\omega, -\varpi)$ can be used to obtain six other cells which are $\{\cdot ab, \cdot ac, \cdot ba, \cdot bc, \cdot ca, \cdot cb\}$ because $\boldsymbol{\iota}(\omega \cdot \omega') = \cdot\omega'\omega$ (see Fig. 5.6).

This set of cells are also obtained as the preimages of the previous cells under the shift of the symbolic dynamics because $\omega \cdot \omega' = \boldsymbol{\sigma}(\cdot\omega\omega')$. The cells $\cdot\omega\omega'$ contain the trajectories on disk ω which will hit the disk ω' at the next collision.

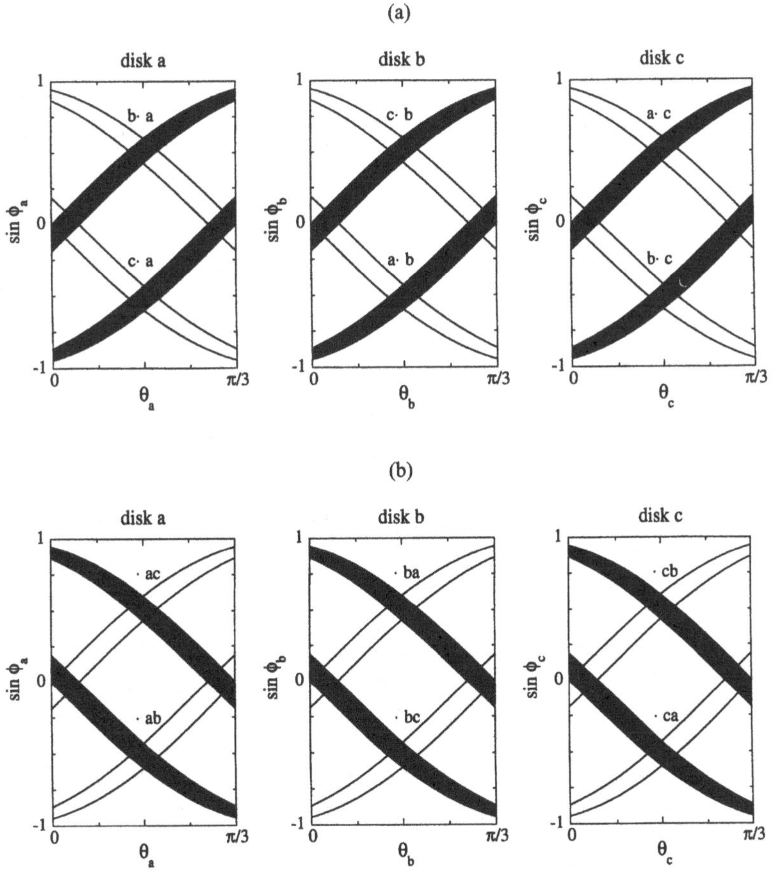

Figure 5.6 (a) The six cells $\omega \cdot \omega'$ of the generating partition and (b) their preimages $\cdot\omega\omega'$ under $\boldsymbol{\sigma}$.

The mapping acts on one of the cells by stretching, which induces dynamical instability on the repeller.

Now, if we consider the intersections between the cells $\omega \cdot \omega'$ and $\cdot\omega'\omega''$ we obtain twelve smaller cells of the type $\omega \cdot \omega'\omega''$ which correspond to trajectories on the current disk ω', which come from the disk ω and go to the disk ω'' (see Fig. 5.7). This structure is reminiscent of a Smale horseshoe in which stretching induces a restriction of the invariant set into smaller and smaller cells (Guckenheimer and Holmes 1983). If we further act by the mapping on the cells $\omega \cdot \omega'$ we obtain six other cells of the type $\omega\omega'\cdot$ which spread over two disks each and which form two very thin strips contained inside the cells $\omega \cdot \omega'$. Continuing this construction we obtain a hierarchy of $3 \times 2^{m+n}$ cells

$$C_{\omega_{-m}\omega_{-m+1}\dots\omega_{-1}\cdot\omega_0\omega_1\dots\omega_n} \, . \tag{5.41}$$

These cells are nested in the sense that the cell (5.41) is always contained inside any cell like (5.41) but with $m' \leq m$ and $n' \leq n$. According to the geometric construction, a unique orbit corresponds to the sequence labelling the cell (5.41) in the limit $m, n \to \infty$. In this way, we may conclude that the trapped trajectories are in one-to-one correspondence with the symbolic dynamics defined by (5.30)–(5.31).

The one-to-one correspondence is valid as long as the cells of the type $\omega \cdot \omega'\omega''$ remain inside the Birkhoff phase space. When r approaches the critical value 2 at which the disks touch each other, we remark that a cell like $b \cdot ac$ approaches the line $\phi = +\pi/2$ which corresponds to some tangency to the disk a. This

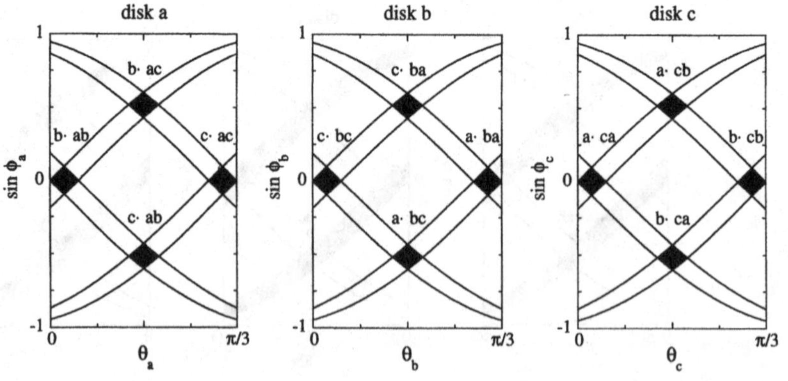

Figure 5.7 Horseshoe-type structures formed in the three-disk scatterer by the repeller as depicted in the Birkhoff phase space $(\theta_\omega, \varpi_\omega)$.

tangency occurs for the first time at $\theta_a = \pi/6$. Accordingly, there is no tangency as long as the condition

$$\tilde{\phi}_{a,1} = \frac{\pi}{6} + \arctan\frac{1/2}{r-1-\sqrt{3}/2} \leq +\frac{\pi}{2}, \tag{5.42}$$

or $r > 1 + 2\sqrt{3}^{-1} \simeq 2.1547$ is satisfied. This value can be taken as the lower bound r_c in the above theorem but we shall see below that the actual value where the symbolic dynamics no longer applies is somewhat smaller.

The dynamical instability can be estimated thanks to the stretching factors (1.141) or the positive Lyapunov exponents (1.138). The stretching factor is the factor by which the widths of the partition cells are reduced at each iteration of the Birkhoff mapping. It is therefore important to have a more detailed estimation. The path lengths and the velocity angles are bounded as follows

$$\begin{aligned} r-2 &\leq \tau_n \leq r-1, \\ 0 &\leq \phi_n < \frac{\pi}{6} + \arctan\frac{1}{2(r-2)}, \end{aligned} \tag{5.43}$$

so that the stretching factors per collision in Eq. (1.141) are bounded by

$$2r + \mathcal{O}(1) \leq \Lambda_n \leq \frac{4r}{\sqrt{3}} + \mathcal{O}(1), \tag{5.44}$$

as $r \gg 2$. Therefore, the stretching factor increases linearly with the size of the scatterer. Iterates of the cells of the partition become extremely thin when $r \gg 2$. On the other hand, the fractal repeller becomes more bulky when $r \simeq 2$. This feature also appears in the escape-time functions in Fig. 5.4. Hence, the dynamics is hyperbolic on the repeller because $|\Lambda_n| > 1$.

Besides, the stable and unstable manifolds of the repeller extend continuously over the Birkhoff phase space without rupture inside this domain as long as the condition (5.42) is satisfied. This condition guarantees therefore that hyperbolicity holds continuously in the neighbourhood of the repeller, which is thus continuously hyperbolic.

Moreover, we remark that the preceding geometric construction holds for deformed disks. Suppose that the shape of the obstacles is given in polar coordinates by $a = 1 + \epsilon f(\theta)$ where f is a smooth function and ϵ is small enough. Then, the cells of the Markov partition would be deformed but they would have the same transverse intersection leading to the same multiplication under iterations. Similarly, the stretching factors would be slightly affected which would not change the fact that they are much larger than one for r large enough. Accordingly, the same conclusion would hold and the conjugacy between the trajectories of the original and the deformed scatterer with the same symbolic dynamics would establish their mutual conjugacy. Therefore, the repeller is structurally stable.

Thanks to the symbolic dynamics, we can conclude that the repeller is of one piece because there exists an orbit which is dense in the repeller. Accordingly, the repeller is transitive and forms a so-called locally maximal nonwandering set like the Smale horseshoe. Finally, the periodic orbits are dense in the repeller. The three-disk repeller is thus of Axiom-A type, which ends the proof of the above theorem.

Symmetry, fundamental domain, and reduced symbolic dynamics

The three-disk scatterer we consider is invariant under actions of the symmetry group C_{3v} which contains six elements (Hamermesh 1962): the identity (ε), two rotations by $\pm 2\pi/3$ (ρ_{abc}, ρ_{cba}), and three reflections ($\sigma_{ab}, \sigma_{bc}, \sigma_{ac}$). Accordingly, each image of a trajectory by a group element is also a trajectory. This observation allows us to reduce the set of trajectories that we need to study, by introducing the fundamental domain which is the smallest subdomain of $\mathcal{Q}$ from which the full billiard $\mathcal{Q}$ can be unfolded by the actions of the group. Here, the fundamental domain is formed by the domain obtained by dividing the plane in six with respect to the centre of the equilateral triangle

$$\mathcal{Q}_F = \{(x,y)|\ 0 \le y \le x\sqrt{3} \quad \& \quad (x - r/\sqrt{3})^2 + y^2 \ge 1\} , \tag{5.45}$$

which is shown in Fig. 5.8a.

This new billiard has itself a fractal repeller which can be constructed by the same method as in the full domain (see Fig. 5.8b). It turns out that the

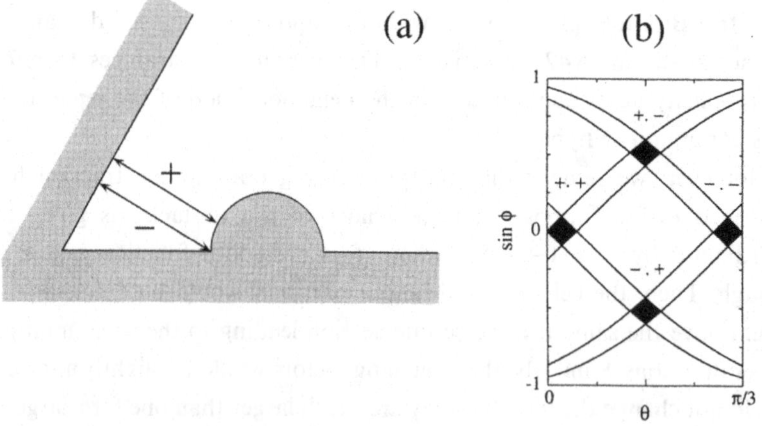

Figure 5.8 (a) The fundamental domain of the three-disk scatterer and (b) its repeller in the corresponding Birkhoff coordinates.

trajectories of this symmetry reduced repeller are in one-to-one correspondence with the following symbolic dynamics proposed by Cvitanović and Eckhardt (1989) under the same conditions as before

$$\tilde{\Sigma}(\tilde{\mathscr{A}}, \tilde{\mathsf{A}}, \mathbb{Z}) \qquad \text{with} \qquad \tilde{\mathscr{A}} = \{+,-\} \qquad \text{and}$$

$$\tilde{\mathsf{A}} = \begin{array}{c} \\ + \\ - \end{array}\!\!\begin{array}{c} \begin{array}{cc} + & - \end{array} \\ \begin{pmatrix} 1 & 1 \\ 1 & 1 \end{pmatrix} \end{array} . \tag{5.46}$$

We remark that there is no longer any constraint on the way the symbols follow each other. The reason is that the symbols are no longer associated with the labels of the disks. This symbolic dynamics is now equivalent to the Smale horseshoe and has the same topological entropy (5.32) as above because

$$|\text{Fix}\ \tilde{\boldsymbol{\sigma}}^n| = \text{tr}\ \tilde{\mathsf{A}}^n = 2^n . \tag{5.47}$$

The correspondence between both symbolic dynamics is not one-to-one because several trajectories of the full domain correspond in general to a single trajectory in the reduced domain. The rule to obtain the symbolic sequence in the fundamental domain from the original symbolic sequence (5.30) is the following (Cvitanović and Eckhardt 1989, Eckhardt 1993). The new symbols are assigned depending on the group elements which are used to bring the trajectory from the full domain to the fundamental one each time the trajectory crosses the boundary of the fundamental domain:

The symbol + is assigned if one reflection is used while the symbol − is assigned if two reflections are used to fold the trajectory between two collisions back into the fundamental domain (ignoring the group element used to bring the initial condition into the fundamental domain).

Table 5.1 gives the rules of translations between the symbolic dynamics Σ and $\tilde{\Sigma}$. An example of reduction is given by

$$\ldots ababcbabcabcbcbcbabcb \ldots \to$$
$$\ldots ++-+-+----+++++-+-+\ldots . \tag{5.48}$$

Table 5.1 Translation rules between the symbolic dynamics Σ with constraint of the three-disk scatterer and the symbolic dynamics $\tilde{\Sigma}$ without constraint of the billiard reduced to its fundamental domain under C_{3v}.

$\{a, b, c\}$	$\{+,-\}$
aba, bab, aca, cac, bcb, cbc	$+$
abc, bca, cab, acb, cba, bac	$-$

We remark that the unfolding on $\{a, b, c\}$ generates six possible sequences for a given reduced sequence in the right-hand member. The correspondence is given in Table 5.2 for the shortest periodic orbits.

The periodic orbits

Because of topological chaos, the periodic orbits proliferate exponentially at a rate determined by the topological entropy (5.32). Thanks to the reduction to the fundamental domain, the task of listing the periodic orbits is greatly simplified. The system is equivalent to a topological Bernoulli shift on two symbols. The periodic orbits of this system have already been described in Chapter 2. It is possible to establish in this way the list of all the shortest prime periodic orbits together with the values of their lengths and of their Lyapunov exponents as shown in Table 5.3. These periodic orbits have been obtained by using the method of Biham and Kvale (1992).

We remark that the higher periodic orbits can be viewed as topological combinations of the two shortest which are $+$ and $-$. These shortest periodic orbits are called the *fundamental periodic orbits* because they generate the whole set of periodic orbits by topological combinations. It turns out that the characteristic properties of the higher periodic orbits are very close to the one obtained by

Table 5.2 Correspondence between the ternary and the dyadic symbolic sequences for the shortest prime periodic orbits of the three-disk scatterer in its fundamental domain. The ternary and dyadic symbolic sequences are denoted by p and $\tilde{p}$ respectively. Note that p is the ternary symbolic sequence of one among several periodic orbits of the full domain which correspond to $\tilde{p}$.

p	$\tilde{p}$
ab	$+$
abc	$-$
abac	$+-$
ababcbcac	$++-$
abcacb	$+--$
abcbcbab	$+++-$
abcacabcbcab	$++--$
abacbabc	$+---$
abcbcbcacacabab	$++++-$
abcacacbab	$+++--$
abcbabacab	$++-+-$
ababc	$++---$
abacabcbabcacbc	$+-+--$
abacbacabc	$+----$

adding the periods and the Lyapunov exponents, especially in the limit $r \gg 2$. Figure 5.9 shows the period–energy diagram for the three-disk scatterer with $r = 6$ where we can observe the clustering of the periods.[2] This property justifies the approximation that the dynamical properties are essentially controlled by the two fundamental periodic orbits when $r \gg 2$. The infinite set of periodic orbits of such a system is therefore ordered by a hierarchy of importance which relates the fundamental periodic orbits to their topological combinations. This property is shared with other chaotic repellers as seen below.

Because of their importance, we calculate the lengths and the Lyapunov exponents of the two fundamental periodic orbits.

Let us first consider the orbit + which bounces between the disk and the

2. Of course, this clustering property cannot hold exactly because each orbit visits a different region of phase space in which a local deformation would lead to a change in the period and exponent of this particular orbit and neighbouring orbits without affecting others.

Table 5.3 The shortest prime periodic orbits of the three-disk scatterer for $r = 6$ in its fundamental domain. They are characterized by their symbolic sequences $\tilde{p}$, their period in number of bounces $n_{\tilde{p}}$, their average length per bounce $\bar{\ell}_{\tilde{p}} = L_{\tilde{p}}/n_{\tilde{p}}$, their Lyapunov exponent per bounce $\lambda_{\tilde{p}}^{(b)} = (1/n_{\tilde{p}}) \ln |\Lambda_{\tilde{p}}|$, as well as their Lyapunov exponent per unit length $\lambda_{\tilde{p}} = (1/L_{\tilde{p}}) \ln |\Lambda_{\tilde{p}}|$.

$\tilde{p}$	$n_{\tilde{p}}$	$\bar{\ell}_{\tilde{p}}$	$\lambda_{\tilde{p}}^{(b)}$	$\lambda_{\tilde{p}} = \lambda_{\tilde{p}}^{(b)}/\bar{\ell}_{\tilde{p}}$
+	1	4.0000000000	2.2924316695	0.57310791739
−	1	4.2679491924	2.4656775496	0.57771951785
+−	2	4.1582647426	2.4105229028	0.57969442834
++−	3	4.1072488720	2.3744347029	0.57810830971
+−−	3	4.1936025803	2.4263350149	0.57858010349
+++−	4	4.0805691185	2.3542513612	0.57694191492
++−−	4	4.1463107265	2.3947849575	0.57757006543
+−−−	4	4.2122679647	2.4364015076	0.57840610523
++++−	5	4.0644660052	2.3419184581	0.57619339296
+++−−	5	4.1171379344	2.3745141318	0.57673902835
++−+−	5	4.1276476772	2.3888451773	0.57874251007
++−−−	5	4.1707143034	2.4091933552	0.57764526169
+−+−−	5	4.1794738776	2.4200334495	0.57902825103
+−−−−	5	4.2233988644	2.4422366475	0.57826331966
+++++−	6	4.0537225726	2.3336737237	0.57568658977
++++−−	6	4.0976224648	2.3608542960	0.57615222397
++−+−+	6	4.1064600416	2.3729827767	0.57786579017
+++−−−	6	4.1423375166	2.3899041845	0.57694578845
++−+−−	6	4.1503611668	2.4002116893	0.57831393291
++−−+−	6	4.1503611668	2.4002116893	0.57831393291
++−−−−	6	4.1869147480	2.4185866984	0.57765367676
+−+−−−	6	4.1942713726	2.4277908389	0.57883494491
+−−−−−	6	4.2308242976	2.4461451969	0.57817224845

symmetry axis between the disks a and b which is the axis of the reflection σ_{ab} (cf. Figs. 5.3 and 5.8a). The monodromy matrix is thus composed of four intermediate monodromy matrices given by Eqs. (1.132)–(1.133): one free flight from the disk to the axis of σ_{ab}, one collision on this flat border, one free flight back to the disk, and one collision on the disk. We notice that the radius of curvature is infinite for collisions on flat borders [i.e., $a = \infty$ in Eq. (1.133)]. The full monodromy matrix is therefore

$$\mathsf{m}_+ = \begin{pmatrix} -1 & 0 \\ -\frac{2}{\cos\phi_+} & -1 \end{pmatrix} \begin{pmatrix} 1 & L_+/2 \\ 0 & 1 \end{pmatrix} \begin{pmatrix} -1 & 0 \\ 0 & -1 \end{pmatrix} \begin{pmatrix} 1 & L_+/2 \\ 0 & 1 \end{pmatrix} = \begin{pmatrix} 1 & L_+ \\ \frac{2}{\cos\phi_+} & 1 + \frac{2L_+}{\cos\phi_+} \end{pmatrix} , \tag{5.49}$$

where $\phi_+ = 0$ and $L_+ = r - 2$. The stability eigenvalue is obtained from the characteristic equation

$$\det(\mathsf{m}_+ - \Lambda_+ \mathsf{1}) = 0 , \tag{5.50}$$

such that

$$\Lambda_+ = 1 + \frac{L_+}{\cos\phi_+} + \sqrt{\frac{L_+}{\cos\phi_+}\left(\frac{L_+}{\cos\phi_+} + 2\right)} , \tag{5.51}$$

where the positive determination of the square root has been chosen in order that $|\Lambda_+| > 1$. We remark that the stability eigenvalue is positive so that this periodic orbit is hyperbolic without inversion.

The other fundamental periodic orbit $-$ is composed of three collisions

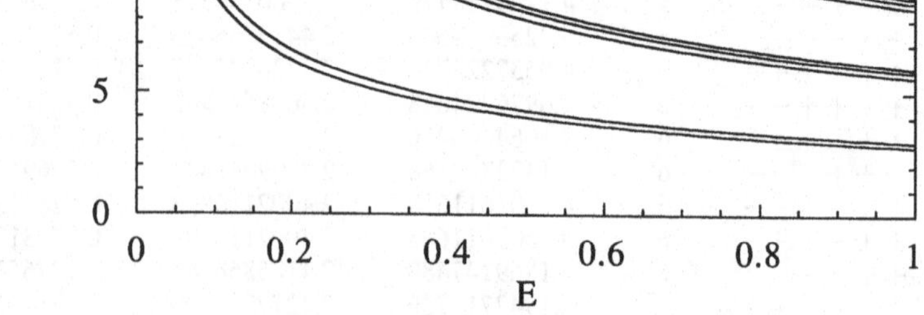

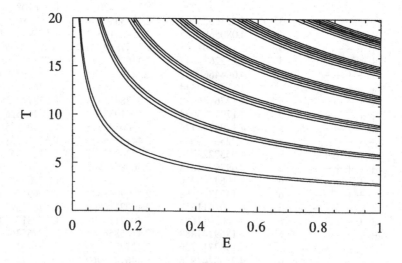

Figure 5.9 Period–energy diagram of the three-disk scatterer with $r = 6$.

(Fig. 5.8a). Indeed, the collision in the corner should be decomposed in two collisions with an infinitesimal free flight between them in order for the velocity direction to come back to its direction at the initial condition. Accordingly, we have successively one free flight from the disk to the symmetry axis of σ_{ab}, one collision on this flat border, one free flight to the symmetry axis of σ_{bc}, one collision on this other flat border, one infinitesimal free flight to the disk, and one collision on the disk so that the full monodromy matrix is

$$\begin{aligned} \mathsf{m}_- &= \begin{pmatrix} -1 & 0 \\ -\frac{2}{\cos\phi_-} & -1 \end{pmatrix} \begin{pmatrix} 1 & 0 \\ 0 & 1 \end{pmatrix} \left[\begin{pmatrix} -1 & 0 \\ 0 & -1 \end{pmatrix} \begin{pmatrix} 1 & L_-/2 \\ 0 & 1 \end{pmatrix} \right]^2 \\ &= \begin{pmatrix} -1 & -L_- \\ -\frac{2}{\cos\phi_-} & -1-\frac{2L_-}{\cos\phi_-} \end{pmatrix} , \end{aligned} \tag{5.52}$$

where $\phi_- = \pi/6$ and $L_- = r - \sqrt{3}$. We remark that the extra collision has the effect of changing the sign of the monodromy matrix so that the stability eigenvalue is negative

$$\Lambda_- = -1 - \frac{L_-}{\cos\phi_-} - \sqrt{\frac{L_-}{\cos\phi_-}\left(\frac{L_-}{\cos\phi_-}+2\right)} , \tag{5.53}$$

here with the negative determination of the square root in order that $|\Lambda_-| > 1$. The stability eigenvalue is thus negative so that the periodic orbit $-$ is inverse hyperbolic. This conclusion is general that each collision in the domain of $-$ induces an inversion. Therefore, a periodic orbit of code $\tilde{\omega}_1\tilde{\omega}_2\ldots\tilde{\omega}_{n_{\tilde{p}}}$ is hyperbolic or inverse hyperbolic if the number of symbol $\tilde{\omega}_n = -$ in its code is even or odd

$$\operatorname{sgn}\Lambda_{\tilde{p}} = \prod_{j=1}^{n_{\tilde{p}}} \tilde{\omega}_j , \qquad \text{for} \quad \tilde{p} = [\tilde{\omega}_1\tilde{\omega}_2\ldots\tilde{\omega}_{n_{\tilde{p}}}]^\infty . \tag{5.54}$$

The Lyapunov exponents per unit length are then given by $\lambda_{\tilde{p}} = (1/L_{\tilde{p}})\ln|\Lambda_{\tilde{p}}|$. In summary, the lengths and the Lyapunov exponents of the two fundamental orbits are

$$\begin{aligned} L_+ &= r - 2 , \\ L_- &= r - \sqrt{3} , \\ \lambda_+ &= \frac{1}{r-2} \ln\left(r - 1 + \sqrt{r(r-2)}\right) , \\ \lambda_- &= \frac{1}{r-\sqrt{3}} \ln\left(\frac{2}{\sqrt{3}}r - 1 + \frac{2}{\sqrt{3}}\sqrt{r(r-\sqrt{3})}\right) . \end{aligned} \tag{5.55}$$

We notice that both Lyapunov exponents decrease similarly as $r \to \infty$ as

$$\lambda_+ \simeq \frac{1}{r} \ln 2r \; ,$$
$$\lambda_- \simeq \frac{1}{r} \ln \frac{4r}{\sqrt{3}} \; , \tag{5.56}$$

except that the exponent of $-$ is larger than that of $+$ by a subleading correction. It should be noted that the average lengths per bounce and the Lyapunov exponents per bounce are the minimum and the maximum values of these quantities for all the periodic orbits of the repeller (see Table 5.3). However, the values of the Lyapunov exponents per unit length fluctuate around λ_- because $\lambda_{\tilde{p}}$ is given by the ratio of the Lyapunov exponent per bounce over the average length per bounce, which both vary.

Pruning in the three-disk scatterer

When the disks come close to each other, the trajectories of the repeller may become tangent to the disks. This occurs for the first time at some critical value which has been calculated by Hansen (1993): $r = r_c = 2.04821419\ldots$. At this critical value, the heteroclinic intersection between the unstable manifold of the period-two orbit $[ab]^\infty$ and the stable manifold of $[ac]^\infty$ reaches the border $\varpi_a = +1$ at $\theta_a = \pi/6$, which corresponds to a tangency for the heteroclinic orbit (see Fig. 5.7). For $r < r_c$, trajectories disappear from the repeller because they would otherwise penetrate the disks. This phenomenon causes a pruning of the symbolic dynamics. The simple symbolic dynamics becomes only a covering symbolic dynamics while the actual one is smaller but more complex. Therefore, the topological entropy per symbol decreases with r.

At the tangencies, it turns out that infinite families of orbits abruptly disappear from the repeller (Sano 1994). In this case, the three-disk scatterer is no longer structurally stable although it is still hyperbolic with all its trajectories being unstable. We notice that the heteroclinic tangencies correspond to trajectories of the flow which are tangent to the disks. Therefore, these tangencies become important in the close configurations of disks where they are responsible for the lack of structural stability of the invariant set. In the regime $r < r_c$, the topological entropy per symbol appears as a devil staircase as a function of r while it was constant at the value $\tilde{h}_{\text{top}} = \ln 2$ for $r > r_c$ (Hansen 1993, Sano 1994).

The decrease of the topological entropy continues even for $r < 2$. When $\sqrt{3} < r \leq 2$, the disks overlap and form a compact billiard with three corners. In this regime, the topology of the trajectories is complicated not only by the trajectories tangent to the disks but also by those which hit the corners.

Bifurcations responsible for the destruction of orbits may occur on both types of singularity lines (Sano 1994). Here also, the billiard is not structurally stable but still hyperbolic.

Let us remark that there exist variants of the three-disk scatterer like the crossroad billiard with three branches (see Fig. 5.1c). We remark that this system has the same repeller as the three-disk scatterer. The only slight difference appears at the level of the periodic orbits *ab*, *bc*, and *ac* which bounce just at the matching points between the disks and the straight walls modelling the wires. Therefore, these periodic orbits are linearly unstable on one of their sides but marginally unstable on the other one. This billiard has been proposed as a simple model of mesoscopic electronic circuits (Eckhardt 1993).

The zeta functions

The scatterer is symmetric under the discrete C_{3v} group which has two one-dimensional representations A_1 and A_2 and one two-dimensional representation E. The Ruelle zeta function factorizes accordingly: $\zeta = \zeta_{A_1}\zeta_{A_2}\zeta_E^2$. The same happens for the Selberg–Smale Zeta function (Cvitanović and Eckhardt 1993).

It is possible to show that the leading zeta functions with $m = 0$ of the three representations are given by the cycle expansions

$$\zeta_{A_1}^{-1} = 1 - t_+ - t_- - (t_{+-} - t_+ t_-) - \cdots , \tag{5.57}$$

$$\zeta_{A_2}^{-1} = 1 + t_+ - t_- + (t_{+-} - t_+ t_-) + \cdots , \tag{5.58}$$

$$\zeta_E^{-1} = 1 + t_- - t_+^2 + t_-^2 + (t_{++-} - t_+^2 t_-) + \cdots , \tag{5.59}$$

where $t_\pm = \exp(-sL_\pm)/|\Lambda_\pm|$ for the energy shell with unit velocity (Gaspard and Alonso Ramirez 1992).

The Pollicott–Ruelle resonances

The resonances are given by the zeros of the inverse Ruelle zeta functions (5.57)–(5.59). They have been calculated from the periodic orbits of Table 5.3. The result is depicted in Fig. 5.10. Several features appear in this figure. The first one is the absence of resonances in a strip below the imaginary axis. This gap is determined by the escape rate γ which appears as the resonance on the real axis which is the closest to the origin. This resonance rules the evolution of statistical ensembles on the longest time scales. It is important for the following to notice that this resonance is determined by the zeta function (5.57) of the fully symmetric A_1-representation. The resonances in the other representations are further below.

These resonances give the generalized eigenvalues of the Liouvillian dynamics

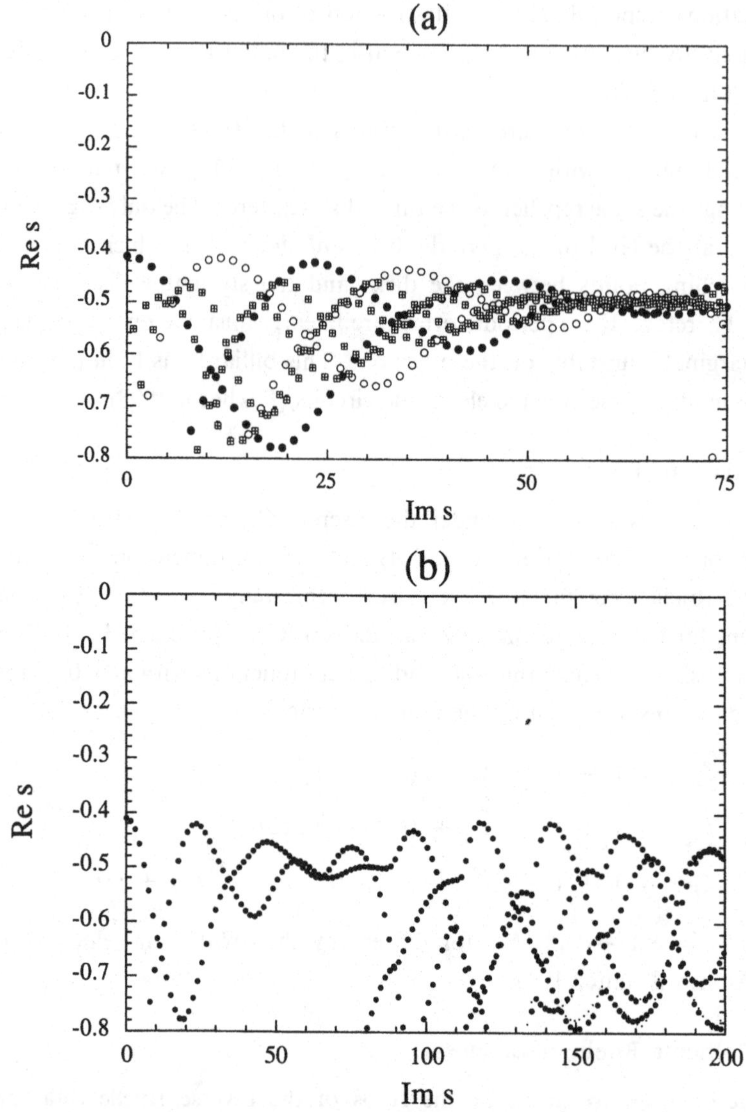

Figure 5.10 (a) Classical resonances for the three-disk scatterer with $r = 6$, $a = 1$, and $v = 1$ obtained with the cycle expansion up to period 6 in the fundamental domain: A_1-resonances are denoted by filled circles, A_2-resonances by open circles, and E-resonances by crossed squares. They are superimposed on the same spectrum calculated up to period 5 (smaller symbols) to show the good convergence. (b) Same as (a) but for the A_1-resonances up to Im $s = 200$ (Gaspard and Alonso Ramirez 1992).

and are therefore involved in the spectral decomposition of the Frobenius–Perron operator, as discussed in Chapter 3. As a consequence, the resonances play an essential role in the escape of an ensemble of particles from the three-disk scatterer. A simulation of this escape has been performed starting from an initial ensemble which is concentrated in a relatively small region (Gaspard and Alonso Ramirez 1992). The decay curve of Fig. 5.11a has been obtained which presents oscillations superposed on the general exponential decay at the escape rate γ. These oscillations have their origin in the resonances of the A_1-representation which are next to the escape rate. In order to show this property, the decay curve has been Fourier transformed according to

$$F(\omega) = \int_0^T e^{i\omega t}\, e^{\gamma t}\, N(t)\, dt\,. \tag{5.60}$$

This Fourier transform is compared with the spectrum of Pollicott–Ruelle resonances in Fig. 5.11b, where we observe the nice agreement of the main peaks at nonvanishing frequencies with the next-to-leading A_1-resonances. This correspondence is an illustration of the Liouvillian resonance theory.

The characteristic quantities of chaos

According to Eqs. (4.102)–(4.103), the pressure function can be calculated from the leading resonance which is given by the leading zero of the inverse zeta function (5.57) where $t_\pm = \exp(-sL_\pm)/|\Lambda_\pm|^\beta$. Using the periodic orbits of Table 5.3 very accurate values of the pressure function and of the characteristic

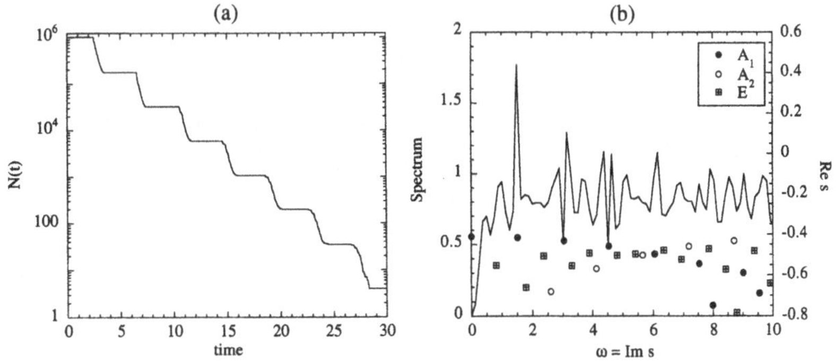

Figure 5.11 (a) Decay curve of an ensemble of 10^6 particles in the three-disk scatterer for intercentre distance $r = 6$, radius $a = 1$, and velocity $v = 1$. The initial velocity is $\mathbf{v} = (1, 0)$ for all particles while the initial position is uniformly distributed between $\mathbf{q} = (0, 0)$ and $\mathbf{q} = (0, 1)$. $N(t)$ is the number of particles which, by the time t, have not had their last bounce before escaping. (b) Power spectrum $|F(\omega)|^2$ of the Fourier transform (5.60) of the oscillations in (a) versus $\omega = \text{Im } s$ and compared with the location of the classical resonances of Fig. 5.10 in the complex plane of s (Gaspard and Alonso Ramirez 1992).

quantities of chaos have been obtained for the scatterer with $r = 6$ and $a = 1$, as given in Table 5.4. The pressure function is given in Fig. 5.12.

In the limit $r \to \infty$, asymptotic expressions can be obtained for the pressure and the different characteristic quantities from the simple approximation $1 \simeq t_+ + t_-$ using (5.55)–(5.56)

$$P(\beta) \simeq \frac{1}{r} \ln\left[(a/2r)^\beta + (a\sqrt{3}/4r)^\beta\right] , \tag{5.61}$$

$$\gamma \simeq \frac{1}{r} \ln(1.072r/a) , \tag{5.62}$$

$$\bar{\lambda} \simeq \frac{1}{r} \ln(2.138r/a) , \tag{5.63}$$

$$h_{\rm KS} \simeq \frac{1}{r} \ln\ 1.995 , \tag{5.64}$$

$$h_{\rm top} \simeq \frac{1}{r} \ln\ 2 , \tag{5.65}$$

$$d_{\rm H} \simeq \frac{\ln\ 2}{\ln(r/a)} , \tag{5.66}$$

for $r \to \infty$. We used respectively, Eqs. (4.63), (4.71), (4.72), and (4.79). We notice that the escape rate and the mean Lyapunov exponent decrease as $(\ln r)/r$ while the entropies have a slower decrease as $1/r$. These behaviours have their origin in the fact that the time of flight between the defocusing collisions is proportional to the interdisk distance r. Since these quantities are evaluated per unit time they decreases as $r \to \infty$. We also notice that the KS entropy is asymptotically very close to the topological entropy because the ratio of the stability eigenvalues is here close to one: $|\Lambda_+/\Lambda_-| \simeq \sqrt{3}/2$. However, the KS entropy remains less than the topological entropy as expected from Chapter 4.

Table 5.4 Characteristic quantities of chaos for the disk scatterers with $r = 6$, $a = 1$ and $v = 1$.

	2-disk scatterer	*3-disk scatterer*	*4-disk scatterer*
γ	0.5731079	0.4103384	0.28909
λ	0.5731079	0.5775526	0.49261
$h_{\rm KS}$	0	0.1672142	0.20351
$h_{\rm top}$	0	0.167232	0.22302
$d_{\rm I}$	0	0.2895220	0.41314
$d_{\rm H}$	0	0.2895376	0.42607

5.2.7 The four-disk scatterer

A direct generalization of the three-disk scatterer is when a disk is added. We may expect that chaos will be increased in this system at least if the four disks form a configuration where each disk is visible from the three others. Therefore, after each collision, the trajectory has the choice of going to three possible disks. The proliferation of orbits in the repeller thus behaves like 3^n with the number of collisions. The topological entropy per bounce may thus be expected to be $\tilde{h}_{top} = \ln 3$.

This is indeed the case for the billiard where the four disks are centered at the vertices of a square of side r (see Fig. 5.1d). In this configuration, the scatterer is symmetric under the group C_{4v} composed of eight elements (four rotations including the identity and four reflections) (Hamermesh 1962). This billiard is very similar to the three-disk system although it presents some structural differences which it is worthwhile to discuss. Moreover, this billiard is directly related to the crossroad billiard which has been considered as a model of a mesoscopic electronic circuit studied in several experiments (Beenakker and van Houten 1989, Roukes and Alerhand 1989, Doron et al. 1991). The relation is that the C_{4v} four-disk scatterer and the crossroad billiard with four connected wires share the same classical repeller. Indeed, the trajectories which are trapped at finite distances in the crossroad billiard are unstable orbits bouncing between the arcs of disks at the crossroad. Most of the other orbits escape in the wires in zig-zag motions. Let us nevertheless observe that the wires with parallel walls sustain continuous families of bouncing ball periodic orbits which are

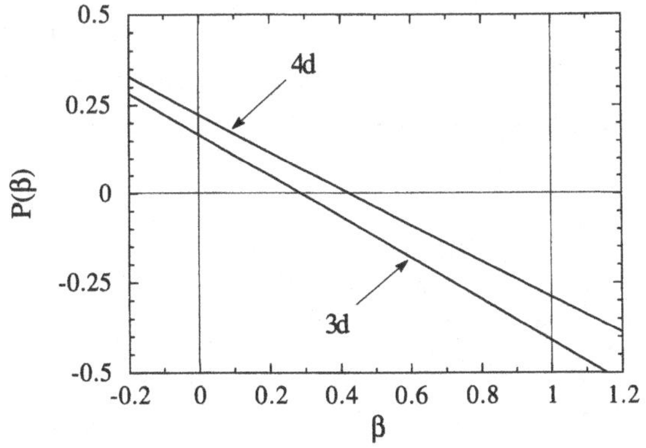

Figure 5.12 Topological pressure functions for the three- (3d) and four-disk (4d) repellers when $r = 6$, $a = 1$, and $v = 1$. The data were obtained using the cycle expansion up to the fundamental-domain periods 6 and 4 for the 3d- and 4d-repellers respectively. The numerical errors on the functions are smaller than the line thickness (Gaspard and Alonso Ramirez 1992).

perpendicular to the directions of the wires. These marginally unstable periodic orbits form a separatrix between the ingoing and the outgoing orbits in the wires. These continuous families end at the matching points between the wires and the arcs of disks where the shortest orbits of the four-disk scatterer exist (Gaspard et al. 1994).

The four-disk square scatterer can be analyzed with the same methods as the three-disk scatterer. The construction of a Markov partition allows us to establish a one-to-one correspondence between the orbits of the repeller and the symbolic dynamics based on the alphabet $\mathscr{A} = \{a, b, c, d\}$ composed of the labels of the disks and the topological transition matrix

$$\mathsf{A} = \begin{array}{c} \\ a \\ b \\ c \\ d \end{array} \begin{array}{c} \begin{array}{cccc} a & b & c & d \end{array} \\ \begin{pmatrix} 0 & 1 & 1 & 1 \\ 1 & 0 & 1 & 1 \\ 1 & 1 & 0 & 1 \\ 1 & 1 & 1 & 0 \end{pmatrix} \end{array} . \qquad (5.67)$$

The partition can be constructed as in the three-disk billiard. One example of such a partition is depicted in Fig. 5.13. Once again, the structure is reminiscent of a Smale horseshoe but here with three branches because the number of cells multiplies by 3 when an extra iteration is considered. The repeller is composed of unstable orbits which are uncountable but occupy a set of zero Lebesgue measure in the Birkhoff phase space so that here also the repeller is a fractal object as long as particles can escape out of the scatterer, i.e., for $r > 2$.

We have here the theorem:

The four-disk scatterer of square geometry such that $r > r_c$ has a continuously hyperbolic repeller which is in one-to-one correspondence with the Markov topological shift $\Sigma(\mathscr{A}, \mathsf{A}, \mathbb{Z})$. This repeller is transitive and of Axiom-A type.

Since the largest eigenvalue of the transition matrix is indeed equal to 3, our

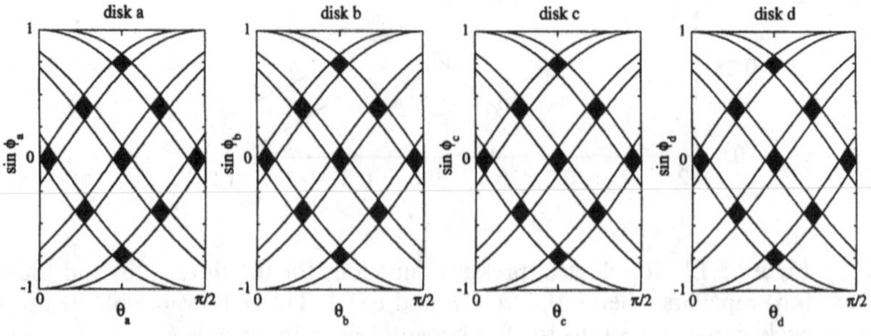

Figure 5.13 Partition of the repeller of the four-disk scatterer in the Birkhoff phase space.

expectation about the value of the topological entropy is confirmed. This result holds as long as the trajectories of the repeller are not tangent to the disks. The first tangency occurs when the heteroclinic point $W_u(ab) \cap W_s(ad)$ on disk a reaches $\varpi_a = \sin \phi_a = +1$. This critical value of the first heteroclinic tangency occurs at $r = r_c = 2.20469453\ldots$ (Hansen 1993). This value is higher than in the three-disk billiard because the angle between the centres of the disks b, a, and d is larger than between the disks b, a, and c of the three-disk system, which favors a possible tangency.

Symmetry and reduction to the fundamental domain

The C_{4v} symmetry can be used to reduce the dynamics to its fundamental domain which is one eighth of the full plane (see Fig. 5.14). In the fundamental domain, the orbits of the repeller can be coded with a reduced symbolic dynamics with only the three symbols $\tilde{\mathscr{A}} = \{0, +, -\}$ and without constraint in the way the symbols follow each other as in the three-disk billiard so that the new transition matrix is

$$\tilde{\mathsf{A}} = \begin{matrix} & \begin{matrix} 0 & + & - \end{matrix} \\ \begin{matrix} 0 \\ + \\ - \end{matrix} & \begin{pmatrix} 1 & 1 & 1 \\ 1 & 1 & 1 \\ 1 & 1 & 1 \end{pmatrix} \end{matrix} . \tag{5.68}$$

The reduction to this symbolic dynamics proceeds in two steps as follows. In the symbolic dynamics based on the four symbols $\{a, b, c, d\}$ which are the labels

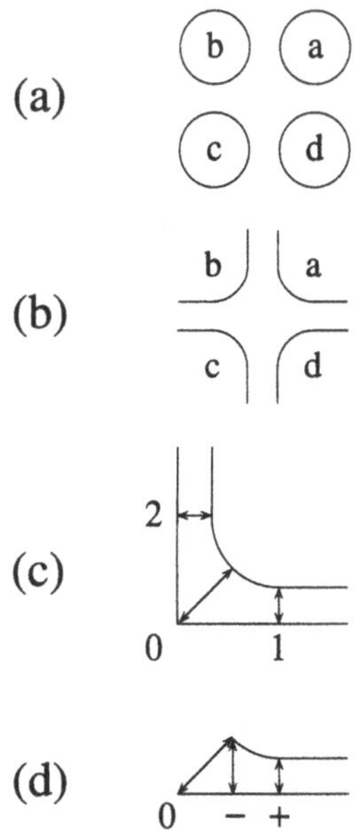

Figure 5.14 Definition of the symbolic dynamics in the C_{4v} billiards and the corresponding symmetry-reduced billiards: (a) four-disk billiard; (b) crossroad billiard; (c) billiard obtained by reduction of the billiard using the horizontal and vertical reflections; (d) fundamental domain of the billiard obtained from (c) using a further diagonal reflection.

of the four disks, there is the constraint that two successive symbols cannot be equal, so that the symbolic dynamics is actually ternary. In a first step, we perform the reduction to a fundamental domain formed by a quarter of the original billiard by using horizontal and vertical reflections (see Fig. 5.14c). In this fundamental domain, there exists a ternary symbolic dynamics without constraint based on the alphabet $\{0,1,2\}$. The symbol 0 corresponds to the periodic orbit along the diagonal while the symbols 1 and 2 correspond to the two shortest periodic orbits which are vertical and horizontal. In these new symbols $\{0,1,2\}$, the symbolic dynamics is free of constraint. The translation rules from $\{a,b,c,d\}$ to $\{0,1,2\}$ are derived by using horizontal and vertical reflections to fold the full orbit onto an orbit of the quarter of the billiard as given in Table 5.5.

In a second step, we can further reduce the symbolic description to the symbols $\{0,+,-\}$ by using a further reflection through the diagonal to go to the fundamental domain formed by an eighth of the plane. The translation rules are given in Table 5.6. In a sequence of symbols $\{0,1,2\}$, the symbols 0 remain unchanged. The symbol 1 (resp. 2) is replaced by + if, ignoring the symbols

Table 5.5 Translation rules between the symbolic dynamics $\{a,b,c,d\}$ with the constraint $\omega_n \neq \omega_{n+1}$ of the four-disk scatterer and the symbolic dynamics $\{0,1,2\}$ without constraint of the quarter of the billiard reduced by horizontal and vertical reflections.

$\{a,b,c,d\}$	$\{0,1,2\}$
ac, ca, bd, db	0
ad, da, bc, cb	1
ab, ba, cd, dc	2

Table 5.6 Translation rules between the symbolic dynamics $\{0,1,2\}$ without constraint and the symbolic dynamics $\{0,+,-\}$ without constraint of the fundamental domain reduced by a further diagonal reflection in the billiard.

$\{0,1,2\}$	$\{0,+,-\}$
10^n1, 20^n2	0^n+
10^n2, 20^n1	0^n-

0, it is preceded by 1 (resp. 2). Conversely, 1 (resp. 2) is replaced by – if it is preceded by 2 (resp. 1), here again ignoring the symbols 0 of the sequence. Accordingly, the two symbolic dynamics are equivalent.

Periodic orbits

The periodic orbits of the repeller can be obtained numerically in a systematic way thanks to the construction of the partition and to the symbolic dynamics. The first few periodic orbits are shown in Fig. 5.15. They appear as topological combinations of three fundamental orbits which are the three shortest ones. The orbits + and – are similar to the fundamental orbits of the three-disk scatterer. The new fundamental orbit 0 bounces along the diagonal of the square so that its length is in an approximate ratio of $\sqrt{2}$ with respect to the two others. This difference in the times of flight among the three fundamental orbits is an important structural difference between the three- and the four-disk scatterers, which increases the irregular character of the dynamics.

A calculation similar to the one performed above gives the following lengths and stability eigenvalues for the three fundamental periodic orbits

$$
\begin{aligned}
L_+ &= r-2\,,\\
L_- &= r-\sqrt{2}\,,\\
L_0 &= r\sqrt{2}-2\,,
\end{aligned}
$$

period 1	+	ab	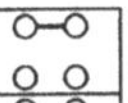
	–	abcd	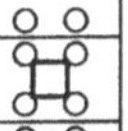
	0	ac	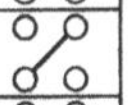
period 2	+ –	abad	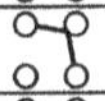
	0 +	abdc	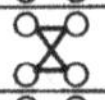
	0 –	abdacdbc	
period 3	+ + –	ababcbcdcdad	
	0 + +	abacdc	
	+ – –	abadcd	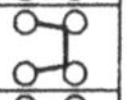
	0 + –	abacbc	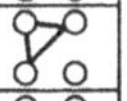
	0 – +	abdcbd	
	0 0 +	abdbac	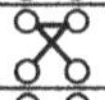
	0 – –	abc	
	0 0 –	abdbcacdbdac	

Figure 5.15 Shortest periodic orbits of the four-disk scatterer (Gaspard et al. 1994).

$$\begin{aligned}\Lambda_+ &= r - 1 + \sqrt{r(r-2)}\,, \\ \Lambda_- &= -r\sqrt{2} + 1 - \sqrt{2r(r-\sqrt{2})}\,, \\ \Lambda_0 &= r\sqrt{2} - 1 + \sqrt{2r(r-\sqrt{2})}\,. \end{aligned} \tag{5.69}$$

We notice the relations $L_0 = \sqrt{2}L_-$ and $|\Lambda_-| = |\Lambda_0|$ between the characteristic quantities of 0 and $-$. For large interdisk distances $r \gg 2$, the Lyapunov exponents behave as

$$\begin{aligned}\lambda_+ &\simeq \frac{1}{r}\ \ln 2r\,, \\ \lambda_- &\simeq \frac{1}{r}\ \ln 2\sqrt{2}r\,, \\ \lambda_0 &\simeq \frac{1}{r\sqrt{2}}\ \ln 2\sqrt{2}r\,. \end{aligned} \tag{5.70}$$

Therefore, the width of the cells of the partition becomes thinner and thinner as the size of the scatterer increases and the fractal repeller is more and more filamentary. On the contrary, when the disks are close to each other $r \simeq 2$, the escape is more and more inhibited because the opening between the disks decreases so that the fractal repeller is more and more bulky as the limit $r = 2$ is approached. For $2 < r < r_c$, the structural stability of the repeller is lost due to the heteroclinic tangencies of the trajectories to the disks, which cause bifurcations and a decrease of the topological entropy per bounce. At $r = 2$, the disks touch each other and the billiard is closed. In this limit, the repeller has filled the whole phase space.

We would like to emphasize the generality of the preceding symbolic description in two-degrees-of-freedom systems which have the C_{4v} symmetry: it is encountered not only in fourfold symmetric billiards but also in the hydrogen atom in a magnetic field (Eckhardt and Wintgen 1990) as well as in the transition state of triatomic molecules as discussed below. In this respect, the four-disk square billiard can be considered as a prototype of a large class of dynamical systems with a chaotic and fractal repeller.

The zeta functions

The scatterer is symmetric under the C_{4v} group which has four one-dimensional representations A_1, A_2, B_1, B_2, and one two-dimensional representation E. Accordingly, the zeta functions factorize as (Gaspard and Alonso Ramirez 1992)

$$\zeta = \zeta_{A_1}\,\zeta_{A_2}\,\zeta_{B_1}\,\zeta_{B_2}\,\zeta_E^2\,. \tag{5.71}$$

with

$$\zeta_{A_1}^{-1} = 1 - t_+ - t_- - t_0 - (t_{+-} - t_+ t_- + t_{0+} - t_0 t_+ + t_{0-} - t_0 t_-) - \cdots, \quad (5.72)$$

$$\zeta_{A_2}^{-1} = 1 + t_+ - t_- + (t_{+-} - t_+ t_- + t_{0+} - t_{0-}) + \cdots, \quad (5.73)$$

$$\zeta_{B_1}^{-1} = 1 - t_+ + t_- + (t_{+-} - t_+ t_- - t_{0+} + t_{0-}) + \cdots, \quad (5.74)$$

$$\zeta_{B_2}^{-1} = 1 + t_+ + t_- - t_0 + (t_{0+} - t_0 t_+ + t_{0-} - t_0 t_- - t_{+-} + t_+ t_-) + \cdots, \quad (5.75)$$

$$\zeta_{E}^{-1} = 1 + t_0 - t_+^2 + t_-^2 + (2\, t_{0++} - 2\, t_{0--} - t_0 t_+^2 + t_0 t_-^2) + \cdots, \quad (5.76)$$

where $t_{\bar{p}} = \exp(-sL_{\bar{p}})/|\Lambda_{\bar{p}}|$. The resonances can be calculated from these zeta functions. In the four-disk scatterer also, these complex resonances determine the time evolution of the scattering process of an ensemble of independent particles. Figure 5.16 shows the number $N(t)$ of particles which are still in the vicinity of the four-disk scatterer after the time t for $r = 12$ and $r = 24$. These decay curves present a gross exponential decay with oscillations superimposed on it. These oscillations are due to the other Pollicott–Ruelle resonances of the

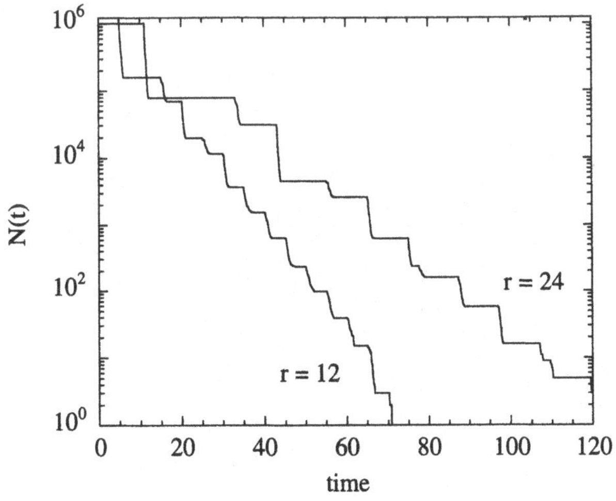

Figure 5.16 Decay curves of two ensembles of 10^6 particles in the four-disk scatterer for intercentre distances $r = 12$ and $r = 24$, radius $a = 1$, and velocity $v = 1$. The initial velocity is $\mathbf{v} = (1, 0)$ for all particles while the initial position is uniformly distributed between $\mathbf{q} = (0, r/2)$ and $\mathbf{q} = (0, r/2 + 1)$. $N(t)$ has the same definition as in Fig. 5.11 (Gaspard and Alonso Ramirez 1992).

systems beside the escape rate as for the three-disk scatterer. Indeed, Fig. 5.17 shows the power spectrum calculated with (5.60) $|\omega F(\omega)|^2$ in comparison with the A_1-resonances. We observe a nice agreement between the locations of the resonances and the peaks from the numerical simulation. This demonstrates the importance of the Pollicott–Ruelle resonances in the study of classical scattering systems.

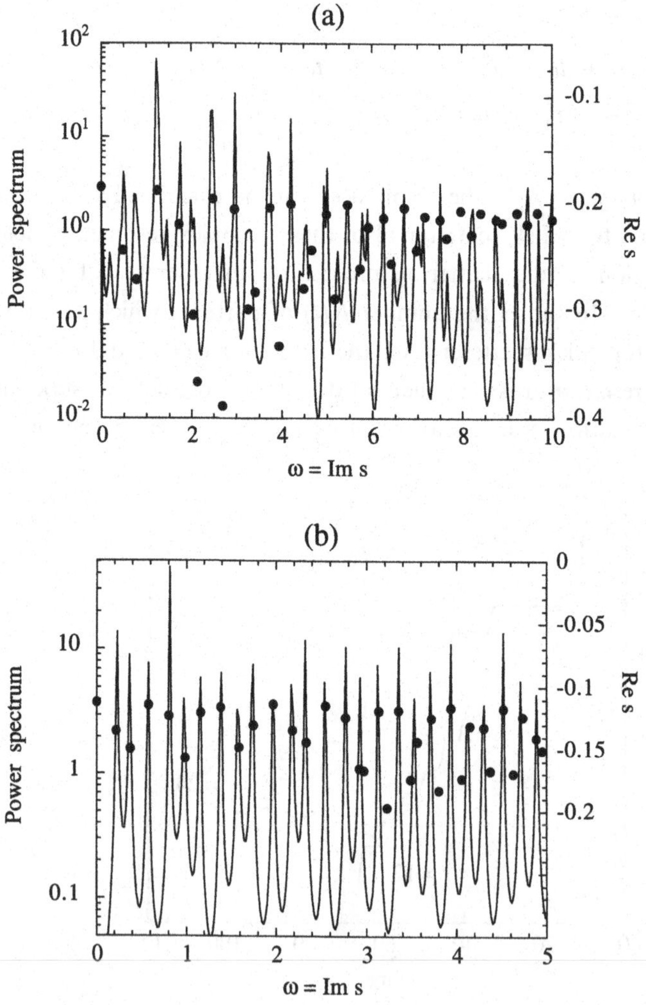

Figure 5.17 (a) Power spectrum of the oscillations appearing in Fig. 5.16 for the four-disk scatterer $r = 12$, versus $\omega = \text{Im } s$ and compared with the location of the classical A_1-resonances for $r = 12$ in the complex plane of s. The power spectrum is here defined as $\omega^2|F(\omega)|^2$ with $F(\omega)$ given by (5.60). (b) Same as (a) for the interdisk distance $r = 24$ (Gaspard and Alonso Ramirez 1992).

Characteristic quantities of chaos

According to Chapter 4, these quantities can be calculated from the leading pole of the zeta function ζ_{A_1} given by (5.72) with $t_{\tilde{p}} = \exp(-sL_{\tilde{p}})/|\Lambda_{\tilde{p}}|^{\beta}$.

In the four-disk scatterer, there are three fundamental periodic orbits. In the limit $r \to \infty$, two of them have comparable lengths while the other has a different length in the ratio $\sqrt{2}$ with respect to the other ones. This difference in the lengths has the consequence that the spectrum of resonances is actually much less regular than in the three-disk scatterer. Another consequence is that, when $\beta \neq 0$, the zeta function (5.72) is dominated at large values of r by the term of the diagonal orbit 0 so that the topological pressure behaves asymptotically as

$$P(\beta) \simeq -\frac{\beta}{2^{1/2}r} \ln 2^{3/2}r \qquad (\beta \neq 0) . \tag{5.77}$$

However, the topological entropy must be evaluated differently at $\beta = 0$. Setting $\xi = \exp(-sr)$, we can obtain an equation for ξ as $\zeta_{A_1}^{-1} \simeq 1 - 2\xi - \xi^{\sqrt{2}} = 0$. Its root gives us the asymptotic behaviour of the topological entropy as

$$h_{\text{top}} \simeq \frac{1}{r} \ln 2.6661789 . \tag{5.78}$$

This result shows how much the times of flight of the different orbits can affect the characteristic quantities of chaos per unit time compared with those evaluated per bounce.

For $r = 6$, the pressure function has been numerically calculated with a total of 32 periodic orbits up to period 4 in the fundamental domain (see Fig. 5.12). The characteristic quantities of chaos calculated with the pressure are given in Table 5.4.

The comparison with the three-disk scatterer shows that, for the same separation between the disks, the KS and the topological entropy are significantly higher in the four-disk scatterer which appears therefore more chaotic for the same interdisk distance. In the same way, the Hausdorff and information dimensions of the fractal repellers are bigger so that the four-disk repeller is more bulky and less filamentary than the three-disk repeller. We may also notice that the topological and KS entropies differ more in the four-disk scatterer than in the three-disk scatterer, which is an effect due to the increase in the number of fundamental orbits. A similar remark holds for the difference between the information and the Hausdorff dimensions.

5.3 Hamiltonian mapping of scattering type

5.3.1 Definition of the model

A further example of chaotic scattering is given by the following periodically kicked Hamiltonian (Rice et al. 1992)

$$H(q,p,t) = \frac{p^2}{2} + v(q) \sum_{n=-\infty}^{+\infty} \delta(t-n) , \tag{5.79}$$

with the function

$$v(q) = d \exp(-q) (q^2 + q + 1) . \tag{5.80}$$

This Hamiltonian describes a particle freely moving on a line except every unit time when it undergoes a kick with an amplitude depending on its current position according to the function (5.80). This function vanishes as $q \to +\infty$ so that the particle is in free motion as $q \to +\infty$. This Hamiltonian system is therefore a model of scattering.

This periodically driven system with one degree of freedom is formally equivalent to a time-independent Hamiltonian system with two degrees of freedom as encountered in the study of unimolecular reactions of van der Waals complexes (Gray et al. 1986, Noid et al. 1986, Tersigni et al. 1990). In these special molecules, a rare-gas atom is weakly bound to a standard molecule by a van der Walls interaction as in HeI_2 (Ewing 1978). When the iodine molecule is excited its vibrational motion periodically drives the motion of the helium atom which soon escapes in a translational motion. The iodine molecule may be so highly excited that its amplitude of vibration does not change much during the dissociation, which can be adequately modelled by Eqs. (5.79)–(5.80) where Eq. (5.80) mimics the dynamical effect of a van der Waals potential.

The above model illustrates a large variety of possible behaviours which can be visualized in the form of phase portraits in the phase plane (q, p). In this plane, the system (5.79) induces a Poincaré map. Taking q and p as the position and momentum before each kick, the map

$$\begin{cases} p_{n+1} = p_n - v'(q_n) , \\ q_{n+1} = q_n + p_{n+1} , \end{cases} \tag{5.81}$$

where $v'(q) = dv/dq$, gives the coordinates before the next kick in terms of the coordinates before the previous kick. This map is area-preserving and has two fixed points:

(1) The origin, $(q = 0, p = 0)$, is a centre when $0 < d < 4$ because the mapping acts as a rotation in its vicinity. Its associated stability eigenvalues

are complex conjugates, $\Lambda_{\pm} = \exp(\pm 2\pi i \rho)$, with the rotation number: $\rho = (1/2\pi)\arccos(1 - d/2)$. At $d = 4$, the origin becomes a saddle with real and negative stability eigenvalues

$$\Lambda_0 = -d/2 + 1 - \sqrt{d(d/4-1)}\,, \qquad \text{and} \qquad 1/\Lambda_0\,. \tag{5.82}$$

(2) The other fixed point, $(q = 1, p = 0)$, is always a saddle for $d > 0$ with positive eigenvalues given by

$$\Lambda_1 = d/2e + 1 + \sqrt{d/e(d/4e+1)}\,, \qquad \text{and} \qquad 1/\Lambda_1\,. \tag{5.83}$$

5.3.2 Metamorphoses of the phase portraits

Figure 5.18 shows several phase portraits as the parameter d increases. These portraits are constructed by plotting several trajectories which remain trapped in the interaction region defined by $U = \{(q,p) : -1 < q < +1.5, -2 < p < +2\}$. The trajectories which exit this domain are not plotted. An invariant set is defined as the set of all the trajectories which remain at finite distance under forward and backward iterations of the map. All initial conditions with $q > 1$ go to infinity under backward or forward iterations. Similarly, for initial conditions with $q < -1$, the kicking amplitude is so strong that the particle is immediately sent at positive q. Therefore, we may expect the invariant set to occur in the neighbourhood U defined above if the parameter d is not too large ($d < 8$). The invariant set undergoes metamorphoses shown in Fig. 5.18. The qualitative features of these portraits can be explained in terms of changes in the stability of the fixed point at the origin.

The centre at the origin undergoes strong classical resonances when the rotation number takes one of the three values, $\rho = 1/4, 1/3, 1/2$, at the parameter values $d = 2, 3$, and 4, respectively (Moser 1958, Arnold 1978, Ozorio de Almeida 1988, MacKay 1993b). At the strong resonances the centre may be nonlinearly unstable with trajectories escaping to infinity even from nearby the centre. On the other hand, for all other values (rational or irrational) of the rotation number the centre is enclosed by concentric KAM tori, which guarantees its stability (Kolmogorov 1954; Arnold 1978, 1988; Moser 1973).

The main elliptic island is delimited by the last of these tori. Beyond this last torus, a chaotic zone extends where unstable trajectories are mixed with very small secondary elliptic islands. All those islands are evidence of the nonhyperbolicity of the system in these regimes. In each elliptic island, there exist uncountably many ergodic components which are concentric irrational tori. These irrational tori are separated by extremely thin chaotic zones resulting from the destruction of rational tori. These chaotic zones are trapped inside

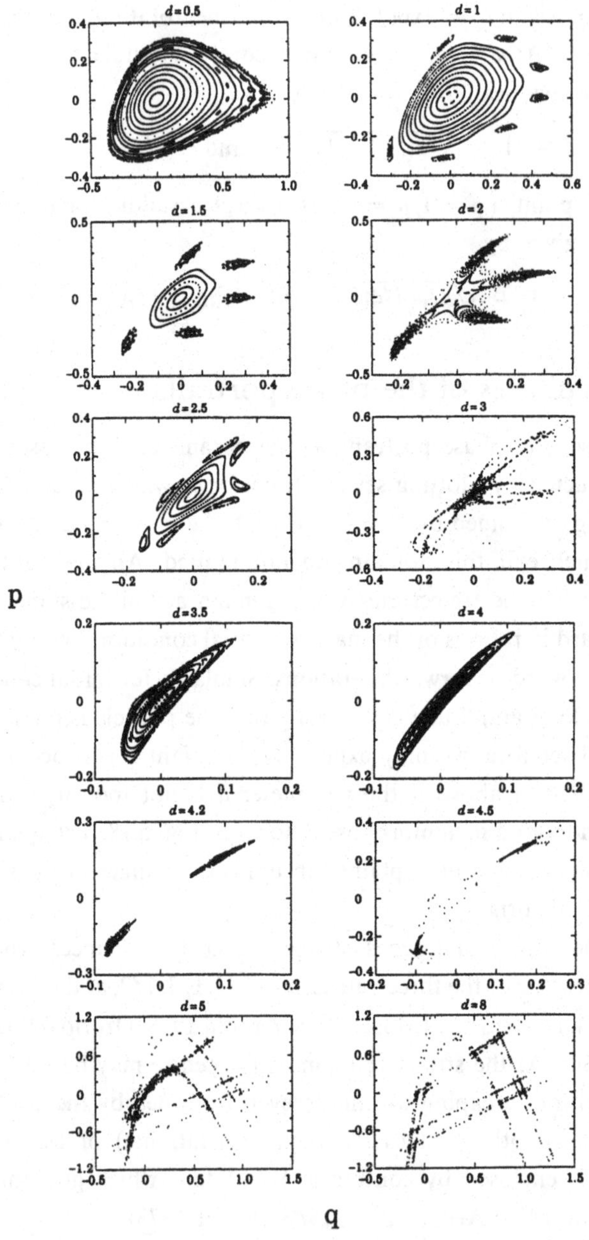

Figure 5.18 Phase portraits of the Hamiltonian map (5.80)–(5.81) for different values of the parameter d (Rice et al. 1992).

the island so that no trajectory can escape from these inner chaotic zones. On the contrary, the outer chaotic zone outside the main island is open with respect to the escape of trajectories to infinity and is therefore a stage of the chaotic scattering process. Indeed, it has been shown that phase-space transport is possible in a chaotic zone due to the homoclinic orbits (MacKay et al. 1984, Bensimon and Kadanoff 1984, Meiss 1992, Wiggins 1992).

In these regimes, the invariant set is nonhyperbolic with uncountably many ergodic components which, moreover, form a very complex hierarchical structure in phase space. The area of the invariant set is positive as shown in Fig. 5.19 so that the Hausdorff dimension is equal to two which is the phase-space dimension. Because of the phenomenon of sticking of trajectories to the tori

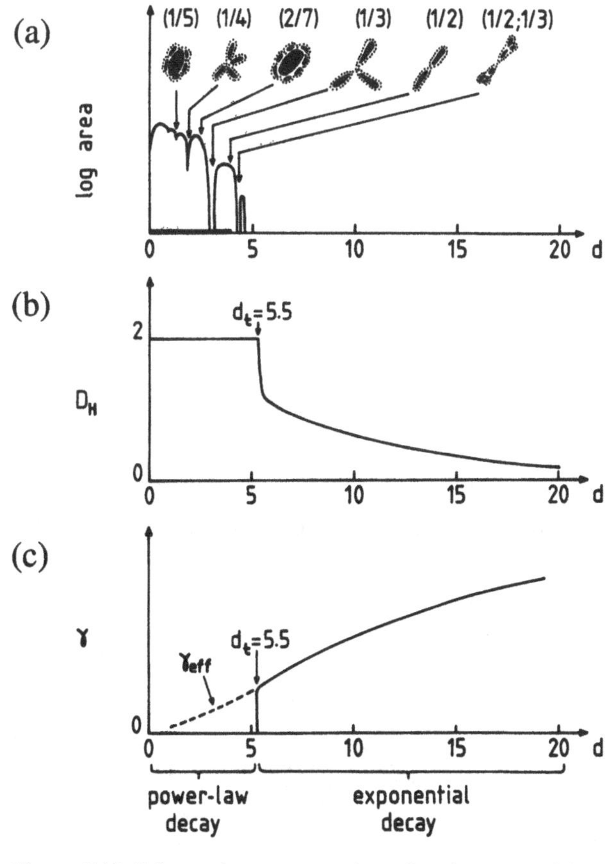

Figure 5.19 Schematic representation of various quantities as a function of the parameter d. (a) The logarithm of the area of the invariant set and the phase portraits at the main resonances. (b) The Hausdorff dimension of the invariant set. (c) The escape rate of the invariant set. The last homoclinic tangency occurs at $d_t \simeq 5.5$ (Rice et al. 1992).

the escape process is very complicated with numerically observed power-law behaviour (Meiss and Ott 1985). Only an effective escape rate could be defined as explained in Chapter 4.

At $d = 4$, the origin undergoes a period-doubling bifurcation and, for $d > 4$, it becomes unstable. The period doubling is followed by a cascade of period doublings, which progressively destroy the main elliptic island. Therefore, the area of the invariant set decreases rapidly. However, there remain tiny elliptic islands as long as the stable and unstable manifolds of the saddle-type orbits have mutual tangencies, called homoclinic tangencies. It is known that these homoclinic tangencies generate periodic orbits of centre type (Guckenheimer and Holmes 1983, MacKay 1993b). Consequently, the invariant set is nonhyperbolic as long as homoclinic tangencies occur. Nevertheless, a last homoclinic tangency occurs at $d_t \simeq 5.5$ between the stable and unstable manifolds of the fixed point $(q = 1, p = 0)$.

After this last homoclinic tangency, the invariant set is a Smale horseshoe and is therefore continuously hyperbolic of Axiom-A type (see Fig. 5.20). It is quite remarkable that such Axiom-A systems occur so easily in scattering systems. A Markov partition of the invariant set in two cells 0 and 1 establishes a one-to-one correspondence between the trapped trajectories of the repeller and a symbolic dynamics based on the alphabet $\mathscr{A} = \{0, 1\}$ and the transition matrix

$$\mathsf{A} = \begin{matrix} & \begin{matrix} 0 & 1 \end{matrix} \\ \begin{matrix} 0 \\ 1 \end{matrix} & \begin{pmatrix} 1 & 1 \\ 1 & 1 \end{pmatrix} \end{matrix} , \tag{5.84}$$

which defines a dyadic Bernoulli shift of topological entropy $h_{\text{top}} = \ln 2$. This horseshoe is the unique transitive nonwandering invariant set in the phase space

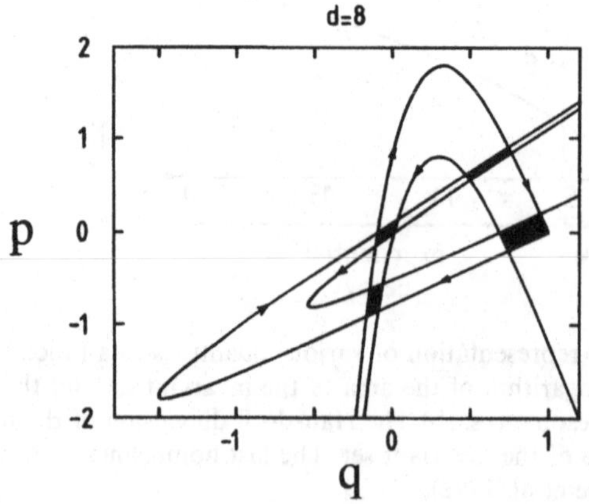

Figure 5.20 Formation of a Smale horseshoe in the Hamiltonian mapping (5.80)–(5.81) at $d = 8$ and construction of a Markov partition in four cells, with the stable and unstable manifolds of the fixed point $(q = 1, p = 0)$ (Rice et al. 1992).

so that a single ergodic component is left after the bifurcation cascade. The Smale horseshoe continues to exist for arbitrarily large values of the parameter d.

This hyperbolic repeller may be characterized by the quantities defined in Chapters 3 and 4 and, in particular, by its Hausdorff dimension. Indeed, the chaotic repeller is uncountable but of zero Lebesgue measure and is thus a typical fractal of Hausdorff dimension lower than two as depicted in Fig. 5.19. Similarly, the escape process is now exponential at long times because the repeller is hyperbolic.

5.3.3 Characteristic quantities of the chaotic repeller

The different characteristic quantities of the Smale horseshoe of the model (5.80)–(5.81) can be obtained asymptotically in the limit $d \to \infty$ as follows. These quantities can be derived from the topological pressure which is the leading pole of the Ruelle zeta function (4.102): $s = P(\beta)$.

In the case of the Smale horseshoe, the symbolic dynamics is binary so that the inverse Ruelle zeta function starts as

$$\begin{aligned}\zeta_0(s;\beta)^{-1} &= \prod_p (1-t_p) = (1-t_0)(1-t_1)(1-t_{01})\cdots \\ &= 1 - t_0 - t_1 - (t_{01} - t_0 t_1) - \cdots = 0\,. \end{aligned} \tag{5.85}$$

where $t_p = \exp(-sn_p)/|\Lambda_p|^\beta$. The terms t_0 and t_1 are completely determined by the stability eigenvalues (5.82) and (5.83) noting that $n_0 = n_1 = 1$. When d is large, the stability eigenvalues behave as $\Lambda_0 \simeq -d$ and $\Lambda_1 \simeq d/e$. On the other hand, the terms in parenthesis in Eq. (5.85) may be neglected as $d \to \infty$, because they involve the differences between quantities of the same order. In this limit, the pressure can thus be calculated from the three first terms, $t_0 + t_1 \simeq 1$, which gives

$$P(\beta) \simeq \ln\left[(1/d)^\beta + (e/d)^\beta\right], \qquad \text{for} \quad d \to \infty\,. \tag{5.86}$$

From this pressure, we deduce the escape rate (4.63), the mean Lyapunov exponent (4.71), the KS entropy (4.72), and the Hausdorff dimension (4.79)

$$\gamma \simeq \ln \frac{d}{3.718}\,, \tag{5.87}$$

$$\bar{\lambda} \simeq \ln \frac{d}{2.077}\,, \tag{5.88}$$

$$h_{\text{KS}} \simeq 0.5822 < h_{\text{top}} = \ln 2 = 0.6931\,, \tag{5.89}$$

$$D_{\text{H}} = 2\, d_{\text{H}} \simeq 2\,\frac{\ln 2}{\ln d}\,, \tag{5.90}$$

as $d \to \infty$. These results have been used to deduce Fig. 5.19. These estimations are comparable to those carried out for the three-disk scatterer in Eqs. (5.61)–(5.66) except that the time between successive passages in the Poincaré surface of section is here constant. As a consequence, both the escape rate and the mean Lyapunov exponent increase logarithmically as $\ln d$. On the other hand, the KS entropy which is the difference between them tends to a constant. Because the stability eigenvalues are in the ratio $|\Lambda_0/\Lambda_1| \simeq e$ at large values of d, the probabilistic Bernoulli process based on the symbols 0 and 1 does not give equal probabilities to both symbols. Since we know that the KS entropy reaches its possible maximum value $h_{\rm top}$ only for equal probabilities we can understand that the KS entropy of the present system is smaller than the topological entropy, as shown by (5.89). We also notice that the Hausdorff dimension decreases to zero as the parameter d increases because the increased dynamical instability causes the repeller to become more and more filamentary, which is the same phenomenon as in the disk scatterers [cf. Eq. (5.66)].

5.4 Application to the molecular transition state

In the theory of unimolecular reactions, the hyperbolic repeller appears as the invariant set classically corresponding to the transition state. Indeed, this repeller controls the classical dissociation dynamics and, in particular, it determines the value of the escape rate which is the classical unimolecular reaction rate. The molecular transition state is a fundamental concept in our understanding of chemical reactions. The transition state is usually considered in a statistical context in the so-called RRKM theory (Holbrook et al. 1996). The present considerations shed a new light on this important problem.

5.4.1 Model of photodissociation of HgI_2

Burghardt and Gaspard (1994, 1995) have studied the photodissociation of HgI_2

$$h\nu + HgI_2(X^1\Sigma_g^+) \rightarrow [I\cdots Hg\cdots I]^\ddagger \rightarrow HgI(X^2\Sigma^+) + I(^2P_{3/2}) . \tag{5.91}$$

This study has been motivated by the femtosecond laser experiments on HgI_2 which have recently been reported by Zewail and coworkers (Dantus et al. 1989, Zewail 1991). The transition state is assumed to be collinear so that the dynamics of the HgI_2 system is modelled according to the following two-degrees-of-freedom Hamiltonian

$$H = T + V = \frac{1}{2\mu_{HgI}} (p_1^2 + p_2^2) - \frac{1}{m_{Hg}} p_1 p_2 + V(r_1, r_2) , \tag{5.92}$$

where $\mu_{\rm HgI} = (m_{\rm Hg}^{-1} + m_{\rm I}^{-1})^{-1}$ is the relative mass of the Hg–I subsystem. The potential energy surface is a damped Morse potential for two degrees of freedom (the two interatomic distances) as proposed by Dantus et al. (1989)

$$V(r_1,r_2) = D\left\{1-\exp\left[-\beta\frac{e^{-\gamma r_1}(r_1-r_e)/r_e+e^{-\gamma r_2}(r_2-r_e)/r_e}{e^{-\gamma r_1}+e^{-\gamma r_2}}\right]\right\}^2 - D\,, \tag{5.93}$$

where the quantities D, β and r_e have further dependencies on r_1 and r_2 according to

$$\begin{aligned} D(r_1,r_2) &= D_0 + D_1 \exp[-(r_1-r_2)^2/\sigma_D^2]\,, \\ r_e(r_1,r_2) &= r_{e0} + r_{e1} \exp[-(r_1-r_2)^2/\sigma_{r_e}^2]\,, \\ \beta(r_1,r_2) &= \beta_0 + \beta_1 \exp[-(r_1-r_2)^2/\sigma_\beta^2-(r_1+r_2-2r_e)^2/\sigma_{\beta_+}^2]\,, \end{aligned} \tag{5.94}$$

with the parameters given in Table 5.7.

This potential is typical of many chemical reactions. It presents a saddle point at an intermediate energy above two exit channels for the products of the reaction. A nontrivial classical invariant set exists at energies above the energy of the saddle point.

As energy increases, the invariant set undergoes bifurcations which are similar to the bifurcations of the Hamiltonian map (5.81) but here with the following function

$$v(q) = \left(a + b\,q + c\,\frac{q^2}{2}\right)\exp\left(-\frac{q^2}{2}\right) \quad \text{with} \quad a,\ c > 0\,. \tag{5.95}$$

The map defined by Eqs. (5.81) and (5.95) can be considered as a model for the Poincaré map of the system (5.92)–(5.94). The symmetry of the molecule I–Hg–I implies that the function (5.95) has to be symmetric under $q \to -q$ so that $b = 0$. An increase in energy corresponds to an increase in the parameter c with $a = 1$ for instance. The bifurcation diagram of the map is depicted in Fig. 5.21 in correspondence with the bifurcations of the flow.

Table 5.7 Set of parameters for the dissociative potential energy surface (5.93)–(5.94) of HgI_2 (Dantus et al. 1989).

$D_0 = 2800\ \text{cm}^{-1}$	$D_1 = -1000\ \text{cm}^{-1}$	$\sigma_D = 1\text{Å}$
$r_{e0} = 2.8\text{Å}$	$r_{e1} = 0.2\text{Å}$	$\sigma_{r_e} = 1\text{Å}$
$\beta_0 = 7.1$	$\beta_1 = 5.5$	$\sigma_\beta = 0.75\text{Å},\ \sigma_{\beta_+} = 2.5\text{Å}$
$\gamma = 2.3\text{Å}^{-1}$		

5.4.2 Transition from a periodic to a chaotic repeller

At energies just above the saddle point, there is a single trapped orbit which is periodic and unstable as in the two-disk scatterer. This periodic repeller is characterized by an escape rate which is equal to the positive Lyapunov exponent of the periodic orbit 0 describing the symmetric stretch of the molecule.

This periodic orbit undergoes an anti-pitchfork bifurcation at the energy E_a. Above, the periodic orbit is of centre type and is surrounded by a main elliptic island. Two new fundamental periodic orbits 1 and 2 are born at this bifurcation. They are unstable and they delimit the phase-space region where the main elliptic island is located. These new periodic orbits correspond to asymmetric stretching of the molecule. In this regime, the invariant set is nonhyperbolic. But the energy range $[E_a, E_{ht}]$ over which the invariant set is nonhyperbolic has a small extension as seen in Fig. 5.22, which shows the escape rate and the Lyapunov exponents of the fundamental periodic orbits. The main elliptic island undergoes a cascade of bifurcations which is very similar to the one depicted in Fig. 5.18. In particular, there is a period doubling at E_d. The cascade ends with a last homoclinic tangency at E_{ht} above which the invariant set is hyperbolic.

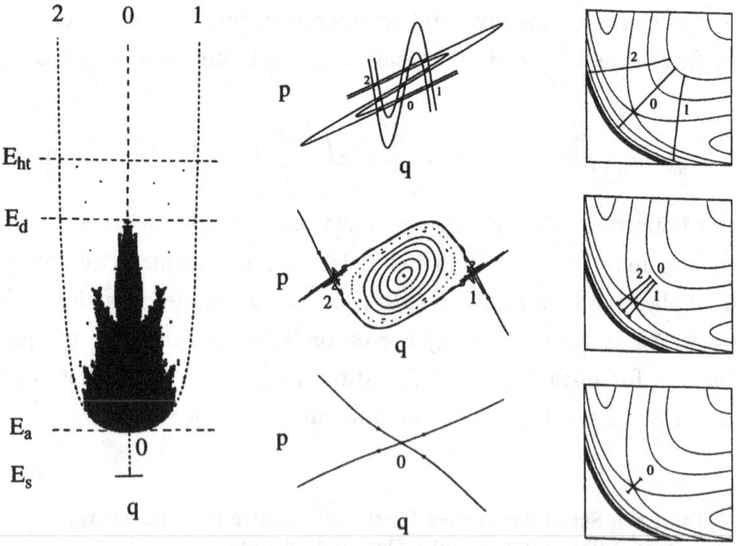

Figure 5.21 Bifurcations of the map (5.81) with (5.95) in correspondence with those of the flow of the model (5.92)–(5.94). The bifurcation diagram is depicted on the left-hand panel as the energy E or the parameter c increases. In the middle panel, several typical phase portraits are presented. The right-hand panel shows the behaviour of the fundamental periodic orbits of the invariant set (Gaspard and Burghardt 1997).

5.4.3 Three-branched Smale repeller and its characterization

A density plot of the sum of the absolute values of the escape-time functions (2.17)–(2.18) is depicted in Fig. 5.23 which reveals that the hyperbolic repeller of the flow is a three-branched Smale repeller (Smale 1980b). Therefore, the trapped trajectories of the repeller form a ternary Bernoulli shift on the symbols $\{0, 1, 2\}$

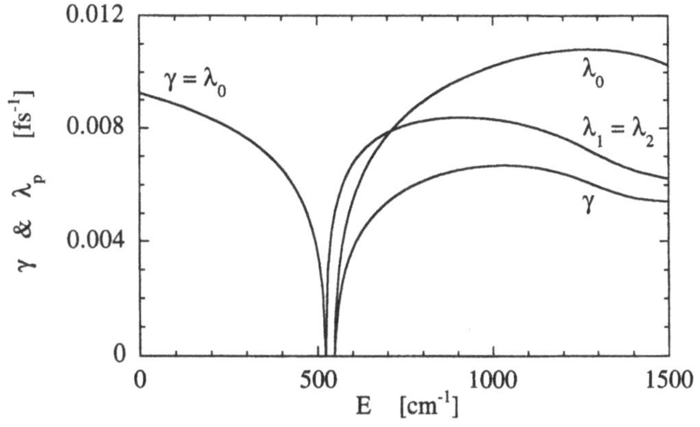

Figure 5.22 Escape rate and the Lyapunov exponents of the fundamental periodic orbits 0, 1, and 2 versus energy above the saddle point in the model (5.92)–(5.94) (Gaspard and Burghardt 1997).

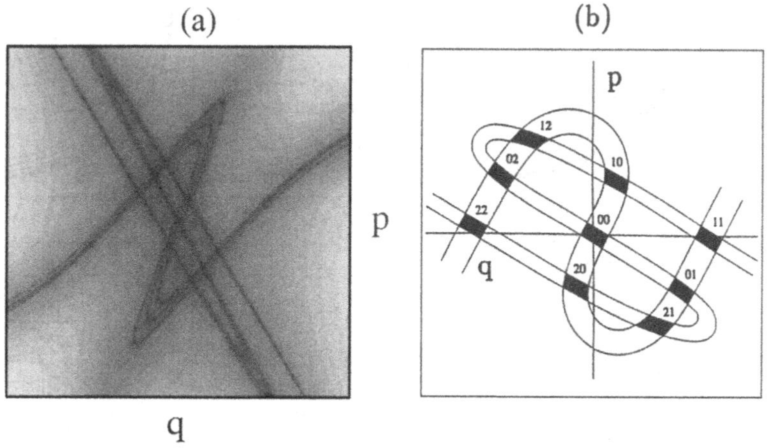

Figure 5.23 (a) Three-branched Smale repeller of the model (5.92)–(5.94) at the energy $E = 600\ \text{cm}^{-1}$ above the saddle point (Burghardt and Gaspard 1995, Gaspard and Burghardt 1997). (b) Schematic Markov partition of the horseshoe with the symbolic assignment of the cells (Burghardt and Gaspard 1994).

of the three fundamental periodic orbits. Figure 5.24 shows the first few periodic orbits of the repeller, where we observe that the periodic orbits are topological combinations of the three fundamental periodic orbits. We notice that the symbolic dynamics in this repeller is the same as in the four-disk scatterer.

The knowledge of these periodic orbits can be used to calculate the characteristic quantities of the fractal repeller. The inverse Ruelle zeta function can here be approximated by a cycle expansion limited to the three fundamental periodic orbits

$$\begin{aligned}\zeta_0(E)^{-1} &= \prod_p [1 - t_p(E)] \\ &= 1 - t_0 - t_1 - t_2 - (t_{01} - t_0 t_1) - (t_{02} - t_0 t_2) - (t_{12} - t_1 t_2) - \ldots \\ &\simeq 1 - t_0 - t_1 - t_2 \,, \end{aligned} \tag{5.96}$$

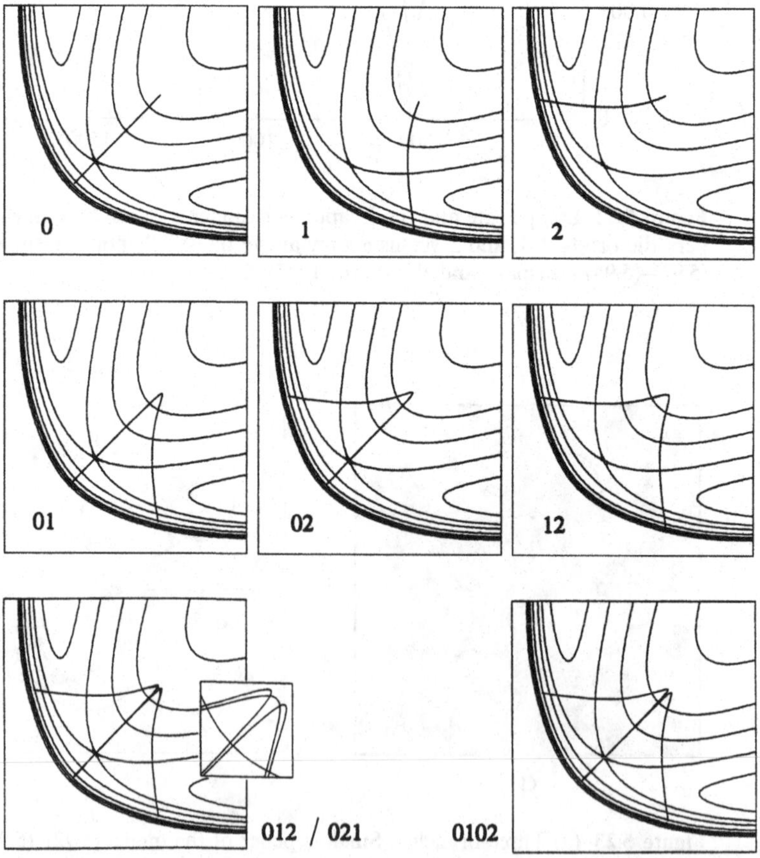

Figure 5.24 First few periodic orbits of the three-branched Smale repeller of the model (5.92)–(5.94) at the energy $E = 1200\ \mathrm{cm}^{-1}$ above the saddle point (Burghardt and Gaspard 1994).

where

$$t_p(E) = \frac{\exp[-sT_p(E)]}{|\Lambda_p(E)|^\beta} \,. \tag{5.97}$$

The other terms in (5.96) may be neglected because the higher periodic orbits are well approximated by combinations of the fundamental periodic orbits. The Lyapunov exponents of the fundamental periodic orbits and the resulting escape rate $\gamma = -P(1;E)$ are shown in Fig. 5.22. We notice that the periods T_p, the stability eigenvalues Λ_p and, consequently, the characteristic quantities of chaos vary with energy in this example.

The escape rate is of particular interest because it corresponds to the classical dissociation rate of the molecule. The classical lifetimes are here of the order of 100–200 fs. We should point out that the quantum scattering resonances can be obtained with a very similar method based on the semiclassical Gutzwiller trace formula (Burghardt and Gaspard 1994, 1995). The quantum resonances are required for the interpretation of the experimental data. In regimes where the density of quantum resonances becomes very high, the classical time evolution provides a good approximation of the quantum one and the above classical quantities become relevant (Gomez Llorente and Taylor 1989, Gaspard and Burghardt 1997, Gaspard and Jain 1997). Therefore, the classical Liouvillian dynamics can be used to characterize the time evolution of quantum density matrices in the quasiclassical limit.

Recent works have shown that the three-branched Smale repeller is very general in such direct unimolecular reactions. This repeller has also been observed in CO_2 and is suspected to exist in other molecules as well (Gaspard and Burghardt 1997). We see that chaotic scattering acquires a particular importance in this perspective.

5.5 Further applications of chaotic scattering

Many different systems with a chaotic repeller have been studied in recent years (Eckhardt 1988, Hénon 1988, Tél 1989, Troll and Smilansky 1989, Smilansky 1991, Jung and Tél 1991, Ott and Tél 1993). In this regard, it turns out that the methods of hyperbolic dynamical systems are applicable to many smooth Hamiltonian and nonHamiltonian systems, which is a most remarkable feature of the theory. Chaotic scattering has therefore many potential applications.

In particular, chaotic repellers exist in the Lagrangian fluid dynamics, i.e., in the flow of the trajectories of the fluid particles themselves (Jung et al. 1993). Recently, experimental evidence has been obtained for chaotic scattering in a fluid wake (Sommerer et al. 1996). In this remarkable experiment, the fractal dimensions, the Lyapunov exponents, and the escape-time statistics of the repeller have been measured by observation of tracer particles and dye in the wake of a moving cylinder, leading to a quantitative test of the escape-rate and Young formulas (4.80). In this hydrodynamic phenomenon, the chaotic repeller turns out to play an unexpected role in the drag properties of the cylinder. Further study would certainly be very interesting in view of the nonhyperbolic character of these hydrodynamic repellers.

Chaotic scattering also arises in celestial mechanics where it may affect the motion of bodies across the Solar System (Petit and Hénon 1986). Moreover, scattering on three attracting Yukawa potentials has been proved to be chaotic and fractal at high enough energy (Knauf 1987, Klein and Knauf 1992).

High-dimensional models have also been considered like the scattering on fixed hard spheres and, in particular, on the four-sphere scatterer in the form of a tetraedron where the symbolic dynamics is *a priori* composed of four symbols (Chen et al. 1990, Ott 1993).

Signatures of chaotic scattering are also observed in quantum systems like the hydrogen atom in a magnetic field (Eckhardt and Wintgen 1990) or the two-electron atoms and ions and, in particular, helium and the hydrogen negative ion (Wintgen et al. 1992, 1994; Gaspard and Rice 1993). The classical dynamics of these systems shows a chaotic invariant set. For helium and the hydrogen negative ion, the invariant set is a hyperbolic repeller with unstable periodic orbits. In contrast with the repeller of unimolecular reactions, the repeller of these Coulomb-type systems extends to large distances and is not compact. Moreover, a simple model by Blümel (1993) shows that algebraic decays may be expected for the chaotic repellers of Coulomb type. These algebraic decays may explain certain recent experiments on Rydberg atoms (Buchleitner et al. 1995). The classical analysis may be relevant to these quantum systems in quasiclassical regimes because the quantum scattering resonances accumulate near thresholds in Coulomb-type systems. On the one hand, a semiclassical quantization of these systems is possible with the Gutzwiller trace formula in terms of the periodic orbits (Gutzwiller 1990). On the other, the accumulation of energy eigenstates may justify the use of classical Liouvillian dynamics to approximate the time evolution of quantum density matrices representing quantum statistical ensembles.

As aforementioned, chaotic scattering is also important in the study of electronic conduction in semiconducting mesoscopic devices (Doron et al. 1991, Smilansky 1991). Here also, the classical Liouvillian resonances may characterize the time evolution of quantum statistical ensembles in quasiclassical regimes.

In this context, billiards with a magnetic field have also been studied as models of electronic circuits in a magnetic field (Beenakker and van Houten 1989, Marcus et al. 1993). The flights between collisions are arcs of circles instead of straight lines. By structural stability, we expect that the hyperbolic repeller keeps its topology at low magnetic fields up to a critical value where the topology changes.

These different examples show the growing importance of chaotic scattering in many fields. In the following chapter, we shall see how chaotic scattering arises in spatially extended systems.

Chapter 6

Scattering theory of transport

6.1 Scattering and transport

When scattering occurs in systems of large spatial extension like solids, a relationship appears between scattering and transport. This is particularly evident in the process of transport of thermal neutrons in nuclear reactors. On the one hand, neutrons may form beams for the study of matter and, on the other hand, as soon as the target presents spatially distributed scattering centres, the scattering process becomes a diffusion process, i.e., a process of transport as studied in kinetic theory. Clearly, both processes are similar and the difference only appears in the number and distribution of scatterers. Therefore, a fundamental connection exists between scattering theory and nonequilibrium statistical mechanics. The scattering approach to diffusion is also natural since diffusion is studied in finite pieces of material in the laboratory. Diffusion is a property of bulk matter which is extrapolated from experiments on finite samples to a hypothetical infinite sample.

Classically, the scattering on a spatially distributed target may be expected to be chaotic because the collisions on spherical scatterers have a defocusing character. Chaoticity will play an important role in such a connection. In the following, we shall elaborate in this direction with the tools developed in the previous chapters to obtain the so-called escape-rate formulas for the transport coefficients, which precisely express such a relationship (Gaspard and Nicolis 1990, Gaspard and Baras 1995).

We should mention here that Lax and Phillips (1967) proposed in the sixties a scattering theory of transport phenomena based on the properties of classical dynamics. These authors have pointed out the importance of trapped orbits and of what is today called the stable and unstable manifolds of the trapped orbits, and developed their theory from such observations. In a sense, we shall follow a very similar way, establishing a connection with dynamical systems theory and its large-deviation formalism in order to obtain the escape-rate formulas.

In the following, we shall first consider the scattering on a large array of disks, which is an open Lorentz gas. On such a large scatterer, the erratic motion of a point particle in elastic collision is of diffusive type so that the escape process out of the scatterer is related, on the one hand, to diffusion and, on the other hand, to the characteristic quantities of chaos in the repeller. After revealing general results, we analyze in detail the periodic Lorentz gas and the multibaker model which both sustain deterministic diffusion. The theory is then generalized to the other transport coefficients as well as to the chemical reaction rates.

6.2 Diffusion and chaotic scattering

6.2.1 Large scatterers and the diffusion equation

Following the idea of Chapter 5, we consider scattering systems of larger and larger size by adding more and more disks to the scatterer. If this construction continued indefinitely we would cover the whole plane with disks, forming what is called a Lorentz gas. The Lorentz gases were originally constructed with a random distribution of disks (Lorentz 1905, van Beijeren 1982). More recent works, in particular by Bunimovich and Sinai (1980), have been devoted to periodic Lorentz gases where the disks form a lattice. The Lorentz gases are models of deterministic diffusion, in which a point particle undergoes elastic collisions on the disks. Conditions for the existence of a positive and finite diffusion coefficient have been determined. As a consequence, we expect that a statistical ensemble of independent point particles will evolve on large spatial scales and at long times according to the diffusion equation

$$\partial_t \rho = D \nabla^2 \rho , \tag{6.1}$$

where $\rho(\mathbf{r}, t)$ is the number of particles per unit volume in the gas and the Laplacian operator acts on the position variable: $\nabla = \partial_{\mathbf{r}}$. The diffusion coefficient is known to be given by the rate of linear growth of the variance of the position, which is Einstein's formula, or equivalently by the time integral of the velocity autocorrelation function

$$D = \lim_{t\to\infty} \frac{1}{2t} \langle [x(\mathbf{\Phi}^t\mathbf{X}) - x(\mathbf{X})]^2 \rangle = \int_0^{\infty} \langle v_x(\mathbf{\Phi}^t\mathbf{X})\, v_x(\mathbf{X}) \rangle \, dt \,, \qquad (6.2)$$

(cf. Subsection 3.5.5). We assume here that the diffusion process is isotropic so that the diffusion coefficient is the same in all directions and the direction x can be used in particular. The average $\langle \cdot \rangle$ is taken over the ensemble of initial conditions.

The dynamics of the Lorentz gases conserves energy so that the motion is identical on each energy shell up to a rescaling of time with the velocity v. We may thus define a diffusion coefficient for each energy shell by taking all the initial conditions of the ensemble in the same energy shell. Consequently, the diffusion coefficient behaves like $D(v) = vD(1)$ where $D(1)$ is the diffusion coefficient in the energy shell with unit velocity.

If the system has the mixing property on each energy shell v (which is the case for the periodic Lorentz gases) the average in Eq. (6.2) is equivalent to an average over the microcanonical invariant measure or equilibrium measure: $\langle \cdot \rangle_v = \langle \cdot \rangle_e$.

Instead of a Lorentz system extending over the whole plane, we consider an open Lorentz gas in which the disks only occupy a large but finite region delimited by a curve C_L. This curve can be defined in polar coordinates ($x = R\cos\theta, y = R\sin\theta$) by the equation $R = Lf(\theta)$, which may be a circle, a hexagon, a square,... . L is a parameter which varies from zero to infinity. Outside the curve C_L, the plane is empty of disks so that the point particle moves in free motion from or to infinity. Accordingly, these open configurations define systems of scattering type. As $L \to \infty$, more and more disks are contained in the scatterer.

6.2.2 Escape-time function and escape rate

Figure 6.1 shows a typical trajectory in such a hexagonal scatterer. The trajectory undergoes many collisions inside the scatterer but, finally, it exits the scatterer and escapes to infinity in free motion. Reversing time, the trajectory would be a typical ingoing trajectory of a scattering experiment. Here, we consider an ensemble of initial conditions taken inside the scatterer and we study the time evolution of all these trajectories.

Although most trajectories escape from the scatterer, there exist trajectories which remain trapped inside the scatterer. They play a very important role because they control the escape dynamics. For instance, the periodic orbits bouncing on the line between the centres of two nearby disks remain trapped in the scatterer for $t \to \pm\infty$. Beside the periodic orbits, there also exist nonperiodic orbits which are forever trapped in the scatterer as soon as the scatterer contains three disks (see Chapter 5). All the trapped orbits are unstable with strictly positive Lyapunov exponents and they form a fractal set of zero Lebesgue measure in phase space. Each trapped orbit has a stable and an unstable manifold which are associated with the negative and positive Lyapunov exponents. Time-reversal symmetry maps the stable manifolds onto the unstable ones. Because the trapped orbits are unstable, the set of trapped orbits is repelling and is accordingly called the fractal repeller $\mathcal{I}_L$. The existence of this fractal repeller is typical of chaotic-scattering processes.

Evidence for the fractal repeller is provided by the escape-time function (see Chapter 2). For each trajectory of initial condition $\mathbf{X}_0$, we can calculate the time at which the border C_L is crossed, which we call the escape time $T_{C_L}(\mathbf{X}_0)$. The escape-time function is finite for almost all trajectories. However, the escape time is infinite for trajectories which are trapped by the fractal repeller. Trapping occurs if the initial condition belongs to an indefinitely trapped orbit or to its stable manifold. The escape-time function therefore has singularities on the fractal set formed by the closure of the stable manifolds of the fractal repeller: $\mathrm{Cl}[W_s(\mathcal{I}_L)]$.

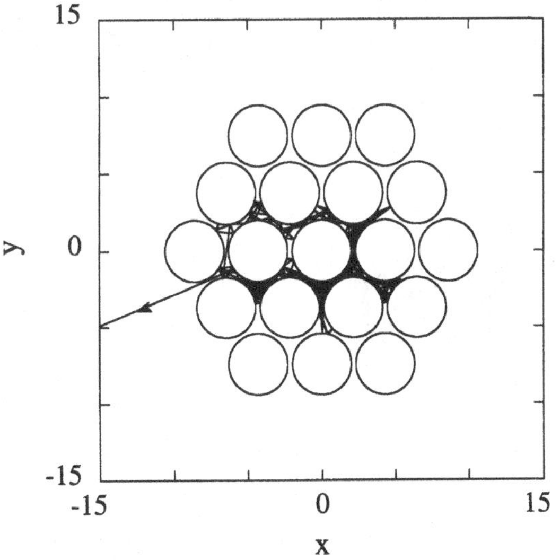

Figure 6.1 Trajectory of a particle escaping after 366 collisions from a hexagonal scatterer of periodic type formed by two shells of disks with $L = 2r$ where $r = 2.15a$ is the distance between the centres of the disks and $a = 2$ is their radius. The initial condition is located on the central disk at an angle $\theta = \pi/4$ (Gaspard and Baras 1995).

For the open Lorentz gas of Fig. 6.1, the escape-time function is depicted in Fig. 6.2 where we observe its self-similarity. We see that the singularities occupy a very important fraction of initial conditions in spite of the fact that they are of zero Lebesgue measure. This phenomenon is due to the fact that the partial Hausdorff dimension of the fractal set is close to one. This observation is fundamental for the following considerations. The deterministic character of the escape process is hidden in the existence of the windows of continuity in the escape-time function where the function is regular in some intervals of initial conditions. For these initial conditions, the trajectories follow nearby paths which bounce on the same disks and exit through the same gap between two next-neighbouring disks. However, we observe that the widths of these windows

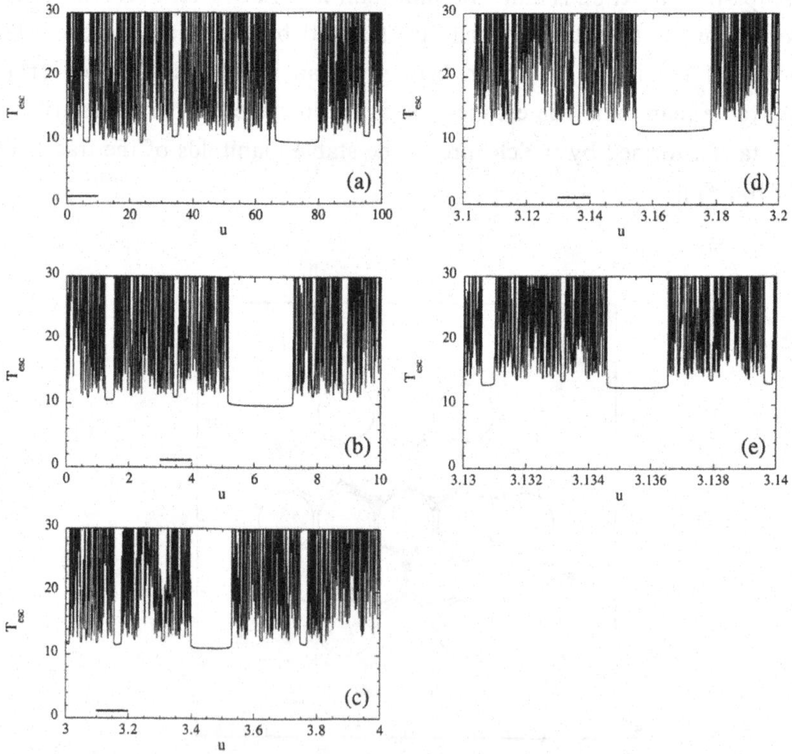

Figure 6.2 Escape-time function for the scatterer of Fig. 6.1 with $L = 2r$, $r = 2.15a$, $a = 1$, for initial conditions $\theta = \pi/4$ and $\phi = \varphi - \theta = \pi/2 - \pi u \times 10^{-10}$, showing the intervals of continuity which provide the signature of the deterministic dynamics on these extremely small scales: (a) $u \in [0, 100]$; (b) $u \in [0, 10]$; (c) $u \in [3, 4]$; (d) $u \in [3.1, 3.2]$; (e) $u \in [3.13, 3.14]$. The comparison between the different intervals makes the scaling behaviour apparent (Gaspard and Baras 1995).

of continuity are extremely small which restricts considerably the range of deterministic considerations.

The escape dynamics can be further characterized by the quantities introduced in the previous chapters and, in particular, by the mean Lyapunov exponent, the KS entropy, and the information and Hausdorff dimensions, all these quantities being defined on the hyperbolic repeller with the nonequilibrium invariant measure (4.61). We have in particular that the escape rate is given as the difference between the mean Lyapunov exponent and the KS entropy according to Eqs. (4.64) or (4.70).

6.2.3 Phenomenology of the escape process

On the other hand, the escape rate can be calculated by the diffusion equation (6.1). Indeed, when the scatterer becomes large enough with respect to the interdisk distance, the motion in the scatterer is essentially controlled by the diffusion equation (6.1). At the border C_L where the particles escape from the scatterer, the density should be considered as being equal to zero because the particles disappear from the diffusion process by escaping to infinity. This is what is called an absorbing boundary. Therefore, the time evolution of the particle density is obtained by solving the diffusion equation with the absorbing boundary condition

$$\rho\Big|_{C_L} = 0 . \tag{6.3}$$

We may expect an exponential decay of the probability density, $\rho \sim \exp(st)$, where $-s$ is a decay rate. We assume that the density evolves in time according to

$$\rho(x, y; t) = \sum_{j=1}^{\infty} c_j \exp(s_j t)\, \psi_j(x, y) , \tag{6.4}$$

where the constants c_j are determined by the initial density. Accordingly, the decay rates $-s_j$ are obtained by an eigenvalue problem for the Helmholtz equation

$$(D\, \nabla^2 - s_j)\, \psi_j = 0 , \qquad \text{with} \qquad \psi_j\Big|_{C_L} = 0 , \tag{6.5}$$

where ψ_j are the associated eigenfunctions.

Using the scaling $(x, y) \to (Lx, Ly)$, the eigenvalue equation (6.5) becomes

$$(\nabla^2 + \chi_j^2)\, \tilde{\psi}_j = 0 , \qquad \text{with} \qquad \tilde{\psi}_j\Big|_{C_{L=1}} = 0 , \tag{6.6}$$

where $\tilde{\psi}_j(x,y) = \psi_j(Lx, Ly)$. The phenomenological decay rates are then given by

$$s_j = -D\left(\frac{\chi_j}{L}\right)^2 . \tag{6.7}$$

We note that the preceding phenomenological description holds only in the limit of a large scatterer ($L \to \infty$) and for the lowest eigenvalues in order to guarantee that the corresponding wavelengths remain large with respect to the mean free path ℓ: $2\pi L/\chi_j \gg \ell$. Indeed, the wavelengths of the higher eigenvalues may become shorter than the lattice cells, in which case we expect a discrepancy between the phenomenological description and the actual Liouvillian dynamics. Hereafter, we shall use the result (6.7) in its domain of validity when $j = 1$ and $L \to \infty$.

Equations (6.4) and (6.7) describe the escape of a statistical ensemble of N_0 particles out of the scatterer. At long times ($t \to \infty$), the escape is dominated by the smallest decay rate, $-s_1$. The number N_t of particles still present inside the scatterer at time t is given in terms of the density according to

$$N_t \simeq N_0 \int_{C_L} dx\, dy\, \rho(x,y;t) \sim N_0 \exp(s_1 t) . \tag{6.8}$$

We obtain the result that the first phenomenological decay rate $-s_1$ gives the escape rate previously defined in Chapter 4 for the scatterer C_L:

$$\gamma = D\left(\frac{\chi_1}{L}\right)^2 + o(L^{-2}) , \tag{6.9}$$

in the limit $L \gg r$, where $o(z)$ denotes a quantity such that $\lim_{z\to 0} o(z)/z = 0$.

For the hexagonal scatterers, the first eigenvalue is

$$\chi_1 = 2.67495 , \quad \text{(hexagon)} , \tag{6.10}$$

(Bauer and Reiss 1978). Figure 6.3 gives the first eigenvalue χ_1 for different shapes of scatterers.

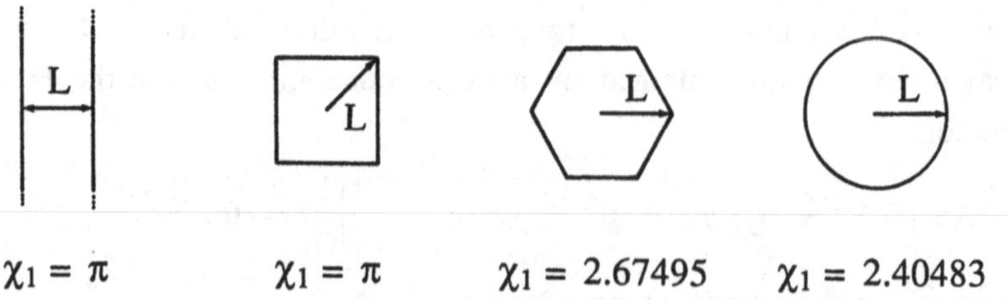

Figure 6.3 The constant χ_1 appearing in Eq. (6.9) for the smallest eigenvalue of the problem (6.6) in different geometries of the absorbing boundary. For the circle, χ_1 is given by the first zero of the Bessel function $J_0(z)$ (Gaspard and Baras 1992).

As a consequence of the previous observations, we can determine the diffusion coefficient by this escape method which is alternative to the methods based on the Einstein and Green–Kubo formulas (6.2). We refer to this method as a first-passage method because the diffusion coefficient is related to the statistics of the times of the first passage for the diffusing particles (Feller 1971, Lindenberg et al. 1975, van Kampen 1981, Weiss 1986).

6.2.4 The escape-rate formula for diffusion

In dynamical systems theory discussed in Chapter 4, the escape rate was obtained by Eq. (4.70) in terms of the Lyapunov exponent and the KS entropy. Combining this result with the relation (6.9) to the diffusion coefficient, we obtain the escape-rate formula

$$D = \lim_{L\to\infty} \left(\frac{L}{\chi_1}\right)^2 [\bar{\lambda}(\mathscr{I}_L) - h_{\rm KS}(\mathscr{I}_L)] , \tag{6.11}$$

where the averages are taken over the invariant measure $\mu_{\rm ne} = \mu_{\beta=1}$ of the fractal repeller $\mathscr{I}_L$ as defined in Chapter 4 (Gaspard and Nicolis 1990). This formula gives the diffusion coefficient from the difference between the Lyapunov exponent and the KS entropy. We know that both should be equal in the spatially extended Lorentz gas (with $L = \infty$) because of Pesin's formula (4.41) which holds in systems without escape. In this limit, both the Lyapunov exponent and the KS entropy reach the same positive value which characterizes the chaoticity of the motion.

In an open system with escape, the difference between the Lyapunov exponent and the KS entropy is nonvanishing. If transport is normal on the scatterer, the difference decreases as

$$\bar{\lambda}(\mathscr{I}_L) - h_{\rm KS}(\mathscr{I}_L) \sim \frac{1}{L^2} , \qquad \text{(normal transport)} , \tag{6.12}$$

from which the diffusion coefficient can be deduced with Eq. (6.11). On the other hand, if the difference behaved as

$$\bar{\lambda}(\mathscr{I}_L) - h_{\rm KS}(\mathscr{I}_L) \sim \frac{(\ln L)^\kappa}{L^\nu} , \qquad \text{(anomalous transport)} , \tag{6.13}$$

with $\nu \neq 2$ or $\kappa \neq 0$, the diffusion coefficient would not exist as is the case in anomalous transport.

In higher-dimensional systems with normal transport, there exist several positive Lyapunov exponents. (Their number is $f - 1$ in a f-dimensional Lorentz gas.) In this case, the escape rate is given in terms of the sum of positive Lyapunov exponents by Eq. (4.64) so that Eq. (6.11) generalizes to

$$D = \lim_{L\to\infty} \left(\frac{L}{\chi_1}\right)^2 \left[\sum_{\bar{\lambda}_i>0} \bar{\lambda}_i(\mathscr{I}_L) - h_{\rm KS}(\mathscr{I}_L)\right] , \tag{6.14}$$

where χ_1 is a constant obtained by solving the multidimensional diffusion equation as in Eq. (6.6) (Gaspard and Nicolis 1990).

In two-dimensional systems, the escape-rate formula (6.11) can be expressed in terms of the partial information codimension (4.84) of the stable (or unstable) manifold $W_{s,u}(\mathscr{I}_L)$ of the repeller. Young's formula (4.80) implies the chain of identities

$$\bar{\lambda} - h_{\rm KS} = \bar{\lambda}\,(1 - d_{\rm I}) = \bar{\lambda}\, c_{\rm I} . \tag{6.15}$$

Hence, Eq. (6.11) becomes

$$D = \lim_{L\to\infty} \left(\frac{L}{\chi_1}\right)^2 \bar{\lambda}(\mathscr{I}_L)\, c_{\rm I}(\mathscr{I}_L) . \tag{6.16}$$

Now, we would like to obtain another formula similar to (6.16) but in which the Hausdorff codimension (4.83) appears instead of the information codimension. The motivation is that the KS entropy (or, equivalently, the information codimension) is difficult to calculate by a direct numerical method although such methods exist for the mean Lyapunov exponent or the escape rate. On the other hand, the Hausdorff codimension can be obtained with Eq. (4.85) as an uncertainty exponent using the Maryland algorithm (McDonald et al. 1985, Ott 1993). Therefore, we want to substitute the calculation of the KS entropy by the calculation of the Hausdorff codimension. The difficulty is that the information and Hausdorff codimensions are in general different and only obey the inequality (4.81). The remarkable point here is that the difference between them vanishes in the limit $L \to \infty$, as follows (Gaspard and Baras 1995).

We assume that the pressure function $P(\beta;\mathscr{I}_L)$ of the hyperbolic repeller has a well-defined limit as $L \to \infty$ so that this limit coincides with the pressure function of the infinitely extended system. This assumption holds in particular for the Lorentz gas with a finite horizon (Gaspard and Baras 1995, see below). We moreover assume that the pressure function $P(\beta;\mathscr{I}_L)$ remains at least twice differentiable in the vicinity of $\beta = 1$ for $L \to \infty$. Expanding the pressure function up to second order we obtain

$$P(\beta) = -\gamma - \bar{\lambda}\,(\beta-1) + \frac{1}{2}\, P''(1)\,(\beta-1)^2 + o[(\beta-1)^2] , \quad \text{(open systems)} , \tag{6.17}$$

where $\lim_{L\to\infty} P''(1)$ exists by our assumption. Since the partial Hausdorff dimension is given by the zero of the pressure function according to Eq. (4.79) and the partial information codimension by $c_I = P(1)/P'(1) = \gamma/\bar{\lambda}$ according to Eqs. (4.80) and (4.84), we obtain

$$c_H = \frac{\gamma}{\bar{\lambda}} - P''(1)\,\frac{\gamma^2}{2\bar{\lambda}^3} + o(\gamma^2) = c_I - \frac{P''(1)}{2\bar{\lambda}}\,c_I^2 + o(c_I^2)\,. \tag{6.18}$$

Using the dependency (6.9) of the escape rate on the distance L, we infer that both the information and the Hausdorff codimensions have a similar leading behaviour as a function of the size L of the scatterer

$$c_H\,,\; c_I = \frac{D}{\bar{\lambda}}\left(\frac{\chi_1}{L}\right)^2 + o(L^{-2})\,. \tag{6.19}$$

Taking the limit $L \to \infty$, we finally obtain the diffusion coefficient in terms of the partial Hausdorff codimension

$$D = \lim_{L\to\infty}\left(\frac{L}{\chi_1}\right)^2 \bar{\lambda}(\mathcal{I}_L)\, c_H(\mathcal{I}_L)\,, \tag{6.20}$$

which is analogous to Eq. (6.16). Similar formulas can be obtained for other generalized codimensions. Equation (6.20) can be rewritten in the form

$$D = \bar{\lambda}_e \lim_{L\to\infty}\left(\frac{L}{\chi_1}\right)^2 c_H(\mathcal{I}_L)\,, \tag{6.21}$$

in terms of the average Lyapunov exponent of the infinite lattice configuration of the Lorentz gas if the limit $\bar{\lambda}_e = \lim_{L\to\infty} \bar{\lambda}(\mathcal{I}_L)$ is finite and positive (Gaspard and Baras 1995). The advantage of this formula over Eq. (6.11) is that it does not require the calculation of the KS entropy but simply of the Hausdorff partial codimension as given, for instance, by the uncertainty exponent of Eq. (4.85).

The above escape-rate formulas establish the connection between diffusion, which takes place at the macroscopic level on the largest scales of the system, and the chaotic dynamics which is generated at the microscopic level on the finer scales of phase space.

Another consequence of the escape out of the system is the formation of a fractal repeller. Its Hausdorff dimension in the full three-dimensional phase space can also be obtained by the methods of Chapter 4 as

$$\begin{aligned}\dim_H(\mathcal{I}_L) &= 2\,d_H + 1 = 3 - 2\,c_H \\ &= 3 - \frac{2D}{\bar{\lambda}_e}\left(\frac{\chi_1}{L}\right)^2 + o(L^{-2})\,,\end{aligned} \tag{6.22}$$

with a similar expression for the information dimension. This formula shows that the fractal repeller fills the whole phase space and its support becomes the plain phase space $\mathcal{M}$ in the limit $L \to \infty$. The fractal properties of the repeller are due to both the dynamical instability and the escape process. Since the escape is actually controlled by diffusion it turns out that the finest scales of the repeller are determined by the motion on the largest scales and vice versa. In this way, the diffusion process is encoded in the fractal repeller. Furthermore, the properties of the fractal repeller determine those of its stable and unstable manifolds. Because these manifolds control the scattering of trajectories into the scatterer we can understand that the scattering process is related to diffusion through the complex object which is the repeller. This fractal repeller thus plays a fundamental role in the transport properties of the system.

In the following, we shall analyze two models of deterministic diffusion from this viewpoint and, thereafter, proceed to the generalization to other transport coefficients.

6.3 The periodic Lorentz gas

6.3.1 Definition

The periodic Lorentz gas is a two-dimensional billiard where a point particle undergoes elastic collisions on hard disks which are fixed in the plane, $\mathbf{q} = (x, y)$. The space between the disks forms the two-dimensional domain $\mathcal{Q}$ of the billiard. All the hard disks have the same radius a. The centres of the disks belong to a triangular lattice

$$\mathbf{q}_c = m_c\, \mathbf{a} + n_c\, \mathbf{b}\,, \tag{6.23}$$

defined in terms of the fundamental translation vectors of the triangular lattice

$$\begin{aligned} \mathbf{a} &= (r,\ 0)\,, \\ \mathbf{b} &= \left(\frac{1}{2}r,\ \frac{\sqrt{3}}{2}r\right), \end{aligned} \tag{6.24}$$

and where $(m_c, n_c) \in \mathbb{Z}^2$ are integers. Different billiards can be constructed depending on the set of pairs of integers which are selected.

The infinite configuration

If all the pairs of integers are selected we fill the whole triangular lattice with hard disks. All the centres are occupied so that the billiard is invariant under the group of spatial translations generated by the vectors (6.23). Accordingly, the whole lattice can be mapped onto a so-called Wigner–Seitz cell with periodic boundary conditions (Bouckaert et al. 1936, Kittel 1976, Harrison 1980). The elementary Wigner–Seitz cell of this triangular lattice is a hexagon of area

$$A_{\rm WS} = \mathbf{a} \times \mathbf{b} = \frac{\sqrt{3}}{2}\, r^2 . \tag{6.25}$$

Under this construction, the position space forms a torus given by the quotient of the plane by the lattice: $\mathbb{T}^2 = \mathbb{R}^2/\mathbb{Z}^2$.

The motion of the point particle in the infinite lattice is unbounded so that transport by diffusion is a priori possible. When the dynamics is reduced to the Wigner–Seitz cell, the position of the particle inside this cell must be supplemented by a lattice vector of the type (6.23) in order to determine the actual position of the particle in the infinite lattice. This lattice vector changes in discrete steps at each crossing of the border of the elementary Wigner–Seitz cell.

The dynamics

Between two successive collisions, the motion of the point particle is determined by the free Hamiltonian $H = (p_x^2 + p_y^2)/(2m)$. At each collision, the velocity or momentum of the particle satisfies the law of elastic collision, as discussed in previous chapters. The system has two degrees of freedom and the coordinates of the phase space are (x, y, p_x, p_y). Energy is conserved so that the system is a three-dimensional flow in each energy shell where the coordinates are (x, y, φ) where φ is the angle of the velocity with respect to the x-axis so that $p_x = p\cos\varphi$ and $p_y = p\sin\varphi$.

6.3.2 The Liouville invariant measure

Since we investigate statistical properties of the billiard dynamics, we introduce the Liouville equilibrium invariant measure given by

$$d\mu_e = I_{\mathscr{D}}(x, y)\ \delta(H - E)\ dx\ dy\ dp_x\ dp_y\ , \tag{6.26}$$

in terms of the indicator function of the domain $\mathscr{D}$. In polar coordinates where $dp_x dp_y = p dp d\varphi$, the invariant measure (6.26) can be rewritten as

$$d\mu_e = m\ I_{\mathscr{D}}(x, y)\ \delta(p - \sqrt{2mE})\ dp\ dx\ dy\ d\varphi\ , \tag{6.27}$$

where we used the property of the Dirac distribution that

$$\delta\left(\frac{p^2}{2m} - E\right) = \frac{m}{p}\ \delta(p - \sqrt{2mE})\ , \tag{6.28}$$

when $p \geq 0$. Equation (6.27) shows that the velocity angles and the positions are uniformly distributed according to the Liouville invariant measure with a density, $d\mathbf{X} = dxdyd\varphi$, on each energy shell. Averages over this invariant measure are denoted by $\langle\cdot\rangle_e$.

The Liouville invariant measure (6.27) applies to every configuration of the Lorentz gas. This measure is not normalizable when the area of the domain $\mathcal{Q}$ is infinite in the case of the infinite and open configurations. However, the measure is normalizable for the reduced dynamics in an elementary Wigner–Seitz cell where the area of the domain $\mathcal{Q}$ takes the finite value (6.25). In this finite case, the Liouville invariant measure is a probability measure which defines the microcanonical ensemble of equilibrium statistical mechanics. For the nonequilibrium mechanics of the open configurations of scattering type, the relevant invariant measures are the nonequilibrium measures described in Chapter 4.

6.3.3 Finite and infinite horizons

Different dynamical behaviours arise depending on the ratio of the distance r between the centres of the disks to the disk radius a. This ratio r/a determines the density of the disk lattice.

(i) Infinite-horizon regime

In a low-density regime where the intercentre distance exceeds the critical value, $r > 4a/\sqrt{3}$, the channels of the lattice are wide enough for an infinite number of disks to be reachable in free flight. In this case, the point particle can travel without any collision through the whole lattice along special trajectories. These special trajectories without collision exist only for velocity angles taking the discrete values $\varphi = 0, \pi/3, 2\pi/3, \pi, 4\pi/3$, and $5\pi/3$ so that they occupy a set of zero Liouville measure and of dimension two in the tridimensional phase space. The collisional horizon of these trajectories is therefore removed to infinity. This situation is referred to as the infinite-horizon configuration (see Fig. 6.4a). As a consequence of this divergence of the horizon, the diffusion coefficient is infinite in this regime ($D = \infty$) (Bunimovich and Sinai 1980).

(ii) Finite-horizon regime

For smaller intercentre distances, $2a < r < 4a/\sqrt{3}$, only a finite set of at most six disks can be reached in free flight by the point particle when its velocity angle φ varies (see Fig. 6.4b). Therefore, the horizon of the particle is now finite. In this finite-horizon regime, the diffusion coefficient in the lattice is known to be positive and finite ($0 < D < \infty$) (Bunimovich and Sinai 1980).

(iii) Localized regime

When the distance between the centres is smaller than the diameter of the disks $\sqrt{3}a < r \le 2a$, the disks partially overlap so that the point particle remains forever localized in compact domains delimited by three neighbouring disks (see Fig. 6.4c). In this case, unbounded motion is no longer possible and the diffusion coefficient vanishes ($D = 0$).

For $r \le \sqrt{3}a$, the disks completely overlap so that the domain of billiard is empty and no dynamics is possible.

6.3.4 Chaotic properties of the infinite Lorentz gas

The mean Lyapunov exponent

Because of the defocusing character of the collisions on disks, the Lorentz gases have one positive Lyapunov exponent which can be calculated as explained in Chapter 1. The periodic Lorentz gas is known to be ergodic and mixing so that an average Lyapunov exponent can be obtained by a time average over the motion in a reduced cell of the periodic lattice as

$$\bar{\lambda}_e = \lim_{T\to\infty} \frac{1}{T} \ln |\Lambda(T, \mathbf{X}_0)| , \tag{6.29}$$

in terms of the stretching factor (1.141) and for almost all initial conditions $\mathbf{X}_0$. The statistics may nevertheless be improved by taking an ensemble of initial conditions distributed according to the ergodic measure.

The average Lyapunov exponent has been calculated for several intercentre distances r and is depicted in Fig. 6.5. We observe that the average Lyapunov exponent is positive in the Lorentz gas even when the horizon becomes infinite at $r = 4a/\sqrt{3} = 2.3094a$. At this critical value, there is no apparent discontinuity

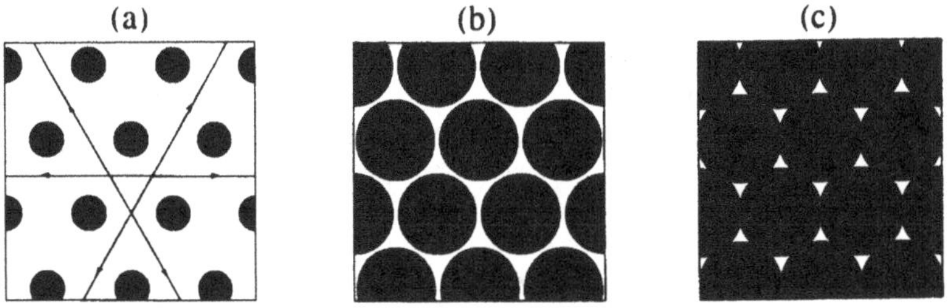

Figure 6.4 (a) Triangular Lorentz gas in the infinite-horizon regime ($4a/\sqrt{3} < r$) with some of the special orbits travelling through the lattice without collisions; (b) Lorentz gas in the finite-horizon regime ($2a < r < 4a/\sqrt{3}$); (c) Lorentz gas in the localized regime ($\sqrt{3}a < r < 2a$) (Gaspard and Baras 1995).

in the behaviour of the Lyapunov exponent as a function of the intercentre distance r.

At large values of r, the average Lyapunov exponent decreases monotonically to zero (see Fig. 6.5a). This behaviour has been explained as follows. When the disks are separated by a large distance r, the Lorentz gas is a dilute gas.

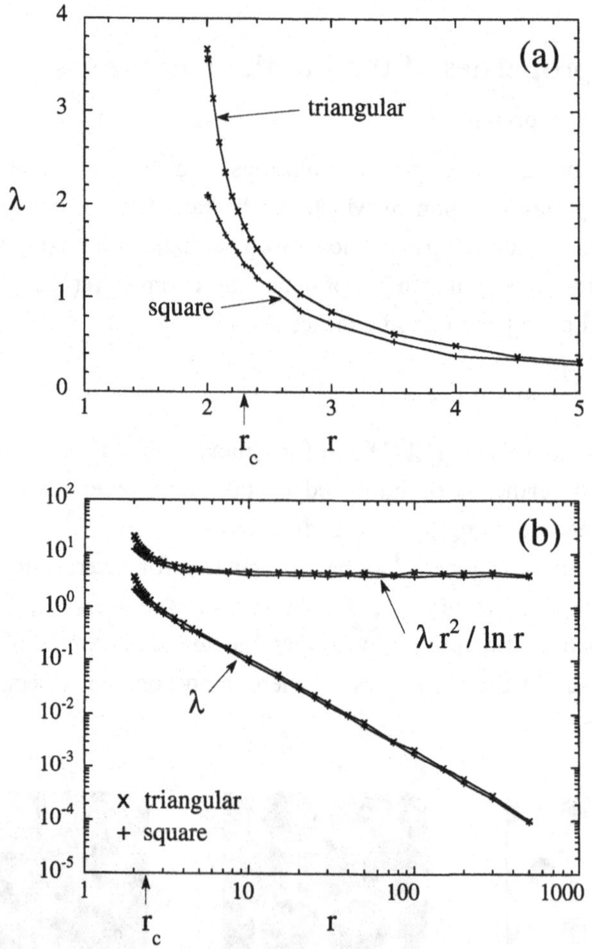

Figure 6.5 (a) Average Lyapunov exponent versus the intercentre distance r for the dynamics in the Wigner–Seitz cell of the infinite Lorentz gas on triangular and square lattices. (b) Plot of $\bar{\lambda}_e r^2/\ln r$ versus r showing the dependency (6.32). The transition from finite to infinite horizon happens at the critical value $r = r_c = 4a/\sqrt{3} = 2.3094a$ for the triangular lattice. This transition does not affect the behaviour of the Lyapunov exponent, as we here observe within the numerical errors (Gaspard and Baras 1995).

The mean free path between collisions is determined by the density of disks, $\rho_{\text{disk}} \sim r^{-2}$, and by the cross section, $\sigma = 2a$, according to

$$\ell \sim \frac{1}{\rho_{\text{disk}}\sigma} \sim \frac{r^2}{a} . \tag{6.30}$$

Over n collisions, the stretching factor (1.141) behaves essentially as

$$\Lambda(T, \mathbf{X}_0) \sim \left(\frac{\ell}{a}\right)^n \sim \left(\frac{r^2}{a^2}\right)^n , \tag{6.31}$$

for large interdisk separations, $r \gg a$. On average, there are about $n \sim T/\ell \sim aT/r^2$ collisions over the path length T (with $v = 1$). Introducing these results in Eq. (6.29), the average Lyapunov exponent is obtained as

$$\bar{\lambda}_e \sim \frac{a}{r^2} \ln \frac{r}{a} . \tag{6.32}$$

Figure 6.5b shows that $\bar{\lambda}_e r^2/\ln r$ is approximately constant as expected from (6.32). Let us mention that more precise formulas for the dependency on r of the Lyapunov exponent have been given (Friedman et al. 1984; Bouchaud and Le Doussal 1985; Chernov 1991, 1997). The case of a random Lorentz gas has been studied by van Beijeren and Dorfman (1995).

For comparison, Fig. 6.5 also shows the average Lyapunov exponent for the Lorentz gas in a square lattice. In this case, the Wigner–Seitz cell is a square. The reduced billiard is also known as the Sinai billiard (Sinai 1970). In a square lattice with $r > 2a$, the horizon is always infinite so that diffusion is anomalous with an infinite diffusion coefficient. This example also shows that the average Lyapunov exponent may have a well-defined value even when the diffusion coefficient is infinite. This result corroborates the previous observation on the triangular lattice that the transition from the finite to the infinite horizon does not induce a change in the behaviour of the average Lyapunov exponent. Besides, Figure 6.5 shows that the average Lyapunov exponent is smaller for the square lattice than for the triangular lattice.

The pressure function in the finite-horizon regime

The pressure function can be calculated using Eq. (4.69). In the present case, there is no escape so that the average is performed over all the trajectories of the ensemble. Figure 6.6 shows the pressure function versus the exponent β for $r = 2.15a$ and $r = 2.2a$ which are two configurations with a finite horizon since $r < 4a/\sqrt{3} = 2.3094a$. We observe that the pressure function here is nearly linear with a very small convexity. This smoothness suggests that at least two derivatives of the pressure function exist at $\beta = 1$. The second derivative of the pressure function can be interpreted as the variance σ_e^2 of the Lyapunov exponents since

$$P_e''(1) = \lim_{T\to\infty} \frac{1}{T} \left\langle \left(\ln|\Lambda_T| - \langle \ln|\Lambda_T| \rangle_e \right)^2 \right\rangle_e = \sigma_e^2 \,, \tag{6.33}$$

where $\ln|\Lambda_T|$ can be considered as a Lyapunov exponent over a time T.

The different characteristic quantities of chaos can be calculated as explained in Chapter 4 thanks to the pressure function. These quantities are given in Table 6.1 for different intercentre distances r.

The pressure function in the infinite-horizon regime

When the horizon becomes infinite a new feature appears in the pressure function which is due to the special orbits with zero Lyapunov exponent. In this regime, a point particle may travel without collision over a distance R if its velocity is directed in the small angles of order R^{-1} in the channels of the lattice. Therefore,

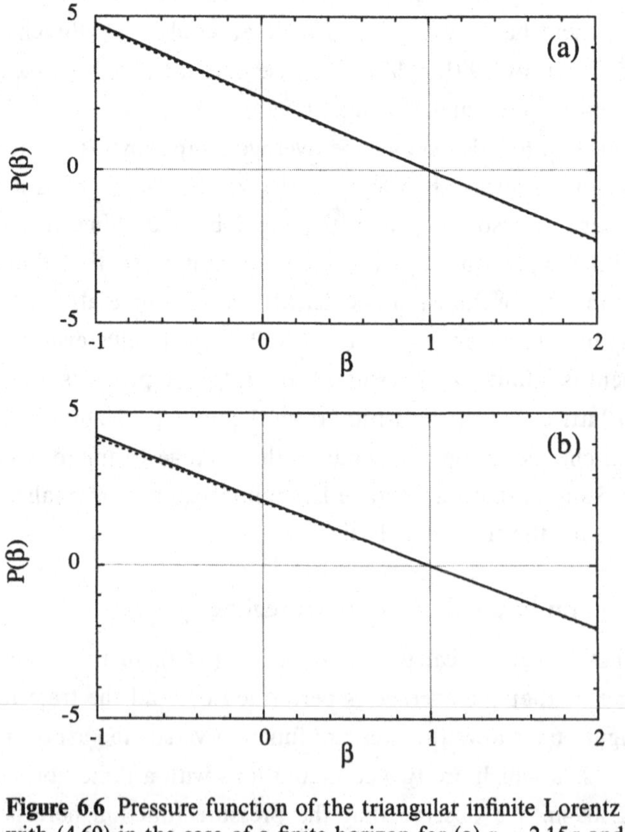

Figure 6.6 Pressure function of the triangular infinite Lorentz gas calculated with (4.69) in the case of a finite horizon for (a) $r = 2.15a$ and (b) $r = 2.2a$. The dashed line is a straight line tangent to the pressure at $\beta = 1$, the slope of which is $-\bar{\lambda}_e$ (Gaspard and Baras 1995).

the probability for a particle of unit velocity to have no collision over the time $T = R$ decreases as R^{-1}. Since this probability is positive for any finite time T, the average in Eq. (4.69) may contain a few terms for which $|\Lambda_T| = 1$. Figure 6.7 shows a histogram of the values of $(1/T)\ln|\Lambda_T|$ obtained from an ensemble of 5000 trajectories of time interval $T = 10000$ when $r = 10a$. Although rare, trajectories with $\ln|\Lambda_T| = 0$ are nevertheless present due to the aforementioned nonvanishing probability, which has important consequences for the pressure function.

Indeed, when $\beta > 1$, the sum in Eq. (4.69) is dominated by the few terms equal to one which are due to the vanishing Lyapunov exponents. The other terms decrease to zero as $T \to \infty$. Accordingly, the pressure function vanishes

Table 6.1 Characteristic quantities for closed triangular Lorentz gas (error of 1–2 %) (Gaspard and Baras 1995).

r/a	$\bar{\lambda}_e = h_{KS}(\mu_e)$	$\sigma_e^2 = P_e''(1)$	h_{top}
2.01	3.56	0.54	3.68
2.15	2.34	0.71	2.38
2.2	2.09	0.67	2.14
3	0.857	–	0.90
7.5	0.170	–	0.19
10	0.107	–	0.12

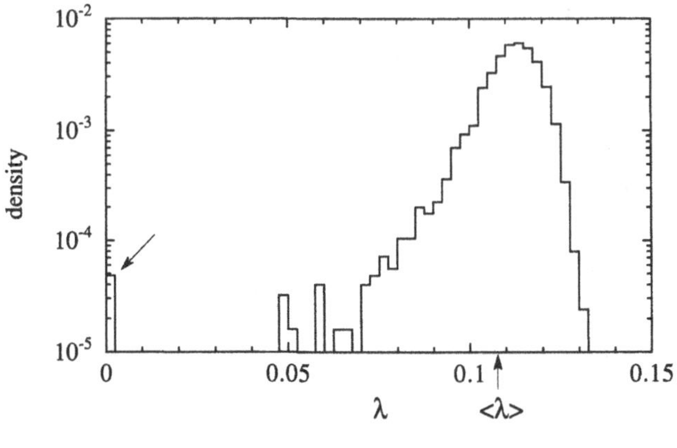

Figure 6.7 Histogram of the Lyapunov exponents of the triangular infinite Lorentz gas $r = 10a$ and $a = 1$, calculated as $\lambda = (1/T)\ln|\Lambda_T|$ with $T = 10000$ for 5000 trajectories. $\langle\lambda\rangle = \bar{\lambda}_e = 0.107$ is the value of the average Lyapunov exponent. The arrow on the left-hand side points to the contribution of the vanishing Lyapunov exponents due to the infinite horizon (Gaspard and Baras 1995).

for $\beta > 1$. On the other hand, for $\beta < 1$, the sum is dominated by the terms associated with unstable orbits with nonvanishing Lyapunov exponents so that the pressure function is then nontrivial. A consequence of the nonhyperbolicity when the horizon is infinite is therefore the appearance of a discontinuity in the derivative of the pressure function (see Fig. 6.8). This discontinuity can be interpreted as a dynamical phase transition as explained in Chapter 4. In the particular case where $\beta = 1$, the invariant measure $\mu_{\beta=1} = \mu_e$ coincides with the equilibrium measure (6.27). For the values $\beta < 1$, the invariant measure μ_β is distributed over the random and unstable trajectories of the system. Accordingly, these states μ_β correspond to a disordered or chaotic phase. On the other hand, for $\beta > 1$, the invariant measure is distributed only over the special regular orbits with zero Lyapunov exponents, so that these states correspond to an ordered phase.

Because of the discontinuity, the average Lyapunov exponent is now given by the left-hand derivative of the pressure at $\beta = 1$ in Eq. (4.71). The right-hand derivative gives the average Lyapunov exponent in the ordered phase, which is vanishing. Since the pressure function is no longer regular at $\beta = 1$ when the horizon is infinite, we remark that the pressure function is no longer differentiable.

The diffusion coefficient

Since the system is mixing, the average in the Einstein and Green–Kubo formulas (6.2) can be taken over any smooth enough density of initial conditions. In the limit $t \to \infty$, the nonequilibrium average is equivalent to the equilibrium average for initial conditions in the Wigner–Seitz cell. Because of energy conservation and of the free motion between collisions, the diffusion coefficient is proportional to the magnitude of the velocity v.

In the triangular lattice, diffusion is isotropic so that the diffusion coefficient can be calculated independently from the x- or y-projection of the position vector $\mathbf{q}_t$. Machta and Zwanzig (1983) have used the Green–Kubo method to calculate numerically the diffusion coefficient of the triangular Lorentz gas as a function of the intercentre distance r.

Moreover, these authors have obtained an approximation for the diffusion coefficient in the limit of small gaps $(r - 2a)$ between the disks. In this case, the particle spends a long time bouncing in the triangular cells of the lattice with rare transitions from cell to cell. Accordingly, the transition rate can be calculated assuming a quasi-equilibrium in each triangular cell, yielding

$$D_{\rm th} \simeq v \, \frac{r^2}{\pi(r^2\sqrt{3} - 2\pi a^2)} \, (r - 2a) \, . \tag{6.34}$$

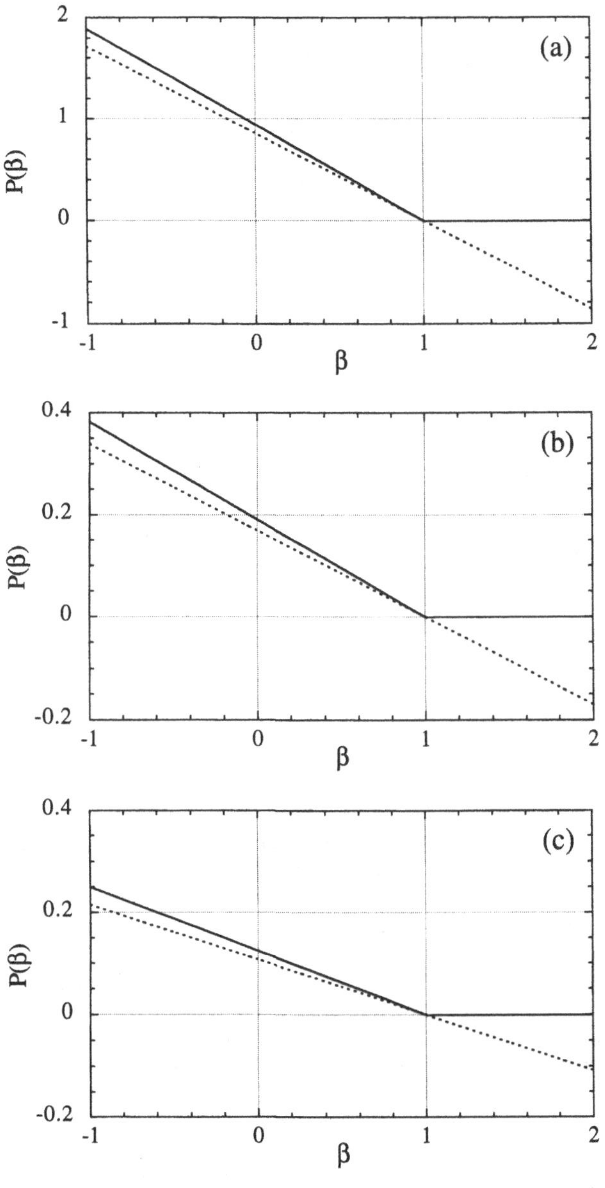

Figure 6.8 Pressure function of the triangular infinite Lorentz gas calculated with (4.69) for (a) $r = 3a$; (b) $r = 7.5a$; (c) $r = 10a$. The dashed line is a straight line tangent to the pressure at $\beta = 1$, the slope of which is $-\bar{\lambda}_e$ (Gaspard and Baras 1995).

According to its derivation, the formula (6.34) applies only when the interdisk gaps are sufficiently small. Equation (6.34) thus provides an excellent approximation of the diffusion coefficient in the limit $r \to 2a$ where it vanishes as $D_{\rm th} \simeq 1.97v(r-2a)$. However, this approximation becomes crude above the value $r \simeq 2.1a$ and its domain of validity is limited in any case to the finite-horizon regime for which $2a < r < 4a/\sqrt{3}$. The values of the diffusion coefficient obtained by Machta and Zwanzig (1983) are reported in Table 6.2 and in Fig. 6.9.

6.3.5 The open Lorentz gas

The open configurations of scattering type

Rather than filling the whole lattice with disks, we can fill only the lattice sides which are inside a closed curve C_L. The number of disks inside the curve C_L increases approximately as

$$N_L \approx \frac{A_L}{A_{\rm WS}} , \tag{6.35}$$

where A_L is the area of the domain enclosed in the curve C_L and $A_{\rm WS}$ is the area (6.25) of the Wigner–Seitz cell.

Table 6.2 Diffusion coefficients of the triangular Lorentz gas versus the intercentre distance r/a in the first column. The second column gives the average Lyapunov exponent $\bar{\lambda}_{\rm e}$ calculated with (6.29). The third column gives the theoretical diffusion coefficient $D_{\rm th}$ of Machta and Zwanzig [Eq. (6.34)]. The fourth column gives the numerical values $D_{\rm GK}$ of Machta and Zwanzig (1983) obtained with the Green–Kubo formula (6.2). The fifth column gives the numerical values $D_{\rm EGK}$ of Morriss and Rondoni (1994) obtained from a combination of extrapolations of Green–Kubo and Einstein formulas. The sixth column gives the diffusion coefficient $D_{\rm esc}$, calculated using the first-passage method and Eq. (6.9). The seventh column gives the diffusion coefficient obtained from the Hausdorff dimension of the fractal repeller thanks to the escape-rate formula (6.21) (Gaspard and Baras 1995).

r/a	$\bar{\lambda}_{\rm e}$	$D_{\rm th}$	$D_{\rm GK}$	$D_{\rm EGK}$	$D_{\rm esc}$	$D_{\rm fractal}$
2.002	–	0.00387	0.0036±0.0003	–	–	–
2.01	3.56	0.0180	0.017±0.002	–	0.0169±0.0009	–
2.04	–	0.0573	0.052±0.002	–	–	–
2.05	3.14	0.0672	–	–	0.0585±0.0004	–
2.06	–	0.0760	0.069	–	–	–
2.1	2.66	0.104	0.10 ±0.01	0.0995±0.0003	0.096±0.007	0.09
2.15	2.34	0.128	0.14	0.1350±0.0005	0.134±0.004	0.13
2.2	2.09	0.147	0.18	0.170±0.001	0.161±0.004	0.17
2.25	1.92	0.162	–	–	0.205±0.003	0.21
2.3	1.76	0.175	0.25 ±0.01	0.2492±0.0003	0.25±0.01	0.25
>2.3094	–	∞	–	–	–	–

Here, we shall especially be concerned with hexagonal scatterers for which the curve C_L is a hexagon reciprocal to the hexagon of the Wigner–Seitz cell. In this case, the number of disk centres inside C_L is

$$N_L = 3\,n^2 + 3\,n + 1\,, \qquad \text{for} \qquad L = nr + \delta\,, \tag{6.36}$$

with $0 < \delta \ll r$. For $L-\delta = r, 2r, 3r, 4r, 5r, 6r, 7r, 8r, \ldots$, we have that the scatterer contains $N_L = 7, 19, 37, 61, 91, 127, 169, 217, \ldots$ disks. If the particle belongs to the interior of the hexagon C_L, its position $\mathbf{q}$ satisfies the condition

$$\max\Big\{ |\mathbf{e}_1 \cdot \mathbf{q}|,\ |\mathbf{e}_2 \cdot \mathbf{q}|,\ |\mathbf{e}_3 \cdot \mathbf{q}| \Big\} < L\,, \tag{6.37}$$

in terms of the vectors which are normal to the sides of C_L: $\mathbf{e}_1 = (0,1)$, $\mathbf{e}_2 = (\sqrt{3}/2, 1/2)$, and $\mathbf{e}_3 = (\sqrt{3}/2, -1/2)$.

The escape rate

The process of escape out of a finite scatterer like the one in Fig. 6.1 can be simulated by taking an ensemble ν_0 of initial conditions. In this ensemble, the positions of the different particles are identical and located at the point $x_0 = y_0 = a \sin(\pi/4)$ on the border of the central disk of the scatterer. The initial velocity angles are uniformly distributed in the interval $-\pi/4 \le \varphi \le 3\pi/4$.

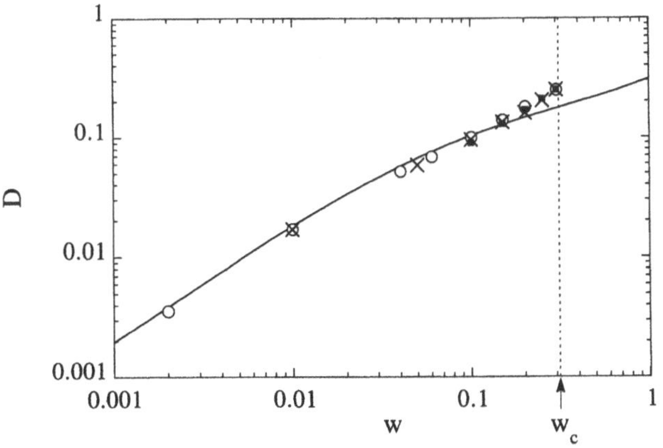

Figure 6.9 Logarithm of the diffusion coefficients versus the logarithm of the gap size $w = (r/a) - 2$ with $a = 1$. The solid line is the theoretical diffusion coefficient D_{th} of Machta and Zwanzig (6.34). The numerical results D_{GK} obtained by Machta and Zwanzig (1983) using the Green–Kubo formula are shown by circles. The diffusion coefficients D_{esc} obtained in the present work with the first-passage method are presented by crosses. The diffusion coefficients D_{fractal} obtained in the present work from the Lyapunov exponents and the Hausdorff codimensions are presented by squares. The transition from finite to infinite horizon at $w_c = (4/\sqrt{3}) - 2 = 0.3094$ is shown by the dashed line (Gaspard and Baras 1995).

Figure 6.10 shows the decay of the number N_T of particles for a scatterer with $L = 2r$ and $r = 2.15a$. We observe without ambiguity that the decay is exponential after the usual transients. The escape rates for different scatterers of increasing sizes L/r but fixed ratio r/a are then collected in Fig. 6.11 where it is possible to determine the diffusion coefficient D_{esc} using the relation (6.9). The diffusion coefficients D_{esc} are obtained by extrapolation from the escape rates for sizes varying from $L = 4r$ to $L = 10r$. We see in Table 6.2 that this

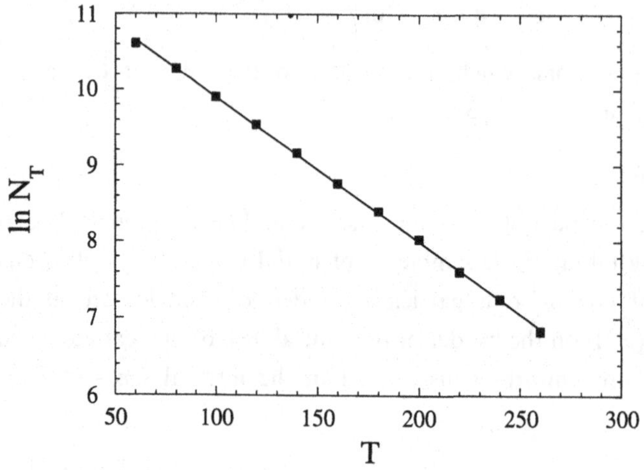

Figure 6.10 $\ln N_T$ versus the time T for a typical escape rate simulation at $L = 2r$, $r = 2.15a$, and $a = 2$ (Gaspard and Baras 1995).

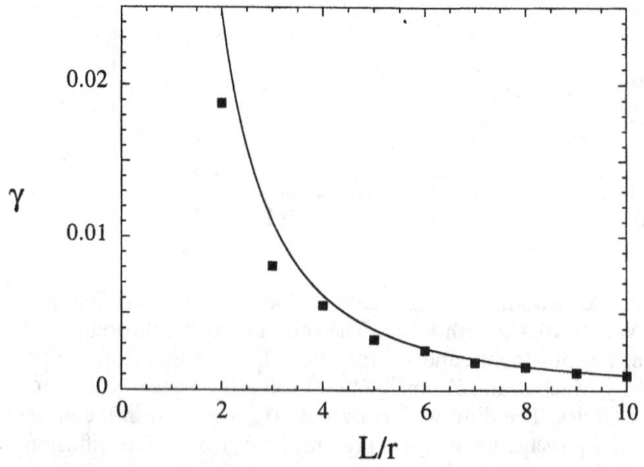

Figure 6.11 Escape rate versus the size L/r for $r = 2.15a$ and $a = 2$ (dots) and compared with the prediction of (6.9) with the theoretical diffusion coefficient (6.34) (Gaspard and Baras 1995).

first-passage method has an accuracy which is comparable to the accuracy of the Green–Kubo method.

Average Lyapunov exponent

For the open configurations, the average Lyapunov exponent over the invariant measure is defined as explained in Chapter 4 with the ensemble of N_T trajectories which are still inside the scatterer by the time T

$$\bar{\lambda} = \lim_{T\to+\infty} \lim_{N_0\to\infty} \frac{1}{T}\frac{1}{N_T}\sum_{i=1}^{N_T} \ln |\Lambda(T,\mathbf{X}_0^{(i)})| \,. \tag{6.38}$$

The numerical calculation shows that the average Lyapunov exponent of the open Lorentz gas is equal to that calculated in the closed configurations up to a numerical error which is of the order of 0.1%. The dependency on the size L of the scatterer is therefore expected to be very small. The values obtained are given in Table 6.2.

Application of the escape-rate formula

By using Eq. (6.21), the diffusion coefficient can be calculated numerically in terms of the Hausdorff dimension of the fractal repeller evidenced by Fig. 6.2. This Hausdorff dimension can be obtained with the Maryland algorithm as an uncertainty exponent [cf. Eq. (4.85)]. To apply the Maryland algorithm, we need to integrate two nearby trajectories and to compare their properties. However, if both trajectories – $(\mathbf{q}_n, \mathbf{v}_n^{(\pm)})$ and $(\mathbf{q}_n', \mathbf{v}_n^{(\pm)\prime})$ – are integrated by the method described in Subsection 5.2.2, we face the difficulty that the differences between their coordinates cannot be smaller than the rounding error of the computer used, namely $\varepsilon > 10^{-15}$ in simple precision and $\varepsilon > 10^{-30}$ in double precision on a CRAY. This restriction would prevent us determining the Hausdorff dimension of large scatterers.

To avoid this restriction, the reference trajectory $(\mathbf{q}_n, \mathbf{v}_n^{(\pm)})$ can be simultaneously integrated together with the difference between the perturbed and the reference trajectories:

$$(\Delta\mathbf{q}_n, \Delta\mathbf{v}_n^{(\pm)}) = (\mathbf{q}_n' - \mathbf{q}_n, \mathbf{v}_n^{(\pm)\prime} - \mathbf{v}_n^{(\pm)}) \,. \tag{6.39}$$

The reference trajectory is integrated by Eqs. (5.10)–(5.16). The difference (6.39) is integrated by equations derived by taking *finite differences* of Eqs. (5.10)–(5.16). In particular, the difference on the reflected velocity is

$$\begin{aligned} \Delta\mathbf{v}_n^{(+)} &= \Delta\mathbf{v}_n^{(-)} - 2\left(\mathbf{n}_n' \cdot \mathbf{v}_n^{(-)\prime}\right)\Delta\mathbf{n}_n \\ &\quad - 2\left(\mathbf{n}_n' \cdot \Delta\mathbf{v}_n^{(-)}\right)\mathbf{n}_n - 2\left(\Delta\mathbf{n}_n \cdot \mathbf{v}_n^{(-)}\right)\mathbf{n}_n \,. \end{aligned} \tag{6.40}$$

The difference in the time of flight between both collisions is

$$\Delta(t_{n+1} - t_n) = -\Delta B_n - \frac{(B'_n + B_n)\Delta B_n - \Delta C_n}{\sqrt{B'^2_n - C'_n} + \sqrt{B^2_n - C_n}} , \tag{6.41}$$

with

$$\begin{aligned} \Delta B_n &= \mathbf{v}_n^{(+)\prime} \cdot \Delta \mathbf{q}_n + \Delta \mathbf{v}_n^{(+)} \cdot (\mathbf{q}_n - \mathbf{q}_c) , \\ \Delta C_n &= (\mathbf{q}'_n + \mathbf{q}_n - 2\mathbf{q}_c) \cdot \Delta \mathbf{q}_n , \end{aligned} \tag{6.42}$$

and where we supposed a velocity of unit magnitude, $A_n = 1$. The difference in the new positions at the next collision is given by

$$\Delta \mathbf{q}_{n+1} = \mathbf{v}_n^{(+)\prime} \, \Delta(t_{n+1} - t_n) + \Delta \mathbf{v}_n^{(+)} \, (t_{n+1} - t_n) + \Delta \mathbf{q}_n . \tag{6.43}$$

The differences in the normal vectors at the points of impact and in the incident velocities are obtained as

$$\begin{aligned} \Delta \mathbf{n}_{n+1} &= \frac{1}{a} \, \Delta \mathbf{q}_{n+1} , \\ \Delta \mathbf{v}_{n+1}^{(-)} &= \Delta \mathbf{v}_n^{(+)} . \end{aligned} \tag{6.44}$$

In this way, both trajectories of the pair can be calculated by recurrence. The integration of Eqs. (6.40)–(6.44) has to be stopped as soon as the reference and

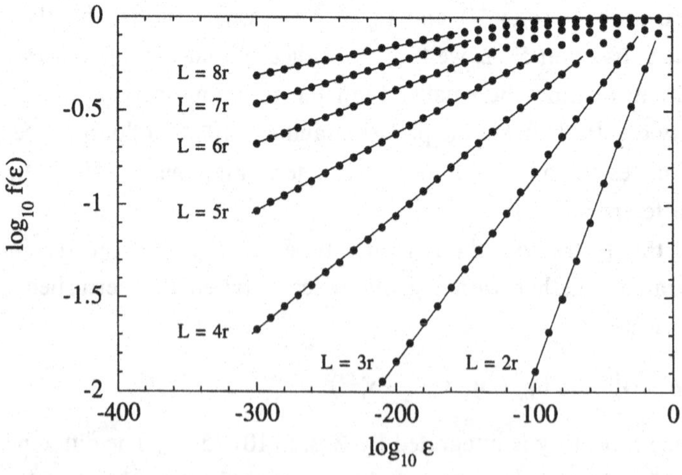

Figure 6.12 Log-log plot of the fraction $f(\varepsilon)$ of uncertain pairs of points, versus the separation ε for the scatterers from $L = 2r$ to $L = 8r$ with an interdisk distance of $r = 2.2a$ ($a = 1$). Note the dramatic shift of the scaling regime towards extremely small separations when L increases. Each point is obtained with a statistical ensemble of 10^4 initial conditions (Gaspard and Baras 1995).

the perturbed trajectories have collisions on different disks, i.e., as soon as one of the square roots in the denominator of Eq. (6.41) becomes undefined.

The differences on the initial conditions may be taken as

$$\begin{aligned} \Delta \mathbf{q}_0 &= a\,[\cos(\alpha_0+\varepsilon)-\cos(\alpha_0),\ \sin(\alpha_0+\varepsilon)-\sin(\alpha_0)] \\ &= 2\,a\,\sin\frac{\varepsilon}{2}\left[-\sin\left(\alpha_0+\frac{\varepsilon}{2}\right),\ \cos\left(\alpha_0+\frac{\varepsilon}{2}\right)\right], \\ \Delta \mathbf{v}_0^{(+)} &= 0\,, \end{aligned} \tag{6.45}$$

where α_0 varies between 0 and $\pi/6$.

This method allows us to reach the value $\varepsilon \simeq 10^{-300}$ which is the minimum decimal power available on standard FORTRAN compilers. Since the dynamics is unstable, the difference $(\Delta\mathbf{q}_n, \Delta\mathbf{v}_n^{(\pm)})$ is expected to grow so that the difference does not decrease below 10^{-300}. In this way, the Hausdorff dimension can be obtained for large scatterers where c_{H} is very small (Gaspard and Baras 1995).

As we can observe in Fig. 6.2, the widths of the intervals of continuity separated by the singularities of the escape-time function may be extremely tiny. A typical width of such an interval of continuity can be estimated as follows. When the size L of the scatterer increases, the time spent by the particles in the scatterer increases on average as $t \sim \gamma^{-1} \sim L^2/D$ and the stretching factors of the individual trajectories as $\Lambda \sim \exp(\lambda t) \sim \exp(cL^2)$ where c is a positive constant. A simple argument moreover shows that the intervals of continuity

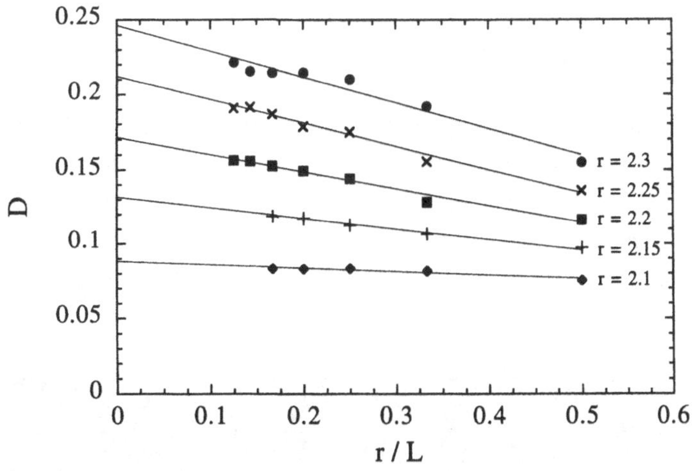

Figure 6.13 Left-hand member of Eq. (6.21) versus r/L. The diffusion coefficient is determined by linear extrapolation assuming that the left-hand member of (6.21) behaves like $D_{\mathrm{fractal}} + C/L$ where C is a constant. The numerical values so obtained are reported in Table 6.2 (Gaspard and Baras 1995).

are of the order of Λ^{-1}. We conclude that the determination of the Hausdorff dimension requires consideration of extremely small differences of the order of $\varepsilon \sim \exp(-cL^2)$ in the fraction (4.85) of uncertain pairs. As a consequence, the above integration method based on finite differences is required in order to calculate the Hausdorff dimension in large scatterers.

Figure 6.12 shows the fraction of uncertain pairs of orbits obtained in the Lorentz gas with $r = 2.2a$. We observe the dramatic decrease in orbit separation that is required to reach the scaling behaviour when the size L increases. This astonishing observation shows how fast the deterministic character of the diffusion process disappears on minute scales as the size of the scatterer increases. A linear regression in the scaling domain where Eq. (4.85) applies provides us with a determination of the Hausdorff codimension c_{H}.

Figure 6.13 shows the values of the right-hand member of the escape-rate formula (6.21) as a function of the inverse size $1/L$, from which the diffusion coefficient D_{fractal} can be determined by linear extrapolation. These values are reported in Table 6.2 where we see the nice agreement with the values of the diffusion coefficient obtained by the other methods.

These results provide important support to the application of the escape-rate formalism to the periodic Lorentz gas. In the following, we shall consider a simpler model which is exactly solvable.

6.4 The multibaker mapping

6.4.1 Definition

The multibaker or bakery map is a model of deterministic diffusion, which may be introduced as follows. We consider the chaotic scattering of a point particle in the billiard of Fig. 6.14 which is made of a sequence of semicircular obstacles in a channel. We call this billiard the open Lorentz channel (Gaspard 1993a, 1995). In the left- and right-hand channels the particle moves into zig-zag

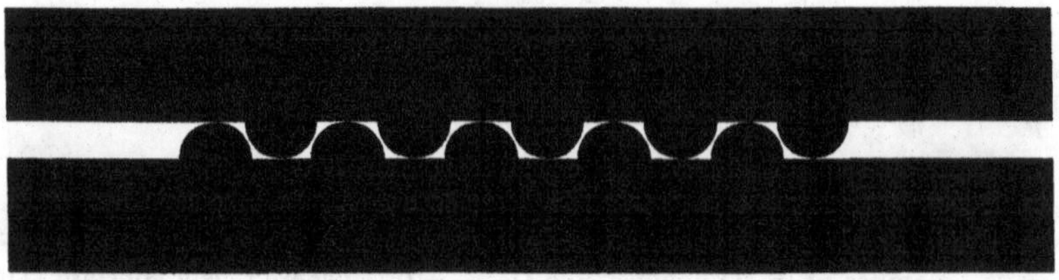

Figure 6.14 The open Lorentz channel billiard (Gaspard 1995).

trajectories with a constant horizontal component of velocity. The time taken by the particle to escape out of the scattering region into the left- and right-hand exit channels is depicted in Fig. 6.15 as a function of the component $\varpi = v_{\parallel}$ of the initial velocity which is parallel to the wall. The initial position is taken on the wall either at the left-hand entrance of the scatterer (Fig. 6.15a) or in the middle (Fig. 6.15b). These plots show that the escape time is a highly singular function of the initial condition, with singularities on the fractal set formed by the stable manifolds of the fractal repeller, as is the case in the open Lorentz gas itself (see Fig. 6.2).

The dynamics of the billiard can be represented by an area-preserving Birkhoff map in terms of the Birkhoff coordinates $\mathbf{x} = (\theta, \varpi)$.

The multibaker model is an area-preserving map which is a caricature of the Birkhoff map of the Lorentz channel. The phase space is composed of a sequence of squares, which correspond to the disks of the Lorentz channel in this analogy. The map acts by stretching and cutting each one of these squares and by redistributing the cut pieces on several neighbouring squares in the same way as the particle collides with successive disks in the Lorentz channel. This redistribution of the cut pieces to neighbouring squares induces a deterministic diffusion along the chain. This map is also a model of the kneading of dough in a bakery.

Originally, a four-adic version of the multibaker map was proposed (Gaspard 1992a). Here, we consider a dyadic version of the multibaker proposed by Tasaki

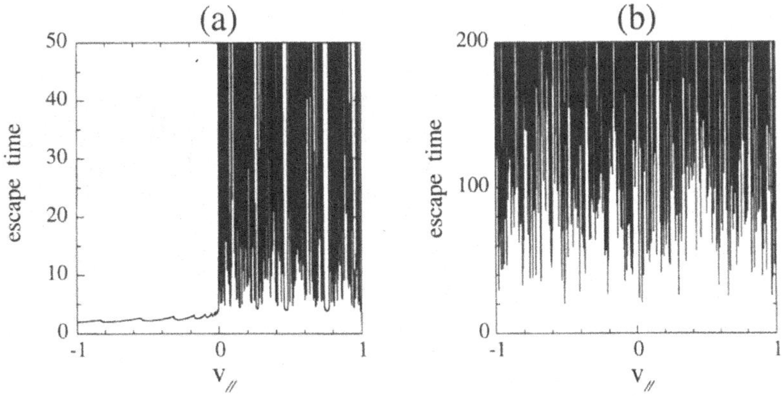

Figure 6.15 Escape-time functions in the open Lorentz channel versus the velocity parallel to the wall: (a) at the left-hand entrance; (b) in the middle (Gaspard 1995).

and Gaspard (1994, 1995; Gaspard 1995), which is defined by

$$\phi(l,x,y) = \begin{cases} \left(l-1, 2x, \frac{y}{2}\right), & 0 \le x \le \frac{1}{2}, \\ \left(l+1, 2x-1, \frac{y+1}{2}\right), & \frac{1}{2} < x \le 1, \end{cases} \tag{6.46}$$

where we recognize the action of the baker on the coordinates (x, y) (see Fig. 6.16). This action is combined with displacements $l \to l \pm 1$.

In order to compare this model with the open Lorentz channel, we consider the escape process out of a finite chain made of the squares $0 \le l \le L$. The corresponding escape-time function is depicted in Fig. 6.17 which shows the analogy with the Lorentz channel. Indeed, a fractal repeller is also underlying the scattering process in the multibaker chain.

The multibaker map is invertible with the following inverse:

$$\phi^{-1}(l,x,y) = \begin{cases} \left(l+1, \ \frac{x}{2}, \ 2y\right), & 0 \le y < \frac{1}{2}, \\ \left(l-1, \ \frac{x+1}{2}, \ 2y-1\right), & \frac{1}{2} \le y < 1. \end{cases} \tag{6.47}$$

Moreover the system is time-reversal invariant, i.e., there exists an involution which corresponds to the velocity inversion in the particle system:

$$\imath(l,x,y) \equiv (l, \ 1-y, \ 1-x), \tag{6.48}$$

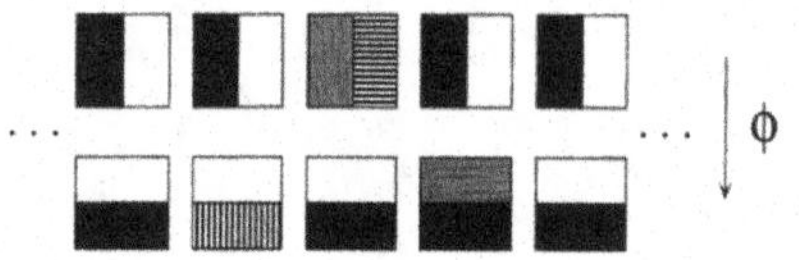

Figure 6.16 Geometric representation of the dyadic multibaker map (Gaspard 1995).

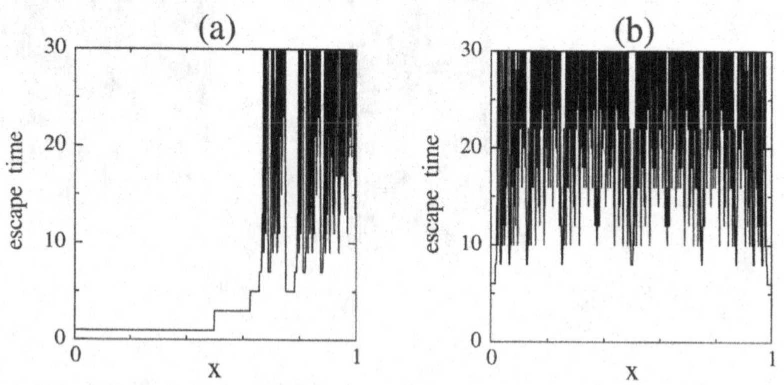

Figure 6.17 Escape-time functions in the multibaker map: (a) at the left-hand entrance; (b) in the middle (Gaspard 1995).

and which satisfies $\iota^2 = 1$ and

$$\phi^{-1} = \iota \circ \phi \circ \iota . \tag{6.49}$$

Since the multibaker is area-preserving the Lebesgue measure $dxdy$ is invariant, which gives equal probabilities $(1/2, 1/2)$ to the displacements to the right or the left. Accordingly, this model is isomorphic to the standard symmetric random walk (Feller 1968) and its diffusion coefficient is equal to

$$D = \frac{1}{2} . \tag{6.50}$$

The map acts by horizontal stretching and vertical contraction so that it is uniformly hyperbolic and its positive Lyapunov exponent is equal to $\lambda = \ln 2$. The dynamics inside a unit cell of the lattice is that of the standard baker map which is isomorphic to a Bernoulli process $B(1/2, 1/2)$ so that its KS entropy is equal to $h_{\rm KS} = \ln 2$.

The free motion of the particle in the exit channels can be modelled by modifying the multibaker map at both ends of a finite chain of length $L+1$. Outside the chain, the particles move in free motion with velocities +1 or −1. This is realized by a simple composition of translations to the left or the right in the half squares extending from both ends to infinity (see Fig. 6.18)

$$\phi_L(l,x,y) = \begin{cases} \left(l-1, 2x, \frac{y}{2}\right) , & 0 \le x < \frac{1}{2} , \quad +1 \le l \le L+1 , \\ \left(l+1, 2x-1, \frac{y+1}{2}\right) , & \frac{1}{2} \le x \le 1 , \quad -1 \le l \le L-1 , \\ (l-1, x, y) , & 0 \le x < \frac{1}{2} , \quad l \le 0 \quad \text{or} \quad L+2 \le l , \\ (l+1, x, y) , & \frac{1}{2} \le x \le 1 , \quad l \le -2 \quad \text{or} \quad L \le l . \end{cases} \tag{6.51}$$

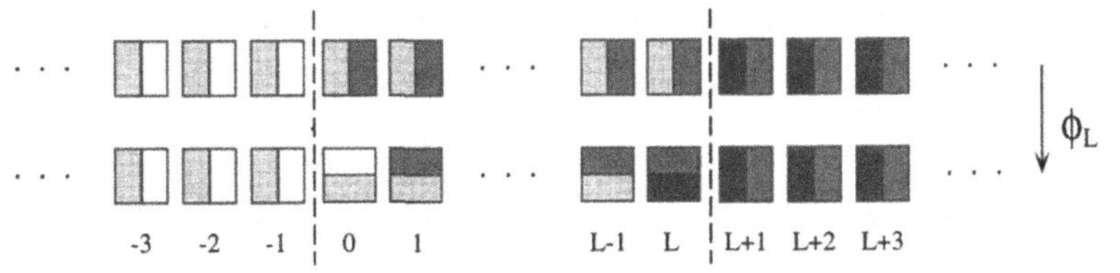

Figure 6.18 Representation of the action of the open multibaker map in its phase space which is composed of an infinity of squares. The map acts like a baker transformation only on the chain of the squares $0 \le l \le L$ and by left or right translations outside the chain up to infinity. This map is a caricature of the Birkhoff map of the open Lorentz channel depicted in Fig. 6.14. See Tasaki and Gaspard (1995).

In this form, the multibaker is a model of chaotic scattering with the typical escape-time functions of Fig. 6.17. Consequently, there exists a hyperbolic repeller in the region $0 \leq l \leq L$ of the system, which is fractal and chaotic. In order to characterize the dynamics on this repeller, we introduce a symbolic dynamics based on the labels of the squares of the finite chain: $\omega \in \mathscr{A} = \{0, 1, 2, \ldots, L\}$. The assignment is carried out so that $\omega_n = l$ iff $\boldsymbol{\phi}^n\mathbf{x}$ belongs to the square of label l. The symbolic dynamics is a topological Markov shift with the $(L+1) \times (L+1)$ transition matrix:

$$
\mathsf{A} = \begin{array}{c}
\\ 0 \\ 1 \\ 2 \\ 3 \\ 4 \\ 5 \\ \vdots \\ L-4 \\ L-3 \\ L-2 \\ L-1 \\ L
\end{array}
\begin{array}{c}
\begin{array}{cccccccccccc}
0 & 1 & 2 & 3 & 4 & 5 & \ldots & L-4 & L-3 & L-2 & L-1 & L
\end{array} \\
\left(\begin{array}{cccccccccccc}
 & 1 & & & & & \ldots & & & & & \\
1 & & 1 & & & & \ldots & & & & & \\
 & 1 & & 1 & & & \ldots & & & & & \\
 & & 1 & & 1 & & \ldots & & & & & \\
 & & & 1 & & 1 & \ldots & & & & & \\
 & & & & 1 & & \ldots & & & & & \\
\vdots & \vdots & \vdots & \vdots & \vdots & \vdots & \ddots & \vdots & \vdots & \vdots & \vdots & \vdots \\
 & & & & & & \ldots & & 1 & & & \\
 & & & & & & \ldots & 1 & & 1 & & \\
 & & & & & & \ldots & & 1 & & 1 & \\
 & & & & & & \ldots & & & 1 & & 1 \\
 & & & & & & \ldots & & & & 1 &
\end{array}\right)
\end{array} . \qquad (6.52)
$$

6.4.2 The Pollicott–Ruelle resonances and the escape-rate formula

The Pollicott–Ruelle resonances of the multibaker map can be obtained from the Fredholm determinant of the Frobenius–Perron operator

$$
Z(s) = \mathrm{Det}(\hat{I} - e^{-s}\hat{P}) = \exp - \sum_{n=1}^{\infty} \frac{\exp(-sn)}{n} \, \mathrm{Tr}\, \hat{P}^n . \qquad (6.53)
$$

Because the multibaker is uniformly hyperbolic the trace of the iterates of the Frobenius–Perron operator can be obtained as

$$
\begin{aligned}
\mathrm{Tr}\, \hat{P}^n &= \sum_{\mathbf{x} \in \mathrm{Fix}\boldsymbol{\phi}^n} \frac{1}{|\det(\mathsf{1} - \partial_{\mathbf{x}}\boldsymbol{\phi}^n)|} \\
&= \frac{|\mathrm{Fix}\boldsymbol{\phi}^n|}{\Lambda^n(1 - \Lambda^{-n})^2} = \mathrm{tr}\, \mathsf{A}^n \sum_{m=0}^{\infty} \frac{m+1}{\Lambda^{n(m+1)}} ,
\end{aligned} \qquad (6.54)
$$

where we used several results from Chapters 2 and 3 and where $\Lambda = 2$. We have therefore reduced the trace of the Frobenius–Perron operator to ordinary traces of the transition matrix (6.52). This latter is here symmetric and can be diagonalized according to

$$\begin{cases} \mathsf{A} \cdot \mathbf{v} = \chi \, \mathbf{v} \, , \\ \mathbf{u}^{\mathrm{T}} \cdot \mathsf{A} = \chi \, \mathbf{u}^{\mathrm{T}} \, , \end{cases} \tag{6.55}$$

where $\mathbf{u} = \mathbf{v}$ since A is symmetric. As a consequence, the Selberg–Smale Zeta function becomes

$$\begin{aligned} Z(s) &= \prod_{m=0}^{\infty} \det\left(1 - \frac{\exp(-s)}{\Lambda^{m+1}} \mathsf{A}\right)^{m+1} \\ &= \prod_{m=0}^{\infty} \prod_{j=1}^{L+1} \left(1 - \frac{\exp(-s)}{\Lambda^{m+1}} \chi_j\right)^{m+1} , \end{aligned} \tag{6.56}$$

where χ_j are the eigenvalues of the topological transition matrix A. The Pollicott–Ruelle resonances are therefore given by

$$s = \ln \chi_j - (m+1) \ln \Lambda \, , \tag{6.57}$$

with the multiplicity $m+1$. Moreover, the pressure function is obtained as

$$P(\beta) = \ln \chi_1 - \beta \ln \Lambda \, . \tag{6.58}$$

Therefore, the characteristic quantities are determined by the largest eigenvalue of the transition matrix (6.52).

These eigenvalues can be calculated by applying the matrix (6.52) on a vector which leads to the second difference equation

$$v_{l-1} + v_{l+1} = \chi \, v_l \, , \tag{6.59}$$

for $0 \leq l \leq L$ with the absorbing boundary conditions $v_{-1} = v_{L+1} = 0$. The solution to this eigenvalue problem is

$$\chi_j = 2 \cos \theta_j \, , \tag{6.60}$$

with the left- and right-eigenvectors

$$u_l^{(j)} = v_l^{(j)} = \sin(l+1)\theta_j \, , \tag{6.61}$$

where

$$\theta_j = \frac{\pi j}{L+2} \, , \qquad \text{with} \quad j = 1, 2, \ldots, L+1 \, . \tag{6.62}$$

The probabilistic Markov chain on the repeller can be constructed in analogy with Eqs. (4.112)–(4.115) where

$$\Pi_{\omega\omega'} = \frac{A_{\omega\omega'} \, v_{\omega'}}{\chi_1 \, v_\omega} \, , \qquad \omega, \omega' = 0, 1, 2, \ldots, L \, , \tag{6.63}$$

with the leading eigenvalue χ_1 and the corresponding eigenvector $\mathbf{v}$ of Eq. (6.55). The characteristic quantities of chaos on the repeller $\mathscr{I}_L$ can here be exactly obtained as

$$\gamma = -\ln\cos\frac{\pi}{L+2} = \frac{1}{2}\left(\frac{\pi}{L+2}\right)^2 + \mathcal{O}(L^{-4}), \tag{6.64}$$

$$\lambda = \ln 2, \tag{6.65}$$

$$\begin{aligned} h_{\text{top}} = h_{\text{KS}} &= \ln\left(2\cos\frac{\pi}{L+2}\right) \\ &= \ln 2 - \frac{1}{2}\left(\frac{\pi}{L+2}\right)^2 + \mathcal{O}(L^{-4}), \end{aligned} \tag{6.66}$$

$$\begin{aligned} d_{\text{H}} = d_{\text{I}} &= 1 + \frac{1}{\ln 2}\ln\cos\frac{\pi}{L+2} \\ &= 1 - \frac{1}{2\ln 2}\left(\frac{\pi}{L+2}\right)^2 + \mathcal{O}(L^{-4}). \end{aligned} \tag{6.67}$$

The KS entropy is very close to the value of the baker map but is slightly smaller due to the escape out of the chain. Similarly, the partial Hausdorff or information dimensions are very close to the unit value so that the fractal repeller occupies the full phase space in the limit $L \to \infty$.

From the asymptotic expression of the escape rate we recover the value (6.50) of the diffusion coefficient. Therefore, the escape-rate formula

$$D = \lim_{L\to\infty}\left(\frac{L}{\pi}\right)^2 [\bar{\lambda}(\mathscr{I}_L) - h_{\text{KS}}(\mathscr{I}_L)], \tag{6.68}$$

holds for this exactly solvable model.

The Pollicott–Ruelle resonances are here given by

$$\begin{aligned} s &= -m\ln 2 + \ln\cos\frac{\pi j}{L+2} \\ &= -m\ln 2 - D\left(\frac{\pi j}{L+2}\right)^2 + \mathcal{O}(L^{-4}), \end{aligned} \tag{6.69}$$

with $j = 1, 2, \ldots, L+1$ and $m = 0, 1, 2, \ldots$. This is the exact spectrum of the relaxation modes of the Liouvillian dynamics, which suggests a comparison with the spectrum of the eigenvalues of the phenomenological equation (6.1) with absorbing boundary conditions at $x = 0$ and $x = L+2$. These phenomenological eigenvalues are

$$s_{\text{ph}} = -D\left(\frac{\pi j}{L+2}\right)^2, \tag{6.70}$$

which reveals that the Pollicott–Ruelle resonances of the leading family $m = 0$ tend to the eigenvalues of the phenomenological diffusion equation as $L \to$

∞. This observation is of course of paramount importance for the theory of transport processes because it shows that the relaxation of the hydrodynamic modes can be formulated in terms of the Pollicott–Ruelle resonances which have a firm mathematical foundation and which do not involve any approximation of Markovian type. Consequences of this observation will be developed in the following chapters. Let us remark that other boundary conditions may be studied as done for the four-adic multibaker map, which lead to the same conclusion (Gaspard 1992a).

Other multibaker models have been analyzed. In particular, Tél et al. (1996) have proposed a biased four-adic multibaker model for diffusion in the presence of an electric field. This model which presents a nonvanishing mean drift has been analyzed from a scattering point of view. In this case, the escape rate contains not only the term decreasing as L^{-2} and proportional to the diffusion coefficient but also an extra term involving the mean drift. This model provides a comparison between the escape-rate formalism described here and another formalism where the transport coefficients are related to the sum of all the Lyapunov exponents in the case of thermostatted dynamical systems (Evans et al. 1990a, 1993) (see below).

6.5 Escape-rate formalism for general transport coefficients

6.5.1 General context

In the previous sections, the diffusion coefficient has been obtained in terms of the escape rate of a particle from a bounded region of the scatterers. This escape rate, in turn, is obtained in terms of the positive Lyapunov exponents and KS entropy that characterize the set of orbits of the moving particle that are trapped forever in the bounded region occupied by the scatterers. This set of trapped orbits forms a fractal set of trajectories of the moving particle, and is referred to as the repeller. In this way, different versions of the escape-rate formula have been derived, which relate the diffusion coefficient to the dynamical quantities on the repeller.

It is the purpose of this section and of the following to extend the scattering theory and, in particular, the escape-rate formalism to include other transport coefficients, and to include a treatment of chemical reaction-rate coefficients. This is accomplished by showing that the basic ideas used to derive the expression for the diffusion coefficient of a moving particle can be extended to apply to other transport and reaction-rate processes (Dorfman and Gaspard 1995). The derivation of the escape-rate expressions for transport and reaction rate coefficients is based on the fact that all of these coefficients can be related to the average mean square displacement of some appropriate dynamical quantity called the Helfand moments (Helfand 1960, 1961).

We consider a large system of N particles governed by Hamilton's equations of classical mechanics or by the dynamical rules of a billiard. We assume that these particles are contained in a rectangular domain of volume V so that the fluid density is $\rho = N/V$. At the borders, we may consider either hard walls of infinite mass or periodic boundary conditions. In the latter case, the total momentum is conserved in addition to the total energy. We suppose that the system is ergodic and mixing on each energy (or energy–momentum) shell. We suppose that almost all the trajectories of the system are unstable of saddle type. Possibly, there is a set of zero Lebesgue measure of trajectories of marginal stability, such as bouncing ball or free orbits. An example of such a system is the dilute hard-sphere gas.

6.5.2 Transport coefficients and their Helfand moments

In the large-system limit ($N, V \to \infty$ with $N/V = \rho$), irreversible processes in such a classical many-body system may be described by hydrodynamic equations such as the Navier–Stokes equations, the diffusion equation, the chemical kinetic equations ruling the time evolution of chemical concentrations, or the equations of electrical conductivity which incorporate Ohm's law. These phenomenological equations contain dissipative terms which are dependent on transport and rate coefficients. The aim of nonequilibrium statistical mechanics is to obtain these coefficients in terms of the microscopic Hamiltonian equations. Since the work of Maxwell and Boltzmann, several methods have been developed to calculate these coefficients. The most general method, developed by Green (1951–60) and Kubo (1957), is the time-correlation function approach whereby the transport and rate coefficients are given as time-integrals of autocorrelations of the fluxes,

$$\alpha = \int_0^\infty \lim_{V\to\infty} \langle J_0^{(\alpha)} \, J_t^{(\alpha)} \rangle \, dt \, , \tag{6.71}$$

where $J_t^{(\alpha)}$ is the flux at time t corresponding to the coefficient α. It is a function of the canonical variables $(\mathbf{q}, \mathbf{p})$ and is obtained by solving the equations of motion for a time t after an initial time so that

$$J_t^{(\alpha)}(\mathbf{q}, \mathbf{p}) = J_0^{(\alpha)}[\mathbf{\Phi}^t(\mathbf{q}, \mathbf{p})] \, , \tag{6.72}$$

where $\mathbf{\Phi}^t$ denotes the flow in phase space induced by Hamiltonian equations. We can express this in terms of an N-particle streaming operator

$$J_t^{(\alpha)} \;=\; \exp(-t\hat{L}) \, J_0^{(\alpha)} \, , \tag{6.73}$$

with the Liouvillian operator given in terms of a Poisson bracket expression $\hat{L} = \{H, \cdot\}$. The average $\langle\cdot\rangle$ in equation (6.71) is taken over a microcanonical ensemble in our case.

It will be useful for us to apply the formulation of the Green–Kubo expressions, as obtained by moments

$$G_t^{(\alpha)} \;=\; G_0^{(\alpha)} \;+\; \int_0^t J_\tau^{(\alpha)} \, d\tau \, , \tag{6.74}$$

such that the fluxes are derivatives as

$$J_t^{(\alpha)} \;=\; \frac{d}{dt} \, G_t^{(\alpha)} \, . \tag{6.75}$$

An integration by parts shows that

$$\begin{aligned} \langle (G_t \;-\; G_0)^2 \rangle \;&=\; \int_0^t \int_0^t \langle J_{t'} J_{t''} \rangle \, dt' \, dt'' \\ &=\; 2\,t \int_0^t \left(1 \;-\; \frac{\tau}{t}\right) \langle J_0 J_\tau \rangle \, d\tau \, . \end{aligned} \tag{6.76}$$

If the time correlation function $\langle J_0 J_t \rangle$ decays fast enough as $t \to \infty$ and, in any case, faster than $\langle J_0 J_t \rangle \sim t^{-1}$, the following condition is satisfied

$$\lim_{t\to\infty} \frac{1}{t} \int_0^t \tau \, \langle J_0 J_\tau \rangle \, d\tau \;=\; 0 \, , \tag{6.77}$$

and we obtain the equality

$$\lim_{t\to\infty} \frac{\langle (G_t \;-\; G_0)^2 \rangle}{2t} \;=\; \int_0^\infty \langle J_0 J_\tau \rangle \, d\tau \, . \tag{6.78}$$

As a consequence, the transport and rate coefficients are also given by Einstein-type formulas

$$\alpha \;=\; \lim_{t\to\infty} \frac{1}{2t} \lim_{V\to\infty} \left\langle \left[G_t^{(\alpha)} \;-\; G_0^{(\alpha)} \right]^2 \right\rangle . \tag{6.79}$$

In Table 6.3, we list moments appropriate for each transport or rate coefficient, where $E_i = \frac{p_i^2}{2m} + \frac{1}{2} \sum_{j(\neq i)} V_{ij}$ is the energy of particle i and V_{ij} is the potential energy of interaction between particles i and j (Helfand 1960, 1961). The following section describes in more detail the case of chemical reactions.

Equation (6.79) shows that Helfand's moments undergo a diffusive type of motion along the axis of the moment $G^{(\alpha)}$ if the transport or rate coefficients are well defined (i.e., are positive and finite). Therefore, the moment may be considered as a random variable having a probability density $p(g)$ obeying a diffusion-type equation, in an equilibrium ensemble

$$\frac{\partial p}{\partial t} = \alpha \frac{\partial^2 p}{\partial g^2} , \tag{6.80}$$

with the transport or rate coefficient α as diffusion coefficient. This equation is the Fokker–Planck equation governing the equilibrium fluctuations of the Helfand moments. Equivalently, one can recover Eq. (6.79) by supposing that the moments satisfy a Langevin stochastic differential equation

$$\dot{G}_t^{(\alpha)} = J_t^{(\alpha)} , \tag{6.81}$$

where the flux $J_t^{(\alpha)}$ is a white noise

$$\langle J_t^{(\alpha)} \rangle = 0 , \qquad \langle J_0^{(\alpha)} J_t^{(\alpha)} \rangle = 2\alpha\, \delta(t) . \tag{6.82}$$

It is important to note that equations (6.80) or (6.81)–(6.82) are to be considered as a simple representation of the results of the time-correlation function method for time scales which are much longer than the time necessary for the time-correlation functions of the microscopic currents to decay to zero. We suppose that this approach applies to situations where, for example, long-time tails in the correlation functions decay sufficiently rapidly for the transport coefficients as defined by Eq. (6.71) to exist. In this way, we can still treat processes

Table 6.3 Helfand's moments.

process	*moment*
self-diffusion	$G^{(D)} = x_i$
shear viscosity	$G^{(\eta)} = \frac{1}{\sqrt{Vk_BT}} \sum_{i=1}^N x_i\, p_{iy}$
bulk viscosity ($\psi = \zeta + \frac{4}{3}\eta$)	$G^{(\psi)} = \frac{1}{\sqrt{Vk_BT}} \sum_{i=1}^N x_i\, p_{ix}$
heat conductivity	$G^{(\kappa)} = \frac{1}{\sqrt{Vk_BT^2}} \sum_{i=1}^N x_i\, (E_i - \langle E_i \rangle)$
charge conductivity	$G^{(e)} = \frac{1}{\sqrt{Vk_BT}} \sum_{i=1}^N eZ_i\, x_i$
chemical reaction rate	$G^{(r)} = \frac{1}{\sqrt{Vk_BT}} (N^{(r)} - \langle N^{(r)} \rangle)$

which are diffusive on long-time scales, but have correlations on shorter time scales. Accordingly, the existence of a flux autocorrelation function which differs from a delta distribution on short-time scales is still perfectly compatible with the validity of the diffusion-type equation, (6.80), on long-time scales.

In this discussion, we have tacitly assumed that we consider a physical system in the proper thermodynamic limit. However, for systems of finite volume, there is an upper limit on the times we can consider because the ranges of variation of the Helfand moments are bounded. For example, if the system consists of hard spheres placed in a cubic box of side length L, the positions of each particle vary only in the interval

$$-\frac{L}{2} \leq x_i,\ y_i,\ z_i \leq +\frac{L}{2}, \tag{6.83}$$

while the momenta can only take values in the interval

$$-\sqrt{2mE_{\text{tot}}} \leq p_{ix},\ p_{iy},\ p_{iz} \leq +\sqrt{2mE_{\text{tot}}}, \tag{6.84}$$

where $E_{\text{tot}} = (3/2)Nk_{\text{B}}T$ is the total energy. Similarly, the energy of a particle can only lie in the range

$$0 \leq E_i \leq \frac{3}{2}\, k_{\text{B}}T\, N\,. \tag{6.85}$$

Because of these bounds, the moments are always of bounded variation in the interval

$$|G^{(\alpha)}| \leq C\, N^{\delta^{(\alpha)}}, \tag{6.86}$$

where C is some constant, and $\delta^{(\alpha)}$ are positive exponents which are respectively $\delta^{(\text{D})} = 1/3$, $\delta^{(\eta)} = 4/3$, $\delta^{(\psi)} = 4/3$, $\delta^{(\kappa)} = 11/6$, $\delta^{(\text{e})} = 5/6$, and $\delta^{(\text{r})} = 1/2$, for three-dimensional gases. Accordingly, the range of variation of the moments grows with the size of the system, when the density and other intensive variables are kept constant. Of course, most variations of the Helfand moments will be due to microscopic motions of the particles and thus will be much smaller than the bounds in Eq. (6.86). At any rate, as the system gets larger, we expect that the diffusive-like behaviour of the moments, described by Eq. (6.80), will be valid over increasingly larger regions of variations of $G^{(\alpha)}$.

After the discussion about the limited range over which the Helfand moments obey the diffusive-type equation (6.80) for finite systems, we may proceed.

6.5.3 Generalization of the escape-rate formula

Within the range of validity of (6.80), we may set up a problem of first passage for the moment $G^{(\alpha)}$ corresponding to the transport or rate processes of coefficient

α (Dorfman and Gaspard 1995). We consider a statistical ensemble formed by copies of the system which we assume to be at equilibrium and microcanonical at total energy E (and eventually at fixed total momentum in the case of periodic boundary conditions: $\mathbf{P}_{\text{tot}} = \sum_{i=1}^{N} \mathbf{p}_i$). For each copy, the motion of the Hamiltonian system is integrated from the initial conditions and the Helfand moment is calculated along the trajectory. At each time, we count the number of copies, $\mathcal{N}^{(\alpha)}(t)$, for which the moment is still in the following interval

$$-\frac{\chi}{2} \leq G_t^{(\alpha)} \leq +\frac{\chi}{2}, \tag{6.87}$$

where the size of the interval χ is sufficiently large to be in the regime of diffusion of the moment but not too large with respect to the total variation interval of the moment allowed by the finiteness of the system. We are here defining a problem of first passage which can be solved using the eigenvalues and eigenfunctions of the diffusion-type equation (6.80) with the absorbing boundary conditions

$$p(-\chi/2) = p(+\chi/2) = 0 . \tag{6.88}$$

The solution of this eigenvalue problem is well known to be

$$p(g,t) = \sum_{j=1}^{\infty} c_j \exp(s_j^{(\alpha)} t) \sin\left[\frac{\pi j}{\chi}\left(g + \frac{\chi}{2}\right)\right] ,$$

$$\text{with} \quad s_j^{(\alpha)} = -\alpha \left(\frac{\pi j}{\chi}\right)^2 , \tag{6.89}$$

where the constants c_j are fixed from the initial probability density $p(g,0)$. The number of copies of the statistical ensemble which are still in the interval (6.87) is then given by

$$\mathcal{N}^{(\alpha)}(t) = \mathcal{N}_0 \int_{-\chi/2}^{+\chi/2} p(g,t)\, dg . \tag{6.90}$$

At long times, the decay is dominated by the slowest decay mode corresponding to the smallest decay rate $-s_1^{(\alpha)}$, which defines the (phenomenological) escape rate of the moment out of the interval (6.87)

$$\gamma_{\text{ph}}^{(\alpha)} = -s_1^{(\alpha)} = \alpha \left(\frac{\pi}{\chi}\right)^2 . \tag{6.91}$$

We are now in a position to establish a relationship with the deterministic dynamics. The Hamiltonian classical motion of the many-body system is chaotic in many cases. This property has been proved by Sinai and coworkers for some simple hard-sphere gas models (Sinai 1970, Bunimovich et al. 1990). Also strong numerical evidence exists which shows that half of the Lyapunov exponents are typically positive in systems of statistical mechanics like the Lennard-Jones gas at room temperatures (Posch and Hoover 1988, 1989; Dellago et al. 1995, 1996). We suppose that the decomposition of phase space into ergodic components is understood and that, beside the decomposition on the known constants of motion, there is only a single main ergodic component.

We consider the set of all the trajectories for which the Helfand moment remains forever within the interval (6.87). Because most of the trajectories are expected to exit this interval, the trapped trajectories must be exceptional and highly unstable forming a set of measure zero with respect to the microcanonical probability measure. On the basis that the trajectories are typically of saddle type in systems of statistical mechanics, we assume that the set of trajectories is a fractal repeller. Indeed, this set is of vanishing measure but may still contain an uncountable infinity of periodic and nonperiodic trajectories. A set satisfying these conditions is necessarily a fractal. Moreover, it is composed of unstable trajectories of saddle type so that it forms a repeller (of saddle type) in phase space. These properties hold for particular models like the multibaker area-preserving map as well as the array of disk scatterers composing the periodic Lorentz gas (see above). Accordingly, it seems reasonable to assume that, for more general systems, such as a gas of hard spheres the set of trajectories for which the moments satisfy Eq. (6.87), form a fractal repeller, with properties to be described in the next paragraph.

A fractal repeller is characterized by different quantities and, especially, by an escape rate which is the deterministic analogue of the escape rate obtained in the preceding first-passage problem. The escape rate of the underlying deterministic Hamiltonian system may thus be identified with the phenomenological escape rate (6.91) when χ is large enough so that

$$\gamma^{(\alpha)} = \alpha \left(\frac{\pi}{\chi}\right)^2 + o(\chi^{-2}) . \tag{6.92}$$

Moreover, in chaotic systems, the escape rate is related to the sum of positive Lyapunov exponents minus the KS entropy per unit time if these quantities are well defined and positive (see Chapter 4). These quantities are evaluated for the natural invariant probability measure (4.61) whose support is the fractal repeller. For the natural invariant measure, each cell of phase space has a weight which is inversely proportional to the local Lyapunov numbers (stretching factors). According to Eq. (4.64), we have

$$\gamma^{(\alpha)} = \sum_{\bar{\lambda}_i>0} \bar{\lambda}_i(\mathscr{I}_\chi^{(\alpha)}) - h_{\mathrm{KS}}(\mathscr{I}_\chi^{(\alpha)}) , \tag{6.93}$$

where we denote by $\mathscr{I}_\chi^{(\alpha)}$ the fractal repeller formed by the trapped trajectories for which the Helfand moment $G_t^{(\alpha)}$ remains forever in the interval (6.87).

Combining the deterministic result (6.93) with the result (6.92) from phenomenological considerations, we obtain the relationship

$$\alpha = \lim_{\chi\to\infty} \left(\frac{\chi}{\pi}\right)^2 \lim_{V\to\infty} \left[\sum_{\bar{\lambda}_i>0} \bar{\lambda}_i(\mathscr{I}_\chi^{(\alpha)}) - h_{\mathrm{KS}}(\mathscr{I}_\chi^{(\alpha)})\right] , \tag{6.94}$$

where the limit $V \to \infty$ denotes the thermodynamic limit to be taken before the limit $\chi \to \infty$ which is internal to the system (Dorfman and Gaspard 1995).

With Table 6.3, Eq. (6.94) shows how a general transport or rate coefficient can in principle be related to the Lyapunov exponents and the KS entropy of a fractal repeller. This fractal repeller is the phase-space object corresponding to the escape process of the Helfand moment associated with the transport or rate coefficient. In this way, a connection is established between statistical and mechanical considerations in phase space.

A remark is now in order about the magnitude of the quantities appearing in (6.94). The sum of positive Lyapunov exponents and the KS entropy per unit time are very large, of the order of the number of particles times the inverse of a typical kinetic time scale (see Introduction). On the other hand, the escape rate, which is the difference between two such large numbers, has a much smaller magnitude given by the time scales characteristic of hydrodynamics. In this way, the kinetic and hydrodynamic levels are naturally connected with an escape-rate formula like (6.94).

6.6 Escape-rate formalism for chemical reaction rates

In this section, we consider in more detail the case of chemical reactions in order to be more explicit on the way the escape-rate formalism may be applied to such a case. We only consider chemical reactions close to equilibrium, where the escape-rate formalism applies. We should point out that the numbers of particles of the chemical species are not conserved quantities. In this respect, the chemical reaction rates differ from the transport coefficients like viscosity and heat conductivity which are derived from the conservation of the total mass, momentum, and energy of the system. However, the escape-rate theory can be formulated in the specific spaces of variations of the different Helfand moments for any kind of exponential relaxations independently of the possible conservation laws (Dorfman and Gaspard 1995).

6.6.1 Nonequilibrium thermodynamics of chemical reactions

In recent decades, many works have been devoted to the nonequilibrium thermodynamics of chemical systems (Glansdorff and Prigogine 1971, Nicolis and Prigogine 1977). Here, we consider a system where a single chemical reaction takes place, namely

$$\sum_{\gamma=1}^{c} \nu_\gamma^+ X_\gamma \leftrightarrow \sum_{\gamma=1}^{c} \nu_\gamma^- X_\gamma \ . \tag{6.95}$$

The numbers of particles of the reactants and products change at each step of the reaction according to

$$\frac{\Delta N_1}{\nu_1} = \frac{\Delta N_2}{\nu_2} = \ldots = \frac{\Delta N_c}{\nu_c} = \Delta N^{(\mathrm{r})} \ , \tag{6.96}$$

where $\nu_\gamma = \nu_\gamma^- - \nu_\gamma^+$ are the stoichiometric coefficients. The degree of advancement of the reaction can be measured in terms of the variation of the number $N^{(\mathrm{r})}$ characterizing the reaction (6.95). We also introduce the chemical concentrations, $\rho_\gamma = N_\gamma / V$. At the phenomenological level of thermodynamics, the velocity of the reaction is defined by

$$\dot{\bar{\rho}}^{(\mathrm{r})} = \frac{\dot{\bar{\rho}}_\gamma}{\nu_\gamma} = w \ , \tag{6.97}$$

where $\bar{\rho}_\gamma$ are the average chemical concentrations. The dependence of the reaction velocity on the concentrations themselves is given by the mass action law

$$w = \kappa_+ \prod_{\gamma=1}^{c} \rho_\gamma^{\nu_\gamma^+} - \kappa_- \prod_{\gamma=1}^{c} \rho_\gamma^{\nu_\gamma^-} \ , \tag{6.98}$$

where $\kappa_\pm$ are the forward and backward reaction rates of (6.95). The affinity of the reaction is defined by

$$A = - \sum_{\gamma=1}^{c} \nu_\gamma \, \mu_\gamma = - k_\mathrm{B} T \, \ln \prod_{\gamma=1}^{c} \left(\frac{\rho_\gamma}{\rho_\gamma^{\mathrm{eq}}} \right)^{\nu_\gamma} \ , \tag{6.99}$$

where $\mu_\gamma = \mu_\gamma^0 + k_\mathrm{B} T \ln \rho_\gamma$ is the chemical potential of the species γ and it is known that the affinity vanishes at thermodynamic equilibrium: $A^{\mathrm{eq}} = 0$. Near the thermodynamic equilibrium, both the reaction velocity w and the affinity A can be expanded in terms of the variations of the chemical concentrations around their equilibrium value, $\rho_\gamma = \rho_\gamma^{\rm eq} + \Delta\rho_\gamma$. In the linear approximation, we infer from Eqs. (6.98) and (6.99) the equality

$$w \simeq w_+^{\rm eq} \frac{A}{k_{\rm B}T} , \qquad \text{with} \qquad w_+^{\rm eq} = \kappa_+ \prod_{\gamma=1}^{c} \left(\rho_\gamma^{\rm eq}\right)^{\nu_\gamma^+} , \tag{6.100}$$

which allows us to obtain the Onsager coefficient of this chemical process as

$$w \simeq \frac{LA}{T} , \qquad \text{with} \qquad L = \frac{w_+^{\rm eq}}{k_{\rm B}} . \tag{6.101}$$

Yamamoto (1960) and, later, Zwanzig (1965) have shown that the Onsager coefficient of the chemical reaction is given by the following integral of the autocorrelation function in a classical system

$$L = \frac{1}{Vk_{\rm B}} \int_0^\infty \langle \dot N_0^{(\rm r)} \dot N_t^{(\rm r)} \rangle \, dt , \tag{6.102}$$

where the corresponding flux is here proportional to the time derivative of the number $N^{(\rm r)}$ of particles which is characteristic of the reaction. Accordingly, we obtain the result given in Table 6.3 for the reaction coefficient

$$\alpha = \frac{w}{A} = \frac{L}{T} = \frac{w_+^{\rm eq}}{k_{\rm B}T} , \tag{6.103}$$

with $\langle N^{(\rm r)} \rangle = N^{(\rm r),\rm eq}$.

6.6.2 The master equation approach

The escape-rate formula can be obtained for chemical reactions by using the master equation approach (Nicolis and Prigogine 1977). As an example, we consider the isomerization

$$A \leftrightarrow B . \tag{6.104}$$

The numbers of particles A and B are the random variables of this process. The total number of particles is conserved, $N_A + N_B = N = N_{\rm tot}$, so that the process is completely determined by the knowledge of the lone variable N_A. The evolution equation of the probability $P(N_A)$ that the system contains N_A particles is

$$\begin{aligned} \frac{d}{dt} P(N_A) &= \kappa_+(N_A+1)\, P(N_A+1) + \kappa_-(N-N_A+1)\, P(N_A-1) \\ &\quad - \kappa_+ N_A\, P(N_A) - \kappa_-(N-N_A)\, P(N_A) , \end{aligned} \tag{6.105}$$

where the constants $\kappa_\pm$ are the reaction rates of (6.104). Introducing the fraction $0 \le x = N_A/N \le 1$ of particles A, it can be shown that the probability density defined according to $p(x) = p(N_A/N) = P(N_A)$ obeys the Fokker–Planck equation

$$\frac{\partial p}{\partial t} + \frac{\partial}{\partial x}\left(\dot{x}\, p\right) = D\, \frac{\partial^2 p}{\partial x^2}\,, \qquad (6.106)$$

in the asymptotic limit where $N \to \infty$ (van Kampen 1981). In Eq. (6.106), the reaction velocity

$$\dot{x} = -\,\kappa_+\, x + \kappa_-(1-x)\,, \qquad (6.107)$$

is the macroscopic rate equation while

$$D = \frac{\kappa_+\kappa_-}{N(\kappa_+ + \kappa_-)}\,, \qquad (6.108)$$

is the diffusion coefficient. This Fokker–Planck equation shows that the Helfand moment associated with the chemical reaction, which is given in Table 6.3 as

$$G^{(\mathrm{r})} = \frac{1}{\sqrt{Vk_{\mathrm{B}}T}}\,(N_A - N_A^{\mathrm{eq}}) = \frac{N}{\sqrt{Vk_{\mathrm{B}}T}}\,(x - x^{\mathrm{eq}})\,, \qquad (6.109)$$

is the random variable of an Ornstein–Uhlenbeck stochastic process around the equilibrium concentration

$$x^{\mathrm{eq}} = \frac{\kappa_-}{\kappa_+ + \kappa_-}\,. \qquad (6.110)$$

On this ground, we can apply the argument of first passage for the Helfand moment (6.109). We look for the first time at which the moment reaches the boundaries of the interval

$$-\frac{\chi}{2} < G^{(\mathrm{r})} < +\frac{\chi}{2}\,. \qquad (6.111)$$

This first-passage problem corresponds to the escape of the variable $x = N_A/N$ out of the interval

$$x^{\mathrm{eq}} - \frac{\xi}{2} < x < x^{\mathrm{eq}} + \frac{\xi}{2}\,, \qquad (6.112)$$

with

$$\xi = \frac{\sqrt{Vk_{\mathrm{B}}T}}{N}\,\chi\,. \qquad (6.113)$$

We consider a statistical ensemble formed by copies of the system and we count the number of copies which remain within the interval (6.112) at time t. The time evolution of this statistical ensemble is the solution of an eigenvalue problem of the same kind as (6.88)–(6.90) but for the Fokker–Planck equation (6.106) which admits the confluent hypergeometric functions as eigenfunctions (Lindenberg et al. 1975). Therefore, the number of copies defined by (6.90) decays here also exponentially

$$\mathcal{N}^{(\mathrm{r})}(t) \sim \exp\left(s_1^{(\mathrm{r})} t\right)\,, \qquad (6.114)$$

where the rate $-s_1^{(r)}$ is the smallest eigenvalue of the Fokker–Planck operator (6.106) solved by requiring that the corresponding eigenfunction vanishes at the boundaries of the interval (6.112). In the limit $\xi \to 0$, it can be shown (Lindenberg et al. 1975) that the eigenvalue is given by

$$s_1^{(r)} \simeq -D\left(\frac{\pi}{\xi}\right)^2, \tag{6.115}$$

in terms of the diffusion coefficient (6.108). We mention that this exponential decay is the slowest decay dominating the time evolution at long times after the faster decay modes have died out. We can replace the size ξ (6.113) of the escape interval of the variable x by the size χ of the escape interval of the corresponding Helfand moment $G^{(r)}$ and we get the escape rate

$$\gamma^{(r)} \simeq -s_1^{(r)} \simeq \alpha\left(\frac{\pi}{\chi}\right)^2, \tag{6.116}$$

with the reaction-rate coefficient

$$\alpha = \frac{N\kappa_+\kappa_-}{k_B T V(\kappa_+ + \kappa_-)}, \tag{6.117}$$

so that we recover the reaction-rate coefficient (6.103) of the macroscopic theory since

$$w_+^{\text{eq}} = \kappa_+ \rho_A^{\text{eq}} = \frac{N\kappa_+\kappa_-}{V(\kappa_+ + \kappa_-)}. \tag{6.118}$$

This result shows the consistency of the first-passage problem applied to the thermodynamic fluctuations described by the master equation with the macroscopic theory as well as with the Green–Kubo formula.

If the chemical reaction (6.104) is simulated by a deterministic molecular dynamics, a first-passage problem can be set up for the chemical fraction x defined in the simulation. This first-passage problem will lead to the escape-rate formula (6.94) for the chemical reaction rate (Dorfman and Gaspard 1995).

6.7 Discussion

6.7.1 Summary

The scattering theory of irreversible phenomena like transports and chemical reactions is based on certain absorbing boundary conditions defined in the full phase space of the N-particle system. A first-passage problem in the space of the Helfand moment turns out to be appropriate for each irreversible process under study. The absorbing boundary condition selects a subset of trajectories in the full phase space, which represents a certain out-of-equilibrium fluctuation.

In chaotic systems, this subset of trajectories may be expected to be a fractal or, at least, to be dominated by a hyperbolic fractal. In nonhyperbolic systems, an effective escape rate – which is related to the Lyapunov exponents and the KS entropy according to (6.93) – may still be defined as explained in Section 4.8.

In this framework, all of the transport coefficients for a simple fluid, as well as the chemical reaction-rate coefficients, can be expressed in terms of the escape rate of an appropriate fractal repeller (Dorfman and Gaspard 1995). This completes a line of argument initiated for the diffusion coefficient of a particle moving in a periodic Lorentz system or a multibaker map (Gaspard and Nicolis 1990).

The invariant measure supported by the fractal repeller can be identified with a nonequilibrium state associated with the corresponding transport process. In the limit where the nonequilibrium constraint is relaxed ($\chi \to \infty$) the nonequilibrium state tends to the equilibrium microcanonical state. In the scattering theory, nonequilibrium processes and out-of-equilibrium fluctuations can thus be defined at the level of the phase space trajectories. In particular, the escape rate of the fractal repeller gives the lifetime of an out-of-equilibrium fluctuation corresponding to the constraint fixed by the parameter χ. The escape rate (6.93) is also the leading Pollicott–Ruelle resonance (generalized eigenvalue) of the Frobenius–Perron (or Liouville) operator corresponding to the Hamiltonian dynamics on the repeller. We have observed with the multibaker map that the Pollicott–Ruelle resonances tend to the eigenvalues of the phenomenological diffusion equation. This further result establishes a connection between nonequilibrium statistical mechanics and the spectral theory of the Liouvillian dynamics of classical statistical ensembles of Chapter 3.

6.7.2 Further applications of the escape-rate formalism

In Chapter 4, we have seen that the large-deviation theory and the relation between the escape rate and the characteristic quantities of chaos extend to several classes of dynamical systems like the lattice gas automata. Therefore, the escape-rate formulas can also be used for the calculation of transport and reaction-rate coefficients in these important models which are the lattice gas automata.

Recent works have been devoted to a particular lattice gas automaton which is the one-dimensional Lorentz lattice gas automaton (Ernst et al. 1995, Dorfman et al. 1995, Appert et al. 1996). In this system, the fluctuations in the density of scatterers are at the origin of a dynamical phase transition in which the pressure function vanishes for $\beta \geq 1$. Such disordered systems therefore present peculiarities which are also encountered in nonhyperbolic deterministic systems.

In a recent work, van Beijeren and Dorfman (1995) have applied the escape-rate formalism to the random Lorentz gas made out of a disordered scatterer of hard disks in the plane and they obtained the characteristic quantities of chaos like the escape rate, the Lyapunov exponent, and the KS entropy versus the size of the scatterer. For this purpose, van Beijeren and Dorfman (1995) developed a method of kinetic theory based on a Boltzmann-type equation generalized to include the curvature of the horocircle accompanying each trajectory (see Chapter 1). In this way, the average Lyapunov exponent can be calculated for such disordered systems as the random Lorentz gas.

The escape-rate formalism has also been applied to calculate the diffusion coefficient in expanding maps of the real line showing that the diffusion coefficient can be a nondifferentiable function of the parameters defining the system (Klages and Dorfman 1995, Klages 1996).

Further applications are possible to the transport of passive scalars in hydrodynamic convection. Here, fractal repellers may be formed due to the trapping of Lagrangian trajectories of the fluid (Jung et al. 1993). The escape-rate formalism may provide a characterization of these special transport phenomena from a dynamical-system point of view.

The formulation of transport processes as scattering processes is also very interesting in the context of quantum systems. First of all, Landauer (1970) formulated the conductance problem as a scattering process, which turned out to be extremely fruitful to conceive electronic conductance in low temperature mesoscopic circuits and to understand the problem of Anderson localization. We should mention here the work by Frensley (1990) who formulated the problem of quantum conductance in mesoscopic circuits at the quantum Liouvillian level. This quantum formulation can be compared to the present classical Liouvillian formulation. Certainly, further progress in this direction may turn out to be particularly fruitful.

6.7.3 Relation to the thermostatted-system approach

Here, we discuss the link between the escape-rate formalism and another approach based on thermostatted systems in which the transport coefficients are also related to the Lyapunov exponents (Hoover 1985, 1991; Evans et al. 1983, 1990a, 1990b; Baranyai et al. 1993; Chernov et al. 1993a, 1993b; Gallavotti and Cohen 1995a, 1995b; Lloyd et al. 1995; Dettmann and Morriss 1996). In the thermostatted-system approach, nonequilibrium systems are defined as a system of particles submitted to interparticle forces, to external forces, but also to a fictitious force modelling the coupling to some hypothetical thermostat. This thermostatting force is defined in such a way that the kinetic energy is conserved instead of the total energy, either exactly in the case of Gaussian thermostats, or on average in the case of Nosé–Hoover thermostats (Nosé 1984a, 1984b; Hoover 1985, 1991). The average thermostatting force vanishes at equilibrium and is automatically switched on in the presence of an external force which would otherwise indefinitely increase the kinetic energy of the system. Further nonequilibrium systems have been considered from a similar viewpoint in which the equations of motion are modified with extra forces modelling nonequilibrium boundary conditions like a gradient of velocity for instance (Evans and Morriss 1990b).

The thermostatting force is nonHamiltonian so that the thermostatted systems do not preserve volumes in phase space, even if they are time-reversal symmetric. As a consequence, on average, the forward time evolution brings the trajectories from the regions which are phase-space expanding toward those which are phase-space contracting. Under nonequilibrium conditions, the system was two distinct ergodic invariant measures μ^+ and μ^-, corresponding to forward and backward time evolutions. There is thus a spontaneous breaking of time-reversal symmetry in such systems. These invariant measures differ from the Liouville measure by their information dimension which is in general smaller than the phase-space dimension (Morriss 1987; Chernov et al. 1993a, 1993b). In hyperbolic systems, these measures can be constructed as the SRB measures presented in Chapter 4. Two regimes exist: (1) Under weak thermostatting forces, the Hausdorff dimension of the support of the measures is still the dimension of the plain phase space, which is called the Anosov regime; (2) Under stronger thermostatting forces, the Hausdorff dimension becomes strictly smaller than the phase-space dimension because the support of the measures becomes a fractal attractor (Gallavotti and Cohen 1995a, 1995b). Contrary to the fractal repeller and the invariant measure μ_{ne} of the escape-rate formalism which are fractal in both the stable and unstable directions, the attractor and the invariant measure μ^+ are absolutely continuous with respect to the Lebesgue measure in the unstable directions and multifractal only in the stable directions (and vice versa for μ^-). Physically, this difference corresponds to the fact that no absorbing boundary is used in thermostatted systems to define the nonequilibrium measures $\mu^\pm$ (although absorbing boundaries could also be considered in these systems as well).

The sum of all the Lyapunov exponents, which measures the expansion of phase-space volumes, is negative when averaged on the invariant measure

μ^+, although it is positive for μ^-. Pairing rules between positive and negative Lyapunov exponents which are similar but different from the Hamiltonian pairing rule have been observed and proved for several thermostatted systems (Dettmann and Morriss 1996).

Because the nonequilibrium constraint has the effect of switching on the average thermostatting force and thus the average phase-space contraction, there is a natural connection which appears in such systems between the sum of all the Lyapunov exponents and the transport coefficients, leading to formulas which are very similar to the escape-rate formulas (Evans et al. 1990a). The characteristic quantities of chaos have been studied in detail for the thermostatted Lorentz gas by Chernov et al. (1993a, 1993b). Since there is no escape in the studied thermostatted systems, Pesin's formula (4.41) applies so that the sum of the *positive* Lyapunov exponents is equal to the KS entropy. Therefore, the sum of *all* the Lyapunov exponents which is related to some transport coefficient splits according to

$$\sum_i \bar{\lambda}_i = \sum_{\bar{\lambda}_i>0} \bar{\lambda}_i + \sum_{\bar{\lambda}_i<0} \bar{\lambda}_i = h_{\text{KS}} - \sum_{\bar{\lambda}_i<0} |\bar{\lambda}_i| \,, \tag{6.119}$$

which shows a formal analogy with the escape-rate formulas. A formula for the information dimension of μ^+ which is very similar to Eq. (6.22) is also given by Chernov et al. (1993a, 1993b).

More recently, Chernov and Lebowitz (1997) have considered energy-conserving systems which are Hamiltonian in the bulk of their phase space but nonHamiltonian at their boundaries. These systems can be maintained out of equilibrium with special reflection rules at their walls, which simulate a shear flow. These boundary-driven systems do not preserve phase-space volumes so that they share several of the aforementioned properties with the thermostatted systems.

6.7.4 The escape-rate formalism in the presence of external forces

We owe to Tél et al. (1996) the physical understanding of the connection between both approaches. As we mentioned earlier, these authors have proposed an area-preserving model of the type of the multibaker for the deterministic diffusion of a particle in the presence of an external force that would sustain a mean drift. In such a model, the density of particles diffuses according to the equation

$$\partial_t \rho = \nabla \cdot (D \, \nabla\rho - \mu \, \mathbf{F} \, \rho) \,, \tag{6.120}$$

where $\mathbf{F}$ is the external force and μ is the mobility. Taking absorbing boundary conditions separated by a distance L, the eigenvalues of Eq. (6.120) are obtained as

$$s_j = -D\left(\frac{\pi j}{L}\right)^2 - \frac{\mu^2 F^2}{4D}, \tag{6.121}$$

with $j = 1, 2, 3, \ldots$. The escape rate of the deterministic system may be identified as the slowest decay rate with $j = 1$ in Eq. (6.121), namely $\gamma \simeq -s_1$. On the other hand, the escape rate is equal to the difference between the sum of positive Lyapunov exponents and the KS entropy calculated on the fractal repeller $\mathscr{I}_{L,F}$ of trajectories selected by the absorbing boundary condition. The following formula is obtained:

$$D\left(\frac{\pi}{L}\right)^2 + \frac{\mu^2 F^2}{4D} \simeq \sum_{\bar{\lambda}_i > 0} \bar{\lambda}_i(\mathscr{I}_{L,F}) - h_{\mathrm{KS}}(\mathscr{I}_{L,F}), \tag{6.122}$$

in the limits $L \to \infty$ and $F \to 0$. If the zero-force limit is taken before the large-system limit, the previous escape-rate formula (6.14) is obtained with $\chi_1 = \pi$. On the other hand, if the large-system limit is taken before the zero-force limit, we get

$$\frac{\mu^2}{4D} = \lim_{F\to 0} \frac{1}{F^2} \lim_{L\to\infty} \left[\sum_{\bar{\lambda}_i > 0} \bar{\lambda}_i(\mathscr{I}_{L,F}) - h_{\mathrm{KS}}(\mathscr{I}_{L,F})\right], \tag{6.123}$$

which is similar to a formula by Baranyai et al. (1993).

This result shows that absorbing boundary conditions can be considered even in the presence of an external force. The specific form (6.123) of the formula relating the transport coefficients to the characteristic quantities of chaos is determined, on the one hand, from dynamical systems theory and, on the other hand, from the hydrodynamic equation like (6.120) holding in the specific system. This result shows that an equation such as (6.123) applies to both Hamiltonian and thermostatted systems.[1] In this respect, the thermostatting hypothesis turns out to be of secundary importance at least in sufficiently large systems where Eq. (6.120) holds in some scaling limit.

In conclusion, we have developed the scattering theory of transport phenomena pioneered in the sixties by Lax and Phillips (1967). In this context, we have investigated the consequences of absorbing boundary conditions imposed on the Liouvillian dynamics of spatially extended systems sustaining transport processes. These assumptions lead to the escape-rate formalism where the irreversible coefficients may be expressed in terms of the characteristic quantities of

1. We notice that absorbing boundary conditions may also be considered for the thermostatted systems.

chaos. The resulting escape-rate formulas turn out to apply to a broad range of irreversible phenomena including transport with or without external forces and chemical reactions. The mathematical structure behind the escape-rate formulas is the large-deviation formalism of chaotic systems presented in Chapter 4, where the slowest decay mode of an escape process can be expressed in terms of a dynamical invariant measure based on the local Lyapunov exponents according to the Ruelle variational principle (4.23) (Ruelle 1978).

Chapter 7

Hydrodynamic modes of diffusion

7.1 Hydrodynamics from Liouvillian dynamics

7.1.1 Historical background and motivation

Hydrodynamics describes the macroscopic dynamics of fluids in terms of Navier–Stokes equations, the diffusion equation, and other phenomenological equations for the mass density, the fluid velocity and temperature, or for chemical concentrations. In nonequilibrium statistical mechanics, these phenomenological equations may be derived from a kinetic equation like the famous Boltzmann equation or other master equations describing the time evolution at the level of one-body distribution functions (Balescu 1975, Résibois and De Leener 1977, Boon and Yip 1980). The kinetic equation itself is derived from Liouvillian dynamics using a Markovian approximation such as Boltzmann's *Stosszahlansatz*. Such approximations may be justified in some scaling limits for dilute fluids or other systems, but the derivation of hydrodynamics is not carried out directly from the Liouvillian dynamics. The only direct link between hydrodynamics and the Liouvillian dynamics – which is used in particular in molecular-dynamics simulations – is established in terms of the Green–Kubo relations.

The recent works in dynamical systems theory have shown that further direct links are possible. In particular, we have observed with the multibaker map in Chapter 6 that the spectrum of the Pollicott–Ruelle resonances actually provides the spectrum of the phenomenological diffusion equation in spatially extended systems (Gaspard 1992a, 1995, 1996). This result suggests that the dispersion relations of hydrodynamics can be obtained in terms of the Pollicott–Ruelle resonances and that the hydrodynamic modes can be constructed as the associated eigenstates. In the simple case of diffusion described by the phenomenological equation

$$\partial_t \rho = D \nabla^2 \rho , \tag{7.1}$$

the hydrodynamic modes are the solutions of the equation of motion like

$$\rho_{\mathbf{k}}(\mathbf{r}, t) = \exp(s_{\mathbf{k}} t) \exp(i\mathbf{k} \cdot \mathbf{r}) , \tag{7.2}$$

which describe periodic profiles of concentration characterized by the wavenumber $\mathbf{k}$. The hydrodynamic modes decay exponentially in time because the corresponding eigenvalues are real and negative

$$s_{\mathbf{k}} = - Dk^2 . \tag{7.3}$$

Accordingly, the concentration becomes spatially uniform and approaches the thermodynamic equilibrium in the long-time limit ($t \to +\infty$). This exponential decay toward the thermodynamic equilibrium seems apparently incompatible with the microscopic Hamiltonian dynamics which is time reversible and which, moreover, preserves phase-space volumes. As a consequence, hydrodynamic modes have for long been considered at the intermediate level of the approximate kinetic equations but not at the fundamental level of the exact Liouvillian equation describing the time evolution of probability densities f_t in phase space.

7.1.2 Pollicott–Ruelle resonances and quasiperiodic boundary conditions

The modern theory of dynamical systems has developed the appropriate methods to define such exponentially decaying modes as we explained in Chapter 3. Resonances and generalized spectral decompositions have been introduced in the theory, in particular, by Pollicott (1985, 1986) and Ruelle (1986a, 1986b, 1987), which can be used to provide the mathematical foundation to such concepts as the hydrodynamic modes and their dispersion relation.

We should mention that such a possibility has long been anticipated by physicists. Liouvillian eigenvalue problems have been introduced in nonequilibrium statistical mechanics in the case of the ideal gas where the spectrum is formed by discrete real frequencies according to Koopman theory (Balescu 1961, Prigogine 1962). That such real frequencies should acquire an imaginary part when particles interact was anticipated on physical grounds (Balescu 1961, 1975; Prigogine 1962). In particular, Kadanoff and Swift (1968) formulated the nonequilibrium statistical mechanics near the liquid–gas critical point as such a Liouvillian eigenvalue problem with complex eigenfrequencies. Today, the complex spectrum of the hydrodynamic modes is known over a large range of wavenumbers, in particular, thanks to experimental studies by neutron scattering (Cohen et al. 1984, Kamgar-Parsi et al. 1987, Cohen and de Schepper 1990). However, the microscopic structure of the Liouvillian eigenstates associated with the complex eigenfrequencies has long remained unclear. It was only with the works on chaotic scattering that it became clear that such eigenstates could only be described in terms of (Schwartz) distributions or linear functionals.

The purpose of the present chapter is to describe the construction of the hydrodynamic modes of diffusion based on this concept of Pollicott–Ruelle resonances and on the resonance spectral theory. In order to formulate this problem, we have first to consider quasiperiodic boundary conditions instead of absorbing boundary conditions as in Chapter 6. Quasiperiodic boundary conditions have been introduced in Chapter 3. The system is supposed to have a dynamics which is periodic in the position space of its particles. On the other hand, the probability density is not assumed to be periodic (otherwise periodic boundary conditions would be defined). Accordingly, the nonperiodic probability density is decomposed by spatial Fourier transforms into a continuum of components labelled by the wavenumber. Consequently, the state space decomposes into a continuum of subspaces (or sectors) with specific wavenumbers. Let us consider the case of a Hamiltonian like (1.7), where the potential is periodic under the spatial translations $\mathbf{q}_i \to \mathbf{q}_i + \mathbf{l}_i$ where $\mathbf{l}_i$ is a vector of a Bravais lattice ℓ of $\mathbb{R}^d$. According to the above discussion, the probability density f admits the Fourier-transform decomposition

$$f(\mathbf{p}_1,\ldots,\mathbf{p}_N,\mathbf{q}_1,\ldots,\mathbf{q}_N) = \frac{1}{|\mathscr{B}|}\int_{\mathscr{B}} d\mathbf{k}_1\ldots d\mathbf{k}_N\, f_{\mathbf{k}_1\ldots\mathbf{k}_N}(\mathbf{p}_1,\ldots,\mathbf{p}_N,\mathbf{q}_1,\ldots,\mathbf{q}_N)\,, \tag{7.4}$$

where $\mathscr{B}$ is the first Brillouin zone corresponding to the lattice $\mathscr{L} = \ell^N$ (Kittel 1976) and where each component is defined by

$$f_{\mathbf{k}_1\ldots\mathbf{k}_N}(\mathbf{p}_1,\ldots,\mathbf{p}_N,\mathbf{q}_1,\ldots,\mathbf{q}_N) = \sum_{(\mathbf{l}_1,\ldots,\mathbf{l}_N)\in\mathscr{L}} \exp(-i\mathbf{k}_1\cdot\mathbf{l}_1 - \cdots - i\mathbf{k}_N\cdot\mathbf{l}_N)\, f(\mathbf{p}_1,\ldots,\mathbf{p}_N,\mathbf{q}_1+\mathbf{l}_1,\ldots,\mathbf{q}_N+\mathbf{l}_N)\,. \tag{7.5}$$

The time evolution acts differently in each of the subspaces $\mathscr{S}_{\mathbf{k}_1\ldots\mathbf{k}_N}$ of functions (7.5). In each subspace, a generalized eigenvalue problem can be formulated and the Pollicott–Ruelle resonances calculated if the system is hyperbolic. The Pollicott–Ruelle resonances would then depend continuously on the wavenumbers.

In the case of the deterministic diffusion of a single particle, the wavenumber can be identified with the wavenumber **k** of the hydrodynamic mode of diffusion (7.2). Accordingly, the Pollicott–Ruelle resonance $s_{\mathbf{k}}$ would be the dispersion relation of diffusion (7.3). The associated eigenstate would be the hydrodynamic mode (7.2) of diffusion.

In the following, we shall show in detail how this program can be formulated. One remarkable result is that the generalized eigenvalues and eigenstates describe in principle the time evolution of the system. Therefore, the transport coefficient can be obtained in the long-wavelength limit as second derivative of the dispersion relations with respect to the wavenumber. In this way, the Green–Kubo relations can be derived from the Pollicott–Ruelle resonances, which establishes a direct link between the Liouvillian dynamics and hydrodynamics (Gaspard 1996).

This chapter is organized as follows. The Liouvillian dynamics of systems of infinite spatial extension is first presented. The general formalism is then applied to deterministic diffusion. A formula obtained by Cvitanović et al. (1992, 1995) is presented which gives the diffusion coefficient in terms of the unstable periodic orbits of the system. The theory is illustrated with the periodic Lorentz gas and the periodic multibaker map. Thereafter, we discuss the possible extension of these results to the hydrodynamic modes of the other transport processes and, in particular, to viscosity in a two-particle fluid. The consequence of spontaneous symmetry breaking on the spectrum of hydrodynamic modes is discussed. In the last part, the problem of the chemio-hydrodynamic modes in near-equilibrium reaction-diffusion systems is discussed and a model is sketched.

7.2 Liouvillian dynamics for systems symmetric under a group of spatial translations

7.2.1 Introduction

If the dynamical system is invariant under a group of spatial translations its Liouvillian dynamics has a decomposition into each of the irreducible representations of the group of translations (Bouckaert et al. 1936, Hamermesh 1962, Robbins 1989, Lauritzen 1990, Cvitanović and Eckhardt 1993). By analogy with the work of Bloch in quantum mechanics, these irreducible representations are labelled by a continuously varying parameter which is the wavenumber. The reduction under a spatial group first proceeds by the reduction based on the subgroup of translations. The subgroup of translations forms a lattice of vectors $\mathbf{l} \in \mathscr{L}$. This lattice may also be symmetric under a point group. But the first reduction by the translations introduces a wavenumber $\mathbf{k}$ which breaks the symmetry under the whole point group. There only remains a little subgroup of the point group which leaves invariant the lattice together with the wavenumber. The little group only contains the identity for most values of the wavenumber. For special wavenumbers on some symmetry axes, the little subgroup is non-trivial which allows a further reduction (Cvitanović et al. 1995). This chapter will be mostly devoted to the reduction under the group of spatial translations.

We proceed by first reducing the flow to a Poincaré–Birkhoff map like (1.5) but here for the dynamics on a lattice. The flow is thus represented as a suspended flow such as (1.6). Consequently, the Frobenius–Perron operator for the flow can be reduced to the Frobenius–Perron operator for the Poincaré–Birkhoff map of the lattice. This reduction is carried out, firstly, by a spatial Fourier transform to go from the lattice to one of its elementary cells, which introduces the wavenumber $\mathbf{k}$. Secondly, a Laplace transform is carried out from the continuous time t to a discrete time n, introducing the variable s conjugate to the time t as already performed in Chapter 3. The chain of reductions is thus

$$\begin{array}{ccccc} (\mathbf{x}, \tau, \mathbf{l}) & \rightarrow & (\mathbf{x}, \tau | \mathbf{k}) & \rightarrow & (\mathbf{x} | s, \mathbf{k}) \\ \hat{P}^t & \rightarrow & \hat{Q}^t_{\mathbf{k}} & \rightarrow & \hat{R}_{\mathbf{k},s} \end{array} \tag{7.6}$$

The last Frobenius–Perron operator $\hat{R}_{\mathbf{k},s}$ depends on the wavenumber associated with the modes in the infinitely extended system as well as on the variable s which turns out to be the generalized eigenvalue of the Liouville operator $\hat{L}$.

7.2.2 Suspended flows of infinite spatial extension

We consider a dynamical system of infinite spatial extension in a phase space defined by the coordinates $\mathbf{X} = (\mathbf{r}, \tilde{\mathbf{r}}) \in \mathbb{R}^M$, where $\mathbf{r} \in \mathbb{R}^L$ are the position coordinates in which the system forms a Bravais lattice $\mathscr{L}$. The coordinates $\tilde{\mathbf{r}} \in \mathbb{R}^{M-L}$ are supplementary coordinates which are necessary to uniquely specify the initial condition of the trajectories. A lattice vector $\mathbf{l_m}$ is centered in each cell of the Bravais lattice, with $\mathbf{m} = (m_1 \cdots m_L) \in \mathbb{Z}^L$. The lattice vectors are given as linear combinations of the basic vectors of the lattice

$$\mathbf{l_m} = m_1\, \mathbf{l}_{100\cdots00} + m_2\, \mathbf{l}_{010\cdots00} + \cdots + m_L\, \mathbf{l}_{000\cdots01} \in \mathscr{L} \,. \tag{7.7}$$

By using the invariance of the lattice dynamics under spatial translations, the flow $\mathbf{\Phi}^t$ over the whole lattice can be reduced to a flow in one and the same elementary cell of the lattice, which may be further reduced to a mapping like (1.5) thanks to a Poincaré hypersurface of section $\mathscr{P}$ with $\dim\mathscr{P} = M - 1$.

As noticed in Chapter 1, the mapping alone does not provide a complete description of the flow because the coordinate along the direction of the flow has been eliminated by introducing the Poincaré hypersurface of section. To restore this information, the current phase-space position along the trajectory can be determined by the interval of time τ elapsed since the last passage in the hypersurface of section. This time τ takes its values in the range $0 \leq \tau < T(\mathbf{x})$ where $T(\mathbf{x})$ is the time of first return in $\mathscr{P}$. Under these assumptions, the complete phase space can be represented in the new coordinates

$$\tilde{\mathbf{X}} = (\mathbf{x}, \tau, \mathbf{l}) \in \mathscr{P} \otimes [0, T(\mathbf{x})[\otimes \mathscr{L} . \tag{7.8}$$

In this phase space, the dynamics on the lattice is described by the following suspended flow $\tilde{\mathbf{\Phi}}^t$, which is a generalization of (1.6). We first observe that the point $(\mathbf{x}, 0, \mathbf{l})$ is a point belonging to the hypersurface of section $\mathscr{P}$ translated to the cell $\mathbf{l}$. As long as the time τ is between 0 and $T(\mathbf{x})$ the trajectory remains on the same segment attached to the position $\mathbf{x}$ and the cell $\mathbf{l}$. When $\tau = T(\mathbf{x})$, the trajectory performs its next passage through the hypersurface $\mathscr{P}$ at the point $\phi(\mathbf{x})$. At this next passage, the trajectory may belong to a different cell $\mathbf{l}'$ of the lattice. We have therefore to introduce a function taking its values in the Bravais lattice $\mathbf{a}(\mathbf{x}) \in \mathscr{L}$ which is the lattice vector of the jump between the cells $\mathbf{l}$ and $\mathbf{l}'$: $\mathbf{l}' - \mathbf{l} = \mathbf{a}(\mathbf{x})$.

At the next passage in the hypersurface of section, we have to identify the point $[\mathbf{x}, T(\mathbf{x}), \mathbf{l}]$ with the point $[\phi(\mathbf{x}), 0, \mathbf{l}+\mathbf{a}(\mathbf{x})]$. The dynamics of the suspended flow is thus controlled by the mapping

$$\begin{aligned} \mathbf{x}_{n+1} &= \phi(\mathbf{x}_n) , \\ t_{n+1} &= t_n + T(\mathbf{x}_n) , \\ \mathbf{l}_{n+1} &= \mathbf{l}_n + \mathbf{a}(\mathbf{x}_n) , \end{aligned} \tag{7.9}$$

instead of (1.5), where $\{t_n\}_{n=-\infty}^{+\infty}$ are the successive times of passage through the hypersurface $\mathscr{P}$ and $\{\mathbf{l}_n\}_{n=-\infty}^{+\infty}$ the centres of the cells successively visited.

The time axis is divided into intervals of length $T(\phi^n\mathbf{x})$ during which the position remains fixed at $\phi^n\mathbf{x}$ and the lattice vector at the value $\mathbf{l}_n$. In the special coordinates (7.8), we note that the flow is defined by the vector field $\tilde{\mathbf{F}}(\mathbf{x}, \tau, \mathbf{l}) = (0, 1, 0)$ for $0 < \tau < T(\mathbf{x})$ (Ruelle 1989c). The flow is thus

$$\tilde{\mathbf{\Phi}}^t(\mathbf{x}, \tau, \mathbf{l}) = \big(\mathbf{x}, \tau + t, \mathbf{l}\big) , \quad \text{for} \quad 0 \leq \tau + t < T(\mathbf{x}) . \tag{7.10}$$

and

$$\tilde{\mathbf{\Phi}}^t(\mathbf{x},\tau,\mathbf{l}) = \left[\phi^n\mathbf{x},\tau+t-\sum_{j=0}^{n-1}T(\phi^j\mathbf{x}),\mathbf{l}+\sum_{j=0}^{n-1}\mathbf{a}(\phi^j\mathbf{x})\right],$$

$$\text{for} \quad 0 \leq \tau+t-\sum_{j=0}^{n-1}T(\phi^j\mathbf{x}) < T(\phi^n\mathbf{x}). \tag{7.11}$$

On the other hand, for the time running backward ($t < 0$), we obtain

$$\tilde{\mathbf{\Phi}}^{-|t|}(\mathbf{x},\tau,\mathbf{l}) = \left[\phi^{-n}\mathbf{x},\tau-|t|+\sum_{j=1}^{n}T(\phi^{-j}\mathbf{x}),\mathbf{l}-\sum_{j=1}^{n}\mathbf{a}(\phi^{-j}\mathbf{x})\right],$$

$$\text{for} \quad 0 \leq \tau-|t|+\sum_{j=1}^{n}T(\phi^{-j}\mathbf{x}) < T(\phi^{-n}\mathbf{x}). \tag{7.12}$$

In the following, we shall denote by $\tilde{\mathbf{\Phi}}_0^t(\mathbf{x},\tau)$ the flow in the elementary cell at the origin $\mathbf{l} = 0$. Thanks to the preceding definitions, we have carried out a reduction of the flow from the infinite phase space to a reduced flow in the fundamental domain obtained by using the translational symmetry on the lattice. This would amount to reducing the dynamics on the infinite space to the dynamics on a torus as if we imposed periodic boundary conditions. However, we emphasize that the position of the particle in the infinite phase space can still be recovered from the lattice vector $\mathbf{l}$ which is driven by the dynamics in the reduced phase space. This possibility allows us to consider quasiperiodic boundary conditions.

To establish the isomorphism with the original coordinates of the system, we have to introduce the vector $\mathbf{c}(\mathbf{x})$ which gives the position of the point $\mathbf{x}$ of intersection with the hypersurface $\mathscr{P}$ with respect to the centre $\mathbf{l}$ of the currently visited cell (cf. Fig. 7.1). At the instant of the intersection with $\mathscr{P}$, the position in the original coordinates is thus

$$\mathbf{r}(\mathbf{X})|_{\mathscr{P}} = \mathbf{r}(\mathbf{x},0,\mathbf{l}) = \mathbf{l} + \mathbf{c}(\mathbf{x}). \tag{7.13}$$

If we denote by $\mathbf{v} = \dot{\mathbf{r}}$ the velocity of the particle as given by the first derivative of the position with respect to the time we obtain the following relation for the segment of trajectory between the points $(\mathbf{x},0,\mathbf{l})$ and $[\phi(\mathbf{x}),0,\mathbf{l}+\mathbf{a}(\mathbf{x})]$:

$$\mathbf{a}(\mathbf{x}) = \mathbf{c}(\mathbf{x}) + \int_0^{T(\mathbf{x})} \mathbf{v}(\mathbf{x},\tau)\, d\tau - \mathbf{c}(\phi\mathbf{x}). \tag{7.14}$$

This equation is of importance for the following because it connects the lattice vector $\mathbf{a}(\mathbf{x})$ with the phase-space quantities $\mathbf{c}$ and $\mathbf{v}$.

7.2.3 Assumptions on the properties of the mapping

For the following developments, we need to assume several properties for the suspended flow:

1. The mapping ϕ is piecewise symplectic.
2. The mapping ϕ is time-reversal symmetric.
3. The mapping ϕ is hyperbolic in the sense that all the trajectories are unstable of saddle type with nonvanishing Lyapunov exponents. We also suppose that the mapping ϕ has the Kolmogorov property (cf. Subsection 3.5.3) which implies ergodicity and mixing. Moreover, the rate of mixing is assumed to be sufficiently fast.

Because of the condition (1), ergodicity holds with respect to the Lebesgue measure ν in the coordinates $\mathbf{x}$ on $\mathscr{P}$. Mixing is considered between functions u and v which are piecewise Hölder continuous. A function v is said to be piecewise Hölder continuous if $|v(\mathbf{x}) - v(\mathbf{x}')| \leq C(v)\|\mathbf{x} - \mathbf{x}'\|^{\beta}$ for both points $\mathbf{x}$ and $\mathbf{x}'$ belonging to a subdomain of a finite union of subdomains of $\mathscr{P}$, for some Hölder exponent $\beta > 0$, and for some positive constant $C(v)$. The class of piecewise Hölder continuous functions is larger than the class of Hölder

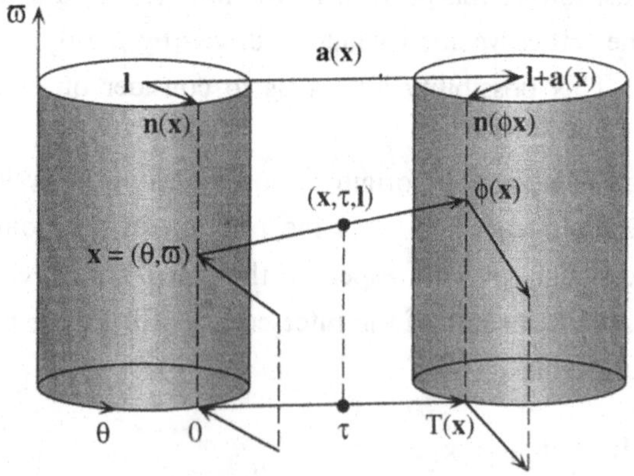

Figure 7.1 Schematic representation of the suspended flow in the case of a planar billiard of the type of the Lorentz gas. The base of the cylinders is the two-dimensional position space available to the particle. The vertical axis of the cylinders corresponds to the projection of the velocity tangent to the walls at collisions. $\mathbf{x} = (\theta, \varpi)$ are the Birkhoff coordinates. $\phi(\mathbf{x})$ is the image of the point $\mathbf{x}$ under the Birkhoff mapping. $\mathbf{l}$ is the centre of the cell to which the point $\mathbf{x}$ is attached. $\mathbf{a}(\mathbf{x})$ is the jump vector so that $\mathbf{l} + \mathbf{a}(\mathbf{x})$ is the centre of the cell to which the image $\phi(\mathbf{x})$ is attached. The vectors $\mathbf{c}(\mathbf{x})$ and $\mathbf{c}(\phi\mathbf{x})$ of Eqs. (7.13)–(7.14) are here identical with the normal vectors from the centres of the cells to the points of collisions: $\mathbf{n}(\mathbf{x})$ and $\mathbf{n}(\phi\mathbf{x})$. $(\mathbf{x}, \tau, \mathbf{l})$ are the coordinates of a current point of the flow with the time coordinate in $0 \leq \tau < T(\mathbf{x})$.

continuous functions defined on the same domain (Chernov 1994). In our context, a stretched-exponential type of mixing may be supposed

$$|\langle u(\mathbf{x})v(\boldsymbol{\phi}^n\mathbf{x})\rangle_\nu - \langle u\rangle_\nu \langle v\rangle_\nu| \leq \exp(-n^\gamma) , \tag{7.15}$$

where $\langle\cdot\rangle_\nu$ denotes the average with respect to ν, n is large enough, and the constant γ is such that $0 < \gamma < 1$. This property of fast decay of correlations is known to imply the central limit theorem (Chernov 1994):

If v is piecewise Hölder continuous and $w = v - \langle v\rangle_\nu$, the following probability approaches the Gaussian error function asymptotically

$$\lim_{n\to\infty} \nu\left\{\mathbf{x} : \frac{1}{\sigma_w\sqrt{n}} \sum_{j=0}^{n-1} w(\boldsymbol{\phi}^j\mathbf{x}) < z\right\} = \frac{1}{\sqrt{2\pi}} \int_{-\infty}^{z} \exp(-y^2/2)\, dy , \tag{7.16}$$

where the variance of fluctuations is given by the sum of the autocorrelation function of w

$$\sigma_w^2 = \sum_{n=-\infty}^{+\infty} \langle w(\boldsymbol{\phi}^n\mathbf{x})\, w(\mathbf{x})\rangle_\nu , \tag{7.17}$$

which is convergent according to (7.15).

The conditions (1) and (2) are not necessary for most of the following considerations which can be extended to dissipative systems and, in particular, to one-dimensional maps as well. The condition (3) is important but may be relaxed. In particular, a slower decay of correlations may also be admitted for part of the following considerations.

7.2.4 Invariant measures

As a consequence of the piecewise symplectic property (1) of the mapping, we find

$$\left|\det \frac{\partial\boldsymbol{\phi}}{\partial\mathbf{x}}\right| = 1 , \tag{7.18}$$

so that volumes are preserved in the Poincaré hypersurface of section $\mathscr{P}$. As aforementioned, the Lebesgue measure $d\mathbf{x}$ is thus invariant under the mapping $\boldsymbol{\phi}$. The average of a function $v(\mathbf{x})$ is defined by

$$\langle v\rangle_\nu = \frac{1}{|\mathscr{P}|} \int_{\mathscr{P}} v(\mathbf{x})\, d\mathbf{x} , \tag{7.19}$$

where $|\mathscr{P}| = \int_{\mathscr{P}} d\mathbf{x}$ is the volume of the hypersurface $\mathscr{P}$.

Consequently, we can define the corresponding invariant measures μ_∞ and μ of the flows $\mathbf{\Phi}^t$ and $\mathbf{\Phi}_0^t$ respectively, with the following averages of the quantity $A(\mathbf{X})$

$$\begin{aligned}\langle A(\mathbf{X})\rangle_{\mu_\infty} &= \int \mu_\infty(d\mathbf{X})\; A(\mathbf{X}) \\ &= \sum_{\mathbf{l}\in\mathscr{L}} \int \mu(d\mathbf{x}d\tau)\; A(\mathbf{x},\tau,\mathbf{l})\,,\end{aligned} \tag{7.20}$$

where

$$\begin{aligned}\int \mu(d\mathbf{x}d\tau)\; A(\mathbf{x},\tau,\mathbf{l}) &= \frac{1}{|\mathscr{P}|}\int_{\mathscr{P}} d\mathbf{x} \int_0^{T(\mathbf{x})} \frac{d\tau}{\langle T\rangle_\nu}\; A(\mathbf{x},\tau,\mathbf{l})\,, \\ &= \frac{1}{\langle T\rangle_\nu}\left\langle \int_0^{T(\mathbf{x})} A(\mathbf{x},\tau,\mathbf{l})\; d\tau\right\rangle_\nu\,,\end{aligned} \tag{7.21}$$

We denote by $\langle\cdot\rangle$ the average over the corresponding measure. The measure μ is equivalent to the Liouville measure describing the microcanonical ensemble in the energy shell $H = E$. The measure μ_∞ is not normalizable because it is defined on an infinite phase space. However, both the measures μ and ν are normalizable.

7.2.5 Time-reversal symmetry

The operation of time reversal is defined by the involution $\imath$ such that

$$\begin{aligned}\imath^2 &= 1\,, \\ \imath\circ\boldsymbol{\phi}\circ\imath &= \boldsymbol{\phi}^{-1}\,, \\ T(\imath\mathbf{x}) &= T(\boldsymbol{\phi}^{-1}\mathbf{x})\,, \qquad \text{and} \\ \mathbf{a}(\imath\mathbf{x}) &= -\mathbf{a}(\boldsymbol{\phi}^{-1}\mathbf{x})\,.\end{aligned} \tag{7.22}$$

The suspended flow itself is time-reversal symmetric under the transformation

$$\begin{aligned}\mathbf{I}(\mathbf{x},\tau,\mathbf{l}) &= \Big[\imath\circ\boldsymbol{\phi}(\mathbf{x}),\, T(\mathbf{x})-\tau,\, \mathbf{l}+\mathbf{a}(\mathbf{x})\Big]\,, \qquad \text{for} \qquad 0<\tau<T(\mathbf{x})\,, \\ &\text{such that} \qquad \mathbf{I}^2 = 1\,, \qquad \text{and} \qquad \mathbf{I}\circ\tilde{\mathbf{\Phi}}^t\circ\mathbf{I} = \tilde{\mathbf{\Phi}}^{-t}\,.\end{aligned} \tag{7.23}$$

This transformation corresponds to a reversal of the velocity of the particle. Since the trajectory is followed in the reversed direction, the lattice vector labelling the cell is also modified (cf. Fig. 7.1).

7.2.6 The Frobenius–Perron operator on the infinite lattice

The phase-space dynamics induces a time evolution in the algebra of classical observables as well as on the probability densities. Let us consider an observable quantity $A(\mathbf{X})$ and the density $f(\mathbf{X})$ of some statistical ensemble defined in the infinite phase space (7.8) of the suspended flow. The statistical ensemble is arbitrary and may be considered as the initial ensemble of a time-evolution process. The average of the observable over the ensemble is given by

$$\begin{aligned} \langle A^*(\mathbf{X}) f(\mathbf{X})\rangle_{\mu_\infty} &= \int \mu_\infty(d\mathbf{X})\, A^*(\mathbf{X})\, f(\mathbf{X}) \\ &= \sum_{\mathbf{l}\in\mathscr{L}} \int \mu(d\mathbf{x}d\tau)\, A^*(\mathbf{x},\tau,\mathbf{l})\, f(\mathbf{x},\tau,\mathbf{l}) \,. \end{aligned} \tag{7.24}$$

The flow induces an evolution of this average according to

$$\begin{aligned} \langle A^*(\mathbf{\Phi}^t\mathbf{X})\, f(\mathbf{X})\rangle_{\mu_\infty} &= \langle A^*(\mathbf{X})\, f(\mathbf{\Phi}^{-t}\mathbf{X})\rangle_{\mu_\infty} \\ &= \langle A^*(\mathbf{X})\, (\hat{P}^t f)(\mathbf{X})\rangle_{\mu_\infty} \,, \end{aligned} \tag{7.25}$$

in which the Frobenius–Perron operator is defined as

$$\hat{P}^t\, f(\mathbf{X}) = f(\mathbf{\Phi}^{-t}\mathbf{X}) \,, \tag{7.26}$$

where we assume that the flow preserves the phase-space volumes.

7.2.7 Spatial Fourier transforms

The symmetry of the system under the group of spatial translations can be used to reduce the representation of the observables and densities onto the phase space of the cell at the lattice origin thanks to Fourier transforms. We define a projection operator by

$$\hat{E}_\mathbf{k} = \sum_{\mathbf{l}\in\mathscr{L}} \exp(-i\mathbf{k}\cdot\mathbf{l})\, \hat{S}^\mathbf{l} \,, \tag{7.27}$$

in terms of the spatial translation operators

$$\hat{S}^\mathbf{l} f(\mathbf{x},\mathbf{l}') = f(\mathbf{x},\mathbf{l}+\mathbf{l}') \,, \qquad \text{for} \qquad \mathbf{l},\, \mathbf{l}' \in \mathscr{L} \,. \tag{7.28}$$

The projection operator (7.27) involves the so-called wavenumber $\mathbf{k}$. This later is defined on the Brillouin zone $\mathscr{B}$ of the lattice $\mathscr{L}$ which is the Wigner–Seitz cell of the reciprocal lattice $\tilde{\mathscr{L}}$ (Kittel 1976). The volume of the Brillouin zone is

$$|\mathscr{B}| = \int_\mathscr{B} d\mathbf{k} = \frac{(2\pi)^L}{|\det(\mathbf{l}_{10\cdots00},\ldots,\mathbf{l}_{00\cdots01})|} \,. \tag{7.29}$$

The operators (7.27) are projection operators since

$$\hat{E}_\mathbf{k}\, \hat{E}_{\mathbf{k}'} = |\mathscr{B}|\, \delta(\mathbf{k}-\mathbf{k}')\, \hat{E}_\mathbf{k} \,, \tag{7.30}$$

which is a consequence of the relation

$$\frac{1}{|\mathscr{B}|} \sum_{\mathbf{l} \in \mathscr{L}} \exp(i\mathbf{k} \cdot \mathbf{l}) = \sum_{\mathbf{k}' \in \tilde{\mathscr{L}}} \delta(\mathbf{k} - \mathbf{k}') . \tag{7.31}$$

The identity operator is recovered by integrating the projection operator over the wavenumber

$$\hat{I} = \frac{1}{|\mathscr{B}|} \int_{\mathscr{B}} d\mathbf{k} \; \hat{E}_{\mathbf{k}} . \tag{7.32}$$

A density f defined on the infinite phase space can thus be decomposed as

$$f(\mathbf{X}) = \frac{1}{|\mathscr{B}|} \int_{\mathscr{B}} d\mathbf{k} \; \hat{E}_{\mathbf{k}} f(\mathbf{X}) , \tag{7.33}$$

in terms of the functions

$$\hat{E}_{\mathbf{k}} f(\mathbf{x}, \tau, \mathbf{l}) = \exp(i\mathbf{k} \cdot \mathbf{l}) \; \hat{E}_{\mathbf{k}} f(\mathbf{x}, \tau, 0) = \exp(i\mathbf{k} \cdot \mathbf{l}) \; f_{\mathbf{k}}(\mathbf{x}, \tau) , \tag{7.34}$$

which are quasiperiodic on the lattice.

The same relation holds for the Fourier components $A_{\mathbf{k}}(\mathbf{x}, \tau)$ of an observable A. We have therefore a decomposition of each observable and each density of the infinite phase space into components defined in the reduced phase space and which depend continuously on the wavenumber **k**.

The average (7.24) of the observable A over the density f can thus be transformed into an integral over the Brillouin zone of an average over the reduced phase space

$$\langle A^* f \rangle_{\mu_\infty} = \frac{1}{|\mathscr{B}|} \int_{\mathscr{B}} d\mathbf{k} \; \langle A^*_{\mathbf{k}} f_{\mathbf{k}} \rangle_{\mu} . \tag{7.35}$$

7.2.8 The Frobenius–Perron operators in the wavenumber subspaces

The time evolution acts in a different way on each one of the **k**-components of the observables or densities so that there is a different Frobenius–Perron operator in each subspace labelled by a given wavenumber.

In order to obtain the time evolution induced by the Frobenius–Perron operator (7.26) on the different components of the Fourier transform, we consider the average of an observable A at time t that we decompose into a Fourier transform using the projection operators (7.27):

$$\langle A^* \hat{P}^t f \rangle_{\mu_\infty} = \langle A^* f^t \rangle_{\mu_\infty} = \frac{1}{|\mathscr{B}|} \int_{\mathscr{B}} d\mathbf{k} \; \langle A^*_{\mathbf{k}} f^t_{\mathbf{k}} \rangle_{\mu} , \tag{7.36}$$

where

$$f^t_{\mathbf{k}}(\mathbf{x}, \tau) = \hat{E}_{\mathbf{k}} \; \hat{P}^t \; f(\mathbf{x}, \tau, 0). \tag{7.37}$$

Using Eq. (7.12), the action of the Frobenius–Perron operator on the density $f(\mathbf{X})$ in the cell at the origin of the lattice is given by

$$\begin{aligned} \hat{P}^t f(\mathbf{x},\tau,0) &= f\left[\tilde{\mathbf{\Phi}}^{-t}(\mathbf{x},\tau,0)\right] \\ &= f\left[\phi^{-n}\mathbf{x},\tau - t + \sum_{j=1}^{n} T(\phi^{-j}\mathbf{x}), -\sum_{j=1}^{n} \mathbf{a}(\phi^{-j}\mathbf{x})\right], \\ &\text{for} \quad 0 \leq \tau - t + \sum_{j=1}^{n} T(\phi^{-j}\mathbf{x}) < T(\phi^{-n}\mathbf{x}) . \end{aligned} \tag{7.38}$$

Applying the projection operator $\hat{E}_{\mathbf{k}}$ and using the property of quasiperiodicity (7.34), we obtain

$$\begin{aligned} \hat{Q}^t_{\mathbf{k}} f_{\mathbf{k}}(\mathbf{x},\tau) &\equiv \hat{E}_{\mathbf{k}}\, \hat{P}^t\, f(\mathbf{x},\tau,0) \\ &= \exp\left\{-i\mathbf{k}\cdot\left[\sum_{j=1}^{n} \mathbf{a}(\phi^{-j}\mathbf{x})\right]\right\} f_{\mathbf{k}}\left[\phi^{-n}\mathbf{x},\tau - t + \sum_{j=1}^{n} T(\phi^{-j}\mathbf{x})\right], \\ &\text{for} \quad 0 \leq \tau - t + \sum_{j=1}^{n} T(\phi^{-j}\mathbf{x}) < T(\phi^{-n}\mathbf{x}) . \end{aligned} \tag{7.39}$$

This equation defines a new Frobenius–Perron operator acting on the different Fourier components of the densities as

$$(\hat{Q}^t_{\mathbf{k}} f)(\mathbf{x},\tau) \equiv \exp\left\{i\mathbf{k}\cdot \mathbf{l}\left[\tilde{\mathbf{\Phi}}^{-t}(\mathbf{x},\tau,0)\right]\right\} f[\tilde{\mathbf{\Phi}}_0^{-t}(\mathbf{x},\tau)] , \tag{7.40}$$

where the vector $\mathbf{l}\left[\tilde{\mathbf{\Phi}}^t(\mathbf{x},\tau,0)\right]$ is the lattice vector corresponding to the path travelled by the point particle on the infinite phase space from the elementary cell at the origin up to the cell reached at the time t. According to Eqs. (7.10)–(7.12), this lattice vector is given by

$$\begin{aligned} \mathbf{l}\left[\tilde{\mathbf{\Phi}}^t(\mathbf{x},\tau,0)\right] &= 0 , \quad \text{for} \quad 0 \leq \tau + t < T(\mathbf{x}) , \\ &= \sum_{j=0}^{n-1} \mathbf{a}(\phi^j\mathbf{x}) , \\ &\text{for} \quad 0 \leq \tau + t - \sum_{j=0}^{n-1} T(\phi^j\mathbf{x}) < T(\phi^n\mathbf{x}) , \end{aligned} \tag{7.41}$$

and by

$$\begin{aligned} \mathbf{l}\left[\tilde{\mathbf{\Phi}}^{-t}(\mathbf{x},\tau,0)\right] &= -\sum_{j=1}^{n} \mathbf{a}(\phi^{-j}\mathbf{x}) , \\ &\text{for} \quad 0 \leq \tau - t + \sum_{j=1}^{n} T(\phi^{-j}\mathbf{x}) < T(\phi^{-n}\mathbf{x}) , \end{aligned} \tag{7.42}$$

if $t > 0$. We emphasize that the suspended flow is defined in such a way that the jumps between the lattice cells occur at the times $\{t_n\}$ of intersections with the hypersurface but not at intermediate times during the free flights between the successive passages. This construction introduces an important simplification in the following developments. We also point out that the Frobenius–Perron operator (7.40) explicitly depends on the wavenumber **k**.

The time evolution of an average value is therefore decomposed as

$$\langle A^* \hat{P}^t f\rangle_{\mu_\infty} = \frac{1}{|\mathscr{B}|} \int_{\mathscr{B}} d\mathbf{k}\, \langle A^*_{\mathbf{k}}\, \hat{Q}^t_{\mathbf{k}} f_{\mathbf{k}}\rangle_\mu \,. \tag{7.43}$$

When $\mathbf{k} = 0$, the Frobenius–Perron operator (7.40) reduces to the usual one for the closed system in an elementary cell with periodic boundary conditions. Since the flow is assumed to be mixing, the spectrum of the operator $\hat{Q}^t_0$ considered in the Koopman approach admits $s = 0$ for the unique discrete eigenvalue corresponding to the invariant probability measure (7.21) (Koopman 1931, Arnold and Avez 1968). We shall here be concerned with the spectral properties for $\mathbf{k} \neq 0$.

7.2.9 Time-reversal symmetry for the k-components

The time-reversal transformation in phase space induces an operation on the observables and densities which is

$$(\hat{K}f)(\mathbf{X}) = f^*(\mathbf{IX}) \,, \tag{7.44}$$

and which has the effect that

$$\hat{K}\, \hat{P}^t\, \hat{K} = \hat{P}^{-t} \,. \tag{7.45}$$

At the level of the spatial Fourier transforms, this operation reverses the wavenumber as expected

$$\hat{K}\, \hat{E}_{\mathbf{k}} = \hat{E}_{-\mathbf{k}}\, \hat{K} \,. \tag{7.46}$$

Hence, the Frobenius–Perron operator of the **k**-component is transformed according to

$$\hat{K}\, \hat{Q}^t_{\mathbf{k}} = \hat{Q}^{-t}_{-\mathbf{k}}\, \hat{K} \,. \tag{7.47}$$

Using these symmetry properties, we obtain the relations

$$\begin{aligned} \left\langle (\hat{K}A)^*_{\mathbf{k}}\, \hat{Q}^t_{\mathbf{k}}\, (\hat{K}f)_{\mathbf{k}} \right\rangle^*_\mu &= \left\langle A^*_{-\mathbf{k}} \hat{Q}^{-t}_{-\mathbf{k}}\, f_{-\mathbf{k}} \right\rangle_\mu \\ &= \left\langle f_{-\mathbf{k}} \big(\hat{Q}^{t\dagger}_{-\mathbf{k}}\, A_{-\mathbf{k}} \big)^* \right\rangle_\mu \,, \end{aligned} \tag{7.48}$$

which show how the forward and backward semigroups in the sectors with the wavenumbers **k** and −**k** are interrelated.

7.2.10 Reduction to the Frobenius–Perron operator of the mapping

The Frobenius–Perron operator of the flow can be reduced to that of the Poincaré map, as presented in Chapter 3. In the case of a spatially extended flow, the reduction proceeds similarly by taking a Laplace transform as in Eq. (3.161), but starting with the operator $\hat{Q}^t_{\mathbf{k}}$ instead of $\hat{P}^t$. The result is given by Eqs. (3.161)–(3.164) in which the operator $\hat{K}_s$ is replaced by the following Frobenius–Perron operator[1]

$$(\hat{R}_{\mathbf{k},s}v)(\mathbf{x}) = \exp\Big[-sT(\boldsymbol{\phi}^{-1}\mathbf{x}) - i\mathbf{k}\cdot\mathbf{a}(\boldsymbol{\phi}^{-1}\mathbf{x})\Big]\; v(\boldsymbol{\phi}^{-1}\mathbf{x})\,. \tag{7.49}$$

This new Frobenius–Perron operator depends not only on the wavenumber $\mathbf{k}$ but also on the complex variable s which should give the relaxation rate of the system. For $\mathbf{k} = 0$, we recover the Frobenius–Perron operator (3.164) of Chapter 3. Through its dependency on the wavenumber $\mathbf{k}$, the new operator also depends on the vectorial function $\mathbf{a}(\mathbf{x})$ giving the motion of the trajectory in the lattice $\mathscr{L}$ by the map (7.9).

The adjoint of the Frobenius–Perron operator (7.49) is defined by the standard condition

$$\langle u^*(\hat{R}_{\mathbf{k},s}v)\rangle_{\mathcal{V}} = \langle(\hat{R}^{\dagger}_{\mathbf{k},s}u)^*v\rangle_{\mathcal{V}}\,, \tag{7.50}$$

so that

$$(\hat{R}^{\dagger}_{\mathbf{k},s}u)(\mathbf{x}) = \exp\Big[-s^*T(\mathbf{x}) + i\mathbf{k}\cdot\mathbf{a}(\mathbf{x})\Big]\; u(\boldsymbol{\phi}\mathbf{x})\,. \tag{7.51}$$

7.2.11 Eigenvalue problem and zeta function

An eigenvalue problem may be posed for the Frobenius–Perron operator (7.49) as explained in Chapter 3 for general Frobenius–Perron operators. The eigenvalue equation here is of the form

$$\hat{R}_{\mathbf{k},s_{\mathbf{k}}}\,\psi_{\mathbf{k}}(\mathbf{x}) = \psi_{\mathbf{k}}(\mathbf{x})\,, \tag{7.52}$$

where the operator is constrained to have one as the eigenvalue corresponding to the eigenstate $\psi_{\mathbf{k}}(\mathbf{x})$. This constraint has the effect of fixing the variable s to take well defined values which vary continuously with the wavenumber $\mathbf{k}$: $s = s_{\mathbf{k}}$. Indeed, as we saw in Chapter 3, the operator (7.49) is equivalent to the original operator divided by its eigenvalue if the return time function is constant. Therefore, it is enough to require that one is an eigenvalue as in Eq. (3.165).

1. Let us recall that the map is here assumed to be volume-preserving [cf. Eq. (7.18)].

For $\mathbf{k} = 0$, the eigenvalue problem (7.52) is equivalent to the problem posed in Chapter 3. For volume-preserving systems, $s = 0$ is always an eigenvalue which corresponds to an invariant measure with a uniform probability density: $\psi_0(\mathbf{x}) = 1$. This eigenvalue is simple and unique if the system is mixing. Other eigenvalues may exist in the half complex plane with $\mathrm{Re}\, s < 0$, which may be interpreted as Pollicott–Ruelle resonances. Branch cuts are also possible.

However, for $\mathbf{k} \neq 0$, the eigenvalue $s = 0$ becomes a nontrivial resonance. When the wavenumber is tuned away from $\mathbf{k} = 0$, we may expect that for $\mathbf{k}$ small enough, there exists an eigenvalue $s_{\mathbf{k}}$ and a corresponding eigenstate $\psi_{\mathbf{k}}(\mathbf{x})$ solution of (7.52), which are given by the continuation from the standard eigenvalue $s = 0$ and eigenstate $\psi_0(\mathbf{x}) = 1$. However, we no longer expect that the eigenstate $\psi_{\mathbf{k}}$ is still a function for $\mathbf{k} \neq 0$. Instead, the results of the previous chapters suggest that the eigenstate is a (Schwartz) distribution or linear functional which acquires a meaning only when it is applied to a smooth enough test function $u(\mathbf{x})$ such that $\langle u^* \psi_{\mathbf{k}} \rangle$ becomes a well-defined complex number. For small values of the wavenumber $\mathbf{k}$, we may even assume that the leading eigenstate is given by a singular (complex) measure, i.e., by a Stieltjes integral.

Assuming the existence of the eigenstate in (7.52), the adjoint (7.51) of the Frobenius–Perron operator also admits an eigenstate

$$\hat{R}^{\dagger}_{\mathbf{k},s_{\mathbf{k}}} \tilde{\psi}_{\mathbf{k}}(\mathbf{x}) = \tilde{\psi}_{\mathbf{k}}(\mathbf{x}) , \tag{7.53}$$

which plays the role of the left eigenstate of the Frobenius–Perron operator itself while $\psi_{\mathbf{k}}$ is the right eigenstate. We may also assume that both eigenstates are mutually orthogonal to fix their normalization

$$\langle \tilde{\psi}^*_{\mathbf{k}}\ \psi_{\mathbf{k}} \rangle_v = 1 . \tag{7.54}$$

In the following, we shall focus on the value $s = s_{\mathbf{k}}$, which is nothing other than a generalized eigenvalue or resonance for the Liouville operator $\hat{L}^+$ of the forward semigroup. As explained in Chapter 3, this eigenvalue can be obtained by considering the Fredholm determinant (3.176) of the Frobenius–Perron operator. This Fredholm determinant is given as a Selberg–Smale Zeta function as in Eq. (3.151). Here, a similar calculation leads to the Zeta function

$$\begin{aligned} Z(s;\mathbf{k}) &= \mathrm{Det}\big(\hat{I} - \hat{R}_{\mathbf{k},s}\big) \\ &= \prod_p \prod_{m_1 \cdots m_{M_u} = 0}^{\infty} \left[1 - \frac{\exp\big(-sT_p - i\mathbf{k}\cdot\mathbf{a}_p\big)}{|\Lambda_{p,1}\cdots\Lambda_{p,M_u}|\Lambda_{p,1}^{m_1}\cdots\Lambda_{p,M_u}^{m_{M_u}}} \right]^{(m_1+1)\cdots(m_{M_u}+1)} , \end{aligned} \tag{7.55}$$

where we assumed that all the periodic orbits are of saddle type and that the stability eigenvalues of the linearized Poincaré section transverse to the orbit come in pairs $\{\Lambda_{p,i}, \Lambda_{p,i}^{-1}\}_{i=1}^{M_u}$ with $M_u = (M-1)/2$. The first product extends over all the prime periodic orbits p of the mapping ϕ. The second product runs over the integers $m_1, \cdots, m_{M_u}$. T_p is the prime period of p.

The new feature with respect to the Zeta functions encountered in Chapter 3 is the dependency on the wavenumber $\mathbf{k}$ and the presence of the vectors $\mathbf{a}_p$ which are the distances travelled by the trajectory on the lattice during the period of the orbit p. If n_p is the number of iterations of the Poincaré map during the period of the prime periodic orbit p and if $\mathbf{x}_p$ is some initial condition, these quantities are respectively given by

$$
\begin{aligned}
T_p &= \sum_{j=0}^{n_p-1} T(\phi^j \mathbf{x}_p) , \\
\mathbf{a}_p &= \sum_{j=0}^{n_p-1} \mathbf{a}(\phi^j \mathbf{x}_p) , \\
\det\left[\frac{\partial \phi^{n_p}}{\partial \mathbf{x}}(\mathbf{x}_p) - \Lambda\, 1\right] &= 0 .
\end{aligned}
\tag{7.56}
$$

Let us now describe a few properties of the Zeta function. For the forward semigroup, no zero is expected when Res is positive and large enough. The zeros are expected for analytic continuation toward negative values of Res. Several kinds of singularities may be encountered like simple zeros, multiple zeros, or branch cuts. When the wavenumber vanishes $\mathbf{k} = 0$, the above discussion shows that the Zeta function admits a simple zero at $s = 0$ if the system is mixing. We may expect that for $\mathbf{k}$ small enough, there exists a zero $s = s_{\mathbf{k}}$ obtained by analytic continuation on the wavenumbers $\mathbf{k}$. This Pollicott–Ruelle resonance would replace the eigenvalue $s = 0$ in the sectors where the wavenumber is nonvanishing. In this way, the dispersion relations of the hydrodynamic modes can be obtained as we shall see in the following.

7.2.12 Consequences of time-reversal symmetry on the resonances

Consequences of the time-reversal symmetry for the eigenvalues can be deduced thanks to the Zeta function (7.55). To each periodic orbit p, a time-reversed periodic orbit $\mathbf{I}p$ is associated such that

$$
T_{\mathbf{I}p} = T_p , \qquad \mathbf{a}_{\mathbf{I}p} = -\mathbf{a}_p , \qquad \Lambda_{\mathbf{I}p,i} = \Lambda_{p,i} \quad (i = 1, \ldots, M_u) . \tag{7.57}
$$

Moreover, we note that all these characteristic quantities are real. Equation (7.57) therefore implies that the product over periodic orbits in the Zeta function $Z(s_{\mathbf{k}};\mathbf{k}) = 0$ can be rewritten to get $Z(s_{\mathbf{k}};-\mathbf{k}) = 0$. On the other hand, taking the complex conjugate of the zeta function implies that $Z(s_{\mathbf{k}}^*;-\mathbf{k}) = 0$. Therefore, the eigenvalue satisfies the relations

$$s_{\mathbf{k}} = s_{-\mathbf{k}} , \qquad s_{\mathbf{k}}^* = s_{-\mathbf{k}}^* . \tag{7.58}$$

These relations can also be obtained from Eq. (7.48).

Introducing the operator

$$\hat{\kappa} v(\mathbf{x}) = v^*(\imath\mathbf{x}) , \qquad \text{such that} \qquad \hat{\kappa}^2 = \hat{I} , \tag{7.59}$$

the Frobenius–Perron operator of the mapping is related to its adjoint by

$$\hat{\kappa}\, \hat{R}_{\mathbf{k},s}\, \hat{\kappa} = \hat{R}^{\dagger}_{-\mathbf{k},s} . \tag{7.60}$$

Therefore, the time-reversal symmetry implies that the left and right eigenstates are related by

$$\tilde{\psi}_{\mathbf{k}}(\mathbf{x}) = \psi^*_{-\mathbf{k}}(\imath\mathbf{x}) . \tag{7.61}$$

The right eigenstates are expected to be smooth along the unstable directions. On the other hand, the left eigenstates are smooth along the stable directions because the former ones are mapped on the latter ones by the time-reversal transformation $\imath$.

7.2.13 Relation to the eigenvalue problem for the flow

The relation between the eigenstates of the flow and those of the map can be treated as in Chapter 3. Carrying out the same deduction as in Eqs. (3.166)–(3.174), we infer that the eigenstate corresponding to the flow, namely

$$(\hat{Q}^t_{\mathbf{k}}\Psi_{\mathbf{k}})(\mathbf{x},\tau,\mathbf{l}) = \exp(s_{\mathbf{k}}t)\, \Psi_{\mathbf{k}}(\mathbf{x},\tau,\mathbf{l}) , \tag{7.62}$$

is given by

$$\Psi_{\mathbf{k}}(\mathbf{x},\tau,\mathbf{l}) = \exp(i\mathbf{k}\cdot\mathbf{l})\, \exp(-s_{\mathbf{k}}\tau)\, \psi_{\mathbf{k}}(\mathbf{x}) , \qquad \text{for} \qquad 0 \le \tau < T(\mathbf{x}) . \tag{7.63}$$

We emphasize that Eq. (7.63) holds for $\tau < T(\mathbf{x})$. At $\tau = T(\mathbf{x})$, the distribution $\Psi_{\mathbf{k}}$ acquires a phase due to the jump in the lattice vector according to

$$\Psi_{\mathbf{k}}[\mathbf{x}, T(\mathbf{x}), \mathbf{l}] = \exp\left[i\mathbf{k}\cdot\mathbf{a}(\mathbf{x})\right] \Psi_{\mathbf{k}}(\phi\mathbf{x}, 0, \mathbf{l}) , \tag{7.64}$$

in agreement with the eigenvalue equation (7.52). We also notice that, since $s_{\mathbf{k}} < 0$, the eigenstate (7.63) presents an exponential growth which only lasts for a finite time interval after which the exponential damping Eq. (7.62) prevails. In this way, the eigenvalue problems of the flow and of the mapping are interconnected.

In the following, we shall apply the general formalism to transport processes in fluids with different kinds of quasiperiodic boundary conditions.

7.3 Deterministic diffusion

7.3.1 Introduction

Deterministic diffusion is a very common phenomenon in which a tagged particle undergoes a diffusive motion in a system. Brownian motion of a colloidal particle in a fluid of atoms or molecules is an example of such a process. Modern methods to measure diffusion coefficients, for instance in solids, make use of radioactive tracers in low concentrations which are diluted in the material under study. Such systems may be modelled by a Hamiltonian system like (1.7) with periodic boundary conditions on all except one particle (for instance the particle 1). The position of the tagged particle is allowed to increment at each passage to a neighbouring lattice cell. Accordingly, the position $\mathbf{q}_1$ belongs to $\mathbb{R}^d$ although each of the other positions $\mathbf{q}_2, \ldots, \mathbf{q}_N$ belongs to $\mathbb{T}^d = \mathbb{R}^d/\mathbb{Z}^d$. The particles 2 to N constitute a deterministic mechanical system which drives the diffusive motion of the tagged particle 1. In this case, we have to make the identification, $\mathbf{r} = \mathbf{q}_1$, while the variable $\tilde{\mathbf{r}}$ regroups the other positions and all the momenta.

Simpler models of deterministic motion have been proposed based on the observation that the deterministic dynamical system should not be too high-dimensional but just high enough for the motion of the particle 1 to be chaotic on the reduced cell of the periodic lattice. This is the case for the periodic Lorentz gas and also for the periodic multibaker map, which both sustain deterministic diffusion under appropriate conditions (see Chapter 6 and below). Another model studied by Knauf (1987, 1989) is the Hamiltonian motion of a nonrelativistic particle in a two-dimensional lattice of attractive Yukawa potentials. This is a model for an electron in a planar solid. The Yukawa potentials model the screened Coulomb interaction of the electron with the ions composing the solid. Energy is conserved and, for high enough energy, Knauf has proved that the motion is strictly hyperbolic and mixing. A positive and finite diffusion coefficient exists. The following theory applies to such hyperbolic systems under conditions that will be enunciated below.

In the case of diffusion, the lattice $\mathscr{L}$ has the dimension of the physical space in which the diffusion process takes place: $L = d$. Quasiperiodic boundary conditions of wavenumber **k** directly describe a fluctuation of the phase-space probability density which is inhomogeneous only in the physical space. Therefore, a **k**-component $f_{\mathbf{k}}$ defined by Eq. (7.34) can be directly interpreted as a hydrodynamic mode of diffusion which is the deterministic analogue of the solution (7.2) of the diffusion equation (7.1). The eigenvalue $s_{\mathbf{k}}$ will thus be the dispersion relation of diffusion.

In view of Eq. (7.3), the diffusion coefficient can be obtained by taking two derivatives with respect to the wavenumber at $\mathbf{k} = 0$. Higher-diffusion coefficients like the Burnett and super-Burnett coefficients can be obtained by taking further derivatives (van Beijeren 1982). In this regard, we shall perform successive derivatives of the eigenvalue and of the right eigenstate at $\mathbf{k} = 0$:

$$\begin{aligned} s_0 &= 0\,,\ \partial_{\mathbf{k}} s_0\,,\ \partial_{\mathbf{k}}^2 s_0\,,\ \partial_{\mathbf{k}}^3 s_0\,,\ \partial_{\mathbf{k}}^4 s_0\,,\ \dots \\ \psi_0 &= 1\,,\ \partial_{\mathbf{k}} \psi_0\,,\ \partial_{\mathbf{k}}^2 \psi_0\,,\ \partial_{\mathbf{k}}^3 \psi_0\,,\ \partial_{\mathbf{k}}^4 \psi_0\,,\ \dots \end{aligned} \tag{7.65}$$

as well as of the left eigenstate. For this purpose, we need to assume *a priori* that both $s_{\mathbf{k}}$ and the eigenstate $\psi_{\mathbf{k}}$ smoothly depend on the wavenumber. We shall see that this condition is equivalent to assuming that the multiple time-correlation functions decay fast enough.

7.3.2 Mean drift

Differentiating the orthogonality equation (7.54) with respect to the wavenumber, we obtain

$$\langle \partial_{\mathbf{k}} \tilde{\psi}_{\mathbf{k}}^* \, \psi_{\mathbf{k}} \rangle_v \ + \ \langle \tilde{\psi}_{\mathbf{k}}^* \, \partial_{\mathbf{k}} \psi_{\mathbf{k}} \rangle_v \ = \ 0\,. \tag{7.66}$$

On the other hand, the orthogonality equation can be rewritten using the eigenvalue equation (7.52) as

$$\langle \tilde{\psi}_{\mathbf{k}}^* \, \psi_{\mathbf{k}} \rangle_v \ = \ \langle \tilde{\psi}_{\mathbf{k}}^* \, \hat{R}_{\mathbf{k},s_{\mathbf{k}}} \, \psi_{\mathbf{k}} \rangle_v \ = \ 1\,, \tag{7.67}$$

which is also differentiated with respect to the wavenumber, to get

$$\begin{aligned} &\langle \partial_{\mathbf{k}} \tilde{\psi}_{\mathbf{k}}^* \, \hat{R}_{\mathbf{k},s_{\mathbf{k}}} \, \psi_{\mathbf{k}} \rangle_v \ + \ \langle \tilde{\psi}_{\mathbf{k}}^* \, \hat{R}_{\mathbf{k},s_{\mathbf{k}}} \, \partial_{\mathbf{k}} \psi_{\mathbf{k}} \rangle_v \\ &\qquad + \langle \tilde{\psi}_{\mathbf{k}}^* \left[-\partial_{\mathbf{k}} s_{\mathbf{k}} \, T(\boldsymbol{\phi}^{-1}\mathbf{x}) - i\mathbf{a}(\boldsymbol{\phi}^{-1}\mathbf{x}) \right] \hat{R}_{\mathbf{k},s_{\mathbf{k}}} \, \psi_{\mathbf{k}} \rangle_v \ = \ 0\,. \end{aligned} \tag{7.68}$$

A simplification using Eqs. (7.49), (7.66), and (7.67) yields

$$-\ \partial_{k_\alpha} s_{\mathbf{k}} \, \langle \tilde{\psi}_{\mathbf{k}}^* \, T(\boldsymbol{\phi}^{-1}\mathbf{x}) \, \psi_{\mathbf{k}} \rangle_v \ - \ i\, \langle \tilde{\psi}_{\mathbf{k}}^* \, a_\alpha(\boldsymbol{\phi}^{-1}\mathbf{x}) \, \psi_{\mathbf{k}} \rangle_v \ = \ 0\,. \tag{7.69}$$

or

$$\partial_{\mathbf{k}} s_{\mathbf{k}} \ = \ -\, i\, \frac{\langle \tilde{\psi}_{\mathbf{k}}^* \, \mathbf{a}(\boldsymbol{\phi}^{-1}\mathbf{x}) \, \psi_{\mathbf{k}} \rangle_v}{\langle \tilde{\psi}_{\mathbf{k}}^* \, T(\boldsymbol{\phi}^{-1}\mathbf{x}) \, \psi_{\mathbf{k}} \rangle_v}\,. \tag{7.70}$$

In the limit $\mathbf{k} = 0$ where $\psi_0 = \psi_0^* = 1$, the mean drift vanishes

$$\partial_{\mathbf{k}} s_0 \; = \; 0 \, , \tag{7.71}$$

if we assume that the motion on the lattice is symmetric under spatial inversion: $\langle \mathbf{a}(\mathbf{x}) \rangle_\nu = 0$.

7.3.3 The first derivative of the eigenstate with respect to the wavenumber

In order to get the derivative of the eigenstate, we differentiate the eigenvalue equation (7.52) with (7.49) directly with respect to $\mathbf{k}$ to obtain

$$\left[- \partial_{\mathbf{k}} s_{\mathbf{k}} \; T(\boldsymbol{\phi}^{-1}\mathbf{x}) \; - \; i \, \mathbf{a}(\boldsymbol{\phi}^{-1}\mathbf{x}) \right] \, (\hat{R}_{\mathbf{k},s_{\mathbf{k}}} \psi_{\mathbf{k}})(\mathbf{x}) \\ + \; (\hat{R}_{\mathbf{k},s_{\mathbf{k}}} \; \partial_{\mathbf{k}} \psi_{\mathbf{k}})(\mathbf{x}) \; = \; \partial_{\mathbf{k}} \psi_{\mathbf{k}}(\mathbf{x}) \, . \tag{7.72}$$

At $\mathbf{k} = 0$, the first derivative of the eigenvalue vanishes because of (7.71) while the eigenstate becomes the microcanonical measure: $\psi_0 = 1$. Therefore, the first derivative of the eigenstate obeys the functional equation

$$\partial_{\mathbf{k}} \psi_0(\boldsymbol{\phi}^{-1}\mathbf{x}) \; - \; i \, \mathbf{a}(\boldsymbol{\phi}^{-1}\mathbf{x}) \; = \; \partial_{\mathbf{k}} \psi_0(\mathbf{x}) \, . \tag{7.73}$$

Similarly, the differentiation of the adjoint eigenvalue equation (7.53) leads to

$$\partial_{\mathbf{k}} \tilde{\psi}_0(\boldsymbol{\phi}\mathbf{x}) \; + \; i \, \mathbf{a}(\mathbf{x}) \; = \; \partial_{\mathbf{k}} \tilde{\psi}_0(\mathbf{x}) \, . \tag{7.74}$$

Equations (7.73) and (7.74) admit the solutions

$$\partial_{\mathbf{k}} \psi_0(\mathbf{x}) \; = \; - \, i \sum_{n=1}^{\infty} \mathbf{a}(\boldsymbol{\phi}^{-n}\mathbf{x}) \, , \tag{7.75}$$

$$\partial_{\mathbf{k}} \tilde{\psi}_0(\mathbf{x}) \; = \; + \, i \sum_{n=0}^{\infty} \mathbf{a}(\boldsymbol{\phi}^{n}\mathbf{x}) \, , \tag{7.76}$$

as can be checked directly. Nevertheless, we notice that the solutions for the derivatives of the left and right eigenstates can be exchanged. In order to resolve this ambiguity, we need to go back to the condition of forward semigroup. The eigenstate $\psi_{\mathbf{k}}$ is obtained by successive applications of the forward evolution operator. It corresponds to a density which is propagated in the future. Therefore, it is smooth along the unstable direction and singular along the stable direction. We notice here that the forward propagation involves the inverse mapping $\boldsymbol{\phi}^{-1}$ as shown by (7.49) while the backward propagation involves the direct mapping $\boldsymbol{\phi}$. Therefore, the solution of (7.73), which is consistent with the forward propagation, must be taken as (7.75) and not as (7.76). In this way, the ambiguity is resolved.

Considered as functions, the derivatives (7.75) and (7.76) are meaningless. However, the assumed property of decay of correlations (7.15) gives them a meaning in the sense of distributions acting on test functions $v(\mathbf{x})$ which are piecewise Hölder continuous. Indeed, if $v(\mathbf{x})$ is such a function we have that

$$|\langle v(\mathbf{x}) a_\alpha(\boldsymbol{\phi}^{-n}\mathbf{x})\rangle_\nu| \leq \exp(-n^\gamma) , \tag{7.77}$$

for n large enough, since $a_\alpha(\mathbf{x})$ is also a piecewise Hölder continuous function of vanishing mean value. As a consequence, the sum

$$\langle v(\mathbf{x}) \partial_{k_\alpha} \psi_0(\mathbf{x})\rangle_\nu = -i \sum_{n=1}^{\infty} \langle v(\mathbf{x}) a_\alpha(\boldsymbol{\phi}^{-n}\mathbf{x})\rangle_\nu . \tag{7.78}$$

is convergent to a finite number so that both (7.75) and (7.76) are distributions or linear functionals, which is all we need for the following.

7.3.4 Diffusion matrix

Differentiating Eq. (7.69) again with respect to the wavenumber, we get

$$\begin{aligned} &- \partial_{k_\alpha}\partial_{k_\beta} s_\mathbf{k} \, \langle \tilde{\psi}_\mathbf{k}^* \, T(\boldsymbol{\phi}^{-1}\mathbf{x}) \, \psi_\mathbf{k}\rangle_\nu \\ &- \partial_{k_\alpha} s_\mathbf{k} \, \langle \partial_{k_\beta}\tilde{\psi}_\mathbf{k}^* \, T(\boldsymbol{\phi}^{-1}\mathbf{x}) \, \psi_\mathbf{k}\rangle_\nu - \partial_{k_\alpha} s_\mathbf{k} \, \langle \tilde{\psi}_\mathbf{k}^* \, T(\boldsymbol{\phi}^{-1}\mathbf{x}) \, \partial_{k_\beta}\psi_\mathbf{k}\rangle_\nu \\ &- i \, \langle \partial_{k_\beta}\tilde{\psi}_\mathbf{k}^* \, a_\alpha(\boldsymbol{\phi}^{-1}\mathbf{x}) \, \psi_\mathbf{k}\rangle_\nu - i \, \langle \tilde{\psi}_\mathbf{k}^* \, a_\alpha(\boldsymbol{\phi}^{-1}\mathbf{x}) \, \partial_{k_\beta}\psi_\mathbf{k}\rangle_\nu = 0 , \end{aligned} \tag{7.79}$$

for $\alpha, \beta = 1, \ldots, d$. Taking the limit $\mathbf{k} = 0$, using Eq. (7.71) and $\psi_0 = \tilde{\psi}_0 = 1$, we obtain

$$\partial_{k_\alpha}\partial_{k_\beta} s_0 = -\frac{i}{\langle T\rangle_\nu} \left[\langle \partial_{k_\beta}\tilde{\psi}_0^* \, a_\alpha(\boldsymbol{\phi}^{-1}\mathbf{x}) \, \rangle_\nu + \langle a_\alpha(\boldsymbol{\phi}^{-1}\mathbf{x}) \, \partial_{k_\beta}\psi_0\rangle_\nu \right] . \tag{7.80}$$

Solving Eq. (7.80) requires the knowledge of $\partial_\mathbf{k}\psi_0$ and of its adjoint which have been obtained in Eqs. (7.75) and (7.76). After substitution, we get the diffusion matrix as

$$D_{\alpha\beta} = -\frac{1}{2} \partial_{k_\alpha}\partial_{k_\beta} s_0 = \frac{1}{2\langle T\rangle_\nu} \sum_{n=-\infty}^{+\infty} \langle a_\alpha(\mathbf{x}) \, a_\beta(\boldsymbol{\phi}^n\mathbf{x})\rangle_\nu . \tag{7.81}$$

If diffusion is isotropic in the lattice, the diffusion matrix is diagonal $D_{\alpha\beta} = D\delta_{\alpha\beta}$ with the diffusion coefficient

$$D = \frac{1}{2d\langle T\rangle_\nu} \sum_{n=-\infty}^{+\infty} \langle \mathbf{a}(\mathbf{x}) \cdot \mathbf{a}(\boldsymbol{\phi}^n\mathbf{x})\rangle_\nu . \tag{7.82}$$

Using the relation (7.14) which connects the map to the flow, we find that

$$\sum_{n=-\infty}^{+\infty} \mathbf{a}(\boldsymbol{\phi}^n\mathbf{x}) = \int_{-\infty}^{+\infty} \mathbf{v}(\boldsymbol{\Phi}^t\mathbf{X}) \, dt . \tag{7.83}$$

Accordingly, the sum in Eq. (7.82) can be transformed as follows

$$\frac{1}{\langle T\rangle_\nu}\left\langle \mathbf{a}(\mathbf{x})\cdot\int_{-\infty}^{+\infty}\mathbf{v}[\tilde{\mathbf{\Phi}}^t(\mathbf{x},0,\mathbf{l})]\,dt\right\rangle_\nu$$

$$= \frac{1}{\langle T\rangle_\nu}\left\langle \int_0^{T(\mathbf{x})}\mathbf{v}(\mathbf{x},\tau,0)\,d\tau\cdot\int_{-\infty}^{+\infty}\mathbf{v}[\tilde{\mathbf{\Phi}}^t(\mathbf{x},0,\mathbf{l})]\,dt\right\rangle_\nu$$

$$= \left\langle \int_{-\infty}^{+\infty}\mathbf{v}(\mathbf{X})\cdot\mathbf{v}(\mathbf{\Phi}^t\mathbf{X})\,dt\right\rangle_\mu , \tag{7.84}$$

where stationarity leads to the cancellation of $\mathbf{c}(\mathbf{x})$ with $-\mathbf{c}(\phi\mathbf{x})$ in the first equality and where the last equality is obtained with the definition (7.21) of the invariant measure μ of the flow. Hence, we obtain the Green–Kubo relation

$$D = \frac{1}{2d}\int_{-\infty}^{+\infty}\langle\mathbf{v}(\mathbf{X})\cdot\mathbf{v}(\mathbf{\Phi}^t\mathbf{X})\rangle_\mu\,dt , \tag{7.85}$$

as a consequence of the eigenvalue problem in terms of the Pollicott–Ruelle resonances (Gaspard 1996). The previous equation (7.82) is particularly interesting for hard-sphere systems because the diffusion coefficient of the flow is given in terms of the discrete-time Poincaré map without the need to integrate the velocity autocorrelation function along the continuous time.

7.3.5 Higher-order diffusion coefficients

The higher-order derivatives of the dispersion relation may be investigated similarly to give the higher-order diffusion coefficients like the Burnett and super-Burnett coefficients (van Beijeren 1982). Assuming that the dispersion relation of diffusion can be expanded in Taylor series around $\mathbf{k}=0$ we obtain

$$s_\mathbf{k} = -D_{\alpha\beta}\,k_\alpha k_\beta + B_{\alpha\beta\gamma\delta}\,k_\alpha k_\beta k_\gamma k_\delta + \mathcal{O}(k^6) . \tag{7.86}$$

$D_{\alpha\beta}$ is the matrix of the diffusion coefficients and $B_{\alpha\beta\gamma\delta}$ is the tensor containing the super-Burnett coefficients, which are thus given by the fourth derivatives of the eigenvalue with respect to the wavenumber at $\mathbf{k}=0$.

To simplify the calculations, we make use of a formal solution of the eigenvalue problem. Thereafter, we observe that taking N derivatives with respect to the wavenumber leads to sums involving at most N-time correlation functions of the characteristic functions $T(\mathbf{x})$ and $\mathbf{a}(\mathbf{x})$ of the system with respect to the mapping ϕ.

Here, the assumption that the mapping has the Kolmogorov property becomes essential. Indeed, the Kolmogorov property is known (Cornfeld et al. 1982) to imply the property of multiple mixing that the multiple correlation functions decay as

$$\langle v_0(\mathbf{x})\, v_1(\boldsymbol{\phi}^{n_1}\mathbf{x})\, v_2(\boldsymbol{\phi}^{n_2}\mathbf{x}) \,\ldots\, v_{N-1}(\boldsymbol{\phi}^{n_{N-1}}\mathbf{x})\rangle_\nu \;\rightarrow\; \langle v_0\rangle_\nu\, \langle v_1\rangle_\nu\, \langle v_2\rangle_\nu \ldots \langle v_{N-1}\rangle_\nu\,, \tag{7.87}$$

for $|n_i - n_j| \rightarrow \infty$ for all $i, j = 0, \ldots, N-1$ and for piecewise Hölder continuous functions v_i. The bound (7.15) on the decay of the 2-time correlation functions suggests a bound like

$$|\langle v_0(\mathbf{x})\, v_1(\boldsymbol{\phi}^{n_1}\mathbf{x}) \,\ldots\, v_{N-1}(\boldsymbol{\phi}^{n_{N-1}}\mathbf{x})\rangle_\nu| \;\leq\; C(v_0, v_1, \ldots, v_{N-1})\ \mathrm{Min}_{0\leq i<j\leq N-1}\Big\{\exp\big(-|n_i - n_j|^\gamma\big)\Big\}\,, \tag{7.88}$$

if $\langle v_i\rangle_\nu = 0$, $n_0 = 0$ and where C is a positive constant. Under such a condition, the sums of such N-time correlation functions over the integers n_i are guaranteed to converge. Hence, higher-order diffusion coefficients like the super-Burnett coefficients may exist in the system.

A formal solution of the eigenvalue equation (7.49) can be obtained by applying successively the Frobenius–Perron operator to the unit function to get

$$\psi_{\mathbf{k}}(\mathbf{x}) \;=\; \lim_{n\to\infty} \prod_{j=1}^{n} \exp\Big[-s_{\mathbf{k}}\, T(\boldsymbol{\phi}^{-j}\mathbf{x}) - i\mathbf{k}\cdot\mathbf{a}(\boldsymbol{\phi}^{-j}\mathbf{x})\Big]\,. \tag{7.89}$$

By Eqs. (7.58) and (7.61), we obtain the adjoint eigenvector as

$$\tilde{\psi}_{\mathbf{k}}(\mathbf{x}) \;=\; \lim_{n\to\infty} \prod_{j=0}^{n-1} \exp\Big[-s_{\mathbf{k}}^{*}\, T(\boldsymbol{\phi}^{j}\mathbf{x}) + i\mathbf{k}\cdot\mathbf{a}(\boldsymbol{\phi}^{j}\mathbf{x})\Big]\,. \tag{7.90}$$

The orthogonality condition becomes an eigenvalue equation to determine $s_{\mathbf{k}}$

$$1 \;=\; \langle \tilde{\psi}_{\mathbf{k}}^{*}\psi_{\mathbf{k}}\rangle_\nu \;=\; \lim_{n\to\infty} \Big\langle \prod_{j=-n}^{n-1} \exp\Big[-s_{\mathbf{k}}\, T(\boldsymbol{\phi}^{j}\mathbf{x}) - i\mathbf{k}\cdot\mathbf{a}(\boldsymbol{\phi}^{j}\mathbf{x})\Big]\Big\rangle_\nu\,. \tag{7.91}$$

The derivatives of the eigenvalue $s_{\mathbf{k}}$ can therefore be obtained by differentiating successively (7.91) with respect to the wavenumber and setting $\mathbf{k} = 0$. The three first derivatives yield the known results that

$$\partial_{k_\alpha} s_0 \;=\; 0\,, \quad \partial_{k_\alpha}\partial_{k_\beta} s_0 \;=\; -2D_{\alpha\beta}\,, \quad \partial_{k_\alpha}\partial_{k_\beta}\partial_{k_\gamma} s_0 \;=\; 0\,. \tag{7.92}$$

An expression for the super-Burnett coefficient can be obtained by differentiating Eq. (7.92) four times with respect to the wavenumber (Gaspard 1996). The calculation is long but straightforward. It is necessary to introduce the quantity $\Delta T(\mathbf{x}) = T(\mathbf{x}) - \langle T\rangle_\nu$ in order for the average $\langle \Delta T\rangle_\nu$ to vanish so that the property (7.88) can be applied. In the calculation, the following correlation functions appear

$$C = \frac{1}{\langle T \rangle_\nu} \sum_{i=-\infty}^{+\infty} \langle \Delta T(\mathbf{x}) \Delta T(\boldsymbol{\phi}^i \mathbf{x}) \rangle_\nu \, , \tag{7.93}$$

$$E_{\alpha\beta} = \frac{1}{\langle T \rangle_\nu} \sum_{i,j=-\infty}^{+\infty} \langle \Delta T(\mathbf{x}) a_\alpha(\boldsymbol{\phi}^i \mathbf{x}) a_\beta(\boldsymbol{\phi}^j \mathbf{x}) \rangle_\nu \, , \tag{7.94}$$

$$\begin{aligned} F_{\alpha\beta\gamma\delta} = \frac{1}{\langle T \rangle_\nu} \sum_{i,j,k=-\infty}^{+\infty} \Big[& \langle a_\alpha(\mathbf{x}) a_\beta(\boldsymbol{\phi}^i \mathbf{x}) a_\gamma(\boldsymbol{\phi}^j \mathbf{x}) a_\delta(\boldsymbol{\phi}^k \mathbf{x}) \rangle_\nu \\ & - \langle a_\alpha(\mathbf{x}) a_\beta(\boldsymbol{\phi}^i \mathbf{x}) \rangle_\nu \, \langle a_\gamma(\boldsymbol{\phi}^j \mathbf{x}) a_\delta(\boldsymbol{\phi}^k \mathbf{x}) \rangle_\nu \\ & - \langle a_\alpha(\mathbf{x}) a_\gamma(\boldsymbol{\phi}^j \mathbf{x}) \rangle_\nu \, \langle a_\beta(\boldsymbol{\phi}^i \mathbf{x}) a_\delta(\boldsymbol{\phi}^k \mathbf{x}) \rangle_\nu \\ & - \langle a_\alpha(\mathbf{x}) a_\delta(\boldsymbol{\phi}^k \mathbf{x}) \rangle_\nu \, \langle a_\beta(\boldsymbol{\phi}^i \mathbf{x}) a_\gamma(\boldsymbol{\phi}^j \mathbf{x}) \rangle_\nu \Big] \, . \end{aligned} \tag{7.95}$$

The fourth derivatives of the eigenvalues are now given by

$$\begin{aligned} \partial_{k_\alpha} \partial_{k_\beta} \partial_{k_\gamma} \partial_{k_\delta} s_0 \; &= \; F_{\alpha\beta\gamma\delta} \; + \; 4 \, C \, (D_{\alpha\beta} D_{\gamma\delta} + D_{\alpha\gamma} D_{\beta\delta} + D_{\alpha\delta} D_{\beta\gamma}) \\ - \; 2 \, (D_{\alpha\beta} E_{\gamma\delta} &+ D_{\alpha\gamma} E_{\beta\delta} + D_{\alpha\delta} E_{\beta\gamma} + D_{\beta\gamma} E_{\alpha\delta} + D_{\beta\delta} E_{\alpha\gamma} + D_{\gamma\delta} E_{\alpha\beta}) \, , \end{aligned} \tag{7.96}$$

and the super-Burnett coefficients by

$$B_{\alpha\beta\gamma\delta} = \frac{1}{4!} \partial_{k_\alpha} \partial_{k_\beta} \partial_{k_\gamma} \partial_{k_\delta} s_0 \, . \tag{7.97}$$

In the case of isochronism ($\Delta T = 0$), only the first term remains in Eq. (7.96) and we obtain a discrete version of a known formula for the super-Burnett coefficient (van Beijeren 1982). However, in the absence of isochronism, correlations between the jump function and the return time function must also be taken into account, which explains the presence of the extra terms with C and $E_{\alpha\beta}$ in Eq. (7.96).

7.3.6 Eigenvalues and the Van Hove function

We have already mentioned that Eq. (7.91) can be used to calculate the eigenvalue. If the return time function $T(\mathbf{x})$ presents bounded deviations around a positive and finite value, we may replace the sum of the times of flight by a given time t and the sum of the jump vectors by the position vector travelled by the particle in the lattice over the time t. In this way, the eigenvalue is given by the alternative expression

$$s_{\mathbf{k}} = \lim_{t \to \infty} \frac{1}{t} \ln F_s(\mathbf{k}, t) \, , \tag{7.98}$$

where

$$F_s(\mathbf{k},t) = \langle \exp[i\mathbf{k}\cdot(\mathbf{r}_t - \mathbf{r}_0)]\rangle_\mu \, , \tag{7.99}$$

with the short notation $\mathbf{r}_t = \mathbf{r}(\mathbf{\Phi}^t\mathbf{X})$ and s stands for 'single-particle' function (Gaspard 1993a). The average is here taken with respect to an ensemble of initial trajectories which may be distributed according to the microcanonical ensemble if the system is mixing. The function (7.99) is known as the Van Hove incoherent intermediate scattering function which has been introduced in the study of diffusion by neutron and light scattering techniques (Van Hove 1954, Boon and Yip 1980). In the form (7.98), we can recognize that the leading eigenvalue of the Frobenius–Perron operator is nothing other than the dispersion relation of diffusion. The diffusion coefficient can also be obtained from (7.98) by taking two derivatives with respect to the wavenumber, which leads to the Einstein formula

$$D_{xx} = \lim_{t\to\infty} \frac{1}{2t} \langle (x_t - x_0)^2 \rangle \, . \tag{7.100}$$

Similarly, the super-Burnett coefficient can here be obtained by four derivatives with respect to the wavenumber, which leads to the known formula (van Beijeren 1982)

$$B_{xxxx} = \lim_{t\to\infty} \frac{1}{4!t} \left[\langle (x_t - x_0)^4 \rangle \; - \; 3 \, \langle (x_t - x_0)^2 \rangle^2 \right] . \tag{7.101}$$

Let us proceed with the interpretation of the super-Burnett coefficients. The diffusion coefficients determine the amplitude of the dynamical fluctuations in the central limit theorem mentioned in Chapter 3. Accordingly, the position of the diffusive particle has a Gaussian distribution after a long enough time. It is known that the central limit theorem only describes the probability distribution of the majority of dynamical fluctuations. On the other hand, the behaviour of the large fluctuations in the tails of the Gaussian distribution requires a refined description within the large-deviation theory. It is at the level of the large deviations that the super-Burnett coefficients play their role. In order to show this connection, we consider the generating function of all the diffusion coefficients, which is given by the analytic continuation of the dispersion relation (7.98)–(7.99) at imaginary wavenumbers $\mathbf{k} = -i\boldsymbol{\eta}$ if it exists:

$$Q(\boldsymbol{\eta}) = \lim_{t\to\infty} \frac{1}{t} \ln \langle \exp[\boldsymbol{\eta}\cdot(\mathbf{r}_t - \mathbf{r}_0)]\rangle \, , \tag{7.102}$$

(Gaspard 1992d, 1993a). The higher-order diffusion coefficients can be obtained from this generating function by successive derivatives with respect to its argument $\boldsymbol{\eta}$ so that we have an expansion like (7.86)

$$Q(\boldsymbol{\eta}) = D_{\alpha\beta}\, \eta_\alpha\eta_\beta \; + \; B_{\alpha\beta\gamma\delta}\, \eta_\alpha\eta_\beta\eta_\gamma\eta_\delta \; + \; \mathcal{O}(\eta^6) \, . \tag{7.103}$$

Travelling through the lattice with a nonvanishing drift velocity is a highly unstable situation since the particle is kicked off its supposed trajectory at each collision so that the mean drift is zero on average. However, we can imagine that there are trajectories for which the particle has a mean velocity υ. These trajectories are necessarily unstable and should form a fractal invariant set $\mathscr{I}_\upsilon$. The rate of escape out of this fractal is defined by $-H(\upsilon)$ according to

$$\mu\left(\frac{\mathbf{r}_t - \mathbf{r}_0}{t} \in [\upsilon, \upsilon + d\upsilon[\right) = C(\upsilon, t)\, e^{tH(\upsilon)}\, d\upsilon\ , \tag{7.104}$$

where $C(\upsilon, t)$ is a slowly varying function and $H(\upsilon)$ is nonpositive. The functions $H(\upsilon)$ and $Q(\boldsymbol{\eta})$ are related to each other by a Legendre transform

$$Q(\boldsymbol{\eta}) = H(\upsilon) + \upsilon \cdot \boldsymbol{\eta}\ , \quad \boldsymbol{\eta} = -\partial_\upsilon H\ , \quad \upsilon = \partial_{\boldsymbol{\eta}} Q\ , \tag{7.105}$$

as can be shown by evaluating the mean value Eq. (7.102) with the distribution (7.104). In particular, the function H vanishes at $\upsilon = 0$ because almost all drift velocities vanish according to the central limit theorem. The Taylor expansion of this function starts as

$$H(\upsilon) = -\frac{\upsilon^2}{4D} + \mathcal{O}(\upsilon^4)\ , \tag{7.106}$$

in the case of a diagonal diffusion matrix. Figure 7.2 depicts these functions for the symmetric random walk on a square lattice. It is well known that the symmetric random walk on a square lattice has an isotropic diffusion in the sense that the diffusion matrix is diagonal. Figure 7.2 reveals that the functions Q and H present some anisotropy which only appears in derivatives higher

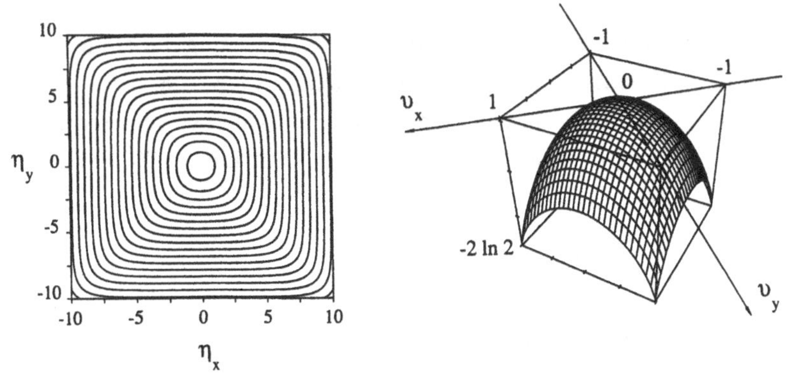

Figure 7.2 The generating function Q and its Legendre transform H for a symmetric random walk on a square lattice with jump probabilities to the next neighbours equal to $1/4$. The generating function Q is defined on the whole plane while its Legendre transform is only defined in the square $|\upsilon_x| + |\upsilon_y| \leq 1$, delimited by the maximum drift velocities (Gaspard 1992d).

than two at origin. In this regard, the super-Burnett coefficients and the higher-order diffusion coefficients are sensitive to the global anisotropy of the lattice. These considerations provide a probabilistic interpretation of the super-Burnett coefficients (Gaspard 1992d, 1993a).

7.3.7 Periodic-orbit formula for the diffusion coefficient

In Chapter 3, we have seen that averages of observables can be evaluated in term of the unstable periodic orbits which are supposed to be dense in the support of the invariant measure. Since the diffusion coefficient is such an average we may expect that diffusion can be estimated in term of the periodic orbits (Cvitanović et al. 1992, 1995).

The starting point is the observation that the diffusion coefficient is given by the second derivative of the dispersion relation $s_{\mathbf{k}}$ with respect to the wavenumber. This dispersion relation is obtained as a Pollicott–Ruelle resonance which is a zero of the Selberg–Smale Zeta function (7.55). This zero is the leading zero of the Zeta function which corresponds to the invariant measure in the limit $\mathbf{k} \to 0$. Therefore, it is also the leading zero of the leading inverse Ruelle zeta function with $m_1 = \ldots = m_{M_u} = 0$:

$$\zeta(s;\mathbf{k})^{-1} = \prod_p \left[1 - \frac{\exp(-sT_p - i\mathbf{k}\cdot\mathbf{a}_p)}{|\Lambda_{p,1}\cdots\Lambda_{p,M_u}|} \right] . \tag{7.107}$$

The dispersion relation is thus given by

$$Z(s_{\mathbf{k}};\mathbf{k}) = 0 , \qquad \text{or} \quad \zeta(s_{\mathbf{k}};\mathbf{k})^{-1} = 0 . \tag{7.108}$$

Similar equations can be obtained for the generating function $Q(\boldsymbol{\eta})$ of the diffusion coefficients by the analytic continuation: $\mathbf{k} = -i\boldsymbol{\eta}$.

The diffusion coefficient can be obtained by taking two successive derivatives of Eq. (7.108) with respect to $\mathbf{k}$ in analogy with Eqs. (3.201)–(3.204). A first derivative yields

$$\partial_{\mathbf{k}}\, s_{\mathbf{k}} = -\frac{\partial_{\mathbf{k}}\zeta^{-1}}{\partial_s\zeta^{-1}} . \tag{7.109}$$

A second derivative evaluated at zero wavenumber gives

$$\partial_{k_\alpha}\partial_{k_\beta} s_0 = -\left.\frac{\partial_{k_\alpha}\partial_{k_\beta}\zeta^{-1}}{\partial_s\zeta^{-1}}\right|_{\mathbf{k}=0} = -2\, D_{\alpha\beta} , \tag{7.110}$$

where we assume the absence of mean drift: $\partial_{\mathbf{k}} s_0 = 0$. Supposing that diffusion is isotropic the diffusion coefficient is related to the trace of the diffusion matrix by: $D = \sum_\alpha D_{\alpha\alpha}/d$. Now, the inverse Ruelle zeta function can be expanded into a sum on pseudo-cycles as

$$\zeta^{-1} = \prod_p (1 - t_p) = \sum_{l=0}^{\infty} (-)^l \sum_{p_1 \cdots p_l}' t_{p_1} t_{p_2} \cdots t_{p_l} , \tag{7.111}$$

where

$$t_p = \exp(-sT_p - i\mathbf{k}\cdot\mathbf{a}_p - U_p) , \quad \text{with} \quad \exp U_p = \prod_{i=1}^{M_u} |\Lambda_{p,i}| , \tag{7.112}$$

U_p being the integral of the local dispersion rate over the period T_p. Inserting this cycle expansion into the expression (7.110), the diffusion coefficient is obtained as

$$D = \frac{1}{2d} \frac{\sum_{l=0}^{\infty}(-)^l \sum_{p_1\cdots p_l}' (\mathbf{a}_{p_1} + \cdots + \mathbf{a}_{p_l})^2 \exp(-U_{p_1} - \cdots - U_{p_l})}{\sum_{l=0}^{\infty}(-)^l \sum_{p_1\cdots p_l}' (T_{p_1} + \cdots + T_{p_l}) \exp(-U_{p_1} - \cdots - U_{p_l})} . \tag{7.113}$$

This formula has been obtained by Artuso (1991) and by Cvitanović, Eckmann, and Gaspard (1995). It has been applied to piecewise-linear maps of the real line by Artuso (1991) and to the periodic Lorentz gas by Cvitanović, Gaspard, and Schreiber (1992).

7.3.8 Consequences of the lattice symmetry under a point group

As we mentioned earlier, a spatial group $\mathcal{G}$ is composed of an infinite abelian subgroup of translations $\mathcal{T}$ and of a finite subgroup of point-like crystallographic transformations $\mathcal{G}_p$: $\mathcal{G} = \mathcal{T} \cdot \mathcal{G}_p$. Above, we have analyzed the consequences of the subgroup of translations. The Liouvillian dynamics has been decomposed into the different irreducible representations of the translation group, which are labelled by the wavenumber $\mathbf{k}$ varying inside the Brillouin zone $\mathcal{B}$. A mode of diffusion corresponds to each wavenumber so that a direction is privileged if a mode of diffusion is present in the system. As a consequence, the symmetry of the object composed of the lattice together with the mode is reduced by the choice of $\mathbf{k}$. This symmetry reduction is formalized by the concept of little group associated with the wavenumber $\mathbf{k}$, which is the subgroup of the lattice point group leaving invariant the vector $\mathbf{k}$ (Bouckaert et al. 1936, Hamermesh 1962). For most values of $\mathbf{k}$ inside the Brillouin zone, the little group is trivial because it only contains the identity. However, the little group is larger when the wavenumber belongs to special symmetry lines or symmetry points in the

Brillouin zone. In particular, the little group coincides with the full point group when $\mathbf{k} = 0$.

These results have the following consequences on the factorization of the zeta function $Z(s;\mathbf{k})$. The little group associated with $\mathbf{k}$ has a single irreducible representation for most values of $\mathbf{k}$ where the Zeta function does not factorize. Only at those special values of $\mathbf{k}$ where the corresponding little group contains several irreducible representations, does there exist a factorization of the Zeta function of the form

$$Z(s;\mathbf{k}) = \prod_R Z_R(s;\mathbf{k}) , \tag{7.114}$$

where R runs over the irreducible representations of the little group of $\mathbf{k}$. For instance, in the triangular Lorentz gas, the Brillouin zone has several lines where the little group contains the identity together with a reflection. We can then simplify the calculation of the diffusion coefficient by following the results of Section 7.2 applied to the factor of (7.114) corresponding to the fully symmetric representation. Nevertheless, the simplification obtained by taking advantage of the invariance under reflection is modest since, in this case, the Zeta function splits into only two factors.

The preceding considerations concern the consequences of the lattice symmetry for the factorization of the Zeta function. There is a different problem which is to express the Zeta function in terms of the prime periodic orbits of the fundamental domain $\mathcal{M}_F$ of the lattice rather than those of the Wigner–Seitz cell $\mathcal{M} = \cup_{g\in\mathcal{G}_p} g(\mathcal{M}_F)$. This problem is a priori independent of the factorization discussed above and presents the following difficulty. The stumbling block appears to be the non-commutativity of translations and rotations. This difficulty can be handled in the following way (Cvitanović et al. 1995).

We can introduce the flow $\mathbf{\Phi}_F^t$ on the fundamental domain $\mathcal{M}_F$ using the full point group of the lattice. Moreover, the density $f(\mathbf{X})$ on which the Frobenius–Perron operator $\hat{Q}_{\mathbf{k}}^t$ acts can always be decomposed using the projectors onto the irreducible representations of the full point group. For an arbitrary wavenumber $\mathbf{k}$, the Frobenius–Perron operator will mix the different components of the density $f(\mathbf{X})$, which we can express by the matrix operator

$$\hat{\mathsf{Q}}_{\mathbf{k}}^t \cdot \mathbf{f}(\mathbf{X}) = \int \delta(\mathbf{X} - \mathbf{\Phi}_F^t\mathbf{Y})\, \mathsf{S}(\mathbf{X},t;\mathbf{k}) \cdot \mathbf{f}(\mathbf{Y})\, d\mathbf{Y} , \tag{7.115}$$

where $\mathbf{f}(\mathbf{X})$ denotes the vector with the different symmetry components of the density and where $\mathsf{S}(\mathbf{X},t;\mathbf{k})$ is a matrix ruling the dynamics of the different components of the density. With Eq. (7.115), the Frobenius–Perron operator is reduced to the flow in the fundamental domain $\mathcal{M}_F$. The Zeta function can then be written as a product over the prime periodic orbits $\tilde{p}$ of the fundamental domain

$$Z(s;\mathbf{k}) = \prod_{\tilde{p}\in\tilde{\mathscr{P}}} \exp\left\{ -\sum_{r=1}^{\infty} \frac{1}{r} \left[\mathrm{tr}\mathsf{S}_{\tilde{p}}(\mathbf{k})^r\right] \frac{\exp(-srT_{\tilde{p}})}{|\det(1 - \mathsf{m}_{\tilde{p}}^r)|} \right\}, \qquad (7.116)$$

where $\mathsf{S}_{\tilde{p}}(\mathbf{k})$ is the matrix $\mathsf{S}(\mathbf{X},t;\mathbf{k})$ associated with the prime periodic orbit $\tilde{p}$. It is only when the little group of the wavenumber $\mathbf{k}$ has nontrivial irreducible representations that the matrices $\mathsf{S}_{\tilde{p}}(\mathbf{k})^r$ split into block-diagonal submatrices which can be assigned to each irreducible representation so that the Zeta function factorizes as explained with Eq. (7.114). The same discussion can be developed with $\boldsymbol{\eta} = i\mathbf{k}$.

We end this section with the remark that the signature of the lattice symmetry appears in the behaviour of the dispersion relation $s_{\mathbf{k}}$ away from $\mathbf{k} = 0$ (see Fig. 7.2).

7.4 Deterministic diffusion in the periodic Lorentz gas

7.4.1 Properties of the infinite Lorentz gas

To illustrate the preceding theory, we consider the Lorentz gas where a point particle undergoes elastic collisions on hard disks which are fixed according to an infinite triangular lattice. This system has been introduced in Chapter 6, where we described the transition between different regimes.

Here, we would like to present more results on the Poincaré–Birkhoff map and on its hyperbolicity, which are at the basis of the decay of correlations.

The disks are assumed to have a radius equal to unity and we denote by r the distance between the centres of the disks. The Birkhoff map $\boldsymbol{\phi}$ was defined in Chapter 5 where we gave an explicit analytic form of the mapping induced by the collisional dynamics from disk to disk. The parameters $r_{n,n+1}$ and $\alpha_{n,n+1}$ in Eqs. (5.21) and (5.22) must be fixed according to the geometry of the disks. The Birkhoff coordinates are the angle θ giving the position of the point of impact on a disk and the component ϖ of the velocity which is tangent to the disk and pointing in the direction of increase of the angle θ.

When the horizon is finite (cf. Subsection 6.3.3) a particle can move from each disk to twelve neighbouring disks of the first and second shells surrounding the current disk, as shown in Fig. 7.3. As a consequence, the domain of definition of the Birkhoff mapping, of the return time function, and of the jump vector in Eq. (7.9) is divided into twelve subdomains

$$\mathscr{P} = [0,2\pi[\otimes]-1,+1[= \cup_{\ell=1,\dots,12} \mathscr{P}_\ell \,. \qquad (7.117)$$

In Eq. (7.9), the function $\mathbf{a}(\mathbf{x})$ for $\mathbf{x} \in \mathscr{P}_\ell$ gives the vector of the jump from the centre of the current disk to the centre of the disk ℓ reached at the next collision. Examples of such subdomains are depicted in Fig. 7.4. On the borders $\partial\mathscr{P}_\ell$ of these subdomains, each one of the functions $\phi(\mathbf{x})$, $T(\mathbf{x})$, and $\mathbf{a}(\mathbf{x})$ defining the mapping (7.9) is discontinuous. Inside the subdomains $\mathscr{P}_\ell$, these functions are smooth, i. e. analytic.

The dynamical instability of the Lorentz gas can be characterized by the largest Lyapunov exponent, as we discussed in Chapter 6. The Lyapunov exponent was calculated with the stretching factor (1.141) according to Eq. (1.139) with a time average in order to obtain the mean Lyapunov exponent (1.138) over the microcanonical ensemble. Here, we would like to emphasize the discontinuous character of the local stretching factor which is due to the discontinuity of the Birkhoff mapping. Figure 7.5 shows the curvature of the expanding horocircle before a collision as a function of the current point $\mathbf{x}$ in the Birkhoff coordinates. This curvature is given by $B_u^{(-)}(\mathbf{x})$ in Eq. (1.136) and can be calculated as the continuous fraction (1.137) extending over the whole past trajectory issued from the current point $\mathbf{x}$: $B_u^{(-)}(\mathbf{x}) = B_u(\tau_{-1})$. This function of $\mathbf{x}$ is defined at almost all points of the Birkhoff surface except for trajectories which are tangent to a disk. For such trajectories, the curvature is discontinuous because an infinitesimal perturbation of the tangency can change the whole past trajectory preceding the tangency. The amplitude of the discontinuity depends on

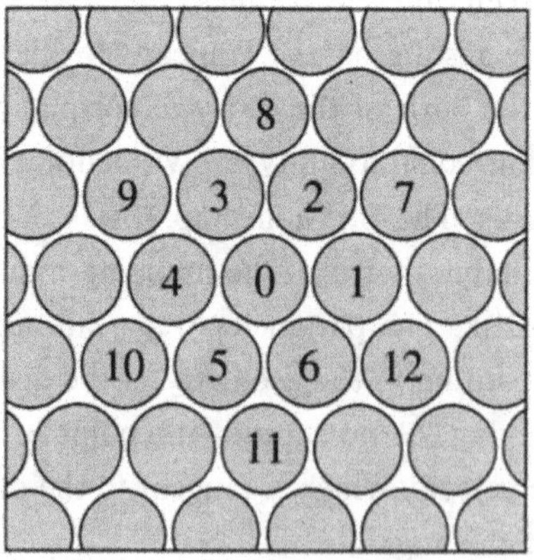

Figure 7.3 Lorentz gas with a finite horizon showing the twelve disks to which transitions are allowed by the geometry starting from disk No. 0.

Figure 7.4 In the coordinates $\mathbf{x} = (\theta, \varpi) \in [0, 2\pi[\otimes]-1, +1[$, representation of the lines of discontinuities of the mapping ϕ separating the subdomains (7.117): (a) for $r = 2.3$; (b) for $r = 2.15$; (c) for $r = 2.01$ (r being the shortest distance between the centres of the disks). The integers are the labels ℓ of the disks reached at the next collision assuming that the particle is on the disk No. 0 in Fig. 7.3 (Gaspard 1996).

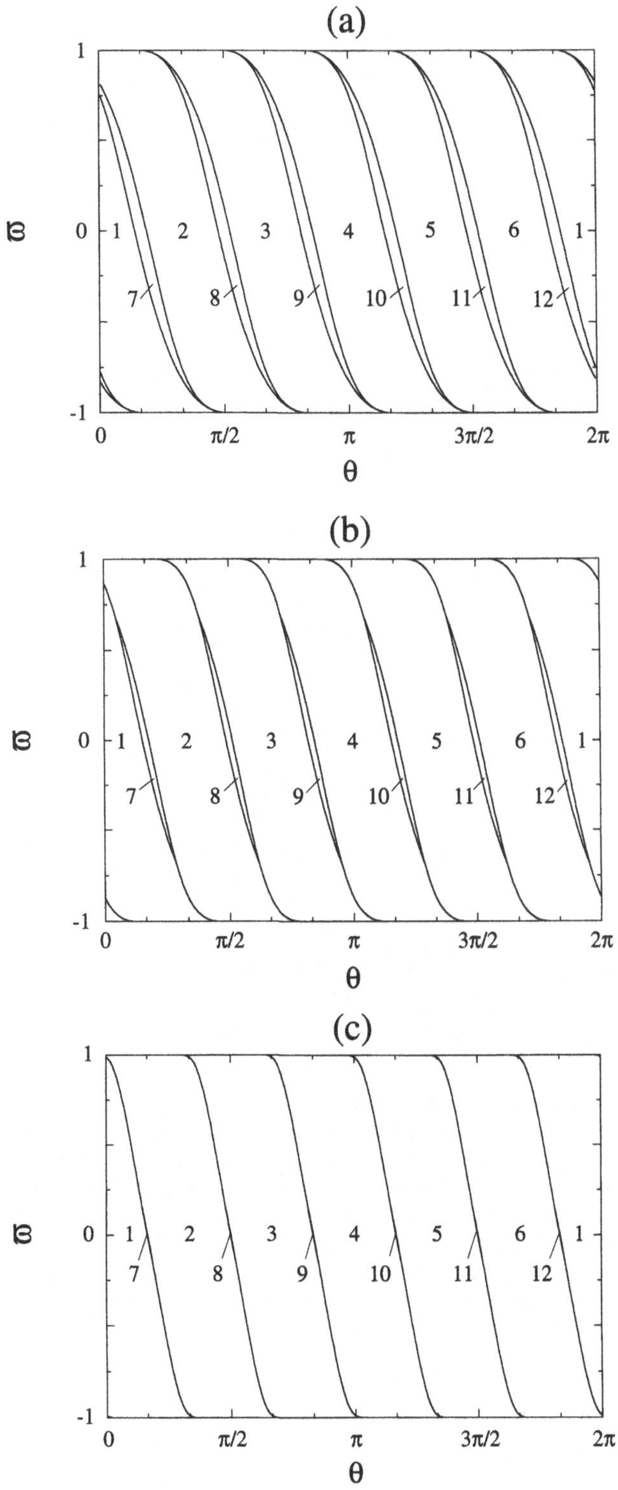

Figure 7.4 For caption, see facing page.

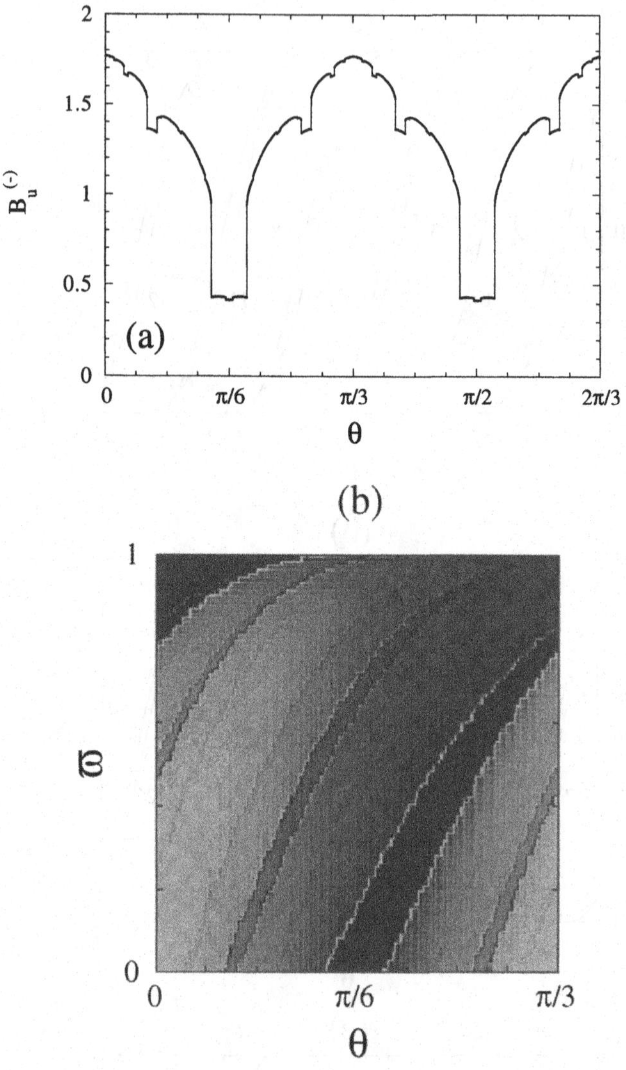

Figure 7.5 In the Lorentz gas with $r = 2.3$: (a) Curvature $B_u^{(-)}(\mathbf{x})$ of the unstable horocircle immediately before the collision along the line $\mathbf{x} = (\theta, \varpi = 0)$ (dots; a solid line joining the dots is used to display the discontinuities). (b) Density plot of $B_u^{(-)}(\mathbf{x})$ in the domain where $0 < \theta < \pi/3$ and $0 < \varpi < 1$. The instability is higher in the light regions than in the dark ones. The preimages of the discontinuity lines appear in this plot. The first preimages are the discontinuities of the inverse mapping ϕ^{-1} which are the lines of Fig. 7.4 mapped by the time-reversal transformation $(\theta, \varpi) \to (\theta, -\varpi)$ (Gaspard 1996).

the time when the tangency happens: the earlier the tangency, the smaller the discontinuity. These discontinuities of the curvature appear on a dense set of zero Lebesgue measure because all the images or preimages of the borders of the subdomains, namely $\phi^n(\partial\mathscr{P}_\ell)$ form discontinuities. Because of these discontinuities, the dynamics is not structurally stable. Indeed, a slight change in the geometry of the Lorentz gas and, in particular, in the interdisk distance r would modify the tangencies and, hence, the topology of the discontinuities. Consequently, the periodic Lorentz gas is not an Axiom-A system.[2]

Figure 7.5 shows that the curvature remains positive so that the dynamics is always hyperbolic due to the defocusing character of the collisions on disks. Therefore, the stable and unstable directions exist at every point $\mathbf{x}$ of the domain of definition of the mapping, even if they depend discontinuously on the point $\mathbf{x}$ because of the discontinuities of the mapping. In spite of these differences with respect to Axiom-A systems, Bunimovich and Sinai (1980) and, more recently, Chernov (1994) proved for the Lorentz gas the series of ergodic properties which we require in condition (3):

(a) The Lorentz gas is a Kolmogorov flow so that it is mixing, and ergodic (Bunimovich and Sinai 1980).

(b) The decay of correlations with respect to the mapping is fast and of stretched exponential type so that (7.15) is satisfied for piecewise Hölder continuous functions (Chernov 1994).

(c) When the Lorentz gas has a finite horizon, both the functions $T(\mathbf{x})$ and $\mathbf{a}(\mathbf{x})$ are bounded and the central limit theorem (7.16)–(7.17) holds (Bunimovich and Sinai 1980; Sinai and Chernov 1987; Bunimovich, Sinai and Chernov 1990, Chernov 1994). Accordingly, the diffusion coefficient is positive and finite and the trajectories follow a standard Brownian process on large scales (Bunimovich and Sinai 1980, Bunimovich 1985).

Since conditions (1)–(3) hold for the regular Lorentz gas with a finite horizon, we can apply the spectral theory of the Frobenius–Perron operator to obtain the dispersion relation of diffusion as a Pollicott–Ruelle resonance, and the hydrodynamic modes as the associated eigenstates.

7.4.2 Diffusion and its dispersion relation

We have seen that the leading Pollicott–Ruelle resonance of the Frobenius–Perron operator gives us the dispersion relation of diffusion. In particular, the

2. It is important to notice that this problem does not occur for the disk scatterer of Chapter 5 when $r > r_c$ because no trajectory of the repeller is tangent to a disk when the interdisk distance is large enough so that the Axiom-A property is satisfied for these disk scatterers in contrast with the Lorentz gases.

diffusion coefficient is obtained from the second derivative of the eigenvalue $s_{\mathbf{k}}$ with respect to the wavenumber at $\mathbf{k} = 0$. Since the triangular lattice is isotropic, the diffusion coefficient is given by Eq. (7.82) with $d = 2$.

The eigenvalue itself can be obtained using the Van Hove function (7.99) and Eq. (7.98), which can be used for a numerical evaluation of the dispersion relation as depicted in Fig. 7.6. We observe that the eigenvalue satisfies $s_{\mathbf{k}} \simeq -Dk^2$ at small values of the wavenumber $\mathbf{k}$. The value of the diffusion coefficient obtained by this method is in agreement with the values previously obtained by other methods (see Chapter 6). No deviation with respect to the quadratic behaviour has been observed at small wavenumbers beyond numerical errors so that the super-Burnett coefficients should be very small. At larger values of $\mathbf{k}$, the eigenvalue $s_{\mathbf{k}}$ seems to encounter other singularities down along the negative real axis which suggests that the dispersion relation becomes either complex or ill defined due to complex singularities, for instance branch cuts.

7.4.3 Cumulative functions of the eigenstates

The properties of the eigenstates (7.52) may also be investigated numerically, by using the formal solution given by Eq. (7.89). However, we expect the eigenstate to be a distribution but not a function. Indeed, the first derivative of the

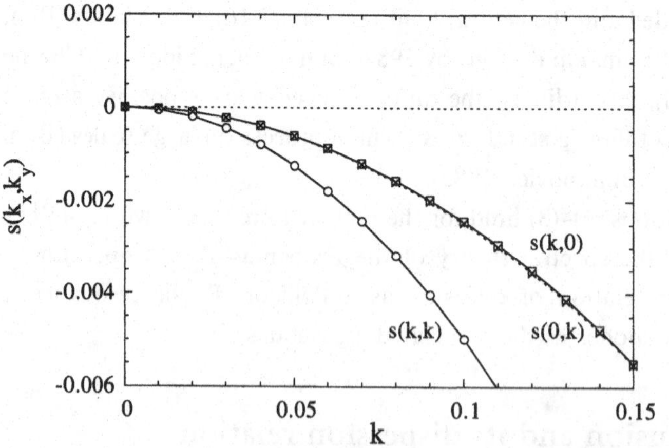

Figure 7.6 In the Lorentz gas with $r = 2.3$, dispersion relation of diffusion $s_{\mathbf{k}} = s(k_x, k_y)$ given by Eq. (7.98)–(7.99) in terms of the Van Hove function. The eigenvalue is evaluated along $\mathbf{k} = (k, 0)$ (squares, solid line); $\mathbf{k} = (0, k)$ (crosses, dashed line); $\mathbf{k} = (k, k)$ (circles, solid line). The three curves are consistent with $s_{\mathbf{k}} \simeq -D(k_x^2 + k_y^2)$ with the diffusion coefficient $D \simeq 0.25$. We remark that the eigenvalue $s_{\mathbf{k}}$ presents a small imaginary part (not visible in the figure) due to numerical errors (Gaspard 1996).

eigenstate with respect to the wavenumber is given by the expression (7.75) which is a distribution because of the rapid decay of correlations (7.77) for piecewise Hölder continuous functions. This result suggests that such a property may extend to the eigenstates at least at small values of the wavenumber. We must be aware that we suppose here that the same property extends to the higher derivatives of the eigenstate with respect to the wavenumber. Under this assumption, we may represent the eigenstate by its cumulative function defined by applying the distribution on the indicator function

$$I_{(\theta,\varpi)}(\mathbf{x}) = \begin{cases} 1\,, & \text{for} \quad \mathbf{x} \in [0,\theta[\otimes]-1,\varpi[\,, \\ 0\,, & \text{otherwise}\,, \end{cases} \tag{7.118}$$

which is also piecewise Hölder continuous. The cumulative function of the eigenstate is thus defined with (7.118) as

$$F_{\mathbf{k}}(\theta,\varpi) = \langle I_{(\theta,\varpi)}\psi_{\mathbf{k}}\rangle_\nu = \frac{1}{4\pi}\int_0^\theta d\theta' \int_{-1}^{\varpi} d\varpi'\ \psi_{\mathbf{k}}(\theta',\varpi')\,. \tag{7.119}$$

Expanding the eigenstate in Taylor series around $\mathbf{k} = 0$, the cumulative function becomes

$$\begin{aligned} F_{\mathbf{k}}(\theta,\varpi) &= \langle I_{(\theta,\varpi)}\rangle_\nu + \mathbf{k}\cdot \langle I_{(\theta,\varpi)}\partial_{\mathbf{k}}\psi_0\rangle_\nu + \frac{1}{2}\,\mathbf{kk} : \langle I_{(\theta,\varpi)}\partial_{\mathbf{k}}^2\psi_0\rangle_\nu + \mathcal{O}(k^3)\,, \\ &= \frac{\theta(\varpi+1)}{4\pi} + i\,k_x\mathscr{T}_x(\theta,\varpi) + i\,k_y\mathscr{T}_y(\theta,\varpi) + \mathcal{O}(k^2)\,, \end{aligned} \tag{7.120}$$

where the first derivatives are given by

$$\mathscr{T}_\alpha(\theta,\varpi) = -\sum_{n=1}^{\infty} \langle I_{(\theta,\varpi)}(\mathbf{x})\ a_\alpha(\boldsymbol{\phi}^{-n}\mathbf{x})\rangle_\nu\,, \tag{7.121}$$

with $\alpha = x, y$. The first term in Eq. (7.120) corresponds to the uniform probability density representing the microcanonical ensemble ν for which the cumulative function is real and linear in the coordinates θ and ϖ ranging in the intervals $\theta \in [0, 2\pi[$ and $\varpi \in]-1, +1[$. We notice that the real part is determined by all the even derivatives while the imaginary part is determined by all the odd derivatives:

$$\begin{aligned} \operatorname{Re} F_{\mathbf{k}}(\theta,\varpi) &= \frac{\theta(\varpi+1)}{4\pi} + \mathcal{O}(k^2)\,, \\ \operatorname{Im} F_{\mathbf{k}}(\theta,\varpi) &= k_x\mathscr{T}_x(\theta,\varpi) + k_y\mathscr{T}_y(\theta,\varpi) + \mathcal{O}(k^3)\,. \end{aligned} \tag{7.122}$$

The real and imaginary parts of these complex functions are depicted in Fig. 7.7. We emphasize that these functions are continuous but nondifferentiable.

7.5 Deterministic diffusion in the periodic multibaker

7.5.1 Properties of the periodic multibaker

This model of deterministic diffusion has been introduced in Chapter 6 by Eq. (6.46). It is an area-preserving map defined in a phase space given by a sequence of squares. The phase-space coordinates are therefore an integer l labelling the squares and two real variables: $(l, x, y) \in \mathbb{Z} \otimes [0, 1] \otimes [0, 1]$. This system is chaotic with a Lyapunov exponent equal to $\lambda = \ln 2$. The direction of

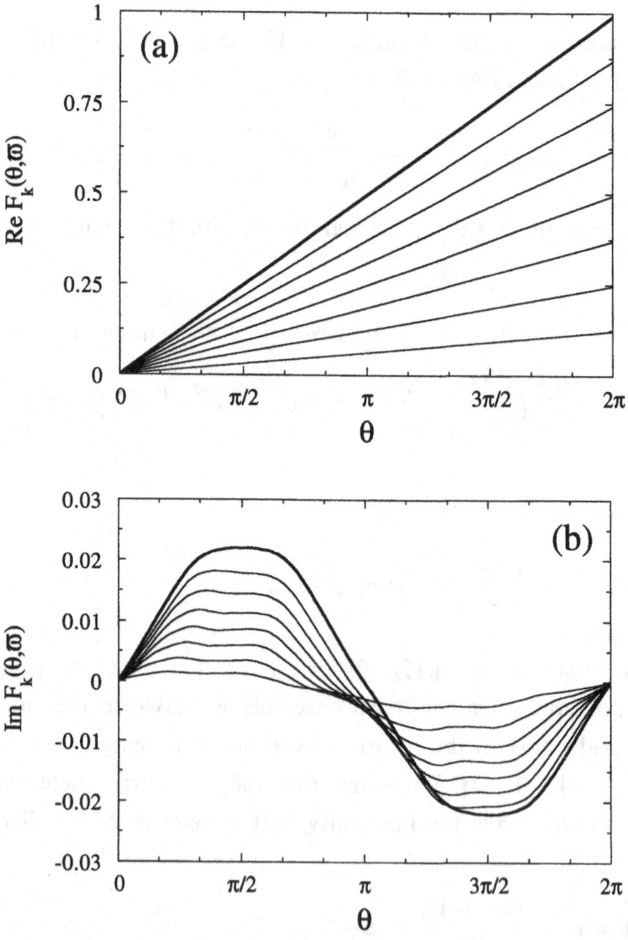

Figure 7.7 In the Lorentz gas with $r = 2.3$, eigenstate $\psi_{\mathbf{k}}$ corresponding to a hydrodynamic mode of wavenumber $\mathbf{k} = (0.1, 0)$ and represented by its cumulative function $F_{\mathbf{k}}(\theta, \varpi)$ given by Eqs. (7.118)–(7.122): (a) real and (b) imaginary parts depicted in the interval $0 < \theta < 2\pi$ for fixed values of $\varpi = -0.75, -0.5, -0.25, 0, 0.25, 0.5, 0.75$, and 1 (Gaspard 1996).

the x-axis is expanding with a uniform stretching factor equal to $\Lambda = 2$ while the y-direction is contracting. An invariant measure is the Lebesgue measure from which we infer that the multibaker is isomorphic to a Bernoulli random process with probabilities $(1/2, 1/2)$ for the dynamics inside each square and isomorphic to a symmetric random walk with a diffusion coefficient $D = 1/2$ for the dynamics along the chain $\mathbb{Z}$.

7.5.2 The Frobenius–Perron operator and its Pollicott–Ruelle resonances

We consider the time evolution of a statistical ensemble represented by the probability density $f_n(l, x, y)$. By taking a Fourier transform, the Liouvillian dynamics along the chain can be reduced to the dynamics in a single cell for components of the density corresponding to a given wavenumber k

$$f_n(l, x, y) = \frac{1}{2\pi} \int_{-\pi}^{+\pi} dk \, \exp(ikl) \, \tilde{f}_n(k, x, y) \, . \tag{7.123}$$

The time evolution is then given by

$$\tilde{f}_n(k, x, y) = \hat{Q}_k^n \, \tilde{f}_0(k, x, y) \, , \tag{7.124}$$

in terms of the reduced Frobenius–Perron operator

$$\begin{aligned} \hat{Q}_k \, f(x, y) \equiv \; & e^{ik} \, \theta\left(\frac{1}{2} - y\right) f\left(\frac{x}{2}, 2y\right) \\ & + e^{-ik} \, \theta\left(y - \frac{1}{2}\right) f\left(\frac{x+1}{2}, 2y - 1\right) . \end{aligned} \tag{7.125}$$

The Pollicott–Ruelle resonances or generalized eigenvalues are given by the zeros of the Fredholm determinant or Selberg–Smale Zeta function

$$\begin{aligned} Z(s; k) &= \mathrm{Det}\big(\hat{I} - e^{-s} \hat{Q}_k\big) \\ &= \exp - \sum_{n=1}^{\infty} \frac{e^{-sn}}{n} \, \mathrm{Tr} \, \hat{Q}_k^n \, . \end{aligned} \tag{7.126}$$

The trace of the Frobenius–Perron operator can be calculated by noting that the number of fixed points of the nth iterate of the map is equal to 2^n and that these fixed points may be labelled by words $\varepsilon_1 \varepsilon_2 \ldots \varepsilon_n$ of length n with the symbols $\varepsilon \in \{+, -\}$ corresponding to jumps to the left or the right. Since the random walk is symmetric all the possible words are present with equal probabilities for left or right jumps on average. This determines the phase factors associated with each fixed point in the trace of the operator (7.125) iterated n times. For the fixed point corresponding to the word $\varepsilon_1 \varepsilon_2 \ldots \varepsilon_n$, the phase factor is given by

$\exp[ik(\varepsilon_1+\varepsilon_2+\cdots+\varepsilon_n)]$. On the other hand, the amplitude is the same for each fixed point so that the trace is finally

$$\begin{aligned}\mathrm{Tr}\ \hat{Q}_k^n &= \frac{\sum_{\varepsilon_1\cdots\varepsilon_n}\exp[ik(\varepsilon_1+\cdots+\varepsilon_n)]}{\Lambda^n(1-\Lambda^{-n})^2}\\ &= \frac{(e^{ik}+e^{-ik})^n}{\Lambda^n(1-\Lambda^{-n})^2} = (\cos k)^n\sum_{m=0}^{\infty}\frac{m+1}{2^{nm}}, \end{aligned}\qquad (7.127)$$

because the sum is unrestricted and because $\Lambda = 2$. Replacing in the Zeta function (7.126) and performing the sum over n, we finally obtain

$$Z(s;k) = \prod_{m=0}^{\infty}\left(1 - e^{-s}\frac{\cos k}{2^m}\right)^{m+1}. \qquad (7.128)$$

For such a simple model as the multibaker, the infinite product over the prime periodic orbits can thus be carried out exactly. The infinite product over the integer m is absolutely convergent so that the Zeta function is an entire function of s as expected. Its zeros give the Pollicott–Ruelle resonances as

$$s_{k,m} = -m\ln 2 + \ln\cos k\,, \qquad m = 0,1,2,3,\ldots\,, \qquad (7.129)$$

of multiplicity $m+1$ (Gaspard 1995). These resonances are depicted in Fig. 7.8. All the eigenvalues have a negative real part as expected for a forward semigroup. They correspond therefore to eigenmodes which relax under the forward time evolution. For $-\pi/2 < k < +\pi/2$, the eigenvalue is real but it becomes imaginary for $|k| > \pi/2$. The imaginary part $+i\pi$ corresponds to a negative sign for the

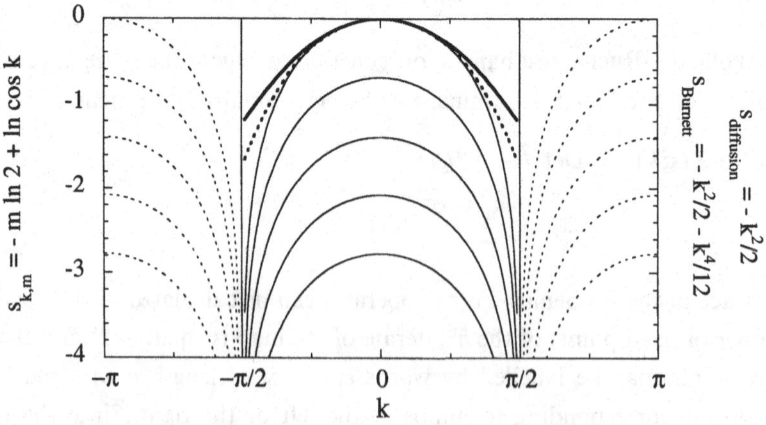

Figure 7.8 Pollicott–Ruelle resonances $s_{k,m}$ of the dyadic multibaker as a function of the wavenumber k (thin solid lines) and compared with the diffusion (thick solid line) and the super-Burnett (thick dashed line) approximations. Resonances with an imaginary part $i\pi$ are depicted by thin dashed lines (Gaspard 1995).

eigenvalue of $\hat{Q}_k$ itself, which comes from an alternate or flip-flop time evolution at wavelengths shorter than $2\pi/k = 4$. Indeed, if we colour one square in white and the next in black successively along the chain the pattern would simply be shifted by one square under the multibaker transformation, which explains that -1 is an eigenvalue of the Frobenius–Perron operator at $k = \pm\pi$.

Figure 7.8 shows that only the resonance for $m = 0$ vanishes at $k = 0$ so that the relaxation dynamics is the slowest for the corresponding mode which we may thus identify as the hydrodynamic mode of diffusion. The other modes have faster decays on time scales of the order of the inverse of the Lyapunov exponent, which may be interpreted as kinetic time scales. We infer that the dispersion relation of diffusion in the multibaker is given by

$$\begin{aligned} s_k &= \ln \cos k = -\frac{k^2}{2} - \frac{k^4}{12} + \cdots \\ &= -D\,k^2 + B\,k^4 + \cdots , \end{aligned} \tag{7.130}$$

which confirms that the diffusion coefficient is $D = 1/2$ and which gives the super-Burnett coefficient as $B = -1/12$. To interpret the full dispersion relation in terms of large deviations, we carry out its analytic continuation at imaginary wavenumber $k = -i\eta$ to obtain the generating function of the diffusion coefficients

$$Q(\eta) = \ln \cosh \eta \, . \tag{7.131}$$

According to Eqs. (7.104)–(7.106), its Legendre transform is given by

$$H(\upsilon) = -\frac{1}{2}(1+\upsilon)\ln(1+\upsilon) - \frac{1}{2}(1-\upsilon)\ln(1-\upsilon) \, , \tag{7.132}$$

where we used the relations

$$\upsilon = \tanh \eta \, , \qquad \text{and} \quad \eta = \frac{1}{2} \ln\frac{1+\upsilon}{1-\upsilon} \, . \tag{7.133}$$

The functions (7.131) and (7.132) are depicted in Fig. 7.9. We observe that the Legendre transform (7.132) is only defined for $-1 \leq \upsilon \leq +1$ because the minimum and maximum drift velocities of the multibaker are $\upsilon = \Delta l = \pm 1$. The function (7.132) gives minus the escape rate out of the fractal $\mathcal{I}_\upsilon$ of trajectories with drift velocity between υ and $\upsilon + d\upsilon$ as $\gamma = -H(\upsilon)$. Using results of Chapter 4, the pressure function on this fractal is given by

$$P(\beta) = H(\upsilon) + (1-\beta)\ln 2 \, , \tag{7.134}$$

because of the uniform hyperbolicity of the multibaker. Therefore, the partial fractal dimensions are given by

$$0 \leq d_{\mathrm{H}} = d_{\mathrm{I}} = 1 + \frac{H(\upsilon)}{\ln 2} \leq 1 \, , \tag{7.135}$$

while the full dimensions are $\dim_q(\mathcal{I}_\upsilon) = 2d_q = 2d_{\rm H}$. The dimension is maximum when $\upsilon = 0$ since almost all trajectories have a zero mean drift. However, the dimension vanishes at $\upsilon = \pm 1$ where $H(\pm 1) = -\ln 2$. Indeed, only one trajectory of length n out of 2^n travels on the multibaker as fast as with a drift velocity ± 1, so that the fractal set reduces to a single trajectory and its dimension to zero.

7.5.3 Generalized spectral decomposition

Since the Zeta function is entire with zeros and without branch cut, the generalized spectrum consists only of Pollicott–Ruelle resonances which are multiple for $m > 0$. This result suggests that the Frobenius–Perron operator (7.125) corresponding to the forward semigroup has a generalized spectral decomposition

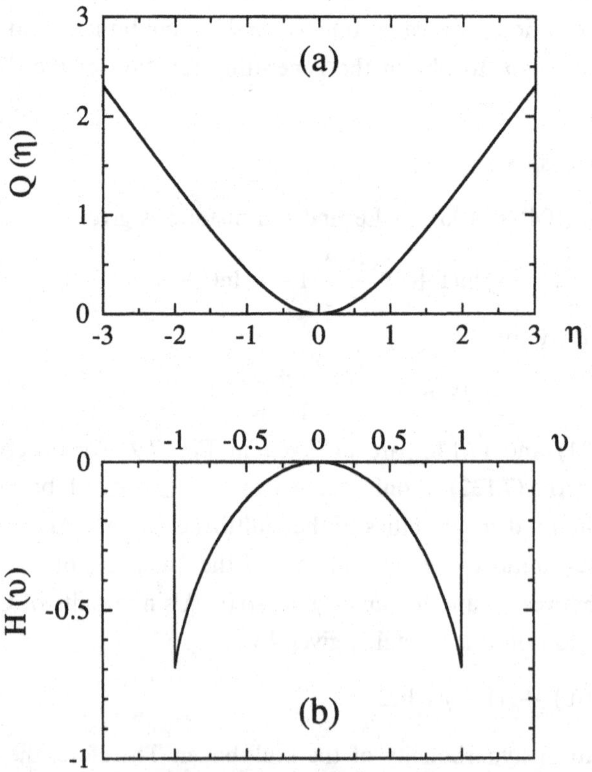

Figure 7.9 (a) Generating function (7.131) of the diffusion coefficients for the dyadic multibaker map. (b) Its Legendre transform (7.132) giving minus the escape rate of the fractal set of trajectories with drift velocity υ.

of the form

$$\langle g|\hat{Q}_k^{+n}|f\rangle = \sum_{m=0}^{\infty} \Big(\langle g|\psi_{m,0}\rangle \ldots \langle g|\psi_{m,m}\rangle\Big) \begin{pmatrix} \chi_m & 1 & 0 & 0 & \ldots & 0 \\ 0 & \chi_m & 1 & 0 & \ldots & 0 \\ 0 & 0 & \chi_m & 1 & \ldots & 0 \\ \vdots & \vdots & \vdots & \vdots & \ddots & \vdots \\ 0 & 0 & 0 & 0 & \ldots & \chi_m \end{pmatrix}^n \begin{pmatrix} \langle\tilde{\psi}_{m,0}|f\rangle \\ \vdots \\ \vdots \\ \langle\tilde{\psi}_{m,m}|f\rangle \end{pmatrix}, \tag{7.136}$$

where

$$\chi_m = \frac{\cos k}{2^m} . \tag{7.137}$$

This spectral decomposition will give the complete time evolution of averages over statistical ensembles distributed on the whole chain as

$$\langle g\rangle_n = \langle g\hat{P}^{+n}f\rangle = \frac{1}{2\pi}\int_{-\pi}^{+\pi} dk\ \langle g_k^*|\hat{Q}_k^{+n}|f_k\rangle ,$$

where f_k and g_k are Fourier transforms of the initial density f and of the observable g.

The Jordan blocks in (7.136) are defined in terms of the right root vectors

$$\hat{Q}_k^+|\psi_{m,0}\rangle = \chi_m\ |\psi_{m,0}\rangle , \tag{7.138}$$

$$\hat{Q}_k^+|\psi_{m,i}\rangle = \chi_m\ |\psi_{m,i}\rangle + |\psi_{m,i-1}\rangle \qquad i = 1, 2, 3, \ldots, m , \tag{7.139}$$

and the left root vectors

$$\langle\tilde{\psi}_{m,m}|\hat{Q}_k^+ = \chi_m\ \langle\tilde{\psi}_{m,m}| , \tag{7.140}$$

$$\langle\tilde{\psi}_{m,i}|\hat{Q}_k^+ = \chi_m\ \langle\tilde{\psi}_{m,i}| + \langle\tilde{\psi}_{m,i+1}| \qquad i = 0, 1, 2, \ldots, m-1 . \tag{7.141}$$

Moreover, the left and right root vectors are biorthonormal

$$\langle\tilde{\psi}_{m,i}|\psi_{m',i'}\rangle = \delta_{m,m'}\ \delta_{i,i'} , \tag{7.142}$$

and they form a complete basis

$$\sum_{m=0}^{\infty}\sum_{i=0}^{m} |\psi_{m,i}\rangle\ \langle\tilde{\psi}_{m,i}| = \hat{I} , \tag{7.143}$$

on appropriate spaces to be determined below. This spectral decomposition is generalized in the sense that both the left and right root vectors are distributions instead of functions. Therefore, the operator $\hat{Q}_k^+$ must be considered as a bidistribution rather than a standard operator mapping functions onto functions.

The bidistribution maps a pair of test functions directly onto a real or complex number according to

$$\left\{ \begin{array}{ccc} f & \in & \mathscr{E} \\ g & \in & \mathscr{E}' \end{array} \right\} \quad \longrightarrow \quad \langle g|\hat{Q}_k^{+n}|f\rangle \in \mathbb{C}\,, \tag{7.144}$$

where the initial densities $\{f\}$ belong to a functional space $\mathscr{E}$ of test functions whereas $\mathscr{E}'$ is a different space of test functions to which the final observables $\{g\}$ belong.

We shall now proceed with the calculation of these distributions (Gaspard 1992c, 1993a, 1995; Hasegawa and Driebe 1992, 1993, 1994; Tasaki et al. 1993, 1994). On the way, we shall determine the functional spaces of test functions on which they are defined.

7.5.4 Analysis of the associated one-dimensional map

We observe that the multibaker map always preserves the unstable and the stable directions which remain parallel to the axis. If we refer to the form (6.47) of the inverse map, we also observe that the variable x is passively driven by the variable y because the two alternatives defining the map are based on a partition at $y = 1/2$ which is parallel to the x-axis. This suggests that a special role is played by the one-dimensional map

$$\varphi(l, y) = \begin{cases} (l+1, 2y)\,, & 0 \leq y \leq \frac{1}{2}\,, \\ (l-1, 2y-1)\,, & \frac{1}{2} < y \leq 1\,. \end{cases} \tag{7.145}$$

This one-dimensional map sustains the same diffusion process and is isomorphic to the symmetric random walk with coefficient $D = 1/2$. The map is uniformly expanding with the Lyapunov exponent $\lambda = \ln 2$. Taking into account the translational invariance of the map, the time evolution of statistical ensembles can be studied in terms of the corresponding Frobenius–Perron operator

$$(\hat{R}_k v)(y) = \frac{e^{-ik}}{2}\, v\left(\frac{y}{2}\right) + \frac{e^{+ik}}{2}\, v\left(\frac{y+1}{2}\right). \tag{7.146}$$

When $k = 0$, this operator reduces to the Perron-Frobenius operator of the dyadic map of the unit interval, which admits the spectral decomposition given in Chapter 3. We are interested here in the extension of this spectral decomposition when the wavenumber k is different from zero.

We first notice that the Fredholm determinant of the present operator is

$$\mathrm{Det}(\hat{I} - e^{-s}\hat{R}_k) = \prod_{m=0}^{\infty} \left(1 - e^{-s}\,\frac{\cos k}{2^m}\right), \tag{7.147}$$

so that the resonances of the Frobenius–Perron operator of the one-dimensional map are identical to those (7.129) of the two-dimensional map, except that the multiplicity is here equal to one (instead of $m+1$).[3] For the one-dimensional map, we should therefore expect the simple spectral decomposition

$$\hat{R}_k = \sum_{m=0}^{\infty} |\Psi_m\rangle \, \chi_m \, \langle\tilde{\Psi}_m| \, , \tag{7.148}$$

with the eigenvalues (7.137). The right eigenvectors,

$$\hat{R}_k \, \Psi_m(x) = \chi_m \, \Psi_m(x) \, , \tag{7.149}$$

may thus be assumed to be polynomials because the operator (7.146) preserves the $(m+1)$-dimensional space of polynomials of degree m. Indeed, we may apply the operator (7.149) to the basis of monomials $\{y^m\}$ to observe that the operator is represented by an infinite triangular matrix which starts as

$$\hat{R}_k \begin{pmatrix} 1 \\ y \\ y^2 \\ y^3 \\ \vdots \end{pmatrix} = \begin{pmatrix} \cos k & 0 & 0 & 0 & \cdots \\ \frac{1}{4}e^{ik} & \frac{\cos k}{2} & 0 & 0 & \cdots \\ \frac{1}{8}e^{ik} & \frac{1}{4}e^{ik} & \frac{\cos k}{4} & 0 & \cdots \\ \frac{1}{16}e^{ik} & \frac{3}{16}e^{ik} & \frac{3}{16}e^{ik} & \frac{\cos k}{8} & \cdots \\ \vdots & \vdots & \vdots & \vdots & \ddots \end{pmatrix} \begin{pmatrix} 1 \\ y \\ y^2 \\ y^3 \\ \vdots \end{pmatrix} , \tag{7.150}$$

where we recognize the eigenvalues (7.137) on the diagonal. Therefore, the diagonalization may be expected to proceed in a straightforward way for such operators. The first few eigenpolynomials are given by

$$\begin{aligned} \Psi_0(y) &= 1 \, , \\ \Psi_1(y) &= y - \frac{1}{2} - \frac{i\tau_k}{2} \, , \\ \Psi_2(y) &= \frac{y^2}{2} - \frac{y}{2} + \frac{1}{12} - \frac{i\tau_k}{2}\left(y - \frac{1}{2}\right) - \frac{\tau_k^2}{6} \, , \\ \Psi_3(y) &= \frac{y^3}{6} - \frac{y^2}{4} + \frac{y}{12} - \frac{i\tau_k}{4}\left(y^2 - y + \frac{1}{7}\right) \\ &\quad - \frac{\tau_k^2}{6}\left(y - \frac{1}{2}\right) + \frac{i\tau_k^3}{21} \, , \\ &\ldots \end{aligned} \tag{7.151}$$

with $\tau_k = \tan k$. When $k = 0$, we observe that they reduce to the Bernoulli polynomials up to a factor: $\Psi_m(y) = B_m(y)/m!$ (as expected according to

3. This is a general feature of area-preserving maps to have resonances of multiplicity $m+1$, although the resonances of one-dimensional maps have in general the multiplicity 1. As a consequence, there is generically no Jordan-block structure in expanding one-dimensional maps. Nevertheless, Jordan blocks may appear in nongeneric one-dimensional maps (Driebe and Ordóñez 1996).

Section 3.10). This may be verified by calculating the generating function of all the eigenpolynomials defined as

$$G(y,t) = \sum_{m=0}^{\infty} \Psi_m(y)\, t^m \, . \tag{7.152}$$

Applying the Frobenius–Perron operator (7.146) on both members of Eq. (7.152) and using Eq. (7.149) with the eigenvalues (7.137), we infer that the generating function must satisfy

$$\begin{aligned}(\hat{R}_k G)(y,t) &= \frac{e^{-ik}}{2}\, G\left(\frac{y}{2},t\right) + \frac{e^{+ik}}{2}\, G\left(\frac{y+1}{2},t\right) \\ &= (\cos k)\, G\left(y,\frac{t}{2}\right) . \end{aligned} \tag{7.153}$$

We observe that the generating function must have an exponential dependence on its first argument y so that

$$G(y,t) = \frac{\exp(yt)}{C(t)} \, , \tag{7.154}$$

where the function $C(t)$ satisfies

$$C(2t) = \frac{\exp(t)+\exp(-2ik)}{1+\exp(-2ik)}\, C(t) \, . \tag{7.155}$$

When $k = 0$, the solution of this equation is $C(t) = (e^t - 1)/t$ so that the generating function is

$$G(y,t) = \frac{t\, e^{yt}}{e^t - 1} \, , \qquad (k=0) \, , \tag{7.156}$$

which we recognize as the generating function of the Bernoulli polynomials since $\Psi_m(y) = B_m(y)/m!$ in this limiting case (Abramowitz and Stegun 1972, Chapter 23).

In the general case for $k \neq 0$, we find generalizations of the Bernoulli polynomials which now depend on the wavenumber. When the generating function is of the form (7.154) with a function $C(t)$ depending only on the parameter t, the corresponding polynomials are known as the Appell polynomials in the theory of polynomial expansions (Boas and Buck 1958).

The function $C(t)$ can be represented by a Taylor series

$$C(t) = \sum_{m=0}^{\infty} c_m\, t^m \, , \qquad c_0 = 1 \, , \tag{7.157}$$

where the coefficients are given by the recurrence formula

$$c_m = \frac{\alpha}{2^m - 1} \sum_{j=1}^{m} \frac{c_{m-j}}{j!} \, , \qquad c_0 = 1 \, , \tag{7.158}$$

with the parameter

$$\alpha = \frac{1}{1 + \exp(-2ik)} = \frac{\exp(ik)}{2\cos k} \, . \qquad (7.159)$$

The first few coefficients are

$$\begin{aligned} c_0 &= 1 \, , \\ c_1 &= \frac{1 + i\tau_k}{2} \, , \\ c_2 &= \frac{(1 + i\tau_k)(2 + i\tau_k)}{12} \, , \\ c_3 &= \frac{(1 + i\tau_k)(7 + 6i\tau_k - \tau_k^2)}{168} \, , \\ &\ldots \end{aligned} \qquad (7.160)$$

We can then show by recurrence that the following inequality holds

$$|c_m| \leq \frac{(|\alpha| + 1)^m}{m!} \qquad \text{for} \qquad m \geq 0 \, , \qquad (7.161)$$

where $|\alpha| = [2|\cos(k)|]^{-1}$. The function $C(t)$ turns out to be a function of exponential type without pole at finite distance if $k \neq \pi(j + 1/2)$

$$|C(t)| \leq \exp\left[(|\alpha| + 1)|t|\right] \, . \qquad (7.162)$$

Moreover, the nonvanishing of c_0 in the Taylor series implies that $C(t)$ does not vanish in the vicinity of $t = 0$. Using (7.155), the function $C(t)$ can then be written as an infinite product (Gaspard 1992c)

$$C(t) = \prod_{n=1}^{\infty} \left[\frac{\exp(2^{-n}t) + \exp(-2ik)}{1 + \exp(-2ik)} \right] , \qquad (7.163)$$

which shows that the zeros of $C(t)$ are

$$t_{n,j} = i\, 2^{n+1} \, [\pi(j + 1/2) - k] \qquad j \in \mathbb{Z} \, , \quad n = 1, 2, 3, \ldots \qquad (7.164)$$

The first zero appears at a radius

$$\rho_k = 2 \left[(\pi - 2|\text{Re}\, k|)^2 + 4\, |\text{Im}\, k|^2 \right]^{1/2} . \qquad (7.165)$$

Let us remark that this radius decreases to zero when $k \to \pm\pi/2$, which is related to the fact that all the eigenvalues (7.137) of $\hat{R}_k$ vanish in this limit.

The null spaces

For an arbitrary wavenumber k, the Frobenius–Perron operator (7.146) admits zero as an eigenvalue. The corresponding eigenfunctions are $\exp(yt_{1,j})$ $(j \in \mathbb{Z})$. We can then define the null spaces $\mathcal{N}_k^{(n)}$ generated by linear combination of the functions $\exp(yt_{n,j})$ with $j \in \mathbb{Z}$. They are mapped successively onto each other under the Frobenius–Perron operator (7.146) according to (Dörfle 1985)

$$0 \xleftarrow{\hat{R}_k} \mathcal{N}_k^{(1)} \xleftarrow{\hat{R}_k} \mathcal{N}_k^{(2)} \xleftarrow{\hat{R}_k} \mathcal{N}_k^{(3)} \xleftarrow{\hat{R}_k} \mathcal{N}_k^{(4)} \cdots . \tag{7.166}$$

Hereafter, we shall see that the null spaces form the border of the functional space on which the expansion into the eigenpolynomials is convergent.

The eigendistributions

The preceding results show that a spectral decomposition like (7.148) is a decomposition of the initial density on the basis formed by the above eigenpolynomials. Such decompositions, which are similar to the Taylor expansions (in which case the polynomials are the monomials y^m), are known as polynomial expansions of analytic functions and a whole theory has been developed in mathematics. In this context, the following theorem has been obtained (Boas and Buck 1958, Theorem 9.2, p. 28)

Theorem. If $\Psi_m(y)$ are the Appell polynomials defined by the generating function (7.152), if $C(t)$ has no zero in the region Ω of the complex plane t, and if Δ is the largest circular disk, with centre at 0, in Ω, and Δ has radius ρ_k, then every entire function of exponential type $\tau < \rho_k$ has the representation

$$f(y) = \sum_{m=0}^{\infty} \langle \tilde{\Psi}_m | f \rangle \, \Psi_m(y) , \tag{7.167}$$

where

$$\langle \tilde{\Psi}_m | f \rangle = \frac{1}{2\pi i} \oint_{\Gamma} t^m \, C(t) \, F(t) \, dt , \tag{7.168}$$

$$F(t) \equiv \int_0^{\infty} e^{-ty} \, f(y) \, dy , \qquad f(y) = \frac{1}{2\pi i} \oint_{\Gamma} e^{yt} \, F(t) \, dt , \tag{7.169}$$

and Γ is a circumference $|t| = \sigma$ with $\tau < \sigma < \rho_k$ on which $C(t)$ is regular.

Proof. Firstly, we consider the Laplace transform (7.169) of (7.167)

$$F(t') = \sum_{m=0}^{\infty} \langle \tilde{\Psi}_m | f \rangle \int_0^{\infty} e^{-t'y} \, \Psi_m(y) \, dy . \tag{7.170}$$

Secondly, we consider the generating function for Appell polynomials as given by (7.152) with (7.154). We multiply both members of Eq. (7.152) by $\exp(-t'y)$ and we integrate over y to get

$$\frac{1}{t'-t} = \sum_{m=0}^{\infty} t^m \, C(t) \int_0^{\infty} e^{-t'y} \, \Psi_m(y) \, dy , \tag{7.171}$$

under the condition of convergence that $\mathrm{Re}t' > \mathrm{Re}t$ for all values of t. We multiply both members of Eq. (7.171) by the Laplace transform $F(t)$ and integrate over the contour Γ. The point t' is outside the contour Γ because of the aforementioned condition of convergence. By Cauchy's formula, we obtain

$$\begin{aligned} F(t') &= \frac{1}{2\pi i} \oint_\Gamma \frac{F(t)}{t'-t}\, dt \\ &= \sum_{m=0}^{\infty} \frac{1}{2\pi i} \oint_\Gamma t^m\, C(t)\, F(t)\, dt \int_0^{\infty} e^{-t'y}\, \Psi_m(y)\, dy\,. \end{aligned} \tag{7.172}$$

The comparison with Eq. (7.170) allows us to identify the coefficient $\langle\tilde{\Psi}_m|f\rangle$ as the contour integral (7.168). *Q.E.D.*

We notice that, if $f(y)$ is an entire function of exponential type τ, the singularities of its Laplace transform $F(t)$ are inside the disk $|t| < \tau$ in the plane of the complex variable t. Therefore, the contour Γ on the circumference $|t| = \sigma$ with $\tau < \sigma$ encloses the singularities of the Laplace transform $F(t)$. In order to obtain a strict polynomial expansion, it is required that the singularities of $F(t)$ are not cancelled by zeros of the function $C(t)$, which imposes the condition that $\tau < \rho_k$, leaving room for a contour such as Γ (cf. Boas and Buck 1958).

The above theorem provides us with the decomposition of the initial density $f(y)$ onto the decaying eigenmodes of the dynamical system. Since $f(y)$ and $F(t)$ are related by a Laplace transform according to (7.169) we have

$$f^{(m)}(0) = \frac{1}{2\pi i} \oint_\Gamma t^m\, F(t)\, dt\,, \tag{7.173}$$

so that the coefficients of the expansion are given by

$$\begin{aligned} \langle\tilde{\Psi}_m|f\rangle &= \langle\tilde{\Psi}_0|f^{(m)}\rangle = \sum_{n=m}^{\infty} c_{n-m}\, f^{(n)}(0) \\ &= \int_{-\infty}^{+\infty} \tilde{\Psi}_m^*(y)\, f(y)\, dy\,, \end{aligned} \tag{7.174}$$

with the distributions

$$\tilde{\Psi}_m(y) \equiv \sum_{n=m}^{\infty} c_{n-m}^*\, (-)^n \delta^{(n)}(y)\,, \tag{7.175}$$

where $f^{(m)}(y)$ and $\delta^{(m)}(y)$ denote respectively the mth derivatives of the function $f(y)$ and of the Dirac distribution $\delta(y)$. We infer that, for every entire function $f(y)$ of exponential type τ such that

$$|f^{(m)}(0)| \leq \kappa_\epsilon\, (\tau + \epsilon)^m \qquad \forall\, \epsilon > 0\,, \tag{7.176}$$

the value of the distribution $\tilde{\Psi}_m$ is finite and bounded as

$$|\langle\tilde{\Psi}_m|f\rangle| \leq \kappa_\epsilon\, (\tau + \epsilon)^m \exp\Big[(|\alpha| + 1)(\tau + \epsilon)\Big]\,, \tag{7.177}$$

according to Eq. (7.161). The functions $f(y)$ for which the expansion (7.167) converges must thus be entire functions of exponential type $\tau < \rho_k$, which form a so-called *Fréchet functional space* $\mathcal{E}(\tau;k)$, i.e., a complete countably normed space (Gaspard 1992b, 1992c). For $k = 0$, Eq. (7.165) gives the critical radius $\rho_k = 2\pi$ and we recover the results known about the Bernoulli polynomials and the Euler–Maclaurin summation formula for which $\tau < 2\pi$ (see Chapter 3). For $k = \pm\pi$, zero is the only eigenvalue and an expansion is possible using the null spaces.

The left eigendistributions $\tilde{\Psi}_m$ can also be represented as finite sums plus a Riemann–Stieljes integral

$$\begin{aligned}
\langle \tilde{\Psi}_m | f \rangle &= \langle \tilde{\Psi}_0 | \frac{d^m f}{dy^m} \rangle , \\
&= \int_0^1 \frac{d^m f}{dy^m}(y)\, dA_0^*(y) , \\
&= \sum_{j=0}^{n-1} (-)^j\, A_j^*(1)\, \frac{d^{j+m} f}{dy^{j+m}}(1) + (-)^n \int_0^1 \frac{d^{m+n} f}{dy^{m+n}}(y)\, dA_n^*(y) ,
\end{aligned} \tag{7.178}$$

where we introduce the functions (Gaspard 1993a)

$$A_j(y) = \begin{cases} \frac{\alpha}{2^j}\, A_j(2y) , & \text{for} \quad 0 < y < \frac{1}{2} , \\ \frac{1-\alpha}{2^j}\, A_j(2y-1) + \alpha\, P_j(y) , & \text{for} \quad \frac{1}{2} < y < 1 , \end{cases} \tag{7.179}$$

with the parameter (7.159). When $j = 0$ and $P_0(y) = 1$, the function $A_0(y)$ is known as a deRham function which is defined only for $|k| < \pi/3$ (deRham 1957; Tasaki et al. 1993, 1994). It gives the cumulative function of the eigendistribution $\tilde{\Psi}_0(y)$ as $A_0(y) = \langle I_y \tilde{\Psi}_0 \rangle \equiv \int_0^y \tilde{\Psi}_0(y')dy'$. In order to extend the interval of wavenumbers k where the integral representation applies, we need to integrate by parts as done in (7.178), which leads to the definition of higher-order functions obtained by successive integrations from the deRham function $A_0(y)$ according to

$$A_j = \frac{dA_{j+1}}{dy} \qquad \text{with} \qquad A_j(0) = 0 , \tag{7.180}$$

with the requirement of continuity at $y = 1/2$. Hence, the polynomials entering Eq. (7.179) are also obtained by successive integrations such that

$$P_j(y) = \frac{dP_{j+1}}{dx}(y) , \qquad \text{from} \quad P_0(y) = 1 . \tag{7.181}$$

For instance, we have $P_1(y) = y + (\alpha - 1)/2$.

The functions A_j are continuous for $|k| < \arccos 2^{-j-1}$ but nondifferentiable for $\arccos 2^{-j} < |k| < \arccos 2^{-j-1}$ (Gaspard 1993a). As a consequence, their smoothness increases with j. Some of them are shown in Figs. 7.10-7.12. Moreover, the Hölder exponent $\nu(k)$ of the function $A_0(y)$ is depicted in Fig. 7.13 as a function of the wavenumber k (Tasaki et al. 1993, 1994). This figure shows that the function is only defined for restricted values of the wavenumber for which $\nu(k) > 0$, while the singular character of the eigenstates increases dramatically near $k = \pm\pi/2$ where $\nu(k) \to -\infty$. As a consequence, it is necessary to integrate the eigenstates several times before obtaining a representation of these distributions as a function.

We may also consider the analytic continuation of the generalized deRham functions for imaginary wavenumbers at $k = -i\eta$ as for the generating function of the diffusion coefficients (Gaspard 1993a). These functions $C_j(y)$ are defined recursively by the same equation as (7.179) but with a real parameter

$$\alpha(-i\eta) = \frac{1}{1 + \exp(-2\eta)} = 1 - \alpha(i\eta) . \tag{7.182}$$

Figure 7.14 shows the function $C_0(y)$ for several values of η where we recognize that this function has previously been considered by Billingsley (1965). Some of its properties are

$$C_0(y) \geq y \quad \text{for} \quad \eta \geq 0 , \tag{7.183}$$

$$C_0(y) = y \quad \text{for} \quad \eta = 0 , \tag{7.184}$$

$$C_0(y) \leq y \quad \text{for} \quad \eta \leq 0 , \tag{7.185}$$

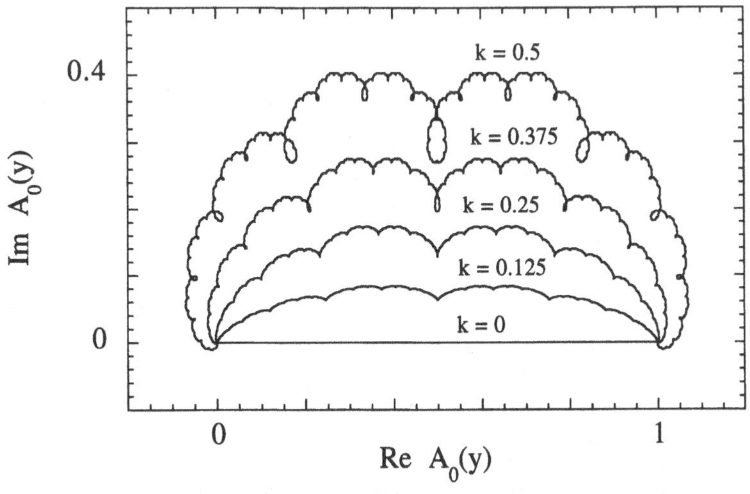

Figure 7.10 Parametric plot in the complex plane of the deRham function $A_0(y)$ when y is varied and for different values of the wavenumber k in the interval $|k| < \pi/3 = 1.0472$ where the deRham function is defined (Gaspard 1993a).

$$\lim_{\eta\to+\infty} C_0(y) = 1 , \tag{7.186}$$

$$\lim_{\eta\to-\infty} C_0(y) = 0 , \tag{7.187}$$

in the interval $y \in [0, 1]$. For $\eta \neq 0$, $C_0(y)$ is continuous in $y \in [0, 1]$ but with zero derivatives almost everywhere and it is nondifferentiable on a set of zero Lebesgue measure. The function $C_0(y)$ is closely related to the fractal set $\mathscr{I}_\upsilon$ of the trajectories with drift velocity υ along the y-axis, which we introduced to interpret Eqs. (7.131) and (7.132).

We notice that, if we take the limit $n \to \infty$ in Eq. (7.178), the preceding definitions allow us to evaluate the eigendistributions either at $y = 0$ or at $y = 1$.

7.5.5 The root states of the two-dimensional map

The above analysis of the associated one-dimensional map (7.145) allows us to construct the spectral decomposition of the area-preserving map itself as follows.

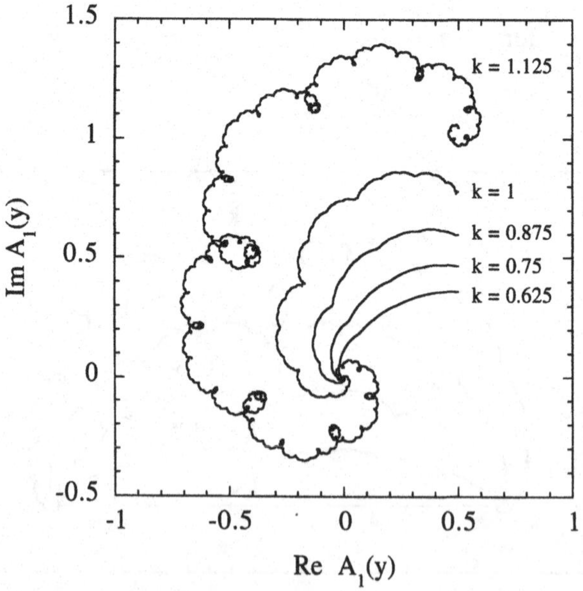

Figure 7.11 Parametric plot in the complex plane of the generalized function $A_1(y)$ when y is varied and for different values of the wavenumber k in the interval $|k| < \arccos(1/4) = 1.3181$ where $A_1(y)$ is defined. We see that this new generalized function $A_1(y)$ remains defined at $k = 1.125$ where the deRham function $A_0(y)$ is no longer defined. We also see the smoothness of $A_1(y)$ at $k = 0.625$ (Gaspard 1993a).

We first observe that, if the distribution $\psi(x,y)$ is a polynomial of degree N in x, Eq. (7.125) implies that $\hat{Q}_k\psi$ is again such a polynomial. Therefore, we may assume that the right root vectors (7.139) are of the form

$$\psi(x,y) = \sum_{\nu=0}^{N} x^\nu \, \mathscr{D}^{(\nu)}(y) , \tag{7.188}$$

where $\mathscr{D}^{(\nu)}(y)$ are distributions of the variable y. This result can be understood by noting that the x-axis is the direction of the unstable manifolds of the mapping where the expanding dynamics tends to smooth the probability density.

Since the root vectors are expected to be distributions we introduce a family of observables

$$I_y^{(j)}(y') = \theta(y-y') \, \frac{(y-y')^j}{j!} , \tag{7.189}$$

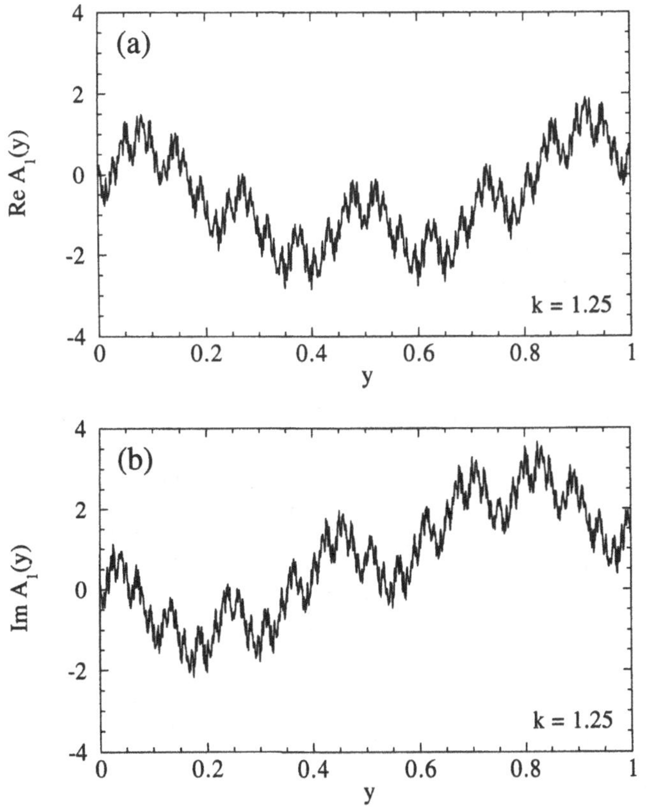

Figure 7.12 The generalized function $A_1(y)$ versus y at the wavenumber $k = 1.25$ which is also in the region where the deRham function $A_0(y)$ is not defined: (a) real part; (b) imaginary part (Gaspard 1993a).

having the role to transform the distributions $|\psi\rangle$ into regular functions $\langle I_x^{(i)} I_y^{(j)}|\psi\rangle$ of x and y if i and j are large enough depending on the singularity of the distribution.

We now apply the Frobenius–Perron operator (7.125) of the multibaker map to a root vector of the form (7.188) in order to obtain separated equations for

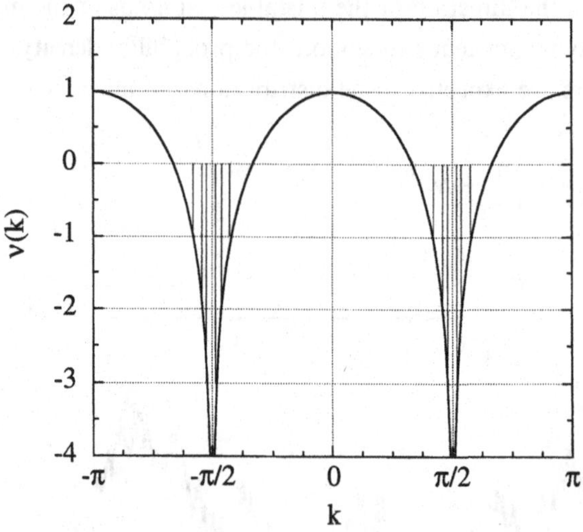

Figure 7.13 Hölder exponent of the function $A_0(y)$, $\nu(k) = -\ln|\alpha(k)|/\ln 2 = 1 + (\ln|\cos k|/\ln 2)$, as a function of the wavenumber k.

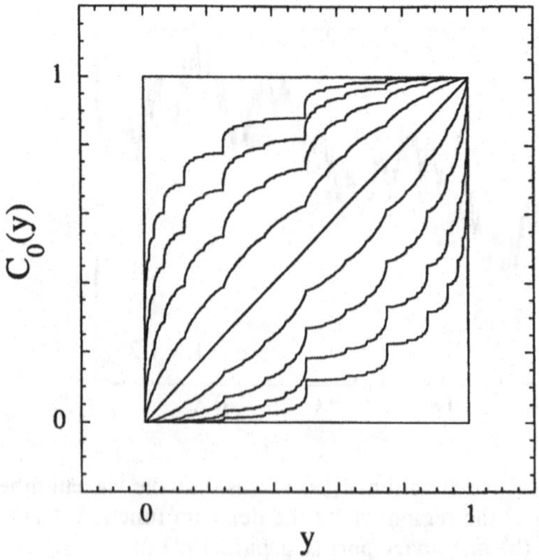

Figure 7.14 The function $C_0(y)$ versus y for different values of the parameter η. When $\eta = -10$, the function is nearly equal to zero whereas it is nearly equal to one when $\eta = 10$. In between, the function is plotted for $\eta = -2, -1.5, -1, -0.5, 0, 0.5, 1, 1.5,$ and 2 (Gaspard 1993a).

the different terms

$$(\hat{Q}_k^+ \psi)(x,y) = \sum_{\nu=0}^{N} x^\nu \left[e^{ik}\, \theta\left(\frac{1}{2}-y\right) \frac{\mathscr{D}^{(\nu)}(2y)}{2^\nu} + e^{-ik}\, \theta\left(y-\frac{1}{2}\right) \sum_{\mu=\nu}^{N} \binom{\mu}{\nu} \frac{\mathscr{D}^{(\mu)}(2y-1)}{2^\mu} \right] . \qquad (7.190)$$

Moreover, we observe that a polynomial of degree N in x is mapped onto another polynomial of degree N in x, a property shared by the operator $\hat{R}_k$ of the preceding Subsection 7.5.4.

However, if $\psi(x,y)$ was a continuous function of y, $(\hat{Q}_k^{+n}\psi)$ would be a discontinuous function of y with discontinuities at $y = j/2^n$ $(j = 1, 2, \ldots, 2^n - 1)$. To avoid this difficulty, we assume that $\mathscr{D}^{(\nu)}(y)$ are the kernels of some distributions as suggested in particular by the above discussion. Our purpose now is to determine these distributions.

The eigenstates

We first look for the eigenstates (7.138) which we assume of the form $\psi_{m,0}(x,y;k) = \mathscr{D}_{m,0}^{(0)}(y)$ independent of x. By Eq. (7.190), these eigendistributions are solutions of the equation

$$e^{ik}\, \theta\left(\frac{1}{2}-y\right) \mathscr{D}_{m,0}^{(0)}(2y) + e^{-ik}\, \theta\left(y-\frac{1}{2}\right) \mathscr{D}_{m,0}^{(0)}(2y-1) = \chi_m\, \mathscr{D}_{m,0}^{(0)}(y) . \qquad (7.191)$$

We multiply both members by a test function $g^*(y)$, we integrate from $y = 0$ to $y = 1$, and we take the complex conjugate. In this way, we obtain the eigenvalue equation for the operator $\hat{R}_k$ of the previous section

$$\langle \mathscr{D}_{m,0}^{(0)} | \hat{R}_k | g \rangle = \chi_m \langle \mathscr{D}_{m,0}^{(0)} | g \rangle , \qquad (7.192)$$

which should hold for an arbitrary test function $g(y)$. As a consequence, the distribution $\mathscr{D}_{m,0}^{(0)}$ may be identified with the left eigendistribution (7.178) of the Frobenius–Perron operator of the one-dimensional map, which corresponds to the eigenvalue χ_m:

$$\mathscr{D}_{m,0}^{(0)}(y) = \tilde{\Psi}_m(y;k) = (-\partial_y)^m \tilde{\Psi}_0(y;k) . \qquad (7.193)$$

According to Eq. (7.175), the eigenvectors of the multibaker are thus obtained as (Gaspard 1993a)

$$\psi_{m,0}(x,y;k) = \mathscr{D}_{m,0}^{(0)}(x) = \tilde{\Psi}_m(y;k) = (-\partial_y)^m \tilde{\Psi}_0(y;k) = \sum_{n=m}^{\infty} c_{n-m}^*(k)\, (-)^n \delta^{(n)}(y) . \qquad (7.194)$$

These eigendistributions can also be represented in different forms as shown previously. In particular, they can be represented in terms of the generalized deRham functions when these distributions are applied to the functions of the type of (7.189)

$$\langle I_x^{(i)} I_y^{(j)} | \psi_{m,0} \rangle = \frac{x^{i+1}}{(i+1)!} \times (-\partial_y)^m A_j(y) , \tag{7.195}$$

where $A_j(y)$ is obtained by the deRham-type iteration (7.179) (Gaspard 1995). For $i = j = 0$, the above solution gives the cumulative function of the eigendistribution when it exists, i.e., for $|k| < \pi/3$ and $|k - \pi| < \pi/3$. Otherwise, for $\pi/3 < |k| < \arccos(1/4)$ and $\pi/3 < |k - \pi| < \arccos(1/4)$, the object $A_0(y)$ does not exist as a function so that we need to integrate it to get the function $A_1(y)$, and so on as $k = \pm\pi/2$ is approached.

The first nontrivial root states

Above, we have obtained the eigenstates of the multibaker under the assumption that these root states are independent of x. Now, we turn to the first right root states (7.139) which satisfy

$$\hat{Q}_k^+ \psi_{m,1} = \chi_m \psi_{m,1} + \psi_{m,0} , \tag{7.196}$$

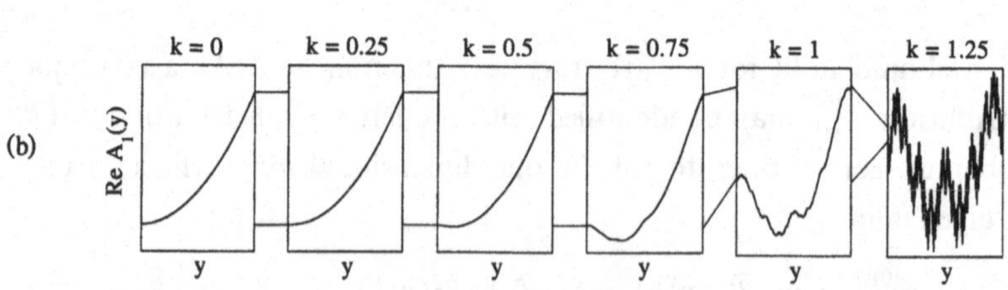

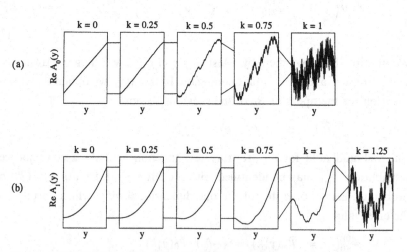

Figure 7.15 (a) Real parts of the cumulative functions $A_0(y;k) = \int_0^1 dx' \int_0^y dy' \psi_{0,0}(x', y'; k)$ of the hydrodynamic modes of diffusion of the multibaker, i.e., of the right eigenstates $\psi_{0,0}(x, y; k)$ corresponding to the Pollicott–Ruelle resonances s_k of the dispersion relation of diffusion. These functions only exist for $|k| < \pi/3$ and $|k - \pi| < \pi/3$. (b) Real parts of the doubly integrated eigendistributions: $A_1(y;k) = \int_0^1 dx' \int_0^y dy'(y - y')\psi_{0,0}(x', y'; k)$, which now exist as functions for $|k| < \arccos(1/4)$ and $|k - \pi| < \arccos(1/4)$ (Gaspard 1995, 1997a).

and, for them, we assume a distribution which is linear in x

$$\psi_{m,1}(x,y;k) \;=\; \mathscr{D}^{(0)}_{m,1}(y) \;+\; x\,\mathscr{D}^{(1)}_{m,1}(y)\,. \tag{7.197}$$

We introduce this distribution in Eq. (7.190) and we identify the coefficients of the terms in x^0 and in x^1 whereupon we get two equations for the two unknown distribution kernels, $\mathscr{D}^{(0)}(x)$ and $\mathscr{D}^{(1)}(x)$,

$$e^{ik}\,\theta\left(\frac{1}{2}-y\right)\,\mathscr{D}^{(1)}_{m,1}(2y) \;+\; e^{-ik}\,\theta\left(y-\frac{1}{2}\right)\,\mathscr{D}^{(1)}_{m,1}(2y-1) \;=\; \chi_{m-1}\,\mathscr{D}^{(1)}_{m,1}(y)\,, \tag{7.198}$$

and

$$e^{ik}\,\theta\left(\frac{1}{2}-y\right)\,\mathscr{D}^{(0)}_{m,1}(2y) + e^{-ik}\,\theta\left(y-\frac{1}{2}\right)\left[\mathscr{D}^{(0)}_{m,1}(2y-1) \;+\; \frac{1}{2}\,\mathscr{D}^{(1)}_{m,1}(2y-1)\right] \;=\; \mathscr{D}^{(0)}_{m,0}(y) \;+\; \chi_m\,\mathscr{D}^{(0)}_{m,1}(y)\,. \tag{7.199}$$

We solve these equations as before using a test function $g(y)$. Equation (7.198) is equivalent to an eigenvalue equation for the operator $\hat{R}_k$ where the eigenvalue is now χ_{m-1} instead of χ_m so that the solution is

$$\mathscr{D}^{(1)}_{m,1}(y) \;=\; \xi\,\tilde{\Psi}_{m-1}(y;k)\,, \qquad \text{with} \quad m \geq 1\,, \tag{7.200}$$

and where ξ is a constant to be fixed in particular by biorthonormality.

Similarly, Eq. (7.199) can be rewritten as

$$\langle \mathscr{D}^{(0)}_{m,1}|\hat{R}_k|g\rangle \;+\; \xi^*\,\frac{e^{ik}}{4}\int_0^1 \tilde{\Psi}^*_{m-1}(y;k)\,g\left(\frac{y+1}{2}\right)\,dy \;=\; \langle\tilde{\Psi}_m|g\rangle \;+\; \chi_m\,\langle\mathscr{D}^{(0)}_{m,1}|g\rangle\,. \tag{7.201}$$

Using the expansion (7.175) of the distributions $\tilde{\Psi}_m$ and assuming an expansion of the type

$$\mathscr{D}^{(0)}_{m,1} \;=\; X_0\,\tilde{\Psi}_{m-1} \;+\; X_1\,\tilde{\Psi}_m \;+\; X_2\,\tilde{\Psi}_{m+1} \;+\; X_3\,\tilde{\Psi}_{m+2} \;+\; \dots\,, \tag{7.202}$$

we can obtain a representation of the distribution $\mathscr{D}^{(0)}_{m,1}$ as a series of derivatives of the Dirac distribution at $y = 0$ in which all the constants X_j are fixed (Gaspard 1993a).

Representing the distributions in terms of Riemann-Stieljes integrals as in Eq. (7.178), we can finally obtain the first root states as (Gaspard 1995)

$$\langle I^{(i)}_x I^{(j)}_y|\psi_{m,1}\rangle \;=\; 2^{m+2}\cos k\,\frac{x^{i+1}}{(i+1)!} \times (-\partial_y)^m\left[e^{-2ik}\,B_j(y) \;-\; \frac{2x}{i+2}\,A_{j+1}(y)\right]\,, \tag{7.203}$$

in terms of further deRham-type functions $B_j(y)$ such that $B_{j-1} = dB_j/dy$. The first of these functions is a solution of another deRham-type equation

$$B_0(y) = \begin{cases} \alpha\, B_0(2y) - \alpha^2 A_0(y)\,, & \text{for} \quad 0 \le y \le \frac{1}{2}\,, \\ (1-\alpha)\, B_0(2y-1) - \alpha\, A_1(2y-1) - \alpha^2 A_0(y) & \\ + 4\alpha^2 \left(y - \frac{1}{2}\right) + \alpha\, B_0(1)\,, & \\ & \text{for} \quad \frac{1}{2} < y \le 1\,, \end{cases} \tag{7.204}$$

with $A_0(y)$ and $A_1(y)$ defined by the iteration (7.179). Figure 7.16 shows the real and imaginary parts of the eigenstates and first root states when $k = 0.75$ in a unit cell (x, y) of the multibaker showing the highly irregular character of these functions along the y-direction of the stable manifolds.

Further root states can be constructed recursively with the same method.

We conclude from this analysis that the root states are, in general, distributions which are smooth functions along the unstable direction x but singular along the stable direction y. These distributions are defined for test functions which are entire functions of exponential type $\tau < \rho_k$ [the Fréchet space $\mathscr{E}_y(\tau;k)$] in the stable direction y and just integrable in the unstable direction x. Since the multibaker is time-reversal symmetric the left root states may be obtained by applying the time-reversal transformation (6.48). Accordingly, the left root states $\tilde{\psi}_{m,i}$ are smooth in the stable direction y and singular in the unstable direction x. This is consistent with the biorthonormality of the left and right root states according to which the smoothness in one direction should be high enough to sustain the singular character in the other direction because of mutual application in the biorthonormality. We may therefore conclude that the Frobenius–Perron operator of the forward semigroup in the decomposition (7.136) is an operator acting on the space $\mathscr{E}_x^{\dagger}(\tau;k) \otimes \mathscr{E}_y(\tau;k)$, where $\mathscr{E}^{\dagger}(\tau;k)$ denotes the dual of the space $\mathscr{E}(\tau;k)$ of entire analytic functions of exponential type, i.e., the space of (anti-)linear functionals on $\mathscr{E}(\tau;k)$. The above analysis reduces to the construction of the spectral decomposition of the baker map when the wavenumber vanishes ($k = 0$) (Antoniou and Tasaki 1992, 1993; Fox 1995, 1997).

The spectral decomposition in terms of the Pollicott–Ruelle resonances can therefore be carried out in full detail in such models. In this way, we can obtain the hydrodynamic modes of diffusion of the multibaker. We observe that, except for the leading eigenstate at $k = 0$, the eigenstates are given as distributions or linear functionals but certainly not as functions. If the eigenfunction assumption is already invalid for such a simple model as the multibaker we may expect that this result extends to more general and more complex systems. When the number of degrees of freedom increases the eigenstates become so complex that they can be approximated by eigenfunctions again, as done with Boltzmann's equation in particular. But such an approximation amounts to ignoring the deterministic point-like character of the motion of the fluid particles.

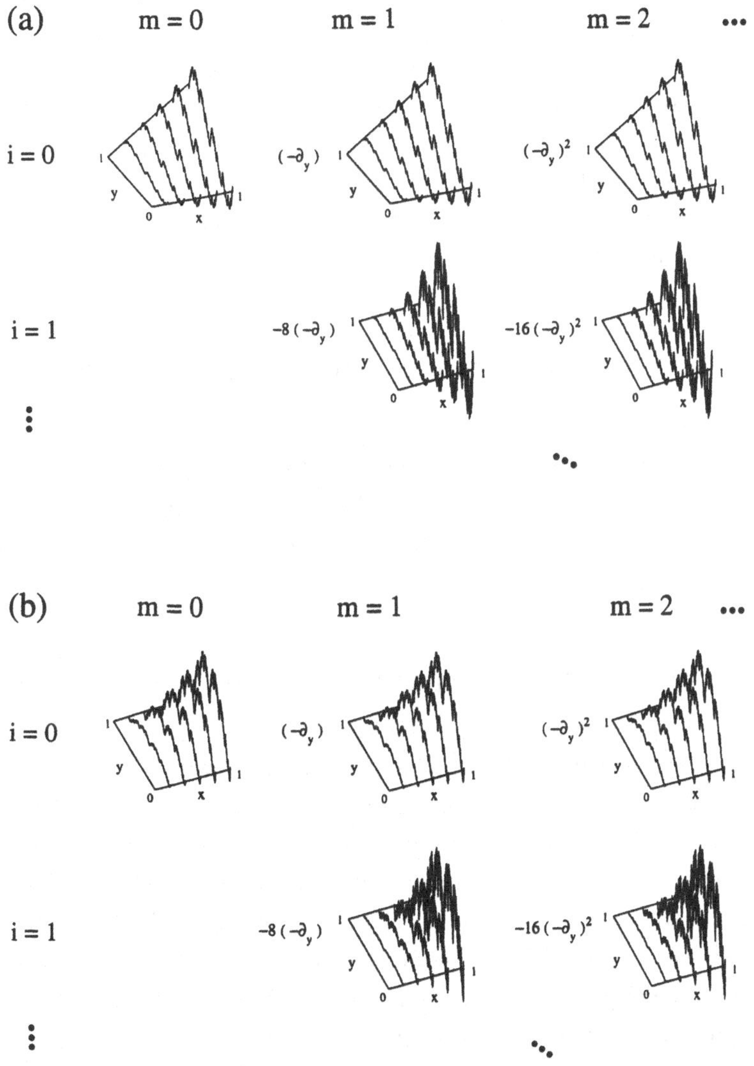

Figure 7.16 (a) Real and (b) imaginary parts of the root vectors, $\langle I_x^{(0)} I_y^{(0)} | \psi_{m,i} \rangle$, of the multibaker for a wavenumber $k = 0.75$ and depicted in a unit cell (x, y) (Gaspard 1995).

7.6 Extensions to the other transport processes

A natural question we may ponder is whether the above dynamical systems theory of diffusion can extend to the other transport processes. In view of the fact that the transport properties are intrinsic to the time evolution of a certain dynamical system we may expect that such extensions are possible. There is probably a difficulty here that the wavenumbers introduced by the quasiperiodic boundary conditions are too numerous to find an immediate physical interpretation as the wavenumber of some collective hydrodynamic mode. Nevertheless, the periodic boundary condition is of importance for numerical simulations of fluids and for the direct calculation of transport properties. Many works have been devoted to this problem and, recently, Bunimovich and Spohn (1996) have considered viscosity in a two-particle fluid from the viewpoint of dynamical systems theory. In this context, the result has been obtained that the viscosity coefficient of the two-particle fluid already has the main features expected for large N-particle systems.

7.6.1 Gaussian fluctuations of the Helfand moments

The formulation of hydrodynamic laws starts from the observation that five quantities are conserved in fluids composed of identical particles: the total mass M, the total momentum $\mathbf{P}$, and the total energy E (Balescu 1975, Résibois and De Leener 1977, Boon and Yip 1980). The conservation laws are then expressed locally in terms of the observables

$$\begin{pmatrix} \rho(\mathbf{r};t) \\ \mathbf{g}(\mathbf{r};t) \\ e(\mathbf{r};t) \end{pmatrix} = \sum_{\mathbf{L}\in\ell} \sum_{i=1}^{N} \begin{pmatrix} 1 \\ \mathbf{p}_i(t) \\ E_i(t) \end{pmatrix} \delta[\mathbf{r} - \mathbf{q}_i(t) - \mathbf{L}] \, , \tag{7.205}$$

which are the densities of mass, momentum, and energy, which involve the momenta $\mathbf{p}_i$ and the energy E_i of each particle i. Here, we consider a fluid defined on a torus with periodic boundary conditions as usually assumed for computer simulations. The periodic boundary conditions are imposed by introducing a lattice ℓ in the physical position space $\mathbb{R}^d$: The vectors $\mathbf{L}$ belong to the lattice ℓ. The observables (7.205) are functions of the phase-space variables $\{\mathbf{q}_i, \mathbf{p}_i\}_{i=1}^N$ and they depend parametrically on the position $\mathbf{r}$. The observables are periodic in the physical position $\mathbf{r}$ and in the particle positions $\{\mathbf{q}_i\}_{i=1}^N$.

Balance equations are derived for the above densities by using Newton's equation of motion for the fluid particles, which leads to the identification of the transport coefficients in terms of the Green–Kubo relations involving the thermodynamic fluxes like that associated with viscosity:

$$J_{\alpha\beta}(t) = \frac{1}{\sqrt{Vk_B T}} \left\{ \sum_{i=1}^{N} \frac{1}{m} p_{i\alpha}(t) p_{i\beta}(t) + \frac{1}{2} \sum_{i \neq j = 1}^{N} [q_{j\alpha}(t) - q_{i\alpha}(t)]\, F_\beta[\mathbf{q}_j(t) - \mathbf{q}_i(t)] \right\}, \tag{7.206}$$

where $F_\alpha(\mathbf{r})$ is the periodic force acting on a particle at $\mathbf{r}$ due to a particle at the origin. A very important point in the derivation of such an expression for a fluid on a torus is that $\mathbf{q}_j - \mathbf{q}_i$ is the shortest distance on the torus, which is called the minimum image convention (Evans and Morriss 1990b, Bunimovich and Spohn 1996).

If the torus is unfolded into a spatially extended fluid, each particle is associated with an infinity of images forming a lattice. All these images move in synchrony following the motion of some reference particle of the lattice. In such interacting lattices, the images separated by the shortest mutual distance change during the time evolution according to the minimum image convention. At each such change, the positions used to locate the particles in Eq. (7.206) should be reset so that the relation between momentum and velocity is therefore modified. This modification explains how the centre of momenta of the particles can be transported in such a system.

Indeed, the transport coefficients may be expressed in terms of the Helfand moments instead of the thermodynamic fluxes as we explained in Chapter 6. The Helfand moment associated with viscosity is proportional to the centre of momenta of the particles. In systems with periodic boundary conditions, the Helfand moments have expressions which differ from those of Table 6.3 as a consequence of the minimum image convention. If the lattice ℓ is cubic the positions undergo jumps by L at times t_n, so that their derivatives are

$$\frac{dq_{i\alpha}}{dt} = \frac{p_{i\alpha}}{m} - L \sum_{n=-\infty}^{+\infty} [\operatorname{sgn} p_{i\alpha}(t_n)]\, \delta(t - t_n), \tag{7.207}$$

where the extra terms are due to the above explained modification. The Helfand moments must therefore be defined according to

$$G_{\alpha\beta}(t) = \frac{1}{\sqrt{Vk_B T}} \left\{ \sum_{i=1}^{N} q_{i\alpha}(t) p_{i\beta}(t) + L \sum_{n=-\infty}^{+\infty} [\operatorname{sgn} p_{i\alpha}(t_n)]\, p_{i\beta}(t_n)\, \theta(t - t_n) \right\}, \tag{7.208}$$

in order for the thermodynamic fluxes (7.206) to be obtained by differentiation: $J_{\alpha\beta} = (d/dt) G_{\alpha\beta}$. In systems with periodic boundary conditions, the fluxes

(7.206) have nonvanishing mean values over the equilibrium measure, $\langle J_{\alpha\beta}\rangle_{\rm e} = \delta_{\alpha\beta}PV/\sqrt{Vk_{\rm B}T}$, where P is a scalar called the dynamical pressure. The viscosity coefficient is obtained from the linear growth of the variance of the Helfand moment, i.e., from the dynamical fluctuations of the Helfand moment around their mean value. This suggests that

$$G_{\alpha\beta}(t) - \langle G_{\alpha\beta}(t)\rangle_{\rm e} \,, \tag{7.209}$$

obeys a central limit theorem and is a random variable with a Gaussian distribution in the limit $t \to \infty$ (cf. Subsection 3.5.5). The viscosity coefficients would be given by the generalized diffusion coefficient of the central limit theorem. This result was proved by Bunimovich and Spohn (1996) for a system of two disks on a torus. This is a way dynamical systems theory may contribute to nonequilibrium statistical mechanics.

7.6.2 The minimal models of transport

The work of Bunimovich and Spohn (1996) is focused on the study of a minimal model for viscosity. We may indeed wonder what is the smallest system in which the variance of the Helfand moment does not vanish, i.e., in which the corresponding transport property is not trivialized. The above sections on deterministic diffusion show that one particle is enough to define the property of diffusion, as illustrated with the Lorentz gases.

In a system with two particles with periodic boundary conditions, the centre of mass is in uniform motion while the relative motion involves only one fictitious particle. We may introduce the coordinates of the centre of mass and the relative coordinates as

$$\begin{cases} \mathbf{q} = \mathbf{q}_1 - \mathbf{q}_2 \,, & \mathbf{p} = \frac{1}{2}(\mathbf{p}_1 - \mathbf{p}_2) \,, \\ \mathbf{Q} = \frac{1}{2}(\mathbf{q}_1 + \mathbf{q}_2) \,, & \mathbf{P} = \mathbf{p}_1 + \mathbf{p}_2 \,. \end{cases} \tag{7.210}$$

In the reference frame of the centre of mass where $\mathbf{P} = 0$, the thermodynamic fluxes (7.206) become

$$J_{\alpha\beta} = \frac{1}{\sqrt{Vk_{\rm B}T}} \left(\frac{2}{m} p_\alpha p_\beta + q_\alpha F_\beta \right) , \tag{7.211}$$

which already leads to nontrivial fluctuations. However, the thermodynamic flux associated with thermal conduction vanishes identically in a two-particle fluid by conservation of momentum so that thermal conductivity does not exist in a two-particle system. It turns out that at least three particles are required for thermal conductivity to exist. In summary, a certain transport process is sustained by a minimum number of particles according to

$$\begin{aligned} \text{diffusion} &: \quad 1 \text{ particle} \,, \\ \text{viscosity} &: \quad 2 \text{ particles} \,, \\ \text{thermal conductivity} &: \quad 3 \text{ particles} \,. \end{aligned} \tag{7.212}$$

This minimal number of particles is related to the fact that the Helfand moment for diffusion is a quantity with only one canonical variable, the one for viscosity is a sum of quantities with two canonical variables, while the one for thermal conductivity is a sum of quantities with three canonical variables (cf. Table 6.3). In each case, a central limit theorem leads to the definition of the transport coefficient.

7.6.3 Viscosity and self-diffusion in two-particle fluids

In the case of a two-disk system, different lattices may be considered in order to impose the periodic boundary conditions. Bunimovich and Spohn (1996) have studied a square lattice but the triangular lattice is also interesting for the property of self-diffusion. Differences appear in the dynamics of the relative motion in these two lattices (cf. Fig. 7.17).

In the case of a *square lattice*, the relative motion is a Sinai billiard (Sinai 1970) in which a fictitious point particle moves in a square domain of side L with periodic boundary conditions forming a torus and undergoes elastic collisions

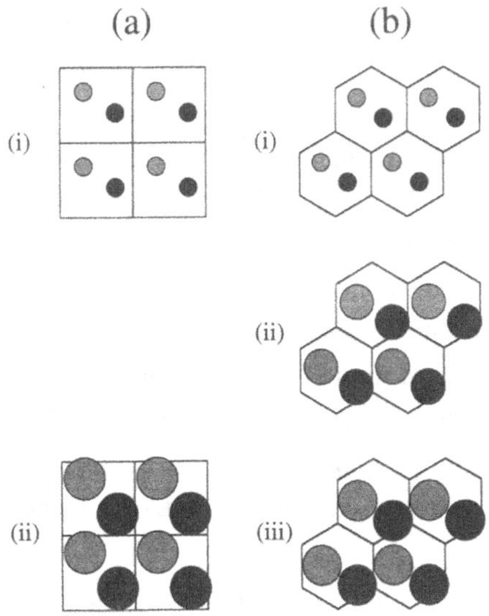

Figure 7.17 Different regimes in the two-disk system (a) on a square lattice and (b) on a triangular lattice.

on a central disk of radius equal to the diameter d of the particles. Two regimes can be distinguished:

(i) At low densities, when the diameter of the particles is small, $d < L/2$, the fictitious point may travel in free motion along special orbits of the Sinai billiard. In this infinite-horizon regime, the central limit theorem does not apply and the viscosity is infinite. Similarly, the self-diffusion coefficient is infinite.

(ii) On the other hand, at high densities for $d > L/2$, the central disk of radius d overlaps with its images and delimits a closed domain in which the fictitious particle is localized. The horizon is finite and the central limit theorem applies to viscosity, yielding a finite and positive shear viscosity coefficient. However, the relative motion remains bounded so that the diffusion coefficient is zero.

An intermediate regime appears in the *triangular lattice.* Here, the relative motion corresponds to the dynamics of a point particle in a Lorentz gas on a triangular lattice of disks of radius equal to the diameter d of the particles with a distance R between the closest disks. The fundamental cell of this billiard is a hexagon with periodic boundary conditions forming a torus. According to the description of this Lorentz gas in Chapter 6, three regimes can be distinguished (cf. Fig. 6.4):

(i) At low densities, when the diameter of the particles is small, $d < R\sqrt{3}/4$, the fictitious point may travel in free motion along special orbits. In this infinite-horizon regime, both the viscosity and the self-diffusion coefficients are infinite.

(ii) At intermediate densities for $(R\sqrt{3}/4) < d < (R/2)$, the horizon is finite but the relative motion is unbounded so that both the viscosity and the self-diffusion coefficients may be finite and positive.

(iii) At high densities for $R < 2d$, the disk overlaps with its images so that the relative motion becomes bounded. Here, the viscosity coefficient is finite and positive but the self-diffusion coefficient vanishes.

Therefore, the two-disk system on a triangular lattice may already simulate a fluid with viscosity and self-diffusion, which is not the case on a square lattice. On the other hand, the square lattice illustrates a system with viscosity but no self-diffusion as may be the case in some solids.

An apparent paradox arises in the way a transport property like viscosity may exist without transport in the relative motion. The resolution of this paradox has been explained above, that a fluid with periodic boundary conditions is a particular system in which the minimum image convention modifies the usual relation between momentum and velocity, creating a nontrivial transport by viscosity, [cf. Eq. (7.208)].

Viscosity has also been considered in another two-particle fluid on a square lattice, where the particles interact by a periodic attractive Yukawa potential (Aumaître 1995). Such attractive Coulomb-type systems with two degrees of freedom are known to be strictly hyperbolic at high enough energy (Knauf 1987). Diffusion has already been studied in such systems and a corresponding central limit theorem was proved by Knauf (1987, 1989). As far as viscosity is concerned, a similar result is expected, that the Helfand moment (7.209) associated with the fluxes (7.206) is asymptotically distributed as a Gaussian random variable. Here, there is always a deflection of the particles because the Yukawa potential has a nonvanishing spatial extension in contrast with the hard-disk potential. Therefore, we should expect normal viscosity and self-diffusion even for a square lattice.

7.6.4 Spontaneous symmetry breaking and Goldstone hydrodynamic modes

It is known that the spontaneous breaking of continuous symmetries introduces extra hydrodynamic modes with vanishing frequency and relaxation rate as the wavenumber becomes very large. This result is known as the Goldstone theorem in the quantum mechanics of many-body systems (Anderson 1984). It is usually considered for the energy eigenvalues of a quantum system. Its extension to statistical mechanics and specially to classical statistical mechanics has been considered in the literature (Forster 1975) but it faces difficulties. In particular, the concept of an energy eigenvalue is no longer adequate in systems at finite temperatures. In this regard, the concept of quasiparticles has been introduced by Landau in order to take into account the possibility of a finite lifetime for the excitations of a many-body system. Such a concept finds a natural expression in the context of the Liouvillian dynamics thanks to the resonances by Pollicott and Ruelle. In this framework, we may reconsider the Goldstone theorem according to which we expect that the spontaneous breaking of continuous symmetries should introduce extra hydrodynamic modes with Pollicott–Ruelle resonances vanishing with their wavenumber.

A dynamical system which may illustrate such a phenomenon is the hard-sphere gas at high densities near close packing. It is known that the hard-sphere gas at high densities presents a solid-type phase and, moreover, that new transverse acoustic modes can appear at the fluid–solid transition, which are the Goldstone hydrodynamic modes. On the other hand, the phase space separates into many different ergodic components near close packing at which point the spheres can no longer move with respect to one another.[4] Therefore, we see

4. In the solid phase, the number of ergodic components is of the order of $N!$.

that the phenomenon of spontaneous symmetry breaking appears especially challenging to consider in dynamical systems theory.

7.7 Chemio-hydrodynamic modes

Systems with reacting chemical species may present a large variety of different modes of relaxation besides the hydrodynamic modes. Because of the chemical reactions, the numbers of particles of the different chemical species are not conserved so that we should not expect that the corresponding modes have relaxation rates which vanish with the wavenumber. Nevertheless, the relaxation rates may be small enough for these modes to be clearly distinct from the faster kinetic modes of hydrodynamics. These questions become particularly important in reaction–diffusion systems with auto- or cross-catalytic reactions which undergo self-organization phenomena far from equilibrium and sustain dissipative structures in space and time such as nonequilibrium patterns (Nicolis and Prigogine 1977). Even in systems with a linear chemistry (without catalysis), the question of hydrodynamic modes is very interesting to investigate in the framework of the theory presented in this chapter. We would like to sketch here a simple model of the type of the Lorentz gas where this question may be investigated.

We consider a Lorentz gas on a periodic triangular lattice in which a point particle undergoes elastic collision. In order to model a chemical reaction like

$$A \leftrightarrow B \,, \tag{7.213}$$

the point particle of the Lorentz gas is allowed to move on 'both sides of the plane'. If the particle is on the upper side it is red (A) and it is green (B) on the lower side. In order to allow transitions between both sides of the plane, two kinds of disks are considered. Most of the disks are simple elastic scatterers at which the particle stays on the same side. Some of the disks are catalysts: the particle changes colour (i.e., side) at collision (see Fig. 7.18). The catalytic disks are like gates from one side to the other. These gates are distributed randomly on the lattice or according to a superlattice. We suppose that one out of N disks is a catalyst on average. Except for the change of colour at the gates, the trajectories are identical with those of the usual Lorentz gas so that energy is conserved and the amplitude of the velocity is constant. Therefore, the diffusion coefficient will be given by the formula (6.34) of Machta and Zwanzig (1983) in the limit $r \to 2a$, where a is the radius of the disks and r the distance between their centres. Several independent particles of both species A and B may be poured in such a system.

The reaction rate can be estimated near equilibrium as follows. We observe that the reaction process is controlled by diffusion. Accordingly, the reaction rate is inversely proportional to the average time for a particle to diffuse toward a catalytic disk. For the described two-dimensional model, we find that the reaction rate at equilibrium is given in the limits $N \to \infty$ and $r \to 2a$ by

$$\kappa^{\text{eq}} \simeq \frac{C\, D_{\text{th}}}{r^2\, N \ln N}, \tag{7.214}$$

where D_{th} is the diffusion coefficient (6.34) and C is a dimensionless quantity depending on the geometry of the system. The expected phenomenological reaction–diffusion equations for the densities ρ_A and ρ_B of species A and B are

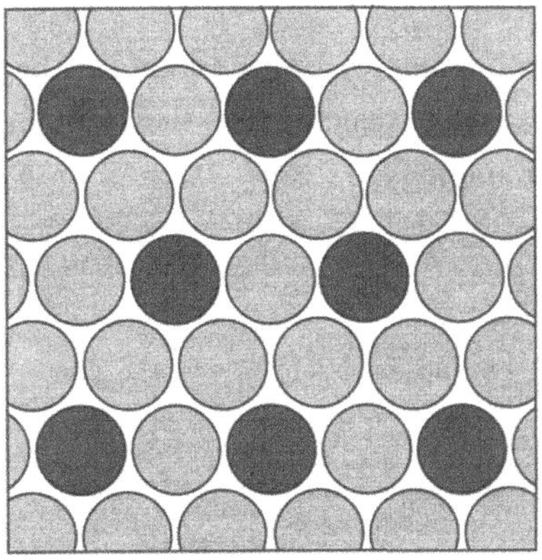

Figure 7.18 Model of a Lorentz gas on a triangular lattice with change of colours at a few catalytic disks.

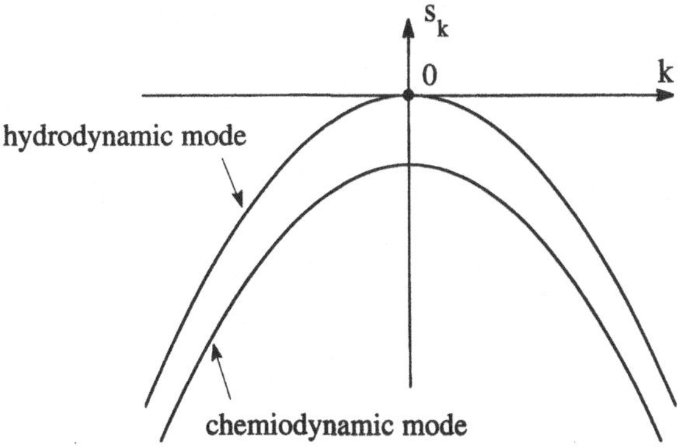

Figure 7.19 The dispersion relations of the chemio-hydrodynamic modes of the reaction–diffusion system (7.215).

often supposed to be

$$\begin{aligned} \partial_t \rho_A &= D \nabla^2 \rho_A + \kappa^{eq} (\rho_B - \rho_A) , \\ \partial_t \rho_B &= D \nabla^2 \rho_B + \kappa^{eq} (\rho_A - \rho_B) . \end{aligned} \tag{7.215}$$

This system admits two different solutions of the type

$$\rho_A(\mathbf{r},t) , \rho_B(\mathbf{r},t) = \exp(i\mathbf{k} \cdot \mathbf{r}) \exp(st) , \tag{7.216}$$

with the dispersion relations

$$\text{hydrodynamic mode of diffusion} : \quad s_{\mathbf{k}} = - D k^2 , \tag{7.217}$$

$$\text{chemiodynamic mode of reaction} : \quad s_{\mathbf{k}} = - 2 \kappa^{eq} - D k^2 . \tag{7.218}$$

These relations are depicted in Fig. 7.19. The mode of diffusion is due to the conservation of the total number of particles. In the uniform limit $\mathbf{k} = 0$, the difference in densities $\Delta\rho = \rho_A - \rho_B$ relaxes exponentially according to the phenomenological equations, $\Delta\dot{\rho} = -2\kappa^{eq}\Delta\rho$, which is a dynamics expected for nonconserved quantities.

The formulation in terms of the Frobenius–Perron operator can be used to study in detail the microscopic phase-space structure of the chemiodynamic modes and confirms or not the phenomenology.

Chapter 8

Systems maintained out of equilibrium

8.1 Nonequilibrium systems in Liouvillian dynamics

Most systems in nature are maintained out of equilibrium either by incident fluxes of particles or by external fields. The earth bathed by sunlight[1] is an illustration of such out-of-equilibrium systems. From this viewpoint, the systems may be considered as subjected to some scattering processes, which leads us to the scattering theory of transport of Chapter 6. In this context, the fact that most classical scattering processes are chaotic has important consequences in our understanding of nonequilibrium states and the methods of the previous chapters are thus required for the investigation of out-of-equilibrium systems.

Works on out-of-equilibrium systems have revealed that such systems remain in a thermodynamic state which is the continuation of the equilibrium state under weak nonequilibrium constraints. Beyond a certain threshold, the thermodynamic branch becomes unstable and new states emerge by bifurcation with spatial or temporal inhomogeneities, called dissipative structures (Prigogine 1961; Glansdorff and Prigogine 1971; Nicolis and Prigogine 1977, 1989). Turing structures in reaction–diffusion systems and convection rolls in fluids are examples of such nonequilibrium structures (DeWit et al. 1992, 1993, 1996; Cross and Hohenberg 1993). The transitions to dissipative structures appear sharp from a macroscopic viewpoint which ignores the thermodynamic fluctuations due to the atomic structure of matter. These fluctuations can be modelled by

1. without talking about meteoritic showers

stochastic dynamical systems like Langevin processes, birth-and-death processes, or lattice-gas automata, which show that transitions may be rounded in systems with finitely many particles (Nicolis and Malek Mansour 1978, Malek Mansour et al. 1981, Dab et al. 1991, Lawniczak et al. 1991, Kapral et al. 1992, Baras and Malek Mansour 1997). However, there is no reason for not considering such phenomena from a Liouvillian point of view, in which the dynamics inside the system is ruled by Hamilton's equations while the nonequilibrium constraints are imposed by boundary conditions on the probability density representing a statistical ensemble. This ensemble is made of infinitely many copies of the original system and evolves in time under Liouville's equation as explained in Chapter 3. This is nothing other than the mathematical formulation of the molecular-dynamics simulations of nonequilibrium systems where particles reaching the walls are reinjected according to Maxwellian distributions fixed by the local temperature of the wall. Such simulations have been able to reproduce the famous Rayleigh–Bénard convection in a fluid of hard disks in a gravitational field and a gradient of temperature (Mareschal et al. 1988).

Boundary conditions at the Liouvillian level have been formulated in Chapter 3. We noticed that the invariant measure depends on the specific boundary condition so that the invariant measure under nonequilibrium constraints differs in general from the equilibrium (Liouville) invariant measure. A system with its nonequilibrium invariant measure $\mu_{\rm ne}$ defines a different dynamical system $(\mathbf{\Phi}^t, \mathcal{M}, \mu_{\rm ne})$ for each nonequilibrium boundary condition. All these dynamical systems can be analyzed for the spectral properties of the time evolution operator with the methods of Chapter 3.

In the present chapter, we shall essentially focus on systems under weak nonequilibrium constraints. Nonequilibrium steady states corresponding to a linear profile of concentration are already very interesting and nontrivial in the Liouvillian formulation. We shall construct their associated invariant measure. Thereafter, we shall consider the limit where the nonequilibrium constraints are imposed at large distances and show that the invariant measure becomes singular with respect to the Liouville invariant measure. These steady-state measures turn out to have been introduced by Lebowitz and MacLennan in 1959 and they are related to Zubarev local integrals of motion. Connection is made with Chapter 7. Fick's law can be derived for such steady-state measures. The steady-state measures are explicitly constructed and represented by their cumulative function for the Lorentz gas and the multibaker map. These measures are extended to the other transport processes in terms of Helfand moments. We then argue that the singular character observed in the large system limit is at the origin of the entropy production known in irreversible thermodynamics

(Prigogine 1961, deGroot and Mazur 1962). The argument is developed for deterministic diffusion in the multibaker map. The extension of the Liouvillian theory to far-from-equilibrium systems is discussed.

8.2 Nonequilibrium steady states of diffusion

8.2.1 Phenomenological description of the steady states

For comparison, we suppose that a gradient of concentration is imposed on a system described by the phenomenological diffusion equation (7.1). Flux boundary conditions are imposed at both sides of a slab of material of width L as

$$\rho_t(x = \pm L/2) = \rho_\pm , \tag{8.1}$$

for all times t. The solution of the diffusion equation for this boundary condition is

$$\rho_t(x) = \frac{\rho_+ + \rho_-}{2} + \frac{\rho_+ - \rho_-}{L}\, x + \sum_{j=1}^{\infty} c_j\, e^{-Dk_j^2 t} \sin k_j(x + L/2) , \tag{8.2}$$

with the wavenumbers $k_j = \pi j/L$ and where the constants c_j are fixed by the initial condition. The gradient of concentration here is $\nabla\rho = (\rho_+ - \rho_-)/L$. At long times, all the transients disappear and there remains the linear profile of concentration which is thus a nonequilibrium steady state.

This linear profile can also be obtained from the hydrodynamic modes of diffusion given by Eq. (7.2). For the hydrodynamic modes, the profile of concentration is sinusoidal as $A_k \sin(kx)$. A linear profile is obtained locally near $x = 0$ in the limit where the wavenumber vanishes while the amplitude A_k of the mode is increased as $A_k = (\nabla\rho)/k$ so that $\lim_{k\to 0} A_k \sin(kx) = (\nabla\rho)x$. A more systematic construction of such linear profiles is obtained by taking the derivative of the hydrodynamic mode with respect to the wavenumber, setting the wavenumber to zero, and multiplying by the gradient:

$$\rho_{\nabla\rho}(\mathbf{r}) = -i(\nabla\rho)\cdot\frac{\partial}{\partial\mathbf{k}}\, \rho_{\mathbf{k}}(\mathbf{r},t)\Big|_{\mathbf{k}=0} = (\nabla\rho)\cdot\mathbf{r} , \tag{8.3}$$

which corresponds to a linear profile of concentration across the system in the direction of the gradient $\nabla\rho$. We notice that the constant $(\rho_+ + \rho_-)/2$ should moreover be added to the solution (8.3) which only gives the concentration profile relative to a mean concentration. This explains why the relative concentration (8.3) may be negative.

8.2.2 Deterministic description of the steady states

We now consider a system of deterministic diffusion as described in previous chapters. To be specific, we take the example of the Lorentz gas with a triangular lattice and in the finite-horizon regime where Bunimovich and Sinai (1980) have shown the existence of normal diffusion. A gas of independent point particles is supposed to undergo elastic collisions on the disks of this system.

In Chapter 6, we considered finite scatterers composed of disks arranged in such triangular lattices. In that chapter, we imposed absorbing boundary conditions on the Liouvillian dynamics. We showed that almost all the trajectories escape out of the scatterer and that there remains a fractal repeller $\mathcal{I}$ with stable and unstable manifolds $W_{s,u}(\mathcal{I})$, which controls the dynamics of escape. With absorbing boundary conditions, any initial density will finally vanish under time evolution so that we should not expect the formation of a nonequilibrium steady state.

For this purpose, we should impose a time-continuous flux of trajectories to the scatterer. At both ends of the system, the density of ingoing trajectories is fixed to certain values which are different on both sides of the scatterer (see Fig. 8.1) so that we have the flux boundary condition

$$f_t(\mathbf{X})\Big|_{\mathbf{X}\in\partial_{\text{in}}\mathcal{M}} = \mathcal{F}(\mathbf{X}) . \tag{8.4}$$

The outgoing density will be fixed by causality, i.e., by the time evolution under the dynamics of trajectories in the scatterer. Therefore, the dynamics takes place on the set of trajectories which is complementary to the repeller: $\mathcal{M}\setminus\mathcal{I}$. Ingoing trajectories may either spend a finite time inside the domain $\mathcal{M}$ before exiting, or they may converge to the repeller if they belong to $W_s(\mathcal{I})$. Similarly, the outgoing trajectories may either have previously entered the domain $\mathcal{M}$, or they may belong to the unstable manifold $W_u(\mathcal{I})$, which form so many discontinuities for the outgoing probability density.

To fix the ideas, we consider a scatterer of size L in the x-direction and of infinite extension in the y-direction. Because of energy conservation, we can focus on the energy shell of unit velocity. The phase space is thus (x, y, φ) with $-L/2 < x < +L/2$, $-\infty < y < +\infty$, and the velocity angle $0 \leq \varphi < 2\pi$, at the exclusion of the regions occupied by the disks.

There is a continuous flux of ingoing trajectories with a uniform phase-space density f_- on the left-hand side at $x = -L/2$ and with a uniform density f_+ on the right-hand side at $x = +L/2$ (see Fig. 8.1). This is expressed by the flux boundary condition

$$
\begin{aligned}
f_t(x=-L/2,y,\varphi) &= f_- , &\text{for}\quad -\pi/2<\varphi<+\pi/2 ,\\
f_t(x=+L/2,y,\varphi) &= f_+ , &\text{for}\quad +\pi/2<\varphi<+3\pi/2 ,
\end{aligned}
\qquad (8.5)
$$

for all admissible values of y and at all times t. We emphasize that $f_t(x,y,\varphi)$ is a density in phase space.[2]

After a long enough time, a steady state will establish itself which can be determined as follows. At each point $\mathbf{X}=(x,y,\varphi)$ inside the scatterer, the density is either f_+ or f_-. In order to determine the density, we have to propagate the trajectory backward in time up to the past instant of time $-T(\mathbf{X})$ when the trajectory entered the phase space $\mathcal{M}$. If the trajectory was on the left-hand side $x=-L/2$ at this time the density is f_- at $\mathbf{X}$. On the other hand, the density is f_+ if the trajectory entered at $x=+L/2$. If the point $\mathbf{X}$ belongs to the unstable manifold $W_u(\mathcal{I})$ of the repeller, we cannot decide about the density because the trajectory of $\mathbf{X}$ remains forever between $x=\pm L/2$ under backward time evolution so that we cannot determine the side from which it comes. This

2. The concentration in Eqs. (8.1)–(8.2) is a density in position space only, which is given by $\rho_t(x,y)=\int_0^{2\pi} f_t(x,y,\varphi)d\varphi$. The relation between $f_\pm$ and $\rho_\pm$ can thus only be calculated after solving the problem and obtaining the outgoing densities.

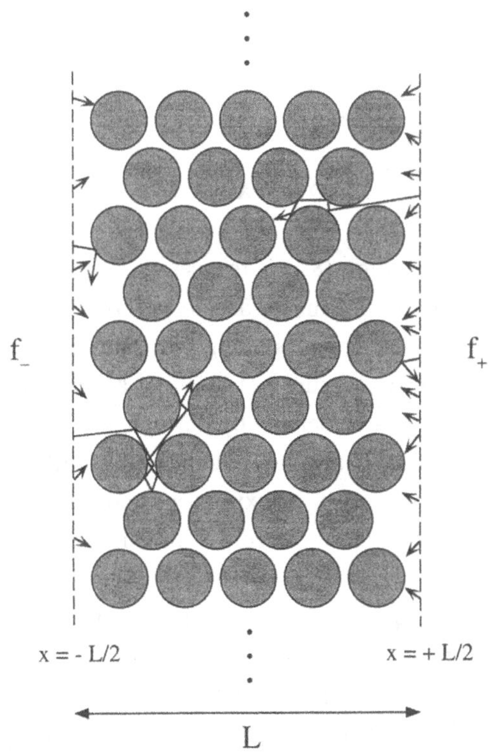

Figure 8.1 Schematic representation of a finite Lorentz gas under flux boundary conditions. The finite Lorentz gas is composed of a slab of width L cut out of the infinite Lorentz gas. This slab of disks forms a scatterer for a gas of independent particles arriving at the left-hand side with a density f_- and at the right-hand side with a density f_+, creating a gradient of density.

process can be visualized by assigning colours (red or green) to the particles coming from the left- or the right-hand sides. If we assume that the motion is free outside the scatterer so that the trajectories are straight lines from infinity before their last collision, the density $f_\pm$ may be decided whether $x \to \pm\infty$ under the full backward time evolution. The invariant density is thus

$$f_{\rm i}(\mathbf{X}) = f_\pm , \qquad \text{if} \quad x\left[\Phi^{-T(\mathbf{X})}\mathbf{X}\right] = \pm\frac{L}{2} , \tag{8.6}$$

where $x(\cdot)$ denotes the projection of $\mathbf{X}$ on the x-axis.

A remarkable result is that the invariant measure (8.6) can be represented by the following expression

$$f_{\rm i}(\mathbf{X}) = \frac{f_+ + f_-}{2} + \mathbf{g}\cdot\left[\mathbf{r}(\mathbf{X}) + \int_0^{-T(\mathbf{X})} \mathbf{v}(\Phi^t\mathbf{X})dt\right] , \tag{8.7}$$

where $\mathbf{g} = \mathbf{e}_x(f_+ - f_-)/L$ is the gradient of density and where $-T(\mathbf{X})$ is the negative time when the boundary ($x = \pm L/2$) is reached from the current point $\mathbf{X}$ under backward evolution. $\mathbf{v} = \dot{\mathbf{r}}$ denotes the particle velocity. The formula (8.7) is obtained by noting that

$$\int_0^{-T(\mathbf{X})} \mathbf{v}(\Phi^t\mathbf{X})dt = \mathbf{r}\left[\Phi^{-T(\mathbf{X})}\mathbf{X}\right] - \mathbf{r}(\mathbf{X}) , \tag{8.8}$$

so that the invariant density simplifies to

$$\begin{aligned} f_{\rm i}(\mathbf{X}) &= \frac{f_+ + f_-}{2} + \mathbf{g}\cdot\mathbf{r}\left[\Phi^{-T(\mathbf{X})}\mathbf{X}\right] \\ &= \frac{f_+ + f_-}{2} + \frac{f_+ - f_-}{L}\, x\left[\Phi^{-T(\mathbf{X})}\mathbf{X}\right] . \end{aligned} \tag{8.9}$$

Since $x\left[\Phi^{-T(\mathbf{X})}\mathbf{X}\right] = \pm L/2$, the definition (8.6) is recovered (Gaspard 1997c).

The invariant density (8.7) gives the mean density of trajectories in the neighbourhood of the point $\mathbf{X}$. The corresponding invariant measure, such that $f_{\rm i} = d\nu_{\rm i}/d\mathbf{X}$, is not a probability measure but gives an average number of trajectories. A probability measure can be defined by normalization as $d\mu_{\rm i}/d\mathbf{X} = f_{\rm i}(\mathbf{X})/\int_{\mathscr{M}} f_{\rm i}(\mathbf{X})d\mathbf{X}$. We notice that $\mu_{\rm i}$ is absolutely continuous with respect to the Liouville measure because the density $f_{\rm i}(\mathbf{X})$ exists for almost all $\mathbf{X} \in \mathscr{M}$ and because $f_{\rm i}(\mathbf{X})$ is undecidable only on a set of zero Liouville measure, i.e., on $W_u(\mathscr{I})$ where $T(\mathbf{X}) = \infty$.

Now, we suppose that the nonequilibrium conditions are imposed at large distances in such a way that $L \to \infty$ and $(f_+ - f_-) \to \infty$ while the density gradient $g = (f_+ - f_-)/L$ is kept constant. In this limit, we must subtract the constant term $(f_+ + f_-)/2$ and only a relative density can be considered. The time to reach the boundaries becomes infinite for all the trajectories, $T(\mathbf{X}) \to \infty$, so that we obtain

$$\Psi_{\mathrm{i}}(\mathbf{X}) = \mathbf{g} \cdot \left[\mathbf{r}(\mathbf{X}) + \int_0^{-\infty} \mathbf{v}(\Phi^t \mathbf{X}) dt \right] . \tag{8.10}$$

The first term describes a mean linear profile of density in the direction of the gradient $\mathbf{g}$. This first term is derived from the phenomenological diffusion equation with flux boundary conditions in Eqs. (8.1)–(8.3). However, the second term is new with respect to the phenomenology and represents fluctuations around the linear density profile. These fluctuations are particularly important because the integral over the past velocity is infinite for almost all the trajectories. Hence, the steady state Ψ_{i} is no longer a function but should be considered as a distribution or singular measure defined by the linear functional

$$\begin{aligned} \langle A \rangle_{\mathrm{ne}} &= \langle A(\mathbf{X}) \Psi_{\mathrm{i}}(\mathbf{X}) \rangle_{\mathrm{e}} = \int d\mu_{\mathrm{e}} \, A(\mathbf{X}) \Psi_{\mathrm{i}}(\mathbf{X}) \\ &= \langle A(\mathbf{X}) \, \mathbf{g} \cdot \mathbf{r}(\mathbf{X}) \rangle_{\mathrm{e}} + \int_0^{-\infty} \langle A(\mathbf{X}) \, \mathbf{g} \cdot \mathbf{v}(\Phi^t \mathbf{X}) \rangle_{\mathrm{e}} \, dt . \end{aligned} \tag{8.11}$$

for an observable $A(\mathbf{X})$ and where μ_{e} is the equilibrium Liouville measure. The nonequilibrium steady states (8.11) have a long history in nonequilibrium statistical mechanics. These steady-state invariant measures were introduced by Lebowitz (1959) and MacLennan (1959). Zubarev (1962, 1974) showed that they are related to local integrals of motion of the underlying mechanical system (see below). Moreover, Procaccia, Oppenheim and coworkers have used such expressions in order to show that the light-scattering spectra of such nonequilibrium steady states may be asymmetric (Procaccia et al. 1979a, 1979b; see also Kirkpatrick et al. 1982). Recently, these nonequilibrium steady states have been considered in a generalization of statistical mechanics outside of local equilibrium (Eyink et al. 1996).

In the limit $L \to \infty$, the steady states (8.11) turn out to be given as invariant measures which are singular along the stable direction but absolutely continuous along the unstable direction. Such steady-state measures may be represented by their cumulative functions by taking $A(\mathbf{X})$ in (8.11) as the indicator function of a rectangular domain in phase space. In order to show this particular property, we shall make the connection with the construction of the hydrodynamic modes of diffusion in Chapter 7.

8.3 From the hydrodynamic modes to the nonequilibrium steady states

Equation (8.3) suggests that the nonequilibrium steady state describing a linear profile of (relative) concentration across the whole system can be obtained directly from the hydrodynamic modes of diffusion by differentiating with respect to the wavenumber. This observation provides an alternative construction of the nonequilibrium steady states. We remark that our derivation of the Green–Kubo relation in Chapter 7 has already required the first derivatives of the generalized eigenstates of the Frobenius–Perron operator with respect to the wavenumber. We can therefore use the calculation carried out in Chapter 7 in order to obtain the nonequilibrium steady states. In doing so, we start from a problem formulated with quasiperiodic boundary conditions to extract the solution of a problem with flux boundary conditions (Tasaki and Gaspard 1994, 1995; Gaspard 1995, 1996).

8.3.1 From the eigenstates to the nonequilibrium steady states

In Chapter 7, we obtained the hydrodynamic modes as the generalized eigenstates (7.62)–(7.63) of the Frobenius–Perron operator (7.40) acting on a subspace of given wavenumber $\mathbf{k}$. The eigenstates (7.63) are given in terms of the eigenstates $\psi_{\mathbf{k}}$ of the Frobenius–Perron operator (7.49) of the Poincaré–Birkhoff mapping of the system. The first derivatives of the right and left eigenstates (7.52)–(7.53) with respect to the wavenumber are given by Eqs. (7.75)–(7.76) in terms of the vector-valued function $\mathbf{a}(\mathbf{x})$ describing the motion of the trajectory on the lattice.

According to Eq. (8.3), the nonequilibrium steady state corresponding to a gradient of concentration in the direction of the constant vector $\mathbf{g}$ is thus given by

$$\psi_{\mathbf{g}}(\mathbf{x}) = -i\mathbf{g}\cdot\partial_{\mathbf{k}}\psi_0(\mathbf{x}) = -\sum_{n=1}^{\infty} \mathbf{g}\cdot\mathbf{a}(\phi^{-n}\mathbf{x}) \, . \tag{8.12}$$

The expression (8.12) is the singular kernel of a linear functional acting on test functions $v(\mathbf{x})$ which are piecewise Hölder continuous so that the rapid decay (7.15) of the correlation functions guarantees the convergence of sums like (7.78). Hence, $\psi_{\mathbf{g}}$ defines a distribution which is a singular measure.

According to the central limit theorem (7.16), the probability for the sum

(8.12) to remain in the finite interval $[-L, +L]$ is equal to

$$\nu\left\{\mathbf{x} : -L < \sum_{n=1}^{N} a_\alpha(\boldsymbol{\phi}^{-n}\mathbf{x}) < +L\right\} \simeq_{N\to\infty} \frac{2L}{\sigma_\alpha\sqrt{2\pi N}} . \tag{8.13}$$

Therefore, this probability vanishes in the limit $N \to \infty$ so that we may expect that the sum (8.12) remains finite only on a set of zero Lebesgue measure. In the following, we shall see that the distribution (8.12) may be represented by its cumulative function and can be visualized in this way.

The nonequilibrium steady state of the flow dynamics is obtained by taking the first derivative of the eigenstate (7.63) of the flow with respect to the wavenumber

$$\Psi_{\mathbf{g}}(\mathbf{x}, \tau, \mathbf{l}) = -i\, \mathbf{g}\cdot\partial_{\mathbf{k}}\Psi_0(\mathbf{x}, \tau, \mathbf{l}) . \tag{8.14}$$

According to Eqs. (7.75) and (8.12) and because $\partial_{\mathbf{k}} s_0 = 0$, we obtain the steady state as

$$\Psi_{\mathbf{g}}(\mathbf{x}, \tau, \mathbf{l}) = \mathbf{g}\cdot\mathbf{l} + \psi_{\mathbf{g}}(\mathbf{x}) = \mathbf{g}\cdot\mathbf{l} - \sum_{n=1}^{\infty} \mathbf{g}\cdot\mathbf{a}(\boldsymbol{\phi}^{-n}\mathbf{x}) . \tag{8.15}$$

A time-reversed nonequilibrium steady state can also be obtained from the adjoint eigenstate $\tilde{\psi}_{\mathbf{k}}$ as

$$\tilde{\Psi}_{\mathbf{g}}(\mathbf{x}, \tau, \mathbf{l}) = \mathbf{g}\cdot\mathbf{l} + \tilde{\psi}_{\mathbf{g}}(\mathbf{x}) = \mathbf{g}\cdot\mathbf{l} + \sum_{n=0}^{\infty} \mathbf{g}\cdot\mathbf{a}(\boldsymbol{\phi}^{n}\mathbf{x}) . \tag{8.16}$$

Now, the quantities of the Poincaré–Birkhoff mapping (7.9) can be related back to the quantities of the flow according to Eqs. (7.13)–(7.14), which allows us to show that the steady states (8.15)–(8.16) are identical to

$$\Psi_{\mathbf{g}}(\mathbf{X}) = \mathbf{g}\cdot\mathbf{r}(\mathbf{X}) + \int_0^{-\infty} \mathbf{g}\cdot\mathbf{v}(\boldsymbol{\Phi}^t\mathbf{X})\, dt , \tag{8.17}$$

and

$$\tilde{\Psi}_{\mathbf{g}}(\mathbf{X}) = \mathbf{g}\cdot\mathbf{r}(\mathbf{X}) + \int_0^{+\infty} \mathbf{g}\cdot\mathbf{v}(\boldsymbol{\Phi}^t\mathbf{X})\, dt , \tag{8.18}$$

where $\mathbf{v}(\boldsymbol{\Phi}^t\mathbf{X})$ is the velocity at time t of the trajectory from the initial condition $\mathbf{X}$.

Equation (8.17) is identical to the relative density (8.10) with the extra singular term involving the integral of the velocity which has a meaning only in the sense of Eq. (8.11). We may here conclude by saying that the spectral theory of the Frobenius–Perron operator not only allows us to derive the Green–Kubo relation from the dispersion relation given by a Pollicott–Ruelle resonance but also to derive the Lebowitz–MacLennan steady-state measures. Therefore, the

Liouvillian theory provides a consistent framework where all the important relations can be systematically derived.

We remark that, contrary to the scaling theories (Spohn 1980), the nonequilibrium steady states are here defined at the microscopic level of phase space. In this regard, such nonequilibrium measures are natural generalizations of the equilibrium measures. A crucial aspect is that the support of the nonequilibrium measures we consider here is the plain phase space of integer dimension. This is in contrast with the chaotic-scattering situation obtained for open systems where the support of the invariant measure is fractal. Nevertheless, the measures (8.17)–(8.18) are singular and are therefore different from the measures of the microcanonical ensembles.

8.3.2 Microscopic current and Fick's law

If the above expressions (8.17)–(8.18) describe nonequilibrium steady states they must satisfy Fick's law according to which the average of the microscopic current is proportional to the gradient with a proportionality constant equal to the diffusion coefficient.

In order to show that is indeed the case, we need to calculate the average current of particles passing through an arbitrary cell. The outgoing and ingoing currents at the cell $\mathbf{l}$ are given respectively by

$$\mathbf{j}^{(\text{out})} = \frac{1}{2}\left\langle \mathbf{v}(\mathbf{x},\tau)\, \Psi_{\mathbf{g}}(\mathbf{x},\tau,\mathbf{l})\right\rangle_{\mu} , \tag{8.19}$$

and

$$\mathbf{j}^{(\text{in})} = \frac{1}{2}\left\langle \mathbf{v}(\iota\mathbf{x},\tau)\, \Psi_{\mathbf{g}}(\mathbf{x},\tau,\mathbf{l})\right\rangle_{\mu} , \tag{8.20}$$

where we take the time-reversal transformation (7.22). The factor $\frac{1}{2}$ is required because each cell of the suspended flow contains particles in transit between two cells.

We first remark that the term linearly increasing with $\mathbf{l}$ in (8.15) is constant in each cell of the lattice so that this term vanishes in the mean currents. Using Eqs. (7.13) and (7.14), $\Psi_{\mathbf{g}}$ can be replaced by $\psi_{\mathbf{g}}$ in Eqs. (8.19) and (8.20). Using the definition (7.21) of the invariant measure, we get

$$\mathbf{j}^{(\text{out})} = -\frac{1}{2\langle T\rangle_{\nu}} \sum_{n=1}^{\infty} \langle \mathbf{a}(\mathbf{x})\, \mathbf{g}\cdot\mathbf{a}(\boldsymbol{\phi}^{-n}\mathbf{x})\rangle_{\nu} + \frac{1}{2\langle T\rangle_{\nu}} \langle \mathbf{c}(\mathbf{x})\, \mathbf{g}\cdot\mathbf{a}(\boldsymbol{\phi}^{-1}\mathbf{x})\rangle_{\nu} , \tag{8.21}$$

$$\mathbf{j}^{(\text{in})} = +\frac{1}{2\langle T\rangle_{\nu}} \sum_{n=0}^{\infty} \langle \mathbf{a}(\mathbf{x})\, \mathbf{g}\cdot\mathbf{a}(\boldsymbol{\phi}^{n}\mathbf{x})\rangle_{\nu} + \frac{1}{2\langle T\rangle_{\nu}} \langle \mathbf{c}(\mathbf{x})\, \mathbf{g}\cdot\mathbf{a}(\boldsymbol{\phi}^{-1}\mathbf{x})\rangle_{\nu} . \tag{8.22}$$

Finally, the total current is

$$\begin{aligned}\dot{\mathbf{j}}^{(\mathrm{tot})} &= \dot{\mathbf{j}}^{(\mathrm{out})} - \dot{\mathbf{j}}^{(\mathrm{in})} \\ &= -\frac{1}{2\langle T\rangle_\nu} \sum_{n=-\infty}^{+\infty} \langle \mathbf{a}(\mathbf{x})\,\mathbf{g}\cdot\mathbf{a}(\phi^n\mathbf{x})\rangle_\nu\,, \\ &= -D\,\mathbf{g}\,. \end{aligned} \tag{8.23}$$

where we used the relation (7.81) for the diffusion matrix which is assumed to be diagonal. Equation (8.23) shows that the current corresponding to the nonequilibrium steady state $\Psi_{\mathbf{g}}$ obeys Fick's law with the diffusion coefficient given precisely by the Green–Kubo relation (7.85) (Gaspard 1996).

8.3.3 Nonequilibrium steady states of the periodic Lorentz gas

Let us illustrate the above result with the periodic Lorentz gas on a triangular lattice with finite horizon as described in Chapter 7, Section 7.4. The nonequilibrium steady state is given by the expression (8.12) as a sum of the jump vectors over the past trajectory. The state is here represented in the Birkhoff coordinates $\mathbf{x} = (\theta, \varpi)$ of the collision events on a disk. That the steady state is a singular measure can here be proved thanks to results by Chernov (1994) showing that the Birkhoff map of the Lorentz gas with finite horizon does present fast decays of correlation functions in the sense of Eq. (7.15) so that sums of correlation functions like (7.78) converge. Therefore, we obtain the following (Gaspard 1996)

Theorem. In the triangular Lorentz gas with a finite horizon, the nonequilibrium steady state $\psi_{\mathbf{g}}(\mathbf{x})$ is a distribution or linear functional defined by Eq. (8.12) on piecewise Hölder continuous functions $v(\mathbf{x})$ defined on the Birkhoff surface of section (7.117).

Because the forward dynamics tends to smooth out the probability densities along the unstable directions the corresponding invariant measure $\psi_{\mathbf{g}}(\mathbf{x})$ is regular along the unstable directions but singular along the stable directions.

A distribution like (8.17) cannot be represented as such but we can overcome this difficulty by considering its cumulative function defined by applying the distribution on the characteristic function (7.118), which is also piecewise Hölder continuous to get

$$\begin{aligned}\mathscr{T}_{\mathbf{g}}(\theta,\varpi) &= \langle I_{(\theta,\varpi)}\psi_{\mathbf{g}}\rangle_\nu = \frac{1}{4\pi}\int_0^\theta d\theta' \int_{-1}^{\varpi} d\varpi'\;\psi_{\mathbf{g}}(\theta',\varpi') \\ &= g_x\,\mathscr{T}_x(\theta,\varpi) + g_y\,\mathscr{T}_y(\theta,\varpi)\,, \end{aligned} \tag{8.24}$$

in terms of the cumulative functions (7.121). These functions which are continuous but nondifferentiable, are depicted in Figs. 8.2 and 8.3 for two different values of the interdisk distance. The comparison with the imaginary part of the cumulative function of the eigenstate in Fig. 7.7b reveals a similarity with the cumulative function of the steady state in Fig. 8.2a. This similarity is due to the relation (7.122) showing that the nonequilibrium steady state is the leading term of the imaginary part of the eigenstate in an expansion in powers of the wavenumber. In Fig. 7.7, the wavenumber was taken in the x-direction so that

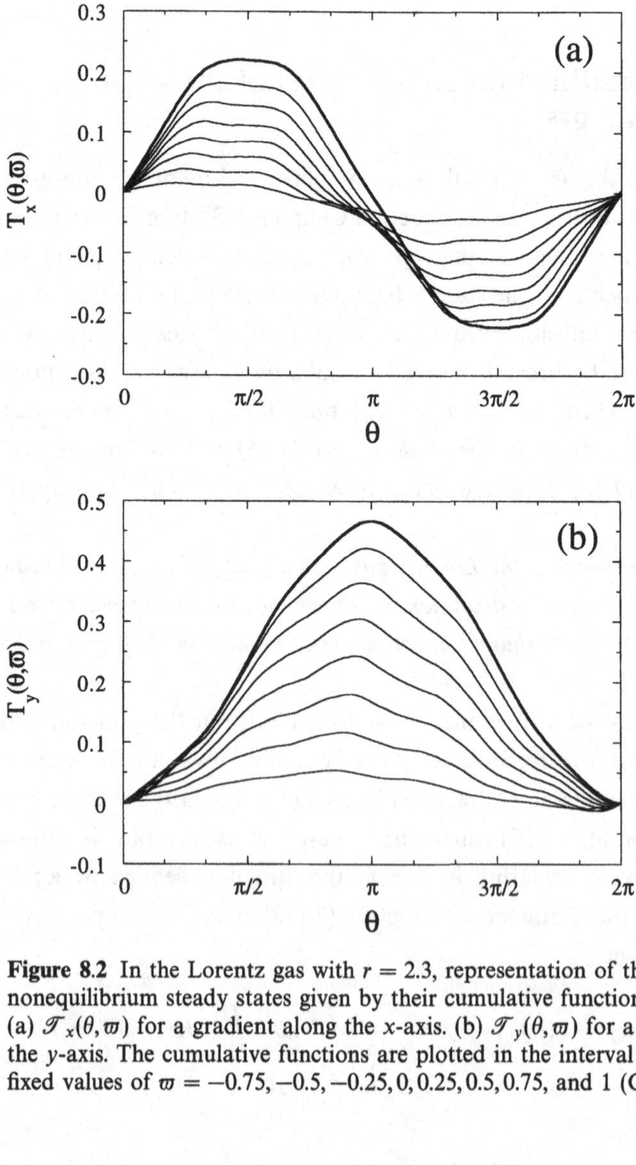

Figure 8.2 In the Lorentz gas with $r = 2.3$, representation of the nonequilibrium steady states given by their cumulative functions (7.121): (a) $\mathcal{T}_x(\theta, \varpi)$ for a gradient along the x-axis. (b) $\mathcal{T}_y(\theta, \varpi)$ for a gradient along the y-axis. The cumulative functions are plotted in the interval $0 < \theta < 2\pi$ for fixed values of $\varpi = -0.75, -0.5, -0.25, 0, 0.25, 0.5, 0.75$, and 1 (Gaspard 1996).

the similarity appears with the steady state with a gradient in the x-direction of Fig. 8.2a.

8.3.4 Nonequilibrium steady states of the periodic multibaker

The same construction can be carried out for the periodic multibaker map defined in Chapter 6 (Tasaki and Gaspard 1994, 1995). According to the expression (8.15), the Lebowitz–MacLennan nonequilibrium steady state is here given by

$$\Psi_{\mathbf{g}}(x, y, l) = g\, l - \sum_{n=1}^{\infty} g\, a[\phi_F^{-n}(x, y)] \,, \tag{8.25}$$

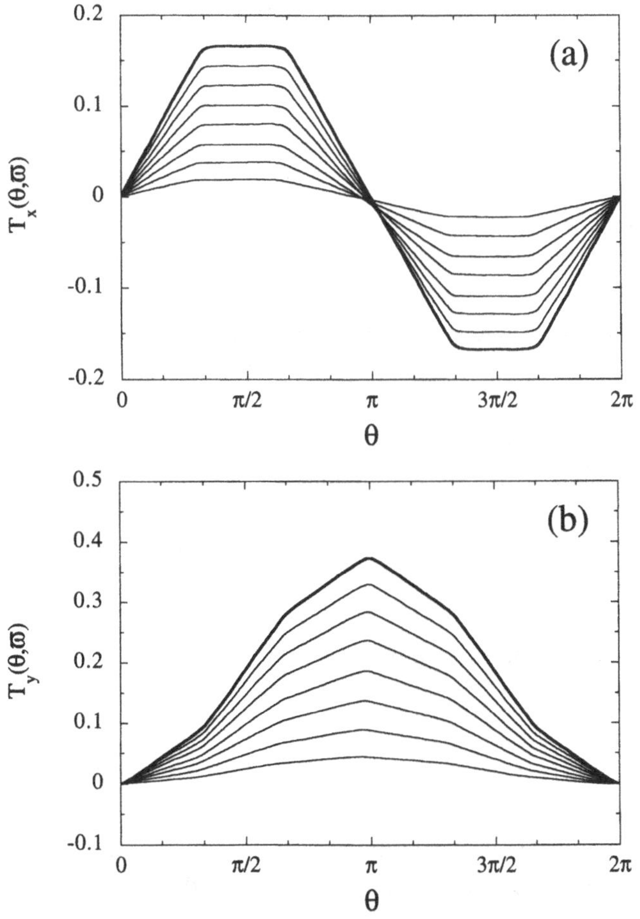

Figure 8.3 Same as Fig. 8.2 in the Lorentz gas with $r = 2.01$ (Gaspard 1996).

where ϕ_F denotes the baker map inside the fundamental domain of the multibaker, which is a square of the chain and where the jump vector $a(x, y) = \mp 1$ for $x < 1/2$ or $x > 1/2$ respectively. We notice that the inverse multibaker map (6.47) is independent of the x variable which is also the case for the jump vector a. The cumulative function of the distribution (8.25) is given by

$$\begin{aligned} \mathcal{T}_g(x, y, l) &= \int_0^x dx' \int_0^y dy' \; \Psi_g(x', y', l) \\ &= g\, l\, x\, y \;+\; g\, x\, T(y) \, , \end{aligned} \tag{8.26}$$

where $T(y)$ is a function such that Eq. (8.25) is recovered by differentiation

$$\frac{dT}{dy}(y) \;=\; -\sum_{n=0}^{\infty} \tilde{a}(\varphi^n y) \, , \tag{8.27}$$

where $\varphi(y) = (2y)$ modulo 1 and $\tilde{a}(y) = \mp 1$ for $y < 1/2$ or $y > 1/2$ respectively. A recurrence relation can be obtained by noting that Eq. (8.27) is equivalent to

$$\frac{dT}{dy}(y) \;=\; \begin{cases} \frac{dT}{dy}(2y) \;+\; 1 \, , & y < \frac{1}{2} \, , \\ \frac{dT}{dy}(2y-1) \;-\; 1 \, , & y > \frac{1}{2} \, . \end{cases} \tag{8.28}$$

The cumulative function should vanish at $y = 0$ by definition: $T(0) = 0$. Moreover, the cumulative function at $y = 1$ is the integral over all the possible trajectories, which is the mean drift of the particle. Since the multibaker is isomorphic to a symmetric random walk the mean drift vanishes so that the integral vanishes: $T(1) = 0$. Integrating Eq. (8.28) and fixing the constants of integration with the two previous conditions, we obtain the recurrence relation

$$T(y) \;=\; \begin{cases} \frac{1}{2}\, T(2y) \;+\; y \, , & y < \frac{1}{2} \, , \\ \frac{1}{2}\, T(2y-1) \;-\; y \;+\; 1 \, , & y > \frac{1}{2} \, . \end{cases} \tag{8.29}$$

This recurrence relation defines a function which is continuous. Indeed, the transformation

$$(\hat{\mathcal{K}} f)(y) \;=\; \begin{cases} \frac{1}{2}\, f(2y) \;+\; y \, , & y < \frac{1}{2} \, , \\ \frac{1}{2}\, f(2y-1) \;-\; y \;+\; 1 \, , & y > \frac{1}{2} \, , \end{cases} \tag{8.30}$$

with $f(0) = f(1) = 0$ is contracting because

$$\|\hat{\mathcal{K}} f - \hat{\mathcal{K}} g\| \;\leq\; \frac{1}{2}\, \|f - g\| \, , \tag{8.31}$$

where the norm is $\|f\| = \mathrm{Sup}_{0 \leq y \leq 1} |f(y)|$. Therefore, the fixed-point equation $\hat{\mathcal{K}} T = T$ which is Eq. (8.29) admits a unique solution. If the function $f(y)$ is continuous and satisfies the conditions $f(0) = f(1) = 0$, the sequence $\hat{\mathcal{K}}^n f$ uniformly converges to a continuous function which is the fixed point (8.29). The limiting function $T(y)$ is continuous but nondifferentiable almost everywhere since its derivative (8.27) is infinite for almost all trajectories. This function is known as the *Takagi function*, which was introduced in 1903 by the Japanese mathematician T. Takagi who proposed this function as a further example of a continuous but nondifferentiable function beyond the example proposed by Weierstrass (Takagi 1903, Hata and Yamaguti 1984, Hata 1986).

Figure 8.4 depicts the first few iterations of (8.30) starting from $f(y) = 0$. We observe that, at the nth iteration, the function has already converged to its limit value at the points $y = j/2^n$ with $j = 0, 1, 2, \dots, 2^n$. These points can be assigned to symbolic sequences with symbols $\omega_n = 0$ or 1 whether $\tilde{a}(\varphi^n y) = +1$ or -1 according to

$$y = y_{\omega_1 \cdots \omega_n} = \frac{\omega_1}{2} + \frac{\omega_2}{4} + \cdots + \frac{\omega_n}{2^n} . \tag{8.32}$$

Using the construction of Fig. 8.4, we deduce the following property of the Takagi function:

$$2\, T\left(y + \frac{\Delta y}{2} \right) - T(y + \Delta y) - T(y) = \Delta y , \tag{8.33}$$

at the points y given by (8.32) and for $\Delta y = 1/2^n$. This is a consequence of the nondifferentiability of this function. Indeed, the second difference (8.33) would vanish as Δy^2 if the Takagi function was twice differentiable. Another property is that

$$\sum_{\omega_1 \cdots \omega_n} \left[T\left(y_{\omega_1 \cdots \omega_n} + \frac{1}{2^n} \right) - T\left(y_{\omega_1 \cdots \omega_n} \right) \right]^2 = \frac{n}{2^n} . \tag{8.34}$$

Both properties will be used in the following (Gaspard 1997b).

The cumulative function of the nonequilibrium steady state (8.26) of the multibaker is thus depicted in Fig. 8.5. This cumulative function is differentiable in the unstable direction x but nondifferentiable in the stable direction y. Indeed, the map tends to smooth out the inhomogeneities in the unstable direction where it is expanding while the inhomogeneities imposed at boundaries by the nonequilibrium constraints tend to build up in the stable direction where the map is contracting, leading to the fractal-like cumulative function given by the Takagi function. We notice that the cumulative function of Fig. 8.5 is for the periodic multibaker while Figs. 8.2 or 8.3 are for the periodic Lorentz gas.

We also remark that the recurrence relation (8.29) of the Takagi function can also be obtained directly as

$$T(y) = -i\, \partial_k\, A_0(y;k)\big|_{k=0} , \tag{8.35}$$

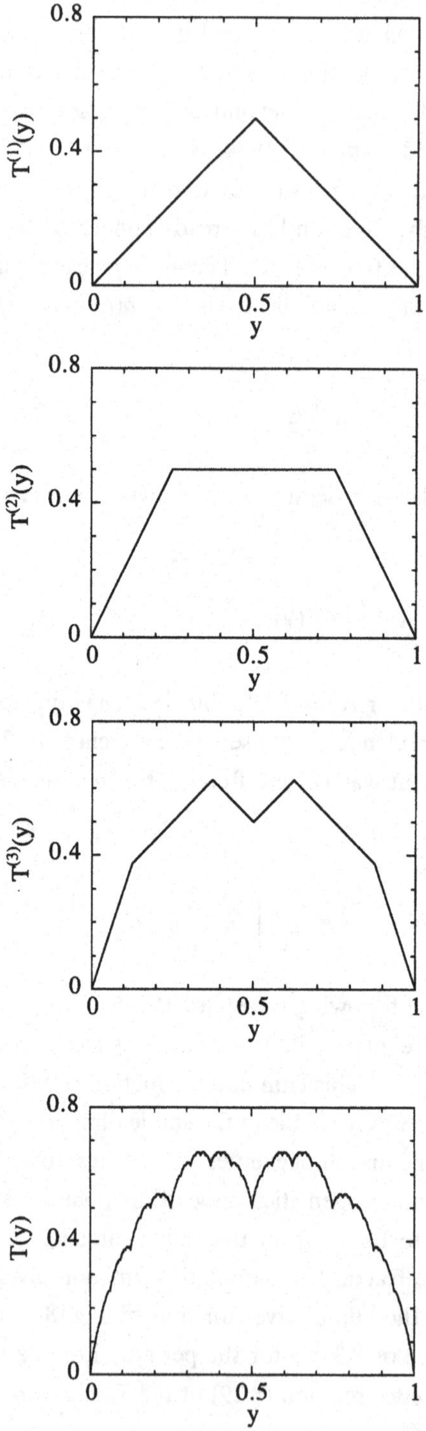

Figure 8.4 Construction of the Takagi function by successive iterations according to Eq. (8.30): $\lim_{n\to\infty} \hat{\mathscr{K}}^n f(y) = T(y)$, the seed function being zero $f(y) = 0$.

by differentiation of the cumulative function of the eigenstate given by the deRham-type function (7.179) for $j = 0$

$$A_0(y;k) = \begin{cases} \alpha(k)\, A_0(2y;k)\,, & y < \frac{1}{2}\,, \\ [1-\alpha(k)]\, A_0(2y-1;k) + \alpha(k)\,, & y > \frac{1}{2}\,, \end{cases} \tag{8.36}$$

with $\alpha(k) = [1 + \exp(-2ik)]^{-1}$, if we use the fact that $A_0(y;k=0) = y$.

8.3.5 Nonequilibrium steady states of the Langevin process

For comparison with the previous systems, it is instructive to obtain the nonequilibrium steady state (8.10) for the Langevin model of Brownian motion. In contrast with the previous models, the Langevin process is stochastic but the cumulative function of the distribution (8.10) is already nontrivial.

In the Langevin model, the position and the velocity of the Brownian particle evolve according to the stochastic equations

$$\frac{dr}{dt} = v\,, \tag{8.37}$$

$$\frac{dv}{dt} = -\,a\,v + c\,W(t)\,, \tag{8.38}$$

where $W(t)$ is a white noise of zero mean $\langle W(t)\rangle = 0$ and of correlation

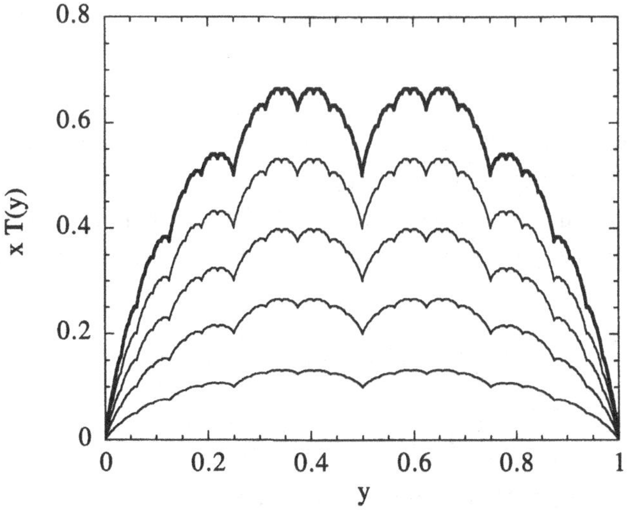

Figure 8.5 Cumulative function (8.26) of the Lebowitz–MacLennan steady-state measure (8.25) for the multibaker map in the cell $l = 0$ and for a unit gradient. This two-dimensional function is given by $xT(y)$ where $T(y)$ is the Takagi function. The cumulative function is depicted versus y for different values of $x = 0, 0.2, 0.4, 0.6, 0.8$, and 1.

$\langle W(0)W(t)\rangle = \delta(t)$ (Wax 1954, van Kampen 1981). Equation (8.38) for the velocity is known to define the so-called Ornstein–Uhlenbeck process. The corresponding Fokker–Planck equation ruling the time evolution of the probability density of the velocity is

$$\frac{\partial f}{\partial t} - a\,\frac{\partial}{\partial v}(vf) = \frac{c^2}{2}\,\frac{\partial^2 f}{\partial v^2}\,. \tag{8.39}$$

The corresponding Green function gives the probability density for the velocity to change from v_0 to v during the time $t > 0$

$$G(v, v_0; t) = \frac{1}{c\sqrt{2\pi b(t)}}\,\exp\left\{-\frac{[v - v_0\exp(-at)]^2}{2c^2 b(t)}\right\}, \tag{8.40}$$

with

$$b(t) = \frac{1}{2a}\,\big[1 - \exp(-2at)\big]\,. \tag{8.41}$$

The stationary probability density of the velocity is obtained as the limit

$$f_{\rm e}(v) = \lim_{t\to\infty} G(v, v_0; t) = \frac{1}{c\sqrt{\pi/a}}\,\exp\left(-\frac{av^2}{c^2}\right), \tag{8.42}$$

which is the Maxwellian distribution of velocities so that the variance of the velocity is

$$\langle v(t)^2\rangle_{\rm e} = \frac{c^2}{2a} = \frac{k_{\rm B}T}{m}\,, \tag{8.43}$$

while the velocity autocorrelation function is

$$\langle v(0)v(t)\rangle_{\rm e} = \frac{c^2}{2a}\,\exp(-a|t|)\,. \tag{8.44}$$

On the other hand, the motion of the position $r(t)$ is a nonstationary process driven by the velocity according to Eq. (8.37).

The cumulative function of the distribution (8.10) describing a nonequilibrium steady state of gradient $g = 1$ can be defined with the indicator function $I_v(v')$ of the semi-axis $v' < v$:

$$\begin{aligned}\langle I_v(v')\,\psi(r, v')\rangle_{\rm e} &= \int_{-\infty}^{+\infty} dv'\; f_{\rm e}(v')\, I_v(v')\left[r + \int_0^{-\infty} v_t(v')\,dt\right]\\ &= r\int_{-\infty}^{v} f_{\rm e}(v')\,dv' + \int_0^{-\infty}\langle I_v(v')\,v_t(v')\rangle_{\rm e}\,dt\,,\end{aligned} \tag{8.45}$$

where $v_t(v')$ is the velocity at time t issued from the initial velocity v'. The first term of Eq. (8.45) is linear in the position and, hence, corresponds to the first term in Eq. (8.26), which describes the linear profile of concentration. On the other hand, the term which may be compared with the Takagi function in Eq. (8.26) is the second one. This term is the time integral of the correlation function between the indicator function $I_v(v')$ and the current velocity $v_t(v')$ so that we introduce

$$T(v) = \int_0^{-\infty} \langle I_v(v')\, v_t(v') \rangle_e \, dt \,. \tag{8.46}$$

The correlation function can be calculated using the Green function (8.40) as

$$\langle I_v(v')\, v_t(v') \rangle_e = \int_{-\infty}^{+\infty} \int_{-\infty}^{+\infty} I_v(v')\, v''\, G(v'',v';t)\, f_e(v')\, dv'\, dv''\,, \tag{8.47}$$

where $v'' = v_t$. Whereupon, we obtain that the interesting part of the steady-state cumulative function is a Gaussian function of the velocity

$$T(v) = \frac{c}{2\sqrt{\pi a^3}} \exp\left(-\frac{av^2}{c^2}\right) . \tag{8.48}$$

This function is for the Langevin model of diffusion what the Takagi-type cumulative functions of Figs. 8.2, 8.3, and 8.5 are for deterministic models such as the periodic Lorentz gas and the multibaker. We observe that the stochastic character of the Langevin process has the effect of smoothing out the nondifferentiability of the cumulative function. We may thus conclude that the aforementioned nondifferentiability has its origin in the deterministic chaos of classical dynamical systems.

8.4 From the finite to the infinite multibaker

The above construction of the Lebowitz–MacLennan steady-state measure for the infinite multibaker immediately leads to the nondifferentiable Takagi function, which is evidence for the singular character of the invariant measure in the limit where the gradient of concentration is imposed at largely separated ends of the chain. It is nevertheless interesting to construct the invariant measures when the nonequilibrium conditions are imposed at finite distances on the multibaker chain as discussed in Section 8.2. This construction has been carried out by Tasaki and Gaspard (1995) for the finite multibaker map of length L extended to infinity with a free-particle dynamics [cf. Chapter 6 and Eq. (6.51)].

A flux of particles is supposed to flow continuously across the chain. The particles in the half squares, $1/2 \leq x \leq 1$ with $l \leq -1$, arriving from infinity at the left-hand end are assumed to be uniformly distributed with the density $\rho_- = f_-$, while those in the half squares, $0 \leq x < 1/2$ with $L+1 \leq l$, arriving at the right-hand end have a density $\rho_+ = f_+$ (see Fig. 6.18).

As a consequence of the chaotic time evolution inside the chain, the measure is very complicated on the half squares which exit the chain. Indeed, there is a fractal repeller in the squares $0 \leq l \leq L$, which has the partial Hausdorff dimension (6.67). Therefore, there are three types of orbits which exit the scatterer at $l \leq -1$ or $L+1 \leq l$:

1. The orbits which entered at the left-hand end. The density is ρ_- in their vicinity.
2. The orbits which entered at the right-hand end. The density is ρ_+ in their vicinity.
3. The orbits of the unstable manifolds of the repeller. The unstable manifolds are segments of horizontal lines which separate the regions of density ρ_+ from those of density ρ_-.

Since the repeller is fractal, it is also the case for its unstable manifolds so that the measure on the exiting half squares is very complicated. However, the density always exists because it is equal to either ρ_+ or ρ_-, except on the set of the unstable manifolds which is of zero Lebesgue measure.

In the limit $L \to \infty$, the partial Hausdorff dimension (6.67) of the unstable manifolds tends to the unit value so that we can understand that the invariant measure becomes singular because the regions of densities $\rho_\pm$ shrink while they continue to alternate across the phase space.

In order to overcome the difficulty of the singular character arising in this limit, we define the cumulative distribution function associated with the measure ν_t at a time t as

$$F_t(l,x,y) = \nu_t\Big(l, [0,x[\otimes[0,y[\Big) , \tag{8.49}$$

which exists as a function even when the density does not exist. The time evolution of the measure ν_t is given by

$$\nu_{t+1}(A) = \nu_t(\phi_L^{-1}A) , \tag{8.50}$$

for any phase-space domain A. According to Eq. (6.51) of the finite multibaker map, we obtain the time evolution of the cumulative function (8.49) as

$$F_{t+1}(l,x,y) = \begin{cases} F_t\Big(l+1, \frac{x}{2}, 2y\Big) , & 0 \leq y < \frac{1}{2} , \\ F_t\Big(l+1, \frac{x}{2}, 1\Big) + F_t\Big(l-1, \frac{x+1}{2}, 2y-1\Big) & \\ \qquad -F_t\Big(l-1, \frac{1}{2}, 2y-1\Big) , & \frac{1}{2} \leq y \leq 1 , \end{cases} \tag{8.51}$$

for $0 \leq l \leq L$, and

$$F_{t+1}(l,x,y) = \begin{cases} F_t(l+1,x,y)\,, & 0 \leq x < \frac{1}{2}\,, \\ F_t\left(l+1,\frac{1}{2},y\right) + F_t(l-1,x,y) - F_t\left(l-1,\frac{1}{2},y\right)\,, & \frac{1}{2} \leq x \leq 1\,, \end{cases} \tag{8.52}$$

for $l \leq -1$ or $l \geq L+1$. The flux boundary conditions are here expressed by requiring that the ingoing particles are uniformly distributed as

$$\begin{aligned} F_0(l,x,y) &= \rho_- xy\,, \qquad \tfrac{1}{2} \leq x \leq 1\,,\quad l \leq -1\,, \\ F_0(l,x,y) &= \rho_+ xy\,, \qquad 0 \leq x < \tfrac{1}{2}\,,\quad l \geq L+1\,. \end{aligned} \tag{8.53}$$

The equations of motion (8.51)–(8.52) can be further simplified because Eq. (8.53) leads to

$$\begin{aligned} F_t(l,x,y) - F_t\left(l,\tfrac{1}{2},y\right) &= \rho_-\left(x-\tfrac{1}{2}\right)y\,, \qquad \tfrac{1}{2} \leq x \leq 1\,,\quad l \leq -1\,, \\ F_t(l,x,y) &= \rho_+ xy\,, \qquad 0 \leq x < \tfrac{1}{2}\,,\quad l \geq L+1\,. \end{aligned} \tag{8.54}$$

Thus, for $l=0$ and $l=L$, we have

$$F_{t+1}(0,x,y) = \begin{cases} F_t\left(1,\frac{x}{2},2y\right)\,, & 0 \leq y < \frac{1}{2}\,, \\ F_t\left(1,\frac{x}{2},1\right) + \rho_- x\left(y-\frac{1}{2}\right)\,, & \frac{1}{2} \leq y \leq 1\,, \end{cases} \tag{8.55a}$$

and

$$F_{t+1}(L,x,y) = \begin{cases} \rho_+ xy\,, & 0 \leq y < \frac{1}{2}\,, \\ \rho_+ x/2 + F_t\left(L-1,\frac{x+1}{2},2y-1\right) & \\ \qquad -F_t\left(L-1,\frac{1}{2},2y-1\right)\,, & \frac{1}{2} \leq y \leq 1\,. \end{cases} \tag{8.55b}$$

We note that the set formed by Eq. (8.51) with $l = 1, 2, \cdots L-1$ together with Eqs. (8.55) is a closed system of equations for the unknown functions $F_t(l,x,y)$ $(l = 0, 1, \cdots L)$.

Provided that the initial state F_0 is twice continuously differentiable in x, the solution of the equations of motion (8.51) and (8.55) asymptotically approaches the steady state given by

$$F_\infty(l,x,y) = x\left\{\frac{\rho_+ - \rho_-}{L+2}\left[(l+1)y + T_l(y)\right] + \rho_- y\right\}\,, \tag{8.56}$$

where $\{T_l(y)\}$ are the so-called incomplete Takagi functions defined by the multiple functional equation

$$T_l(y) = \begin{cases} \frac{1}{2}\,T_{l+1}(2y) + y\,, & 0 \leq y < \frac{1}{2}\,, \\ \frac{1}{2}\,T_{l-1}(2y-1) + 1 - y\,, & \frac{1}{2} \leq y \leq 1\,, \end{cases} \tag{8.57}$$

with the boundary conditions, $T_{-1}(y) = T_{L+1}(y) = 0$ (Tasaki and Gaspard 1995). Equation (8.57) defines a contracting operator which admits a unique continuous solution satisfying

$$\mathrm{Max}_{0\leq l\leq L}\mathrm{Sup}_{0\leq y\leq 1}\ |T_l(y)| \leq 1 , \tag{8.58}$$

and

$$\mathrm{Sup}_{0\leq y\leq 1}|T_l(y) - T(y)| \leq 2\left(\frac{1}{2}\right)^{\min(l,L-l)} . \tag{8.59}$$

The inequality (8.59) implies that, if the lattice site l is far from the ends of the chain, the incomplete Takagi function T_l tends to the Takagi function. Hence, the steady state (8.26) is recovered from the middle part of the asymptotic state (8.56) in the limit $L \to \infty$ because $T_l(y) = T(y) + \mathcal{O}(2^{-L/2})$ in the middle of the chain at $l = [L/2]$. For a multibaker chain of length $L = 10$, the incomplete Takagi functions $\{T_l(y)\}$ are depicted in Fig. 8.6, where we observe that the distribution $T_5(y)$ at the middle site practically coincides with the Takagi function shown in Fig. 8.4. The incomplete Takagi functions are almost everywhere differentiable. The reason is that the corresponding invariant measure has a well-defined density almost everywhere except on the unstable manifolds of the fractal repeller as discussed in Section 8.2.

The average current in the nonequilibrium steady state ν_∞ is given by:

$$\begin{aligned} J_{l|l+1} &= \nu_\infty\big(l, [1/2, 1[\otimes[0,1]\big) - \nu_\infty\big(l+1, [0,1/2[\otimes[0,1]\big) \\ &= F_\infty(l,1,1) - F_\infty\left(l,\tfrac{1}{2},1\right) - F_\infty\left(l+1,\tfrac{1}{2},1\right) \\ &= -\tfrac{1}{2}[F_\infty(l+1,1,1) - F_\infty(l,1,1)] = -D\ (\nabla\rho) . \end{aligned} \tag{8.60}$$

Accordingly, the state ν_∞ obeys Fick's law since the diffusion coefficient of the dyadic multibaker is $D = 1/2$ while the gradient is

$$\nabla\rho = \frac{\rho_+ - \rho_-}{L+2} . \tag{8.61}$$

In the limit $L \to \infty$, the gradient can be kept fixed by increasing the concentration difference $\rho_+ - \rho_-$ at both ends of the chain. In this limit, the incomplete Takagi function in the middle of the chain tends to the Takagi function itself so that the cumulative function of the invariant measure tends to

$$F_\infty(l,x,y) = x\Big[\rho_l\ y + (\nabla\rho)\ T(y)\Big] , \tag{8.62}$$

with the mean density in the lth square of the chain given by

$$\rho_l = (\nabla\rho)(l+1) + \rho_- . \tag{8.63}$$

We can recognize in Eqs. (8.62)–(8.63) the cumulative function (8.26) for a gradient $g = \nabla\rho$ up to a constant. Because the density exiting the chain is a very complicated mixture of the densities $\rho_\pm$ the mean density on the left-hand squares for $-\infty < l \leq -1$ is not equal to the density ρ_- on the ingoing half-squares but actually to $\rho_l = (1/2)(\rho_- + \rho_0) = \rho_- + (1/2)(\nabla\rho)$. A similar moderation holds at the right-hand side.

We conclude that the local density given by

$$f_\infty(x, y, l) = \partial_x\, \partial_y\, F_\infty(l, x, y) = \rho_l - \sum_{n=0}^{\infty} (\nabla\rho)\, \tilde{a}(\varphi^n y)\,, \tag{8.64}$$

fluctuates between the values $\rho_\pm$ in such a way that a mean linear profile is

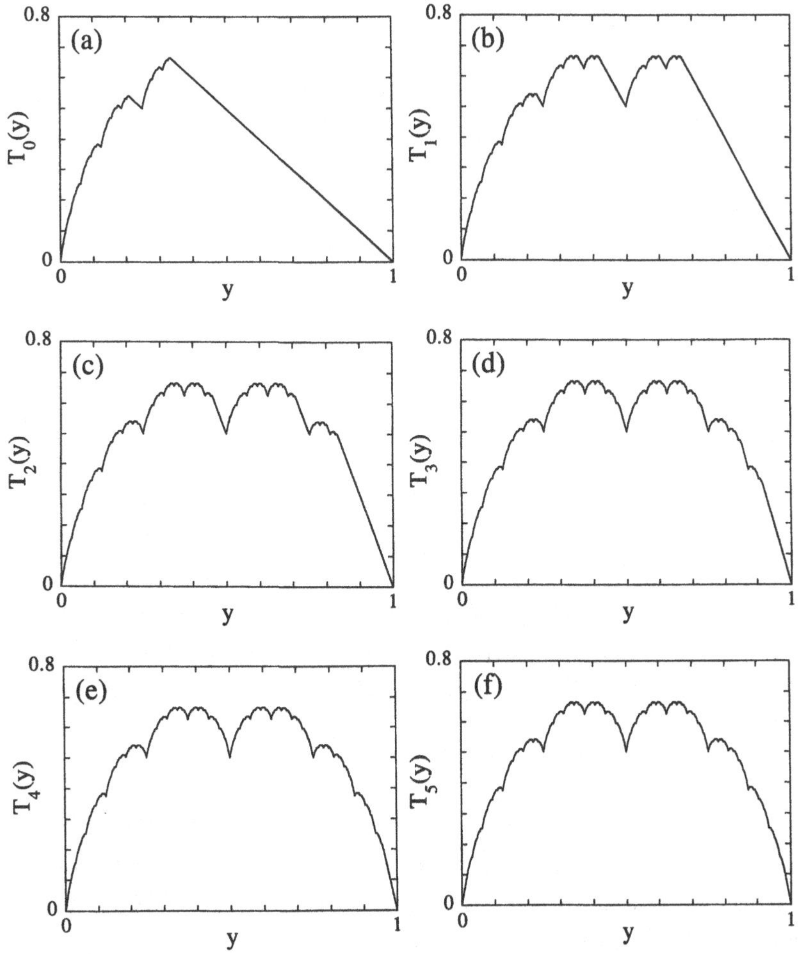

Figure 8.6 The incomplete Takagi functions $\{T_l(y)\}$ given by Eqs. (8.57) along an open multibaker chain of length $L = 10$: (a) $l = 0$; (b) $l = 1$; (c) $l = 2$; (d) $l = 3$; (e) $l = 4$; (f) $l = 5$. On the right-hand side of the chain, the same functions are obtained up to the reflection $y \to 1 - y$ because of the symmetry $T_l(y) = T_{L-l}(1 - y)$.

formed along the chain. However, the fluctuations are so wild that the invariant measure becomes singular in the large-system limit. The cumulative function is nevertheless well defined in this limit, which allows us to represent the limiting nonequilibrium steady state as a function. We shall see below that this singular character of the invariant measure has important consequences for the production of entropy of irreversible thermodynamics.

Let us remark that the infinite multibaker chain may have other invariant measures than the one given in terms of the Takagi function (Tasaki and Gaspard 1994, 1995). However, these other invariant measures have exponential profiles of concentration instead of a linear profile. Moreover, these other invariant measures are singular in *both* the stable and the unstable directions in contrast with the invariant measure presented here. In order to produce such other invariant measures in a physical system the ingoing particles should be distributed according to a singular measure instead of a measure which is absolutely continuous with respect to the Lebesgue measure as with the flux boundary conditions (8.53). Therefore, these other invariant measures appear to be physically excluded because they are not selected by smooth physical boundary conditions.

8.5 Generalization to the other transport processes

The simplicity of the expression (8.17) of the nonequilibrium steady states of diffusion suggests the generalization to the other transport and reaction-rate coefficients in N-particle systems. Let us suppose that a gradient $\mathbf{g}$ inducing a transport process α is imposed on a system by some nonequilibrium constraint. The microscopic current corresponding to the process α is denoted by $J_{\mathbf{g}}^{(\alpha)}$. A Helfand moment $G_{\mathbf{g}}^{(\alpha)}$ is associated with the microscopic current such that

$$J_{\mathbf{g}}^{(\alpha)} = (d/dt)G_{\mathbf{g}}^{(\alpha)} = \{G_{\mathbf{g}}^{(\alpha)}, H\} , \tag{8.65}$$

where $\{\cdot,\cdot\}$ denotes the Poisson bracket and H the Hamiltonian of the system. In the case of diffusion: $G_{\mathbf{g}}^{(\mathrm{D})} = \mathbf{g}\cdot\mathbf{r}$ and $J_{\mathbf{g}}^{(\mathrm{D})} = \mathbf{g}\cdot\mathbf{v}$ where $\mathbf{r}$ and $\mathbf{v}$ are the particle position and velocity according to Table 6.3. The nonequilibrium steady state and its time-reversed state are thus given by

$$\Psi_{\mathbf{g}}^{(\alpha)}(\mathbf{X}) = G_{\mathbf{g}}^{(\alpha)}(\mathbf{X}) + \int_0^{-\infty} J_{\mathbf{g}}^{(\alpha)}(\mathbf{\Phi}^t\mathbf{X})\, dt , \tag{8.66}$$

$$\tilde{\Psi}_{\mathbf{g}}^{(\alpha)}(\mathbf{X}) = G_{\mathbf{g}}^{(\alpha)}(\mathbf{X}) + \int_0^{+\infty} J_{\mathbf{g}}^{(\alpha)}(\mathbf{\Phi}^t\mathbf{X})\, dt , \tag{8.67}$$

for a trajectory of initial condition at the phase space point $\mathbf{X} = \{\mathbf{q}_i, \mathbf{p}_i\}_{i=1}^N$. We should notice that these distributions are invariants of motion in the sense that

$$\Psi_{\mathbf{g}}^{(\alpha)}(\mathbf{X}) = \Psi_{\mathbf{g}}^{(\alpha)}(\mathbf{\Phi}^t\mathbf{X}) , \tag{8.68}$$

as can be verified using the definitions of $G_{\mathbf{g}}^{(\alpha)}$ and $J_{\mathbf{g}}^{(\alpha)}$. Their invariance is rendered compatible with the presence of chaotic behaviour thanks to their singular character. Indeed, a theorem by Moser asserts that there are no analytic invariant functions in the presence of transverse homoclinic orbits as is the case here (Moser 1973). The presence of chaotic behaviour thus opens the way to new types of invariants of motion given by distributions.

The average current in the direction $\mathbf{e}$ is given as one-half the outgoing flux minus one-half the ingoing flux which can be calculated with the time-reversed steady state in analogy with the case of diffusion treated here above

$$\begin{aligned} j_{\mathbf{e}}^{(\alpha)} &= \frac{1}{2} \langle J_{\mathbf{e}}^{(\alpha)}(\mathbf{X})\, \Psi_{\mathbf{g}}^{(\alpha)}(\mathbf{X})\rangle_\mu - \frac{1}{2} \langle J_{\mathbf{e}}^{(\alpha)}(\mathbf{X})\, \tilde{\Psi}_{\mathbf{g}}^{(\alpha)}(\mathbf{X})\rangle_\mu \\ &= -\frac{1}{2} \int_{-\infty}^{+\infty} \langle J_{\mathbf{e}}^{(\alpha)}(\mathbf{X}) J_{\mathbf{g}}^{(\alpha)}(\mathbf{\Phi}^t\mathbf{X})\rangle_\mu \, dt \\ &= -\alpha\, \mathbf{e}\cdot\mathbf{g} , \end{aligned} \tag{8.69}$$

where $\langle\cdot\rangle_\mu$ denotes the average over the equilibrium state and where we used a change of variable $t \to -t$. This result is obtained for an isotropic system under the condition of validity of the Green–Kubo relations. The result (8.69) shows that the average current is constant and given by the unit gradient multiplied by the transport coefficient in agreement with Fick's and Fourier's laws, or the laws ruling viscosity. In this way, the above method can be used to construct nonequilibrium steady states associated with viscosity, in particular, in the two-disk fluid mentioned in Chapter 7.

The above nonequilibrium steady states were first introduced by Lebowitz (1959) and MacLennan (1959). They are also closely related to local integrals of motion introduced by Zubarev (1962)

$$P^{(\alpha)}(\mathbf{x},t) = \rho^{(\alpha)}(\mathbf{x},t) + \int_{-\infty}^{0} \nabla\cdot\mathbf{j}^{(\alpha)}(\mathbf{x},t+\tau)\, d\tau , \tag{8.70}$$

where $\rho^{(\alpha)}$ and $\mathbf{j}^{(\alpha)}$ are the local density and current associated with the property α and which obey the local conservation equation $\partial_t\rho^{(\alpha)} + \nabla\cdot\mathbf{j}^{(\alpha)} = 0$. In the case of diffusion, $\rho^{(\mathrm{D})} = \delta[\mathbf{x}-\mathbf{r}(t)]$ and $\mathbf{j}^{(\mathrm{D})} = \mathbf{v}(t)\delta[\mathbf{x}-\mathbf{r}(t)]$ so that Zubarev's local integral of motion turns out to be related to the steady-state distribution (8.17) according to

$$\Psi_{\mathbf{g}}^{(\mathrm{D})}(\mathbf{r},\mathbf{v}) = \int (\mathbf{g}\cdot\mathbf{x})\, P^{(\mathrm{D})}(\mathbf{x},0)\, d^3x , \tag{8.71}$$

as shown by a straightforward calculation using properties of the derivative of the Dirac distribution, as well as $\mathbf{r}(0) = \mathbf{r}$ and $\mathbf{v}(0) = \mathbf{v}$. In this sense, Zubarev's singular integrals of motion replace Poincaré–Moser's analytic integrals of motion when the system becomes chaotic. The above steady-state measures have been considered to extend statistical mechanics outside of equilibrium (Spohn 1991, Eyink et al. 1996).

8.6 Entropy production

The problem of entropy production is one of the oldest problems of nonequilibrium statistical mechanics (Boltzmann 1896; Gibbs 1902; Prigogine 1961, 1980). It originates from the confrontation between the thermodynamics of irreversible processes and the classical and later the quantum mechanics which are supposed to describe the motion of atoms and molecules in matter. Classical mechanics is reversible and preserves volumes in phase space (cf. Liouville's theorem). Both properties are in apparent contradiction with irreversible thermodynamics. However, the above properties of the nonequilibrium steady states and, in particular, their singular character in the large-system limit provide a way to reconcile classical mechanics with irreversible thermodynamics.

In previous sections, the nonequilibrium steady states corresponding to gradients of concentration have been constructed for flux boundary conditions. In finite systems, such a nonequilibrium steady state is given by an invariant measure which is different from the Liouville equilibrium invariant measure but which remains absolutely continuous with respect to the Liouville measure. In the limit of large systems where the concentration gradient is maintained to a fixed value, the nonequilibrium invariant measure becomes singular with respect to the Liouville measure. For instance, in the multibaker map which is a model of deterministic diffusion, the invariant measure corresponding to a gradient of concentration is given in terms of the nondifferentiable Takagi function in the limit where the gradient is imposed at borders which are more and more separated while keeping the gradient constant. The convergence of the invariant measure – which remains absolutely continuous as long as the borders are finitely separated – to the singular measure is very rapid in the large-system limit because it is determined by the Lyapunov exponential instability. In this way, the absolute continuity disappears exponentially fast below tiny scales in phase space. This result is of crucial importance for the following arguments.

The approach toward the thermodynamic equilibrium can be understood in terms of statistical properties like the Pollicott–Ruelle resonances as explained in

Chapter 7. These properties are intimately related to the usual explanation of the approach to equilibrium in terms of the mixing property introduced by Gibbs. The mixing property assumes the decrease of the two-time correlation functions of the physical observables, so that the approach to equilibrium can be directly formulated in the context of the Liouvillian dynamics of statistical ensembles of trajectories. In this explanation of the approach to equilibrium, the concept of entropy is not required and the connection to the second law of thermodynamics can be eluded: the time evolution toward equilibrium is only formulated in terms of statistical averages of physical observables: $\langle A \rangle_t = \int A(\mathbf{X}) f_t(\mathbf{X}) d\mathbf{X} \rightarrow \langle A \rangle_e$ for $t \rightarrow \infty$.

Recently, several works have been devoted to the problem of the possible connections between the aforementioned results on chaotic dynamical systems and the time evolution of a quantity of the type of the thermodynamic entropy (Nicolis and Nicolis 1988; Evans 1989; Chernov et al. 1993a, 1993b; Ruelle 1996, 1997; Breymann et al. 1996; Nicolis and Daems 1996). These works are concerned with our understanding of the entropy production and of the second law in the context of dynamical systems theory. It is a very old and difficult problem because it is well know that an entropy defined like the Gibbs entropy as

$$S_{\mathrm{G}}(t) = - \int_{\mathscr{M}} d\mathbf{X} \; f_t(\mathbf{X}) \; \ln \; f_t(\mathbf{X}) \, , \tag{8.72}$$

remains constant during the time evolution of the probability density $f_t(\mathbf{X})$ in closed volume-preserving systems. However, if the invariant measure becomes singular in some limit as is the case in open volume-preserving systems, we are no longer allowed to use the Gibbs entropy. Indeed, the Gibbs entropy is given in terms of the probability density which exists only if the associated measure is absolutely continuous with respect to the Liouville measure: $f_t(\mathbf{X}) = d\nu_t/d\mathbf{X}$. (Let us recall here that the Liouville measure $d\mathbf{X}$ is invariant in volume-preserving systems.) Therefore, the constancy of the entropy is in question in large open volume-preserving systems (Gaspard 1997b).

As a consequence, the problem of entropy production must be reconsidered in the light of the above result showing the singularity of the invariant measure of nonequilibrium states in open volume-preserving systems of large spatial extension. If the invariant measure becomes singular in some limit, the Gibbs entropy should be replaced by a coarse-grained entropy or ε-entropy which is essentially $-\sum p \log p$ where the p's are the probabilities for the trajectories to visit cells of size ε in phase space. Indeed, the work by Kolmogorov and Tikhomirov has shown that an ε-entropy diverges for $\varepsilon \rightarrow 0$ in a way which is characteristic of the type of singularities of the measure (Kolmogorov 1956,

Kolmogorov and Tikhomirov 1959, Tikhomirov 1963). Such an ε-entropy is thus required if the type of the measure happens to change in some limit, especially, if we want to keep the operational interpretation of entropy as a measure of disorder.

In the following, we first describe the problem of entropy production by going back to the original definition of entropy production in the thermodynamics of irreversible processes. For systems with flux boundary conditions, a probability measure is defined in such open systems with infinitely many particles in terms of a Poisson suspension over the dynamical system. The time evolution is introduced in this formulation. We then define the ε-entropy and we show that the definition is consistent with the standard equilibrium entropy per unit volume. Hence, the time evolution of this ε-entropy and the corresponding ε-entropy production are obtained. The definitions are applied to diffusion in the multibaker map where the ε-entropy production is determined by the nondifferentiable Takagi function in the large system limit and gives precisely the entropy production expected from irreversible thermodynamics (Gaspard 1997b).

8.6.1 Irreversible thermodynamics and the problem of entropy production

The concept of entropy production is introduced in the thermodynamics of irreversible processes (Prigogine 1961, deGroot and Mazur 1962, Glansdorff and Prigogine 1971, Nicolis and Prigogine 1977). In order to identify in deterministic dynamical systems a quantity like the entropy production, we shall first present the phenomenological entropy production and discuss its properties which should be recovered in the deterministic approach. We only discuss here the case of diffusion which is the process observed in the Lorentz gas and the multibaker map.

A first remark is that the thermodynamics of irreversible processes is a macroscopic theory where the quantities are defined as averages over volumes of sizes larger than the mean free path of the fluid particles. In the case of the diffusion of tracer particles in a fluid, the density evolves in time according to the phenomenological equation (7.1). The phenomenological diffusion equation is obtained from the conservation law of tracer particles,

$$\partial_t \rho + \nabla \cdot \mathbf{j} = 0 , \tag{8.73}$$

where the tracer current is given by Fick's law,

$$\mathbf{j} = -D \, \nabla \rho . \tag{8.74}$$

If the tracer concentration is not too high we may suppose that the tracer particles and the fluid form an ideal solution so that the entropy S is given as the integral of the entropy per unit volume or entropy density s:

$$S = \int_V s \, d\mathbf{r} \, , \qquad \text{with} \qquad s = \rho \, \ln \frac{\rho_0}{\rho} \, , \tag{8.75}$$

where ρ_0 is a constant defining a reference density for which $s(\rho_0) = 0$. An evolution equation for the entropy density can be derived from its definition (8.75) and the diffusion equation (8.73)–(8.74) as

$$\partial_t \, s + \nabla \cdot \mathbf{J}_s = \sigma_s \, , \tag{8.76}$$

with the following entropy current and entropy source

$$\mathbf{J}_s = \mathbf{j} \, \ln \frac{\rho_0}{e\rho} = - D \nabla \rho \, \ln \frac{\rho_0}{e\rho} \, , \tag{8.77}$$

$$\sigma_s = D \, \frac{(\nabla \rho)^2}{\rho} \geq 0 \, . \tag{8.78}$$

The balance equation (8.76) for the entropy density can be expressed for the global entropy S as

$$\frac{dS}{dt} = - \int_{\partial V} \mathbf{J}_s \cdot d\mathbf{A} + \int_V \sigma_s \, d\mathbf{r} = \frac{d_e S}{dt} + \frac{d_i S}{dt} \, . \tag{8.79}$$

$d_e S/dt$ is the *entropy flow* at the borders ∂V of the system and $d_i S/dt$ is the so-called *entropy production* inside the system due to the irreversible process of diffusion. This entropy production is always nonnegative according to the second law of thermodynamics:

$$\frac{d_i S}{dt} = \int_V \sigma_s \, d\mathbf{r} \geq 0 \, . \tag{8.80}$$

The entropy production vanishes at equilibrium and is positive away from equilibrium. In contrast, the entropy flow may take positive or negative values depending on the gradient of concentration imposed at borders.

8.6.2 Comparison with deterministic schemes

We now compare irreversible thermodynamics with the deterministic dynamics for the motion of atoms and molecules in the fluid. We suppose that $\mathbf{X}$ denotes the positions and momenta of these particles, which define the phase space $\mathcal{M}$ of the dynamical system. The motion is supposed to be governed by a set of differential equations of first order in time, given by a vector field $\mathbf{F}(\mathbf{X})$ in phase space

$$\dot{\mathbf{X}} = \mathbf{F}(\mathbf{X}) \, . \tag{8.81}$$

We consider a statistical ensemble of copies of the system which is defined by a probability measure ν_t, we suppose for the moment to be absolutely continuous with respect to the Lebesgue measure so that the corresponding probability density exists $f_t(\mathbf{X}) = d\nu_t/d\mathbf{X}$. This probability density evolves in time according to the (generalized) Liouville equation which expresses the conservation of probability in phase space:

$$\partial_t f(\mathbf{X}) + \partial_{\mathbf{X}} \cdot [\mathbf{F}(\mathbf{X}) f(\mathbf{X})] = 0 , \tag{8.82}$$

and which has to be solved with appropriate boundary conditions corresponding to the nonequilibrium constraints. The time evolution of the probability density induces a time evolution for the Gibbs entropy density

$$s_{\mathrm{G}} = -f \ln f , \tag{8.83}$$

and, consequently, for the Gibbs entropy (8.72). The balance equation for the local evolution of this density is given by

$$\partial_t s_{\mathrm{G}} + \nabla \cdot \mathbf{J}_{s_{\mathrm{G}}} = \sigma_{s_{\mathrm{G}}} , \tag{8.84}$$

with the following Gibbs entropy current and entropy source

$$\mathbf{J}_{s_{\mathrm{G}}} = (-f \ln f)\, \mathbf{F} = s_{\mathrm{G}}\, \mathbf{F} , \tag{8.85}$$

$$\sigma_{s_{\mathrm{G}}} = f\, \partial_{\mathbf{X}} \cdot \mathbf{F} . \tag{8.86}$$

Therefore, the variation of the Gibbs entropy is obtained as

$$\frac{dS_{\mathrm{G}}}{dt} = -\int_{\partial\mathcal{M}} \mathbf{J}_{s_{\mathrm{G}}} \cdot d\mathbf{A} + \int_{\mathcal{M}} \sigma_{s_{\mathrm{G}}}\, d\mathbf{X} . \tag{8.87}$$

The first term is the flow of entropy at the boundary of the phase space and may be identified with the entropy flow in Eq. (8.79). The second term is the average value of the divergence of the vector field (8.81). In a conservative system, this divergence vanishes, $\partial_{\mathbf{X}} \cdot \mathbf{F} = 0$, and $\sigma_{s_{\mathrm{G}}} = 0$. Therefore, the second term vanishes in Eq. (8.87) so that the time variation of the Gibbs entropy dS_{G}/dt is only due to boundary conditions. At a steady state, the variation of entropy vanishes so that the flow of Gibbs entropy is also zero.

These properties of Gibbs entropy are in contradiction with the properties expected for an entropy in view of the thermodynamics of irreversible processes. If we had to identify the Gibbs entropy (8.72) with the thermodynamic entropy (8.75), as is the case in equilibrium statistical mechanics, the famous problem would arise that the source and thus the production of Gibbs entropy vanishes for the class of conservative systems. Recently, this problem has been revisited from different viewpoints.

Mackey has suggested that a positive entropy production should have its origin in the property of exactness of dynamical systems (Mackey 1989). Exact dynamical systems are defined as discrete-time systems, $\mathbf{X}_{t+1} = \mathbf{\Phi}(\mathbf{X}_t)$, which are expansive $|\partial_{\mathbf{X}}\mathbf{\Phi}| \geq 1$. The expansivity is compatible with a finite phase space if the mapping $\mathbf{\Phi}$ sends several different points $\mathbf{X}$ onto the same point $\mathbf{\Phi}(\mathbf{X})$ (Lasota and Mackey 1985). If entropy is conceived as a measure of disorder in phase space we understand that there is a loss of information and thus disorder production in such systems. In flows, the property of exactness corresponds to the assumption that $\partial_{\mathbf{X}} \cdot \mathbf{F} \geq 0$ which means that the flow is expansive. According to Eqs. (8.86)–(8.87), the entropy production would then be positive. However, Hamiltonian systems are not expansive.

In thermostatted systems (Hoover 1991, Evans et al. 1990a), the trajectories are attracted toward phase-space regions of contractivity where $\partial_{\mathbf{X}} \cdot \mathbf{F} \leq 0$ on average. In such systems, there is a negative entropy production. Indeed, since trajectories converge to a strange attractor which has an information dimension lower than the total phase-space dimension the probability distribution $f_t(\mathbf{X})$ is more disordered at the initial time than at following times. As a consequence, the Gibbs entropy – which is a measure of disorder – decreases! To overcome this problem, a hypothetical mechanism of entropy conservation between the system and the thermostat has been proposed: $S_{\text{G, total}} = S_{\text{G, system}} + S_{\text{G, thermostat}} = \text{constant}$ (Chernov et al. 1993a, 1993b). If we now consider the so-defined entropy of the thermostat there is a change of sign in its time variation and the entropy production of the thermostat should thus be positive. Although useful for the computations of thermodynamic quantities, this reasoning is unsatisfactory as an explanation of the origin of entropy production.

8.6.3 Open systems and their Poisson suspension

The previous difficulties can be overcome in the approach we now present (Gaspard 1997b). We consider volume-preserving systems which are open. The openness of the system is very important if we want to conceive a process of the kind described by irreversible thermodynamics. As an example, we consider a finite Lorentz gas as described in Section 8.2. A statistical ensemble of such systems is introduced which corresponds to filling space with a gas of infinitely many particles (which are independent of each other). The time evolution of the statistical ensemble is governed by the Liouville equation, or alternatively, by a Frobenius–Perron operator as shown in Chapter 7. Different boundary conditions can be considered to solve the Liouvillian dynamics in such systems:

(1) Absorbing boundary conditions

If we suppose that the particle density is zero at the borders of the finite lattice almost all the particles of the gas escape out of the lattice so that the density falls to zero inside the system according to

$$\rho_t \simeq \psi \exp(-\gamma t) \qquad (t \to \infty)\,, \tag{8.88}$$

where ψ is the first eigenfunction of the diffusion equation (8.73)–(8.74) as explained in Chapter 6. If we replace this solution into the phenomenological entropy source (8.78) we get

$$\sigma_s = D\frac{(\nabla\rho)^2}{\rho} \simeq \exp(-\gamma t)\, D\frac{(\nabla\psi)^2}{\psi} \to 0\,, \tag{8.89}$$

which vanishes when $t \to \infty$. As a consequence, this situation does not allow us to directly identify the thermodynamic entropy production as a stationary property because the entropy source vanishes together with the density itself due to the escape of all the particles.

In order to properly identify the thermodynamic entropy production, we consider the following boundary conditions:

(2) Flux boundary conditions

We suppose that the Lorentz-type scatterer is submitted to a continuous flux of particles as explained in Section 8.2 (see Fig. 8.1). In the phenomenological description, the steady state shows the linear profile of Eq. (8.2) after transients have decayed. Because of the asymptotic stationarity, the phenomenological entropy (8.75) in any part of the system reaches a constant value so that the time variation (8.79) of the entropy vanishes. On the other hand, the entropy production (8.78) takes the form

$$\sigma_s = D\,\frac{\left(\frac{\rho_+-\rho_-}{L}\right)^2}{\frac{\rho_++\rho_-}{2} + \frac{\rho_+-\rho_-}{L}x} \sim \frac{1}{x}\,, \tag{8.90}$$

which is everywhere positive. By stationarity, the entropy flow is equal to minus the entropy production.

This result suggests that such a steady state is an appropriate physical situation to proceed with an identification of the entropy production with some microscopic quantities. With this aim, we have first to introduce an adequate microscopic description of an open system which is maintained in a nonequilibrium steady state by flux boundary conditions. In such open and infinite systems, we can define a measure ν_t at time t which gives the local density of particles in phase space such that $\nu_t(B)$ is the number of particles in the phase-space region B at time t. We notice that this measure is not normalizable because there is an infinity of particles in the whole system extending to infinity. A normalizable measure μ_t can nevertheless be defined in these open systems, as follows.

The gas of independent particles is a many-body system which is called a Poisson suspension over the dynamical system of measure ν_t (Cornfeld et al. 1982). This Poisson suspension is a dynamical system on a phase space which is a direct product of infinitely many copies of the original phase space $\mathcal{M}$: $\mathcal{M}^\infty = \otimes_{i=1}^{\infty}\mathcal{M}_i$. A point in this phase space $\mathcal{M}^\infty$ defines an ensemble of copies of the system: $Y = \{\mathbf{X}_i\}_{i=1}^{\infty} \subset \mathcal{M}$. We can define subsets of the phase space such that the number of copies $\mathbf{X}_i$ inside the region B is fixed to the integer N

$$C_{BN} = \{Y \subset \mathcal{M} \ : \quad \text{Number}(Y \cap B) = N\} . \tag{8.91}$$

A probability measure μ_t can be defined for the Poisson suspension corresponding to the measure ν_t as

$$\begin{aligned} \mu_t(C_{BN}) &= \frac{[\nu_t(B)]^N}{N!} \exp[-\nu_t(B)] , \qquad \text{and} \\ \mu_t(C_{B_1N_1} \cap C_{B_2N_2}) &= \mu_t(C_{B_1N_1})\, \mu_t(C_{B_2N_2}) \quad \text{if} \quad B_1 \cap B_2 = \emptyset . \end{aligned} \tag{8.92}$$

To show that this measure is a probability, we consider a partition $\{B_i\}$ of some subset A of the phase space: $A = \cup_i B_i \subset \mathcal{M}$ with $B_i \cap B_j = \emptyset$ for $i \neq j$. Each cell B_i of the partition contains a certain number of points of Y given by $\text{Number}(Y \cap B_i) = N_i \in \{0, 1, 2, 3, \ldots\}$. The partition induced by $\{B_i\}$ in the phase space $\mathcal{M}^\infty$ is given by

$$\begin{aligned} &\cup_{\{N_i\}}\, C_{\{B_iN_i\}} \qquad \text{with} \\ C_{\{B_iN_i\}} &= C_{B_1N_1} \cap C_{B_2N_2} \cap C_{B_3N_3} \cap \cdots \cap C_{B_mN_m} , \end{aligned} \tag{8.93}$$

where $\{N_i\}$ denotes a configuration in which the number of particles in each cell B_i is equal to a given integer N_i. The measure of one element of the induced partition, i.e., the measure of a given configuration is given by applying the definition (8.92). Summing over all the configurations $\{N_i\}$, we get that

$$\sum_{\{N_i\}} \mu_t(C_{\{B_iN_i\}}) = \prod_{i=1}^{m} \sum_{N=0}^{\infty} \frac{[\nu_t(B_i)]^N}{N!} \exp[-\nu_t(B_i)] = 1 , \tag{8.94}$$

so that the measure μ_t is normalized to unity at all times t and is thus a probability measure in this sense.

The time evolution of the measure μ_t under a specific dynamical system (with discrete or continuous time) is induced by the following time evolution of the measure ν_t:

$$\nu_{t+1}(B) = \nu_t(\mathbf{\Phi}^{-1}B) . \tag{8.95}$$

Since the flux of particles at boundaries is continuous in time an invariant measure ν_∞ will established itself after some time at the level of the statistical ensemble (Tasaki and Gaspard 1995). We have already obtained the invariant density of this measure which is given by (8.6) or (8.7). Hence, the invariant measure μ_∞ of the Poisson suspension, i.e., of the gas of infinitely many particles, can be constructed with the above definition (Gaspard 1997b).

Let us remark here that the flux boundary conditions break the time-reversal invariance of the steady-state measure ν_∞ under nonequilibrium conditions, in contrast to the Liouville equilibrium measure which is time-reversal invariant.

8.6.4 The ε-entropy

Thanks to the probability measure of the Poisson suspension, we are now able to define an ε-entropy for open systems with an infinite number of particles.

The entropy of the measure $\mu = \mu_t$ corresponding to the partition $\{B_i\}$ of A in the original phase space $\mathscr{M}$ is defined in terms of the probabilities of the elements of the induced partition in $\mathscr{M}^\infty$. It is therefore a coarse-grained entropy which characterizes the disorder of the probability measure μ in the sense of Boltzmann. When all the cells B_i have the same given size ε we shall speak of an ε-entropy S_ε. Our definition is thus

$$S_\varepsilon = S(\{B_i\}) = -\sum_{\{N_i\}} \mu(C_{\{B_iN_i\}}) \ln \mu(C_{\{B_iN_i\}}) \, . \tag{8.96}$$

More systematic definitions of ε-entropy may be given as for instance

$$S_\varepsilon = \mathrm{Inf}_{\mathrm{diam}B_i \le \varepsilon} S(\{B_i\}) \, , \qquad \text{or} \qquad S_\varepsilon = \mathrm{Sup}_{\mathrm{diam}B_i \ge \varepsilon} S(\{B_i\}) \, , \tag{8.97}$$

but, for simplicity, we shall use the definition (8.96) where all the cells are identical.

Using the definitions (8.92) and (8.93), Eq. (8.96) becomes

$$S_\varepsilon = -\sum_i \sum_{N=0}^{\infty} \mu(C_{B_iN}) \ln \mu(C_{B_iN}) \, , \tag{8.98}$$

and

$$S_\varepsilon = \sum_i \nu(B_i) \ln \frac{e}{\nu(B_i)} + \mathscr{R}(\varepsilon) \, , \tag{8.99}$$

where the rest $\mathscr{R}(\varepsilon) = \sum_i \mathscr{O}[\nu(B_i)^2]$ is important only if $\nu(B_i) \gg 1$ but is negligible for $\varepsilon \to 0$ even if the measure is singular. This rest plays no role in the following argument and may be considered negligible but we shall keep it for rigor.

In the case where the measure ν is absolutely continuous with respect to the Liouville measure, the associated density exists: $f(\mathbf{X}) = d\nu/d\mathbf{X}$. The measures of the cells are given by $\nu(B_i) = f(\mathbf{X}_i)\Delta\mathbf{X}$ where $\mathbf{X}_i$ is a point inside B_i according to the mean theorem of Riemann integration theory and where $\Delta\mathbf{X}$ is the Liouville measure of the cells B_i. The ε-entropy is given by

$$S_\varepsilon = \left(\ln \frac{e}{\Delta\mathbf{X}}\right) \int_A f(\mathbf{X})\, d\mathbf{X} \\ - \int_A d\mathbf{X}\, f(\mathbf{X}) \ln f(\mathbf{X}) + \mathcal{O}(\Delta\mathbf{X}) + \mathcal{R}(\Delta\mathbf{X}) . \tag{8.100}$$

The second term is nothing other than the Gibbs entropy (8.72). The first term diverges as $\Delta\mathbf{X} \to 0$. Since $\int_{\mathcal{M}} f d\mathbf{X} = 1$ for a closed system with a normalized measure the first term remains constant in time and may be disregarded. This term fixes the famous constant of entropy according to the third law of thermodynamics (Fermi 1936). This term is very important to establish the correspondence with the entropy of quantum statistical mechanics where ε should be fixed according to $\Delta\mathbf{X} = \Delta^f q \Delta^f p = (2\pi\hbar)^f$. Thanks to the previous definition (8.96), we recover the usual expression of the equilibrium entropy per unit volume for instance in an ideal gas where

$$S_\varepsilon^{(\text{space})} = \rho \ln \frac{e^{5/2}(2\pi m k_B T)^{3/2}}{\rho\, \Delta^3 q \Delta^3 p} + \mathcal{O}(\Delta^3 q \Delta^3 p) . \tag{8.101}$$

The previous definition is therefore entirely consistent with standard equilibrium statistical mechanics (Pathria 1972).

The ε-entropy (8.99) of a domain A evolves in time and we are interested in its time variation, i.e., in the difference in its values between two successive instants of time separated for instance by a unit time:

$$\Delta S_\varepsilon = S_\varepsilon(t+1;A) - S_\varepsilon(t;A) \\ = \sum_{B_i \subset A} \left[\nu_t(\mathbf{\Phi}^{-1}B_i) \ln \frac{e}{\nu_t(\mathbf{\Phi}^{-1}B_i)} - \nu_t(B_i) \ln \frac{e}{\nu_t(B_i)} \right] + \mathcal{R}(\varepsilon) . \tag{8.102}$$

If the measure ν_t is absolutely continuous with respect to the Liouville measure we obtain

$$\Delta S_\varepsilon = \int_{\mathbf{\Phi}^{-1}A} d\mathbf{X}\, f_t(\mathbf{X}) \ln \left| \frac{\partial \mathbf{\Phi}}{\partial \mathbf{X}} \right| \\ + \left(\int_{A_{\text{in}}} - \int_{A_{\text{out}}} \right) d\mathbf{X}\, f_t(\mathbf{X}) \ln \frac{e}{f_t(\mathbf{X})\Delta\mathbf{X}} + \mathcal{O}(\Delta\mathbf{X}) + \mathcal{R}(\Delta\mathbf{X}) , \tag{8.103}$$

where we used the identity

$$\int_{\mathbf{\Phi}^{-1}A} - \int_A = \int_{A_{\text{in}}} - \int_{A_{\text{out}}} , \tag{8.104}$$

according to which the difference between the integrals over the preimage of A and over A itself is equal to the difference between the integrals over the domain $A_{\rm in}$ which enters A and the domain $A_{\rm out}$ which exits A.

In a volume-preserving system, the Jacobian of the mapping $\mathbf{\Phi}$ is equal to unity so that there is no term in (8.103) which could be identified with the entropy production. However, this holds as long as the density exists so that the terms $\mathcal{O}(\Delta\mathbf{X})$ may be neglected. This is no longer the case for singular measures.

Let us now separate the ε-entropy variation into an ε-entropy flow and an ε-entropy production in analogy with Eq. (8.79). The ε-entropy flow can be naturally defined as the difference between the ε-entropies of the domains ingoing and outgoing A

$$\Delta_e S_\varepsilon \;=\; S_\varepsilon(t;A_{\rm in}) \;-\; S_\varepsilon(t;A_{\rm out}) \;=\; S_\varepsilon(t;\mathbf{\Phi}^{-1}A) \;-\; S_\varepsilon(t;A)\,, \tag{8.105}$$

where the last identity follows from Eq. (8.104).

The ε-entropy production can now be defined as

$$\Delta_i S_\varepsilon \;=\; \Delta S_\varepsilon \;-\; \Delta_e S_\varepsilon\,. \tag{8.106}$$

In the next section, we shall apply the previous definitions to a simple model of diffusion.

8.6.5 Entropy production in the multibaker map

We proceed with the calculation of the entropy production (8.106) in the multibaker on a finite chain of length L as defined with Eq. (6.51). The nonequilibrium steady states have been explicitly constructed for this system as explained in previous sections (Tasaki and Gaspard 1994, 1995).

We first calculate the time variation (8.102) of the entropy. We consider the cells B_i of size $(\Delta x, \Delta y)$ in the square $A = (l, [0,1] \otimes [0,1])$ as well as their preimages $\phi^{-1}B_i$ of size $(\Delta x/2, 2\Delta y)$ which belong to the half squares $\phi^{-1}A = (l+1, [0,1/2[\otimes[0,1]) \cup (l-1, [1/2,1] \otimes [0,1])$ (see Fig. 8.7). We take the size $\Delta x = 1/(2m)$ with an integer m while $\Delta y = 1/2^n$. The time variation of the ε-entropy at the steady state is thus given by

$$\Delta S_\varepsilon \;=\; \left[\left(\frac{\Delta x}{2}, 2\Delta y\right)\text{—entropy of } \phi^{-1}A\right] \\ -\;\left[(\Delta x, \Delta y)\text{—entropy of } A\right]. \tag{8.107}$$

The number of cells $(\Delta x, \Delta y)$ in the unit square A is equal to $(1/\Delta x) = 2m$ in the unstable direction x and to $(1/\Delta y) = 2^n$ in the stable direction y. Equation (8.56) shows that the cumulative function of the steady state is of the form

$$F_\infty(l,x,y) \;=\; x\,[\rho_l\; y \;+\; (\nabla\rho)\; T_l(y)] \;\equiv\; x\; g_l(y)\,, \tag{8.108}$$

in terms of the incomplete Takagi functions (8.57), of the mean density (8.63) in the lth square, and of the concentration gradient (8.61). This measure is uniform in the x-direction, but complicated in the y-direction. We infer that the measure of the cell

$$B_i \;=\; [x, x+\Delta x[\otimes[y_{\omega_1\cdots\omega_n}, y_{\omega_1\cdots\omega_n}+\Delta y[\,, \tag{8.109}$$

is

$$\begin{aligned} \nu(B_i) \;&=\; \Delta x\;\Delta g_l(\omega_1\cdots\omega_n) \qquad \text{with} \\ \Delta g_l(\omega_1\cdots\omega_n) \;&=\; g_l(y_{\omega_1\cdots\omega_n}+\Delta y) \;-\; g_l(y_{\omega_1\cdots\omega_n})\,, \end{aligned} \tag{8.110}$$

where $y_{\omega_1\cdots\omega_n}$ is defined by Eq. (8.32). Substituting in Eq. (8.107), the entropy variation becomes explicitly

$$\begin{aligned} \Delta S_\varepsilon \;=\; &\sum_{\omega_1\cdots\omega_{n-1}}\left[\frac{1}{2}\;\Delta g_{l+1}(\omega_1\cdots\omega_{n-1})\;\ln\frac{2e}{\Delta x\Delta g_{l+1}(\omega_1\cdots\omega_{n-1})}\right.\\ &\left.+\;\frac{1}{2}\;\Delta g_{l-1}(\omega_1\cdots\omega_{n-1})\;\ln\frac{2e}{\Delta x\Delta g_{l-1}(\omega_1\cdots\omega_{n-1})}\right]\\ &-\;\sum_{\omega_1\cdots\omega_n}\;\Delta g_l(\omega_1\cdots\omega_n)\;\ln\frac{e}{\Delta x\Delta g_l(\omega_1\cdots\omega_n)}\;+\;\mathcal{O}(\Delta x)\,, \end{aligned} \tag{8.111}$$

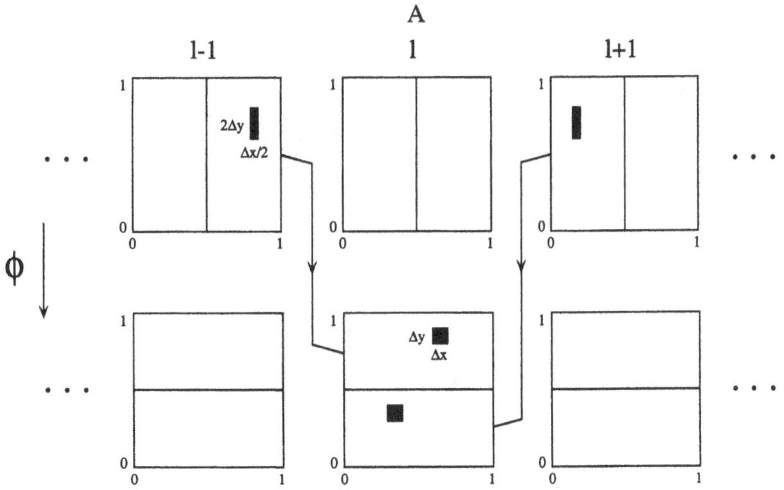

Figure 8.7 Action of the multibaker map on three successive squares of the chain and on the cells $\{B_i\}$ of a $(\Delta x, \Delta y)$-partition of the lth square taken as the domain A in Eq. (8.107).

where we used the uniformity in the x-direction so that the sum over all the cells of size Δx in the interval $[0, 1]$ contributes by a factor $1/\Delta x$, which cancels the factor Δx. A similar cancellation happens for the cells of size $\Delta x/2$ in both half-squares composing $\boldsymbol{\phi}^{-1}A$. Moreover, we notice that the factors $e/\Delta x$ in the logarithms can be eliminated using the properties that

$$\sum_{\omega_1 \cdots \omega_n} \Delta g_l(\omega_1 \cdots \omega_n) = g_l(1) - g_l(0) = \rho_l , \tag{8.112}$$

and that $\rho_{l+1} + \rho_{l-1} = 2\rho_l$.

On the other hand, the ε-entropy flow (8.105) is given by

$$\Delta_e S_\varepsilon = \left[(\Delta x, \Delta y)\text{—entropy of } \boldsymbol{\phi}^{-1}A\right] - \left[(\Delta x, \Delta y)\text{—entropy of } A\right] . \tag{8.113}$$

As a consequence, we obtain the ε-entropy production (8.106) as

$$\Delta_i S_\varepsilon = \left[\left(\frac{\Delta x}{2}, 2\Delta y\right)\text{—entropy of } \boldsymbol{\phi}^{-1}A\right] - \left[(\Delta x, \Delta y)\text{—entropy of } \boldsymbol{\phi}^{-1}A\right] , \tag{8.114}$$

or, by using the stationarity that implies $\Delta S_\varepsilon = 0$ in Eq. (8.107), we get

$$\Delta_i S_\varepsilon = \left[(\Delta x, \Delta y)\text{—entropy of } A\right] - \left[\left(2\Delta x, \frac{\Delta y}{2}\right)\text{—entropy of } A\right] . \tag{8.115}$$

We carry out a calculation similar to that leading to Eq. (8.111). We observe that the cell $\omega_1 \cdots \omega_n$ is composed of both the cells $\omega_1 \cdots \omega_n 0$ and $\omega_1 \cdots \omega_n 1$, so that their probability weights add like $\Delta g_l(\omega_1 \cdots \omega_n) = \Delta g_l(\omega_1 \cdots \omega_n 0) + \Delta g_l(\omega_1 \cdots \omega_n 1)$. Accordingly, the ε-entropy production of the multibaker is explicitly given by

$$\Delta_i S_\varepsilon = \sum_{\omega_1 \cdots \omega_n} \left[\Delta g_l(\omega_1 \cdots \omega_n 0) \ln \frac{2\Delta g_l(\omega_1 \cdots \omega_n 0)}{\Delta g_l(\omega_1 \cdots \omega_n 0) + \Delta g_l(\omega_1 \cdots \omega_n 1)} + \Delta g_l(\omega_1 \cdots \omega_n 1) \ln \frac{2\Delta g_l(\omega_1 \cdots \omega_n 1)}{\Delta g_l(\omega_1 \cdots \omega_n 0) + \Delta g_l(\omega_1 \cdots \omega_n 1)}\right] , \tag{8.116}$$

where we have taken the limit $\Delta x \to 0$ to eliminate the term of $\mathcal{O}(\Delta x)$ which plays no role in our argument because the measure remains regular in the unstable direction x.

The first remarkable property of the ε-entropy production (8.116) is its positivity

$$\Delta_i S_\varepsilon = \sum \left(a \ln \frac{2a}{a+b} + b \ln \frac{2b}{a+b} \right) \geq 0 , \tag{8.117}$$

which follows from the concavity of the function $z \ln z$ [i.e., $(d/dz)^2 (z \ln z) \geq 0$].

A numerical calculation of the entropy production $\Delta_i S_\varepsilon$ given by Eq. (8.116) is depicted in Fig. 8.8 for a multibaker chain of length $L = 50$, as a function of $n = \log_2(1/\Delta y)$. We observe that $\Delta_i S_\varepsilon$ is approximately constant with respect to n. Therefore, the ε-entropy production displays a plateau at a positive value which depends on the position l along the chain. In the case of Fig. 8.8, we have taken $\rho_- = 1$ and $\nabla\rho = 1$ so that the mean density in the lth cell is $\rho_l = l + 2$. Figure 8.9 shows $\Delta_i S_\varepsilon$ as a function of the position l, for different values of ε. We then compare with the behaviour expected from the phenomenological entropy production which is

$$\Delta_i S_{\text{phenom}} = D \, \frac{(\nabla\rho)^2}{\rho_l} = \frac{1}{2(l+2)} , \tag{8.118}$$

for a square of unit length because the diffusion coefficient is $D = 1/2$ in the multibaker. We observe a remarkable agreement between both curves except at the ends of the chain. The decrease of the ε-entropy production at the ends is explained by the fact that the density there is constant at the values $\rho_\pm$ over large parts of the square so that $\Delta_i S_\varepsilon$ tends to zero more rapidly at the ends

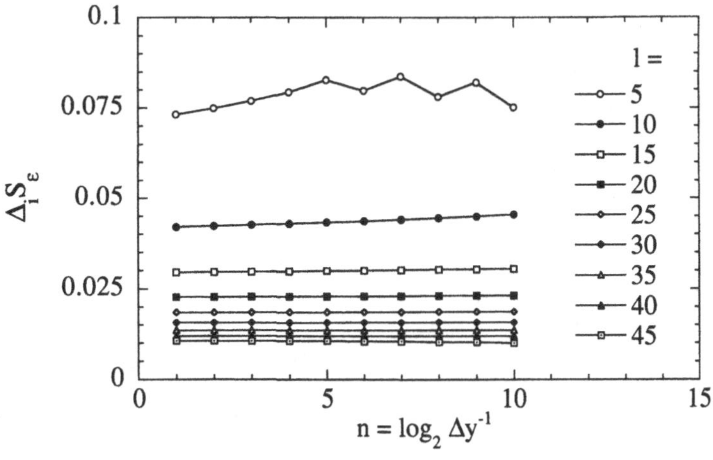

Figure 8.8 The ε-entropy production (8.116) calculated numerically for a multibaker chain of length $L = 50$, a unit gradient $\nabla\rho = 1$, and $\rho_- = 1$. The ε-entropy production is depicted as a function of $n = \log_2(1/\Delta y)$, for different positions l along the chain (Gaspard 1997b).

than in the middle of the chain as $\varepsilon \to 0$. The critical value of Δy below which $\Delta_i S_\varepsilon$ tends to zero depends on the position l along the chain in a way which is determined by Eq. (8.59) as

$$\Delta y_c \sim 2^{-\min(l,L-l)} . \tag{8.119}$$

Hence, the critical scale decreases exponentially fast as $L, l \to \infty$ due to the Lyapunov instability of the dynamics. For a fixed value of Δy, we should thus observe the vanishing of the ε-entropy production only in some boundary layers of the order of the inverse Lyapunov distance: $l_c \sim (\ln 2)^{-1}$. Away from these small boundary layers, the ε-entropy production reaches a positive value which we shall now calculate.

We suppose that we are in the middle of the chain at values of Δy above the critical value (8.119) so that the incomplete Takagi functions in (8.108) can be replaced by the limiting Takagi function. In Eqs. (8.116)–(8.117), we set $a = m + \delta/2$ and $b = m - \delta/2$ and we expand in Taylor series of δ/m to get

$$\Delta_i S_\varepsilon = \sum m \left[\left(\frac{\delta}{2m}\right)^2 + \frac{1}{6}\left(\frac{\delta}{2m}\right)^4 + \mathcal{O}\left(\frac{\delta^6}{m^6}\right) \right] , \tag{8.120}$$

with

$$m = \rho_l \frac{\Delta y}{2} + (\nabla\rho)\frac{\Delta T_a + \Delta T_b}{2} , \qquad \text{and} \qquad \delta = (\nabla\rho)(\Delta T_a - \Delta T_b) , \tag{8.121}$$

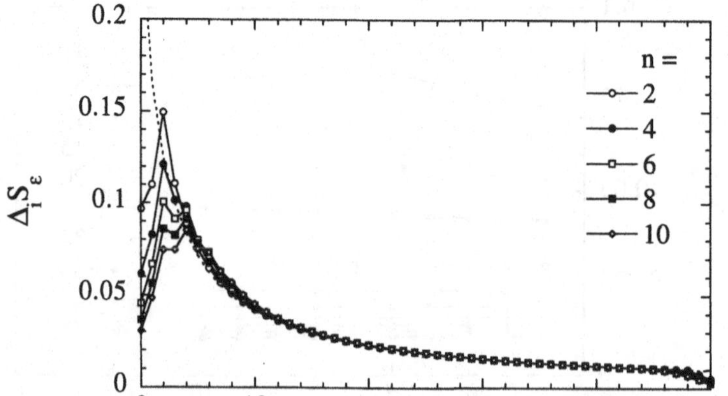

Figure 8.9 The same ε-entropy production as in Fig. 8.8 for a multibaker chain of length $L = 50$ but depicted here as a function of the position l, for different values of $n = \log_2(1/\Delta y)$. The dashed line represents the phenomenological entropy production (8.118) (Gaspard 1997b).

where

$$\Delta T_a = T\left(y_{\omega_1\cdots\omega_n} + \frac{1}{2^{n+1}}\right) - T(y_{\omega_1\cdots\omega_n}) ,$$

$$\Delta T_b = T\left(y_{\omega_1\cdots\omega_n} + \frac{1}{2^n}\right) - T\left(y_{\omega_1\cdots\omega_n} + \frac{1}{2^{n+1}}\right) . \tag{8.122}$$

We can here use the properties (8.33) and (8.34) of the Takagi function which imply that

$$\Delta T_a - \Delta T_b = \frac{1}{2^n} = \Delta y , \tag{8.123}$$

$$\sum(\Delta T_a + \Delta T_b)^2 = \frac{n}{2^n} = \Delta y \log_2 \frac{1}{\Delta y} , \tag{8.124}$$

and, moreover, the property that $\sum(\Delta T_a + \Delta T_b) = T(1) - T(0) = 0$. Expanding in series of $(\nabla\rho)/\rho_l$, we finally obtain

$$\Delta_i S_\varepsilon = \frac{(\nabla\rho)^2}{2\rho_l} + \frac{(\nabla\rho)^4}{2\rho_l^3}\left(\frac{1}{6} + \log_2 \frac{1}{\Delta y}\right) + \mathcal{O}\left[\frac{(\nabla\rho)^6}{\rho_l^5}\right] , \tag{8.125}$$

for $\Delta y > \Delta y_c$. We notice that, if the Takagi function was twice differentiable, Eq. (8.33) would behave like Δy^2 instead of Δy, so that the leading term of Eq. (8.125) would vanish as $\Delta y \to 0$. Therefore, the property (8.33) of nondifferentiability of the Takagi function directly implies the positivity of this leading term and its particular expression.

The most remarkable result is that the leading term is precisely the entropy production expected from irreversible thermodynamics. The next term is a correction which is small like $(\nabla\rho)^4$ and which slowly increases as $\Delta y \to \Delta y_c$. This behaviour is observed in Figs. 8.8 and 8.9. Near the left-hand end of the chain where ρ_l is small enough, we observe a slow linear increase of $\Delta_i S_\varepsilon$ versus $\log_2(1/\Delta y)$ in Fig. 8.8. This increase becomes negligible where ρ_l is larger. We emphasize that these results are entirely due to the Tagaki function and to its nondifferentiability, which therefore controls the entropy production.

8.6.6 Summary

We may summarize the situation about the entropy production as follows.

A system is supposed to sustain a diffusion process and to be submitted to flux boundary conditions. With such boundary conditions, the microscopic invariant measure of density (8.7) is no longer the uniform Liouville measure. In the limit where fixed nonequilibrium gradients are imposed at arbitrarily large distances, arbitrarily small subsets of phase space contain points coming from almost every point of the boundary $\partial\mathcal{M}$ so that the density varies infinitely fast and the measure becomes singular, which is the central result of this chapter.

These asymptotic steady-state measures (8.11) or (8.66) have been known since works by Lebowitz (1959) and MacLennan (1959). The singular character of these steady-state measures forces us to use a coarse-grained entropy or ε-entropy instead of the Gibbs entropy which only applies to regular measure. The ε-entropy production turns out to have precisely the behaviour expected from irreversible thermodynamics for diffusion in the multibaker map, which can be formulated as

$$\lim_{\varepsilon\to 0} \lim_{(\nabla\rho)/\rho\to 0} \lim_{L\to\infty} \frac{\rho}{(\nabla\rho)^2} \Delta_i S_\varepsilon = D > 0 , \tag{8.126}$$

where the limits are not commutative. Because the limit $L \to \infty$ of a large chain has to be taken before the fine-grained limit $\varepsilon \to 0$ we should understand the entropy production as an emerging property appearing in the limit of large systems. We remark that the ε-entropy is the entropy obtained after a coarse graining of the phase space with an ε-partition. It has been known since Gibbs that coarse graining defines an entropy which tends to its equilibrium value because of the mixing property. However, the previous result is independent of the particular coarse graining because the entropy production defined in (8.126) does not vanish in the fine-grained limit. This is in contrast to the usual coarse-graining considerations which depend on the particular partition. We should therefore emphasize that the origin of the result (8.126) is the singular character of the underlying microscopic steady-state measure but not the coarse graining. The entropy production (8.126) would indeed vanish if the limiting steady-state measure was absolutely continuous with respect to the Liouville measure. Coarse graining is only a procedure to define an appropriate characteristic quantity of the dynamical system (see Chapter 4). Hence, the new and fundamental result is that the limiting steady-state measure is singular, which has its origin in the mixing and chaotic properties combined with the nonequilibrium conditions.[3]

We notice that the singular character of the nonequilibrium invariant measure appears very rapidly in large systems because of the convergence property (8.59). We can translate this rapid convergence for a fluid of particles of diameter d as follows. From the analogy with a Lorentz gas, the transition from one square of the multibaker to neighbouring squares corresponds to the free flight of a particle from a collision to the next one, i.e., to a mean free path $\ell \sim 1/(\rho\sigma)$ where ρ is the particle density and $\sigma = \pi d^2$ the collision cross-section. At each collision, a perturbation $\delta\varphi$ on a velocity angle is amplified as $\delta\varphi \to (4\ell/d)\delta\varphi$. At a distance $z = n\ell$ of n mean free paths from the wall, about n collisions

3. This singular character of the measure is not due to the average phase-space volume contraction of thermostatted systems since the systems we here consider are volume-preserving.

have occurred so that the critical scale below which the absolute continuity is hidden is $\Delta\varphi_c \sim (d/4\ell)^n$. Expressed in terms of the distance z from the wall, the critical scale would be $\Delta\varphi_c \sim \exp(-z\lambda)$ where $\lambda \sim (1/\ell)\ln(4\ell/d)$ is the Lyapunov exponent per unit distance. The width of the boundary layer where the ε-entropy production should be smaller than its bulk thermodynamic value should thus be $z_c \sim \ell/\ln(\ell/d)$. It is only beyond this boundary layer that we may expect the ε-entropy production to reach its thermodynamic value. We notice that the scale where the absolute continuity of the nonequilibrium invariant measure is hidden becomes exponentially small as $z \gg z_c$ toward the bulk of the fluid because of the Lyapunov dynamical instability. Therefore, the bulk behaviour of the nonequilibrium steady state is essentially determined by the nondifferentiability of the Tagaki function or, equivalently, by the singularity of the steady-state measures (Gaspard 1997b).

We may conclude that we have here obtained the fundamental mechanism at the origin of the thermodynamic entropy production, which is compatible with a *reversible and volume-preserving* microscopic dynamics. This understanding is based on the hypothesis of microscopic chaos and on the general expressions (8.66) of the limiting steady-state measures, so that our arguments extend *mutatis mutandis* to the other transport and reaction-rate processes of irreversible thermodynamics. Quantum generalizations can also be envisaged since the steady states (8.66) are defined in terms of physical observables.

8.7 Comments on far-from-equilibrium systems

In this chapter, we have been concerned with a Liouvillian approach to nonequilibrium states of the thermodynamic branch, i.e., to states which remain close to the thermodynamic equilibrium. We have shown that it is possible to define the nonequilibrium steady states in terms of invariant measures defined in the microscopic phase space. These invariant measures may become singular when the nonequilibrium constraints are imposed at large distances, which explains the phenomenological entropy production of nonequilibrium thermodynamics.

Here, we would like to point out that the construction of invariant measures to describe nonequilibrium systems in a Liouvillian approach is not restricted to near-equilibrium states, but also extends to far-from-equilibrium systems. Such far-from-equilibrium systems are usually described at the macroscopic level but a microscopic description may be required in certain conditions, as for instance in rarefied gases. We already mentioned that nonequilibrium states can be defined with appropriate boundary conditions on the probability density at the border of the microscopic phase space. We may expect that there exists an invariant measure which is unique for the purpose of averaging over typical trajectories in such systems. All the methods described in particular in Chapter 3 apply to such dynamical systems. The behaviour of the time-correlation functions and of the associated spectral functions can be characterized in terms of Pollicott–Ruelle resonances, in particular. Lyapunov exponents may also be defined as recently shown by Ruelle (1997).

We should expect that the Pollicott–Ruelle resonances of a nonequilibrium system differ from those of the equilibrium system. Indeed, observation reveals that nonequilibrium systems are the stage of new processes like spatial structurations or temporal activities on time scales specific to the nonequilibrium conditions (Nicolis and Prigogine 1977). Nevertheless, we may here also expect that some Pollicott–Ruelle resonances tend to the eigenvalues of the phenomenological equations ruling the time evolution of the macroscopic densities and fields, in the limit of large systems. For instance, the phenomenologically expected resonances associated with far-from-equilibrium oscillations of period T would be $s_n = in(2\pi/T) - \gamma_n$, where γ_n are the decay rates of the time correlations which should be damped due to the statistical fluctuations (Nicolis and Malek Mansour 1978, Baras and Malek Mansour 1997). The inverses of these decay rates define in a sense the lifetimes of the dissipative structure, and they constitute an important aspect of order in nonequilibrium systems (Nicolis and Gaspard 1994). It is interesting to notice that all these resonances of the forward time evolution of the Liouvillian dynamics have a negative real part corresponding to the decay of the time correlations even if the macroscopic description may have unstable trajectories (cf. Chapter 3). Therefore, the Liouvillian description has the stability that is required for the consistency of the macroscopic description in the presence of bifurcations.

Chapter 9

Noises as microscopic chaos

9.1 Differences and similarities between noises and chaos

In the previous chapters, we have shown how to characterize the random processes induced by deterministic dynamical systems. In many cases, the trajectories of Newton's equations are dynamically unstable, which generates chaotic behaviour. This chaos turns out to play an important role in transport phenomena like diffusion in spatially extended systems.

These results suggest that many stochastic models, like the ones by Langevin or Fokker and Planck, which have been developed to study transport and reaction-rate phenomena, can be interpreted in terms of deterministic chaotic processes. The advantage of such an interpretation is that there is no logical gap introduced in the theoretical scheme between the atomic hypothesis that fluid particles obey Newton's equations and the description of random phenomena. Moreover, with this chaotic hypothesis, the origin of randomness is not attributed to the accumulation of many particles but to the intrinsic dynamical instability of the system.

Nevertheless, a distinction should be made between deterministic chaos and stochastic noises beyond the fact that both are probabilistic random processes. Indeed, such processes display clear differences as illustrated with the two following examples.

In Fig. 9.1, a trajectory is plotted of the Rössler deterministic chaotic attractor

(Rössler 1979, Gaspard and Nicolis 1983)

$$\begin{aligned} \dot{X}_1 &= -X_2 - X_3 , \\ \dot{X}_2 &= X_1 + a X_2 , \\ \dot{X}_3 &= b X_1 - c X_3 + X_1 X_3 . \end{aligned} \tag{9.1}$$

For comparison, Fig. 9.2 depicts a trajectory from the Ornstein–Uhlenbeck stochastic process given by the Langevin equation

$$\dot{X} = -a X + c W(t) , \tag{9.2}$$

where $W(t)$ is a δ-correlated white noise of zero mean and unit variance. This process may be viewed as a model of the velocity of a Brownian particle immersed in a fluid.

Both trajectories appear irregular in time. However, there are important differences between them and, in particular, the second is often recognized as

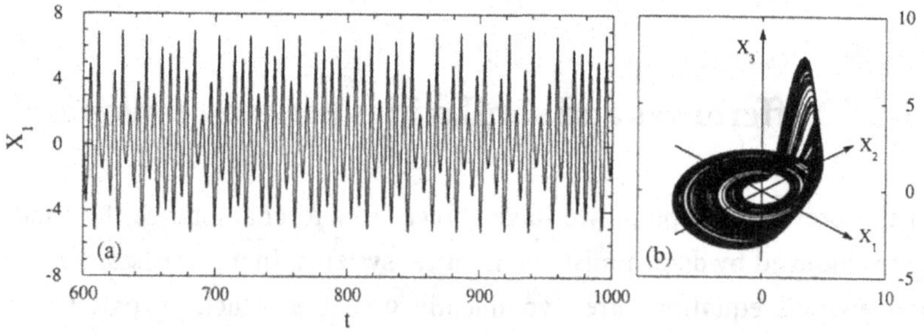

Figure 9.1 (a) Trajectory of the Rössler chaotic system (9.1) with the parameter values $a \simeq 0.32$, $b \simeq 0.3$, and $c \simeq 4.5$, integrated from the initial conditions $X_1 = 0$, $X_2 = -4$, and $X_3 = 0$. (b) The corresponding chaotic attractor depicted in the phase space (X_1, X_2, X_3) (Gaspard and Wang 1993).

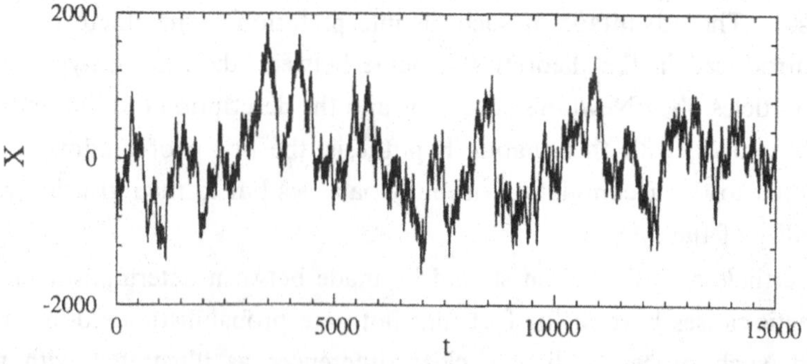

Figure 9.2 Typical trajectory of the Ornstein–Uhlenbeck process of stochastic equation (9.2) with the parameter values $a = 2^{-8}$ and $c = 50$ as an example of a noise signal (Gaspard and Wang 1993).

being more irregular than the first. In this regard, the chaotic trajectory is a differentiable curve according to Cauchy's theorem; while the stochastic signal is nowhere differentiable, irregular on arbitrarily small scales if we refer to the strict mathematical construction of the process. Already, the visual comparison reveals that stochastic noises are qualitatively more random than the chaotic signals. From this observation, we may wonder if there exists a quantitative measure of randomness that would capture our visual intuition.

Such a question is of crucial importance in the natural sciences where the time evolution of many physico-chemical and biological phenomena is described by random processes. Brownian motion has been known for more than a century and was modelled at the beginning of this century by stochastic differential equations like the Langevin equations. The foundations of stochastic processes were firmly established in the thirties and the forties by Wiener and others who showed the importance of white noise (Wax 1954).

It was largely recognized only recently that deterministic nonlinear differential equations like Newton's equations also admit random solutions in the phenomenon of chaos (Hao Bai-Lin 1984, 1990). In this context, dynamical randomness is characterized by the Kolmogorov–Sinai (KS) entropy per unit time as explained in Chapter 4. The system is considered as a source of information in the sense of Shannon and the KS entropy is the data accumulation rate which is minimal to avoid redundancy in the recorded data (Shannon and Weaver 1949). This general reasoning is not limited to the chaotic systems but can be extended to more general stochastic processes. The concept of KS entropy per unit time can be generalized to a concept called (ε, τ)-entropy per unit time, which applies to a very large class of processes that includes both chaotic and stochastic ones (Gaspard and Wang 1993).

The idea is the following. In the definition of the KS entropy per unit time, the phase space is partitioned into small cells. The entropy per unit time is then calculated for the probabilities to visit different cells successively at times separated by a lapse τ. Afterwards, the supremum of the entropy is taken for smaller and smaller partitions. This supremum is finite only in deterministic systems, but it is infinite for stochastic processes because the trajectories of such processes display randomness on arbitrarily small scales. It is therefore necessary to suppose that the cells of a partition have a diameter ε. The entropy per unit time of the processes for such ε-partitions is essentially what we call the (ε, τ)-entropy per unit time. For stochastic noises, the (ε, τ)-entropy is thus a function that diverges as $\varepsilon \to 0$ or as $\tau \to 0$. The dependency of the divergence on ε or τ characterizes the stochastic process as will be shown below. In continuous processes, the dependency on the time lapse τ can be eliminated to recover the concept of ε-entropy per unit time which has been much studied in the mathematical literature and, in particular, by Kolmogorov and Tikhomirov (1959) (see also Kolmogorov 1959, Tikhomirov 1963). In this chapter, one of our purposes will thus be to characterize different time and spacetime random processes in terms of this (ε, τ)-entropy per unit time.

Thereafter, we come back to the comparison between Brownian motion and microscopic chaos of a system of particles, as already described in the Introduction. We argue that Brownian motion can be used to provide evidence for the hypothesis of microscopic chaos.

9.2 (ε, τ)-entropy per unit time

9.2.1 Dynamical processes

A random process is a sequence of random variables defined at successive times. We shall denote the random variables by upper case letters $X, Y, Z, \ldots$. The random variables take their values – denoted by lower case letters $x, y, z, \ldots$ – in $\mathbb{R}^d$.

The process may be discrete or continuous in time. When it is continuous we proceed to a discretization of the signal at small time intervals $\tau = \Delta t$ according to the model of observation described in Subsection 4.1.1. A new multi-time random variable is then defined by

$$\mathbf{X} = [X(t_j)\,,\ X(t_j+\tau)\,,\ X(t_j+2\tau)\,,\ \ldots\,,\ X(t_j+N\tau-\tau)]\,, \tag{9.3}$$

which belongs to $\mathbb{R}^{Nd}$ and which corresponds to the signal during the time period $T = N\tau$. We use bold face letters to emphasize their vectorial character, having N components. Each component is itself a d-dimensional vector corresponding to one of the N successive times. t_j is the initial time of the sequence.

From the point of view of probability theory, the process is defined by the N-time joint probability

$$P(\mathbf{x}; d\mathbf{x}, \tau, N) = \text{Prob}\{\mathbf{x} < \mathbf{X} < \mathbf{x} + d\mathbf{x}\} = p(\mathbf{x})\, d\mathbf{x}\,, \tag{9.4}$$

where $p(\mathbf{x})$ is the probability density for $\mathbf{X}$ to take the value $\mathbf{x} \in \mathbb{R}^{Nd}$. If the process is stationary, the joint probability (9.4) does not depend on the initial time t_j.

In the following sections, we shall also use the probability for the random variable $\mathbf{X}$ to be within a distance ε of the value $\mathbf{x}$

$$P(\mathbf{x}; \varepsilon, \tau, N) = \text{Prob}\{\ \text{dist}[\mathbf{x}, \mathbf{X}] < \varepsilon\}\,. \tag{9.5}$$

This other N-time probability involves the definition of a distance dist$[\cdot,\cdot]$ in $\mathbb{R}^{Nd}$, which will be specified later. The joint probabilities (9.4) and (9.5) are closely related and their knowledge provides in principle a full characterization of the random process over the time interval $T = N\tau$. In this regard, the integer N is equivalent to the time T during which the random process is observed by the measuring device and we shall often replace N by T and vice versa.

When the random variables of the process take discrete values as in birth-and-death processes where the random variables are numbers of particles – $X(t) \in \mathbb{N}^d$ or $\mathbb{Z}^d$ – the signal is not continuous any more and there is no need to introduce the infinitesimal quantities $d\mathbf{x}$ or ε. Similarly, when the process is discrete in time the infinitesimal time τ is not necessary and can be dropped. In that case the total time of observation T is related to the number N of random variables in (9.3) by the average time $\bar{\mathscr{T}}$ between the occurrences of these random variables: $T = N\bar{\mathscr{T}}$.

In physics, chemistry, or biology, the joint probabilities are not the quantities which are primarily available in the definition of a dynamical process. More often, the dynamical process is defined in terms of ordinary or stochastic differential equations from which we need to calculate the joint probability which is invariant under the time evolution, as shown below.

9.2.2 Entropy of a process over a time interval T and a partition $\mathscr{C}$

We now consider a dynamical process defined by a joint probability like (9.4). The time signal is a function $X(t)$ in the phase space $\mathbb{R}^d$.

We partition the phase space into cells, $\mathscr{C} = \{C_1, \ldots, C_L\}$. Each cell is labelled by an integer $\omega \in \{1, 2, \ldots, L\}$. We calculate the probabilities to visit successively the cells $\omega_0\omega_1 \ldots \omega_{N-1}$ at the times $0, \tau, \ldots, (N-1)\tau$. These probabilities are given by

$$\begin{aligned}\mu(\omega_0\omega_1 \ldots \omega_{N-1}) &= \text{Prob}\Big\{X(0) \in C_{\omega_0},\ X(\tau) \in C_{\omega_1},\\ &\qquad \ldots, X(N\tau - \tau) \in C_{\omega_{N-1}}\Big\}\\ &= \int_{\mathbb{R}^{Nd}} I_{C_{\omega_0}}(x_0)\ \ldots\ I_{C_{\omega_{N-1}}}(x_{N-1})\ p(\mathbf{x})\ d\mathbf{x}\,, \qquad (9.6)\end{aligned}$$

where $p(\mathbf{x})$ is the joint probability density in (9.4) and $I_{C_\omega}(x)$ denotes the characteristic function of the cell C_ω.

The entropy of the process with respect to the partition $\mathscr{C}$ – also called the block entropy – is then defined as usual by

$$H(\mathscr{C},\tau,T) \;=\; -\sum_{\omega_0\ldots\omega_{N-1}} \mu(\omega_0\ldots\omega_{N-1})\,\log\,\mu(\omega_0\ldots\omega_{N-1})\,, \tag{9.7}$$

where we recall that $T=\tau N$.

The entropy $H(\mathscr{C},\tau,T)$ grows at most linearly with time

$$H(\mathscr{C},\tau,T) \;\leq\; T\,\log L\,, \tag{9.8}$$

where L is the number of cells in the partition $\mathscr{C}$.

If $\mathscr{C}$ is the partition of the algorithm for the reconstruction (4.2) of the random process, the entropy (9.7) gives an evaluation of the algorithmic complexity of Chapter 4:

$$H(\mathscr{C},\tau,T) \;\simeq\; K(\omega_T)\,, \tag{9.9}$$

for almost all trajectories [cf. Eqs. (4.6)–(4.9)]. According to the results (4.3)–(4.5), the entropy (9.7) can thus distinguish between the regularly random processes and the other processes where information is produced more slowly than proportionally with time:

$$H(\mathscr{C},\tau,T)$$

$$\sim\; \log\,T\,, \qquad \text{(quasiperiodicity)}, \tag{9.10}$$

$$\sim\; T^{\alpha}\;(0<\alpha<1)\;\text{ or }\;\frac{T}{(\log T)^{\beta}}\;(\beta>0)\,, \qquad \text{(sporadic randomness)}, \tag{9.11}$$

$$\sim\; T\,, \qquad \text{(regular randomness)}. \tag{9.12}$$

In the case of regular randomness, the entropy per unit time of the system with respect to the partition $\mathscr{C}$ is then defined by the limit

$$h(\mathscr{C},\tau) \;=\; \lim_{T\to\infty}\,\frac{1}{T}\,H(\mathscr{C},\tau,T)\,. \tag{9.13}$$

Let us make the remark that the basis of the logarithms used in Eq. (9.7) fixes the unit of the entropy per unit time according to

$\log_2$:	bits/second ,
$\log_e = \ln$:	nats/second ,
$\log_{10}$:	digits/second .

9.2.3 Partition (ε,τ)-entropy per unit time

However, as we shall see in the next section, no supremum exists for stochastic processes which appear to have a higher degree of randomness than chaotic systems. There is no supremum because the entropy per unit time $h(\mathscr{C},\tau)$ grows indefinitely as the cells of the partition become smaller and smaller. To control this growth, let us suppose that the diameter of the cells $\{C_i\}$ is smaller or larger than ε and define partition (ε, τ)-entropies per unit time according to

$$\underline{h}_{\rm P}(\varepsilon,\tau) = {\rm Inf}_{\mathscr{C}:\ {\rm diam}(C_i)\leq\varepsilon}\ h(\mathscr{C},\tau)\ , \qquad \text{and} \tag{9.14}$$

$$\bar{h}_{\rm P}(\varepsilon,\tau) = {\rm Sup}_{\mathscr{C}:\ {\rm diam}(C_i)\geq\varepsilon}\ h(\mathscr{C},\tau)\ , \tag{9.15}$$

which are functions of ε and τ and where the index P refers to the word partition. An estimation of such quantities can be obtained by using specific partitions into identical cells like cubic cells.

In deterministic systems, the KS entropy per unit time is obtained from the partition (ε, τ)-entropy (9.15) by taking the limit

$$h_{\rm KS} = \lim_{\varepsilon\to 0} \bar{h}_{\rm P}(\varepsilon,\tau)\ , \tag{9.16}$$

which realizes the supremum over all possible partitions in Eq. (4.9). Equation (9.16) holds in principle for arbitrary nonvanishing values of the time interval $\tau = \Delta t$, as in Chapter 4.

9.2.4 Cohen–Procaccia (ε, τ)-entropy per unit time

A numerical method to evaluate an (ε, τ)-entropy per unit time has been provided by Cohen and Procaccia (1985). A realization of the process $X(t)$ over a very long time interval $M\tau >> T = N\tau$ is given by the time series $\{x(j\tau)\}_{j=0}^{M-1}$. Within this long time series, sequences of length N are compared with each other. A set of $R << M$ reference sequences, which are also of length N, is considered

$$\mathbf{x}_i = [x(i\tau),\ldots,\ x(i\tau+N\tau-\tau)]\ , \qquad i\in\{j_1,\ldots,j_R\}\ . \tag{9.17}$$

The distance between a reference sequence and another sequence of length N is defined by

$$\begin{aligned}{\rm dist}_N[\mathbf{x}_i,\mathbf{x}_j] &= \\ \max\Big\{|x(i\tau)-x(j\tau)|,\ldots,\ &|x(i\tau+N\tau-\tau)-x(j\tau+N\tau-\tau)|\Big\}\ ,\end{aligned} \tag{9.18}$$

for $j = 1,\ldots, M' = M - N + 1$.

The probability (9.5) for this distance to be smaller than ε is then evaluated by

$$P(\mathbf{x}_i;\varepsilon,\tau,N) = \frac{1}{M'}\ {\rm Number}\Big\{\mathbf{x}_j\ :\ \ {\rm dist}_N[\mathbf{x}_i,\mathbf{x}_j]\ \leq\ \varepsilon\Big\}\ , \tag{9.19}$$

where $M' = \text{Number}\{\mathbf{x}_j\}$. The average of the logarithm of these probabilities over the different reference points $\mathbf{x}_i$ is then calculated

$$\mathscr{H}(\varepsilon, \tau, N) = -\frac{1}{R} \sum_{\{\mathbf{x}_i\}} \log P(\mathbf{x}_i; \varepsilon, \tau, N) , \tag{9.20}$$

where $R = \text{Number}\{\mathrm{x}_i\}$. The Cohen–Procaccia (CP) (ε, τ)-entropy per unit time is then

$$h_{\mathrm{CP}}(\varepsilon, \tau) = \frac{1}{\tau} \lim_{N\to\infty} \lim_{R,M'\to\infty} \left[\mathscr{H}(\varepsilon, \tau, N+1) - \mathscr{H}(\varepsilon, \tau, N)\right] . \tag{9.21}$$

For chaotic systems, it is known that

$$h_{\mathrm{KS}} = \lim_{\varepsilon\to 0} h_{\mathrm{CP}}(\varepsilon, \tau) , \tag{9.22}$$

for values of τ which are neither too large nor too small (Eckmann and Ruelle 1985). Prescriptions have been given in the literature concerning the choice of τ (Fraser and Swinney 1986, Ravani et al. 1988).

We shall show in the following that the Cohen–Procaccia method can also be applied to noise signals where $h_{\mathrm{CP}}(\varepsilon, \tau)$ may diverge as $\varepsilon, \tau \to 0$, in a similar way as the partition (ε, τ)-entropy per unit time but with a different prefactor.

9.2.5 Shannon–Kolmogorov (ε, τ)-entropy per unit time

Other criteria than before may be used to introduce the quantity ε. A particularly important variant is the Shannon–Kolmogorov ε-entropy per unit time which was originally called the rate distortion function by Shannon (1960) and was called the ε-entropy by Kolmogorov (1956). This quantity has been introduced for continuous-amplitude stationary sources.

Let us suppose that we want to reproduce the time signal of a process up to a precision ε. The exact time signal $X(t)$ is then approached by an approximate copy $Y(t)$. The distortion between them is measured by a cost function

$$\rho_N(\mathbf{X}, \mathbf{Y}) = \frac{1}{N} \sum_{j=0}^{N-1} \rho[X(j\tau) - Y(j\tau)] \simeq \frac{1}{T} \int_0^T \rho[X(t) - Y(t)]\, dt , \tag{9.23}$$

where the distance $\rho(X, Y)$ is chosen to depend only on the difference $(X - Y)$. It can be for instance the absolute error function, $\rho(z) = |z|$, or the squared-error function, $\rho(z) = z^2$. The cost function measures the distance between the copy $Y(t)$ and the actual process $X(t)$. We require that their separation measured with the cost function is not larger than $\rho(\varepsilon)$. There remains the important question: what is the general method used to construct the copy $Y(t)$ from our knowledge of the actual process $X(t)$? The following scheme is adopted.

The probability density for the actual signal $X(t)$ to take successively the values

$$\mathbf{x} = [x(0),\ x(\tau),\ \dots\ ,\ x(N\tau-\tau)]\ , \tag{9.24}$$

is given by $p(\mathbf{x})$. The approximate signal $Y(t)$ is constructed according to a joint probability for $X(t)$ to take the values $\mathbf{x}$ while $Y(t)$ takes the values

$$\mathbf{y} = [y(0),\ y(\tau),\ \dots\ ,\ y(N\tau-\tau)]\ . \tag{9.25}$$

Its conditional probability density is a general function $q(\mathbf{y}|\mathbf{x})$. The average mutual information between $\mathbf{x}$ and $\mathbf{y}$ is then

$$J[q] = \int_{\mathbb{R}^{Nd}} \int_{\mathbb{R}^{Nd}} d\mathbf{x}\ d\mathbf{y}\ p(\mathbf{x})\ q(\mathbf{y}|\mathbf{x}) \log \frac{q(\mathbf{y}|\mathbf{x})}{q(\mathbf{y})}\ , \tag{9.26}$$

where

$$q(\mathbf{y}) = \int_{\mathbb{R}^{Nd}} d\mathbf{x}\ p(\mathbf{x})\ q(\mathbf{y}|\mathbf{x})\ . \tag{9.27}$$

The average cost due to the distortion of $\mathbf{y}$ with respect to $\mathbf{x}$ is evaluated by

$$\bar{\rho}_N[q] = \int_{\mathbb{R}^{Nd}} \int_{\mathbb{R}^{Nd}} d\mathbf{x}\ d\mathbf{y}\ p(\mathbf{x})\ q(\mathbf{y}|\mathbf{x})\ \rho_N(\mathbf{x},\mathbf{y})\ , \tag{9.28}$$

using the definition (9.23). This average cost depends on the way the process $Y(t)$ is constructed. This dependency appears through the conditional probability density $q(\mathbf{y}|\mathbf{x})$ which has been arbitrary up till now in a similar way as the partition $\mathscr{C}$ was arbitrary in (9.7). To deal with this arbitrariness, we consider the set $\mathscr{Q}(\varepsilon)$ of all conditional probabilities such that the average cost is less than $\rho(\varepsilon)$

$$\mathscr{Q}(\varepsilon) = \Big\{ q(\mathbf{y}|\mathbf{x}) \ :\ \ \bar{\rho}_N[q]\ \leq \rho(\varepsilon) \Big\}\ . \tag{9.29}$$

The Shannon–Kolmogorov (SK) (ε,τ)-entropy is then defined by

$$H_{\mathrm{SK}}(\varepsilon,\tau,T) = \mathrm{Inf}_{q\in\mathscr{Q}(\varepsilon)}\ J[q]\ . \tag{9.30}$$

Whereupon, the SK (ε,τ)-entropy per unit time is

$$h_{\mathrm{SK}}(\varepsilon,\tau) = \lim_{T\to\infty} \frac{1}{T}\ H_{\mathrm{SK}}(\varepsilon,\tau,T)\ . \tag{9.31}$$

An important property of the (ε,τ)-entropy per unit time is that it is monotonically non-decreasing in ε. The entropy may or may not depend on τ according to the specific choice of the cost function (9.23). The entropy no longer depends on τ if the cost function is taken as an integral over a time interval T in the case of a time-continuous process.

When the process is stationary and Gaussian it is possible to calculate explicitly the ε-entropy per unit time thanks to a formula obtained by Kolmogorov (1956).

For stationary Gaussian processes in one dimension ($d = 1$), the joint probability density is given by

$$p(\mathbf{x}) = \frac{\exp\left(-\frac{1}{2}\,\mathbf{x}^{\mathrm{T}} \cdot \mathbf{C}_N^{-1} \cdot \mathbf{x}\right)}{(2\pi)^{N/2}(\det \mathbf{C}_N)^{1/2}} , \tag{9.32}$$

where $\mathbf{C}_N$ is the matrix of the correlations

$$[\mathbf{C}_N]_{ij} = \langle X_i X_j \rangle = C(|i-j|) \qquad \text{with} \quad \langle X_i \rangle = 0 . \tag{9.33}$$

For time-continuous processes, X_i denotes the random variable $X(t_i)$ at the discretized time $t_i = i\tau$. Since the process is stationary, the correlation function depends only on $|i-j|$ and $\mathbf{C}_N$ is a symmetric Toeplitz matrix. For Gaussian processes, the amplitude is always a continuous random variable and the dependency on τ of the (ε, τ)-entropy per unit time disappears in the limit $\tau \to 0$.

For time-discrete processes, the spectral density is defined by

$$S(\omega) = \sum_{n=-\infty}^{+\infty} \exp(-i\omega n)\, \langle X_n X_0 \rangle , \tag{9.34}$$

which is defined in the frequency interval $\omega \in]-\pi, +\pi]$. On the other hand, for time-continuous processes, the spectral density is

$$S(\omega) = \int_{-\infty}^{+\infty} \exp(-i\omega t)\, \langle X(t) X(0) \rangle \, dt , \tag{9.35}$$

where $\omega \in \mathbb{R}$.

The Shannon–Kolmogorov ε-entropy can be estimated in the case of the squared-error cost function. This fidelity criterion requires that the approximate signal $Y(t)$ is close to the exact signal $X(t)$ according to

$$\lim_{T\to\infty} \frac{1}{T} \int_0^T \langle [X(t) - Y(t)]^2 \rangle \, dt \leq \varepsilon^2 . \tag{9.36}$$

The following Kolmogorov formula then gives an exact evaluation of $h_{\mathrm{SK}}(\varepsilon)$ for stationary Gaussian processes

$$\varepsilon^2 = \frac{1}{2\pi} \int_{-\infty}^{+\infty} \min[\theta \, , \, S(\omega)] \, d\omega , \tag{9.37}$$

$$h_{\mathrm{SK}}(\varepsilon) = \frac{1}{4\pi} \int_{-\infty}^{+\infty} \max\left[0 \, , \, \log \frac{S(\omega)}{\theta}\right] d\omega , \tag{9.38}$$

in terms of the spectral density (9.35). For time-discrete processes the integration should be carried out from $-\pi$ to $+\pi$ and the spectral density (9.34) used. The geometry involved in the calculation of the integrals (9.37)–(9.38) is schematically depicted in Fig. 9.3. A derivation of this fascinating formula can be found in the book by Berger (1971).

9.3 Time random processes

9.3.1 Deterministic processes

For random process, the (ε, τ)-entropy reduces to the KS entropy defined in Chapter 4, in the limit $\varepsilon \to 0$ according to Eqs. (9.16) or (9.22). As a consequence, we should expect that the (ε, τ)-entropy presents a plateau as a function of ε, where the value of the KS entropy is reached. According to Pesin's theorem (cf. Chapter 4), the (ε, τ)-entropy should then saturate at a value given by the sum of the positive Lyapunov exponents $\{\bar{\lambda}_i\}$. This is indeed observed in numerical calculation of the (ε, τ)-entropy with the Cohen–Procaccia method (see Fig. 9.4).

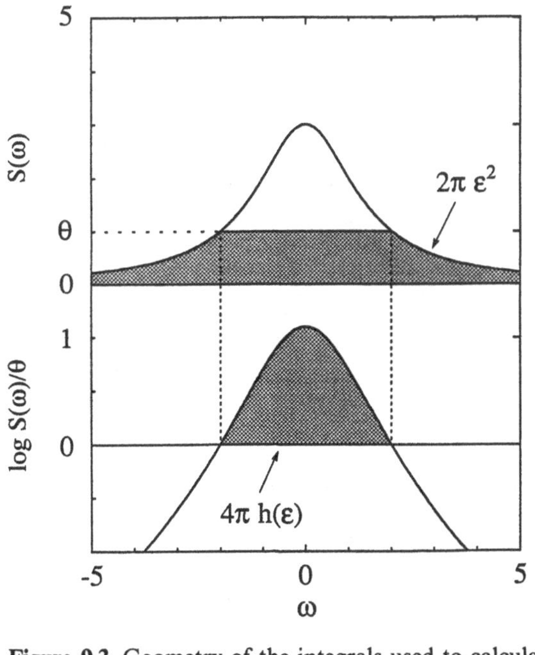

Figure 9.3 Geometry of the integrals used to calculate the Shannon–Kolmogorov ε-entropy from the spectral density $S(\omega)$ of a stationary Gaussian process with the Kolmogorov formula (9.37)–(9.38). The upper diagram corresponds to the first integral (9.37) which gives ε in terms of the intermediate quantity θ. The lower diagram corresponds to the second integral (9.38) giving the ε-entropy (Gaspard and Wang 1993).

9.3.2 Bernoulli and Markov chains

Bernoulli and Markov chains are time- and amplitude-discrete random processes defined by a matrix of transition probabilities $P_{\alpha\beta}$ satisfying Eq. (4.105). The random events occur at time intervals separated by the non-infinitesimal average time constant $\bar{\mathcal{T}}$.

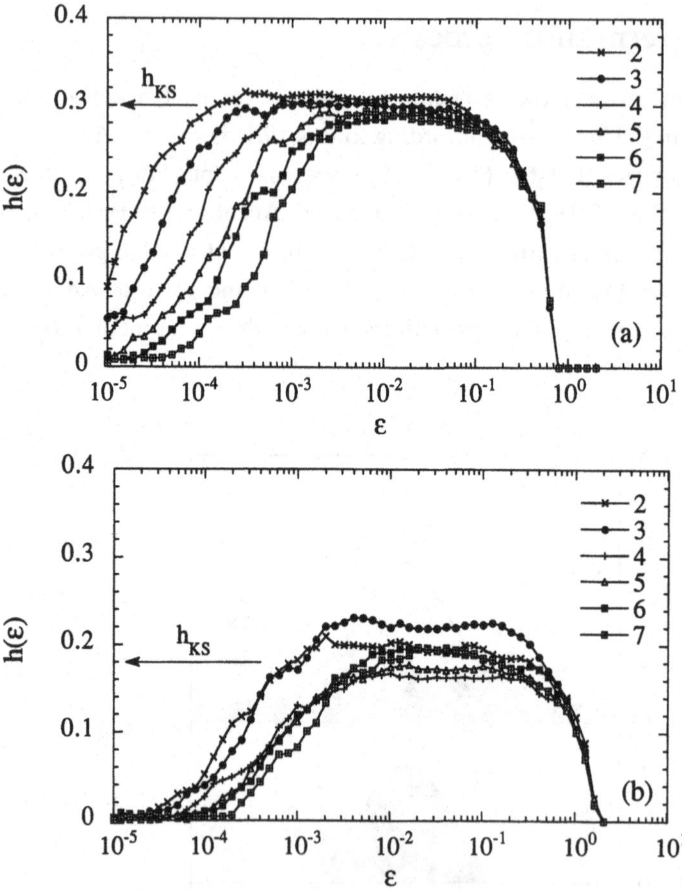

Figure 9.4 (a) Numerical evaluation of the KS entropy per unit time for the one-dimensional logistic map $X_{t+1} = 4X_t(1 - X_t)$ using the Cohen–Procaccia method. The known value of the KS entropy is $h_{KS} = \bar{\lambda} = \log 2 =$ 0.30 digits/iteration. (b) The same for the X-component of the two-dimensional Hénon map $X_{t+1} = 1 + Y_t - 1.4X_t^2$, $Y_{t+1} = 0.3X_t$ where $h_{KS} = \bar{\lambda} = 0.18$ digits/iteration. The different curves correspond to the calculation of the entropy for time sequences of lengths $N = 2 - 7$ [cf. (9.17)–(9.21)]. Since the total length of the time series is limited to $M = 131072$, the probability of finding a string of length N decreases as N increases. As a consequence, the statistics of long strings diminishes at small ε and the plateau of the corresponding curve shrinks as seen in the figures (Gaspard and Wang 1993).

Because of the Markov property of the statistical independence of events separated by times larger than $\bar{\mathcal{T}}$, the joint probability to visit different states successively factorizes as expressed by Eq. (4.106). Accordingly, the (ε, τ)-entropy per unit time is a constant giving the KS entropy per unit time

$$h_{KS} = -\frac{1}{\bar{\mathcal{T}}} \sum_{\alpha} p_\alpha \log p_\alpha \qquad \text{(Bernoulli chains)}, \tag{9.39}$$

$$h_{KS} = -\frac{1}{\bar{\mathcal{T}}} \sum_{\alpha\beta} p_\alpha P_{\alpha\beta} \log P_{\alpha\beta} \qquad \text{(Markov chains)}. \tag{9.40}$$

The finiteness of the entropy per unit time has its origin in the discreteness of these random processes in time: the random choices between the different states generate a finite amount of information per time step $\bar{\mathcal{T}}$. The (ε, τ)-entropy per unit time is bounded because Markov chains belong to the same class as the deterministic systems with a continuous phase space as discussed in Section 4.6. Both the random Markov chains and the deterministic chaotic systems are therefore sources of information of the same degree of dynamical randomness.

For long, a fundamental distinction was made between discrete Markov chains and deterministic systems. In particular, it was known that Markov chains obey an H-theorem (Schnakenberg 1976). Recent works have shown that Markov partitions of a deterministic system can be used to prove the H-theorem for deterministic systems (Nicolis and Nicolis 1988). Such an H-theorem can be established for arbitrarily small Markov partitions into cells corresponding to the multiplet states $\omega_0\omega_1 \cdots \omega_{N-1}$. However, the decay of the H-theorem is further and further delayed in the point limit $N \to \infty$.

From the preceding discussion, we see that a totally new point of view must be adopted. Both Markov chains and deterministic systems must be considered as random processes, if both have a positive KS entropy per unit time. We shall be able to say that one process is more random than the other if there is a difference between their KS entropies per unit time. On the other hand, there exists an isomorphism between them if their KS entropies take the same value, i.e., if they produce precisely the same dynamical randomness. Whether the dynamical process is defined by a deterministic map or by a Markov chain therefore becomes irrelevant from the point of view of probability theory.

9.3.3 Birth-and-death processes

These processes are often encountered in kinetic theories (Schnakenberg 1976, Nicolis and Prigogine 1977, van Kampen 1981). Certainly, the simplest and most well-known among them is the Poisson random process which is defined as follows.

The time axis is divided into small time intervals τ. During each small interval, several events may occur. There is a random variable N_j which is the number of events in the jth time interval $t \in [j\tau, (j+1)\tau[$ and which is defined by the probabilities $p_n = \text{Prob}\{N_j = n\}$

$$\begin{aligned} p_0 &= 1 - w\,\tau + \mathcal{O}(\tau^2)\,, \\ p_1 &= w\,\tau + \mathcal{O}(\tau^2)\,, \\ p_2 &= p_3 = \ldots = \mathcal{O}(\tau^2)\,, \end{aligned} \tag{9.41}$$

where w is the rate at which the events occur. Finally, it is assumed that the random variables N_j are mutually independent so that the discretized process forms a Bernoulli chain. Since there is a random variable for each infinitesimal time interval τ, the process has a τ-entropy per unit time which will diverge as $\tau \to 0$. We can calculate it using (9.40) with $\bar{\mathcal{T}}$ here replaced by τ

$$h(\tau) = -\frac{1}{\tau} \sum_{n=0}^{\infty} p_n \ln p_n\,, \tag{9.42}$$

and we obtain

$$h(\tau) = w \ln \frac{e}{w\tau} + \mathcal{O}(\tau)\,, \tag{9.43}$$

where e is the basis of natural logarithms.

The same result can be obtained in another way using the property that the random variable $\mathcal{T}_k$, which is the time between two successive events, has an exponential distribution of parameter w

$$\text{Prob}\{\mathcal{T}_k > t\} = \exp(-wt)\,. \tag{9.44}$$

The different random times $\mathcal{T}_k$ are independent. The corresponding probability density is $p(t) = w\exp(-wt)$ and the expectation value is $\bar{\mathcal{T}} = 1/w$. A τ-entropy per unit time can be evaluated using a discretization of the t-axis of this continuous variable distribution. Each cell centered on $t_i = i\tau$ has a probability mass equal to $P_i = p(t_i)\tau$ and we can use the following definition for the τ-entropy per unit time

$$\tilde{h}(\tau) = -\frac{1}{\bar{\mathcal{T}}} \sum_{i=0}^{\infty} P_i \ln P_i\,, \tag{9.45}$$

where the tilde is there to recall that this definition is a priori different from the preceding one (9.42). If we introduce the definitions we made in Eq. (9.45) it is easy to see that the same leading term as (9.43) is recovered in the limit $\tau \to 0$.

The previous analysis can be extended to general birth-and-death processes governed by the Pauli master equation

$$\frac{dp_\beta}{dt} = \sum_\alpha p_\alpha W_{\alpha\beta} , \tag{9.46}$$

where $p_\alpha(t)$ is the continuous time probability to be in the state α while $W_{\alpha\beta}$ is the rate for a transition between the states α and β. These rates are here assumed to be constant in time. The conservation of probability implies that

$$W_{\alpha\alpha} = - \sum_{\beta(\neq\alpha)} W_{\alpha\beta} . \tag{9.47}$$

Accordingly, the solution of the master equation is

$$\sum_\alpha p_\alpha(0) P_{\alpha\beta}(t) = p_\beta(t) , \qquad \text{with} \quad P_{\alpha\beta}(t) = [\exp \mathbf{W}t]_{\alpha\beta} , \tag{9.48}$$

where **W** is the matrix composed of the elements $W_{\alpha\beta}$. Figure 9.5 shows a typical trajectory of a birth-and-death process.

For the purpose of calculating the τ-entropy per unit time, we shall follow the first method above. If we discretize the time axis the probability to be in the state α at the time $t_j = j\tau$ obeys a Markov chain according to (9.48). The transition probabilities in (9.48) are expanded in Taylor series

$$P_{\alpha\beta}(\tau) = \delta_{\alpha\beta} + W_{\alpha\beta} \tau + \mathcal{O}(\tau^2) . \tag{9.49}$$

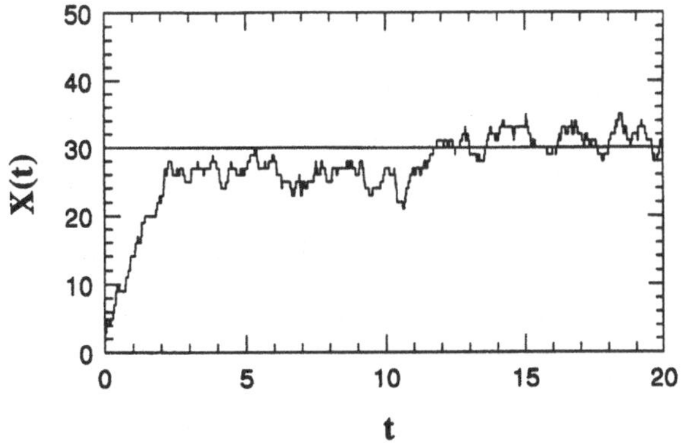

Figure 9.5 Typical trajectory of the birth-and-death process $X \leftrightarrow Y$ with reaction rates $\kappa_\pm$. The total number of particles is assumed to be constant $X + Y = N$. The master equation is $\dot{p}_X = \kappa_+(X+1)p_{X+1} + \kappa_-(N - X + 1)p_{X-1} - [\kappa_+ X + \kappa_-(N - X)]p_X$. The constants were chosen to be $\kappa_+ = 0.2$, $\kappa_- = 0.3$, and $N = 50$. The horizontal line indicates the mean value $\langle X \rangle = N\kappa_-/(\kappa_- + \kappa_+) = 30$ (Gaspard and Wang 1993).

As for the Poisson process, the τ-entropy per unit time is defined using (9.40) with $\bar{\mathscr{T}}$ replaced by τ

$$h(\tau) = -\frac{1}{\tau} \sum_{\alpha\beta} p_\alpha^0 \, P_{\alpha\beta}(\tau) \ln P_{\alpha\beta}(\tau) \, , \tag{9.50}$$

where p_α^0 are the stationary probabilities, solutions of

$$\sum_\alpha p_\alpha^0 \, W_{\alpha\beta} = 0 \, . \tag{9.51}$$

In the limit of small τ, we get

$$h(\tau) = \left(\sum_{\alpha\neq\beta} p_\alpha^0 \, W_{\alpha\beta}\right)\left(\ln \frac{e}{\tau}\right) - \sum_{\alpha\neq\beta} p_\alpha^0 \, W_{\alpha\beta} \ln W_{\alpha\beta} + \mathscr{O}(\tau) \sim \ln \frac{1}{\tau} \, , \tag{9.52}$$

so that the divergence is logarithmic in τ as for the Poisson process.

Hence, we are in the presence of a τ-entropy per unit time. The discreteness of the random variables suppresses the need for any dependency on ε. However, the process being continuous in time, we need to keep the dependency on the small time interval τ. As $\tau \to 0$, the entropy per unit time slowly increases toward infinity so that the birth-and-death processes are more random than the Markov chains and the chaotic systems.

9.3.4 Time-discrete, amplitude-continuous random processes

Let us now suppose that the random process is discrete in time but continuous in space. The random variables $\{\ldots, X_{-1}, X_0, X_1, X_2, \ldots\}$ take their values in $\mathbb{R}^d$. If we suppose that the variables are independent in time the joint probability is

$$\text{Prob}\Big\{x_0 < X_0 < x_0 + dx_0, \; \ldots, \; x_{N-1} < X_{N-1} < x_{N-1} + dx_{N-1}\Big\} = p(x_0)\, p(x_1) \ldots p(x_{N-1}) \, dx_0 \ldots dx_{N-1} \, . \tag{9.53}$$

The ε-entropy per unit time (9.14)–(9.15) can be estimated by taking a partition of $\mathbb{R}^d$ into small hypercubes C_i of side ε

$$h_{\tilde{\mathrm{P}}}(\varepsilon) = -\sum_i \left[\int_{C_i} p(x)\, dx\right] \ln \left[\int_{C_i} p(x)\, dx\right] , \tag{9.54}$$

where $\tilde{\mathrm{P}}$ refers to the particular partition into hypercubes. We see that the (ε, τ)-entropy per unit time is here independent of τ because time is discrete.

Assuming that the density $p(x)$ admits second derivatives

$$\int_{C_i} p(x)\, dx = \varepsilon^d\, p(x_i) + \mathcal{O}(\varepsilon^{d+2}) , \tag{9.55}$$

where x_i is the centre of the hypercube C_i and d is the dimensionality of the space where the random variables X_j take their values. The partition ε-entropy per unit time is therefore

$$h_{\mathfrak{p}}(\varepsilon) = d \ln \frac{1}{\varepsilon} - \int_{\mathbb{R}^d} dx\, p(x) \ln p(x) + \mathcal{O}(\varepsilon) . \tag{9.56}$$

Compared with the chaotic processes, we see that the time-discrete amplitude-continuous processes are more random since their ε-entropy per unit time diverges as $\varepsilon \to 0$. This divergence originates from the fact that the probability distribution is continuous. Indeed, the outcome of each new random event can arise in any small interval of $\mathbb{R}^d$ irrespective of the previous outcome. This is in contrast with chaotic systems which are constrained by the deterministic differential equation so that if the trajectory is passing in an infinitesimal ball of size ε of a Poincaré surface of section, the next passage is constrained to occur within a ball which is determined by the flow and which is still of infinitesimal size $\varepsilon \exp(\lambda \bar{\mathcal{T}})$. The Lyapunov exponent λ limits the ε-entropy to a finite value, independent of ε.

The result (9.56) can be generalized to processes composed of independent random variables X_j whose probability measure has a fractal set for support. If we suppose that d_{I} is the information dimension of this measure the partition ε-entropy per unit time is then given by

$$h_{\mathfrak{p}}(\varepsilon) \simeq d_{\mathrm{I}} \ln \frac{1}{\varepsilon} . \tag{9.57}$$

On the other hand, the Cohen–Procaccia ε-entropy per unit time is given by Eq. (9.56) with ε replaced by 2ε because a hypercubic cell such that $\mathrm{dist}[x, X] \leq \varepsilon$ is of side 2ε according to the distance (9.18).

When the process is moreover assumed to be Gaussian, the Shannon–Kolmogorov ε-entropy can be calculated for a squared-error cost function $\rho(z) = z^2$ to get

$$h_{\mathrm{SK}}(\varepsilon) = d \ln \frac{1}{\varepsilon} - \int_{\mathbb{R}^d} dx\, p(x) \ln p(x) - d \ln \sqrt{2\pi e} + \mathcal{O}(\varepsilon) , \tag{9.58}$$

where $p(x)$ is the d-dimensional Gaussian density (Tikhomirov 1963). For correlated Gaussian random variables, the Shannon–Kolmogorov ε-entropy can be calculated with (9.37)–(9.38) as

$$h_{\mathrm{SK}}(\varepsilon) = \ln \frac{1}{\varepsilon} + \frac{1}{4\pi} \int_{-\pi}^{+\pi} \ln S(\omega)\, d\omega , \tag{9.59}$$

in terms of the spectral density (9.34) for one-dimensional processes (Tikhomirov 1963).

For independent Gaussian processes, we find therefore that the different ε-entropies obey the inequalities

$$h_{\mathrm{SK}}(\varepsilon) \;\leq\; h_{\mathrm{CP}}(\varepsilon) \;\leq\; h_{\mathfrak{P}}(\varepsilon)\,, \tag{9.60}$$

for small ε. In spite of their differences, they display the same dominant behaviour as $\varepsilon \to 0$

$$h_{\mathfrak{P}}(\varepsilon) \;\simeq\; h_{\mathrm{CP}}(\varepsilon) \;\simeq\; h_{\mathrm{SK}}(\varepsilon) \;\simeq\; d\,\log\frac{1}{\varepsilon}\,. \tag{9.61}$$

In this sense, the different entropies are equivalent. As a consequence, we may conclude that the logarithmic divergence (9.61) is characteristic of time-discrete amplitude-continuous random processes. To illustrate the previous result, Fig. 9.6 depicts the ε-entropy for independent Gaussian random variables where we observe the expected divergence (9.61).

The same logarithmic divergence appears in deterministic mappings which are perturbed by some noise described by a time-discrete amplitude-continuous Gaussian random process as for

$$Z_t \;=\; X_t \;+\; a\,Y_t\,, \qquad \text{where} \quad X_{t+1} \;=\; 4\,X_t\,(1-X_t)\,, \tag{9.62}$$

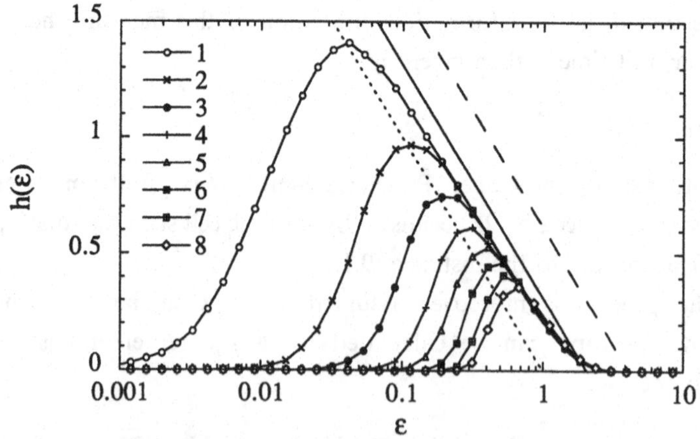

Figure 9.6 Numerical evaluation of the CP ε-entropy in digits per unit time, for the random process of independent one-dimensional Gaussian variables of variance $\langle X^2\rangle = 1$. The numerical CP ε-entropy is compared with its theoretical value (solid line) as well as with the P (long dashed line) and the SK (short dashed line) ε-entropies respectively given by (9.56) and (9.58) ($d = 1$). The different curves correspond to the calculation of the entropy for time sequences of lengths $N = 1 - 8$ [cf. (9.17)–(9.21)] (Gaspard and Wang 1993).

while the random variables Y_t are independent Gaussians of zero mean and unit variance. When $a = 0$, the random process is purely deterministic so that the ε-entropy per unit time presents a plateau at the constant value of the KS entropy $h_{KS} = \log\ 2$ (see Fig. 9.4a). However, when a is not vanishing but small, the signal becomes noisy at small scales ε where the ε-entropy per unit time now increases as $\log(a/\varepsilon)$ according to (9.56). Nevertheless, at values of ε which are larger than the size a of the fluctuations, the noise cannot be resolved so that there remains the plateau at $h(\varepsilon) \simeq h_{KS}$. For larger noise amplitudes a, the rise in the ε-entropy starts at larger values of ε so that the plateau shrinks (see Fig. 9.7).

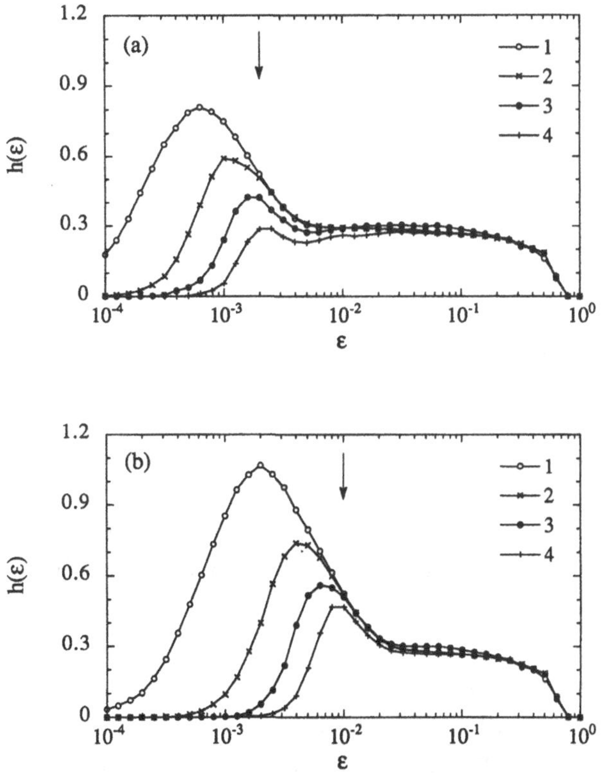

Figure 9.7 Numerical CP ε-entropy in digits per unit time, for the noisy maps (9.62) with (a) $a = 0.002$; (b) $a = 0.01$. The arrows show the corresponding values of the noise amplitude a. Note the plateau at the value of the KS entropy of the logistic map ($h_{KS} = \log 2 = 0.30$ digits/iteration) (cf. Fig. 9.4a) followed by the rise as $\log(1/\varepsilon)$ for smaller ε. The different curves correspond to the calculation of the entropy for time sequences of lengths $N = 1 - 4$ [cf. (9.17)–(9.21)] (Gaspard and Wang 1993).

9.3.5 Time- and amplitude-continuous random processes

Langevin processes

Stochastic differential equations like the Langevin equations define random processes which are continuous in both time and amplitude (van Kampen 1981). These stochastic differential equations are of the form

$$\frac{dX_i}{dt} = F_i(\mathbf{X}) + \sum_{j=1}^{d} c_{ij}\, W_j(t)\,, \tag{9.63}$$

where $W_i(t)$ are white noises of zero mean $\langle W_i(t)\rangle = 0$ and of correlations $\langle W_i(0)W_j(t)\rangle = \delta_{ij}\delta(t)$.

An example of such a stochastic process is the stationary Ornstein–Uhlenbeck process (8.38) which describes the behaviour of the velocity of a Brownian particle. For this process, the correlation function of the velocity $X(t) = v(t)$ is

$$\langle X(t)X(0)\rangle = \frac{c^2}{2a}\exp(-a|t|)\,, \tag{9.64}$$

so that the spectral density (9.35) is

$$S(\omega) = \frac{c^2}{\omega^2 + a^2}\,. \tag{9.65}$$

The joint probability is given here in terms of the Green function (8.40)–(8.41) according to

$$\begin{aligned} P(\mathbf{x};d\mathbf{x},\tau,N) &= \text{Prob}\Big\{x_0 < X(0) < x_0 + dx_0\,,\,\ldots\,, \\ &\qquad x_{N-1} < X(N\tau-\tau) < x_{N-1} + dx_{N-1}\Big\} \\ &= G(x_{N-1},x_{N-2};\tau)\,\ldots\,G(x_1,x_0;\tau)\,p_{\text{st}}(x_0)\,dx_0\ldots dx_{N-1}\,, \end{aligned} \tag{9.66}$$

where the factorization of the joint probability results from the Markovian character of the Ornstein–Uhlenbeck process.

According to the Cohen–Procaccia method, the (ε,τ)-entropy is based on the consideration of the following probabilities

$$\text{Prob}\Big\{|X_0 - x_0| \le \varepsilon,\,\ldots,\,|X_{N-1} - x_{N-1}| \le \varepsilon\Big\} \simeq \frac{(2\varepsilon)^N}{(2\pi)^{N/2}(\det\mathbf{C}_N)^{1/2}}\exp\left(-\frac{1}{2}\,\mathbf{x}^{\mathrm{T}}\cdot\mathbf{C}_N^{-1}\cdot\mathbf{x}\right), \tag{9.67}$$

where the variables $x_i = x(i\tau)$ represent a given realization of the process, while $X_i = X(i\tau)$ are the random variables. The correlation matrix is of the form

$$\mathbf{C}_N = \frac{c^2}{2a} \begin{pmatrix} 1 & \alpha & \alpha^2 & \alpha^3 & \dots & \alpha^{N-1} \\ \alpha & 1 & \alpha & \alpha^2 & \dots & \alpha^{N-2} \\ \alpha^2 & \alpha & 1 & \alpha & \dots & \alpha^{N-3} \\ \vdots & \vdots & \vdots & \vdots & \ddots & \vdots \\ \alpha^{N-1} & \alpha^{N-2} & \alpha^{N-3} & \alpha^{N-4} & \dots & 1 \end{pmatrix} , \tag{9.68}$$

where $\alpha = \exp(-a\tau)$ (Wax 1954). The inverse matrix is

$$\mathbf{C}_N^{-1} = \frac{2a}{c^2(1-\alpha^2)} \begin{pmatrix} 1 & -\alpha & 0 & 0 & \dots & 0 & 0 \\ -\alpha & 1+\alpha^2 & -\alpha & 0 & \dots & 0 & 0 \\ 0 & -\alpha & 1+\alpha^2 & -\alpha & \dots & 0 & 0 \\ \vdots & \vdots & \vdots & \vdots & \ddots & \vdots & \vdots \\ 0 & 0 & 0 & 0 & \dots & 1+\alpha^2 & -\alpha \\ 0 & 0 & 0 & 0 & \dots & -\alpha & 1 \end{pmatrix} . \tag{9.69}$$

The determinant of (9.68) is therefore

$$\det \mathbf{C}_N = \left(\frac{c^2}{2a}\right)^N (1 - \alpha^2)^{N-1} , \tag{9.70}$$

so that, for small τ,

$$\det \mathbf{C}_N \simeq \frac{(c^2\tau)^N}{2a\tau} . \tag{9.71}$$

We then obtain the following estimation for the CP (ε, τ)-entropy

$$h_{\mathrm{CP}}(\varepsilon, \tau) = \frac{1}{\tau}\, \psi\left(\frac{c^2\tau}{\varepsilon^2}\right) \simeq \frac{1}{\tau} \ln \frac{c\sqrt{2\pi e\tau}}{2\varepsilon} \qquad \text{for} \quad \tau >> \frac{\varepsilon^2}{c^2} , \tag{9.72}$$

where $\psi(x)$ is a certain function which can be approximated by a logarithmic function for $x \gg 1$ according to the above calculation. For the Ornstein–Uhlenbeck process, the Cohen–Procaccia (ε, τ)-entropy depends on both the arguments ε and τ in the function $\psi(x)$. Nevertheless, in a plot of the (ε, τ)-entropy versus ε, the different curves obtained by varying τ form an envelope[1] which is

$$h_{\mathrm{CP}}(\varepsilon, \tau) \leq h_{\mathrm{envelope}}(\varepsilon) = A\, \frac{c^2}{2\varepsilon^2} , \tag{9.73}$$

with the positive constant $A = 2\psi'(x_0)$ where x_0 is a solution of the equation $\psi(x_0) = x_0\psi'(x_0)$. Numerically, the value $A \simeq 3.1$ nats per unit time is observed.

This phenomenon is illustrated in Fig. 9.8 where the Cohen–Procaccia (ε, τ)-

1. The envelope of a parametric family of curves is calculated by eliminating the parameter τ between the equations $F(\varepsilon, h; \tau) = 0$ and $\partial_\tau F(\varepsilon, h; \tau) = 0$, with $F(\varepsilon, h; \tau) = \psi(c^2\tau/\varepsilon^2) - \tau h$.

entropy has been numerically calculated for a deterministic map which simulates the Brownian motion of a particle [i.e., the nonstationary process defined by the stochastic equation $dX/dt = cW(t)$]. We consider the one-dimensional sine map of the real line

$$X_{t+1} \;=\; X_t \;+\; p \;\sin 2\pi X_t \;, \tag{9.74}$$

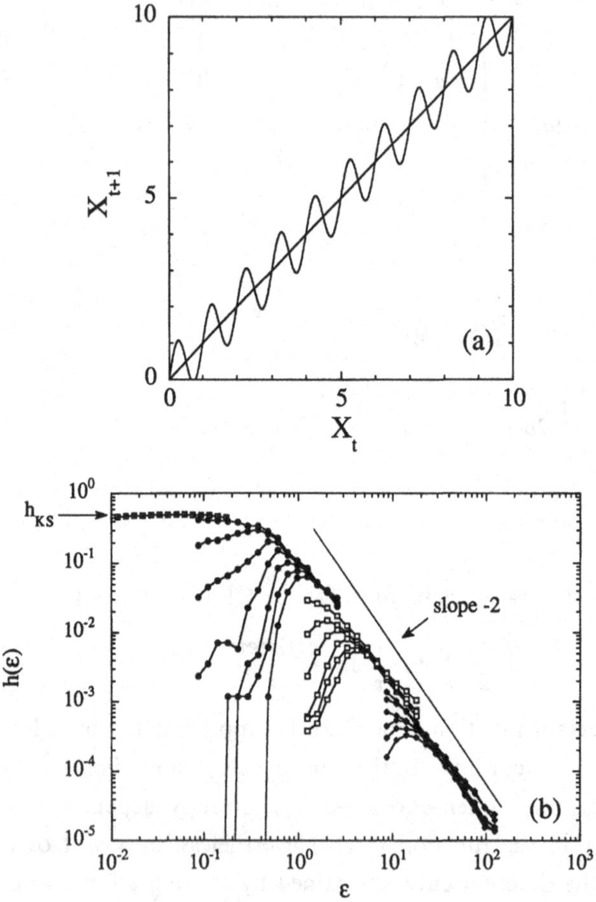

Figure 9.8 (a) The map (9.74) as a model of deterministic diffusion. (b) Numerical evaluation of the (ε, τ)-entropy of the nonstationary process on the real line for different values of $\tau = 1$ (filled circles), $\tau = 10$ (squares), and $\tau = 100$ (filled diamonds). For each value of τ, the different curves correspond to time sequences of lengths $N = 1, 4, 7, 10, 13, 16, 19$ [cf. (9.17)–(9.21)]. The crossed squares give the entropy calculated with periodic boundary conditions over 50 lattice cells ($0 \leq X < 50$) which coincides with the value of the mean Lyapunov exponent $h_{KS} = \bar{\lambda} = 0.49$ digits/iteration. The envelope of the different curves decreases as $(1/\varepsilon)^2$ at large ε but saturates at the value of the KS entropy at small ε (Gaspard and Wang 1993).

which presents diffusion, in particular, at the value $p = 0.8$ with a diffusion coefficient $D = 0.18$ (Schell et al. 1982). Figure 9.8b depicts the (ε, τ)-entropy per unit time calculated for different values of the time interval τ where we observe the formation of an envelope (9.73). Moreover, the saturation toward the KS entropy $h_{\mathrm{KS}} = \bar{\lambda} = 0.49$ digits/iteration is visible at small values of ε.

Another estimation of the entropy is provided by the Shannon–Kolmogorov ε-entropy (9.37)–(9.38) which turns out to be independent of τ. In this respect, this entropy is comparable to the aforementioned envelope except that the difference between the distances used in the Cohen–Procaccia and the Shannon–Kolmogorov methods precludes an immediate quantitative comparison. Since the Ornstein–Uhlenbeck process is Gaussian, the Shannon–Kolmogorov ε-entropy can be estimated thanks to the Kolmogorov formula (9.37)–(9.38). We find that

$$h_{\mathrm{SK}}(\varepsilon) \simeq \frac{2\, c^2}{\pi^2\, \varepsilon^2} , \tag{9.75}$$

in nats per unit time.

Such dependencies in ε^{-2} can be explained as follows. The ε-entropy probes the fine scales of the noise corresponding to the high frequencies where the spectral density decreases as $S(\omega) \simeq c^2/\omega^2$. Thus, for small ε, the ε-entropy does not depend on the relaxation rate a. Einstein's relation $\langle(\Delta X_t)^2\rangle \simeq 2Dt$ holds on arbitrarily small scales. Cells of size ε are crossed on average in a time $t_\varepsilon \sim \varepsilon^2/(2D) = \varepsilon^2/c^2$. Since the ε-entropy per unit time has the physical unit of the inverse of a time and measures the number of transitions from cell to cell during one unit of time, we have that $h(\varepsilon) \sim (1/t_\varepsilon) \sim c^2/\varepsilon^2$.

The Ornstein–Uhlenbeck process and the related Brownian motion can therefore be characterized by the divergence in ε^{-2} of their ε-entropies.

Yaglom noises

These random processes form a whole family of time- and amplitude-continuous stationary Gaussian processes which embed and generalize the Ornstein–Uhlenbeck process (Yaglom 1987). They are the stationary analogues of Mandelbrot's fractional Brownian motions (Mandelbrot and Van Ness 1968). Their correlation function is taken as

$$\langle X(t)X(0)\rangle = \frac{c^2}{a\sqrt{2\pi}}\, |at|^H\, K_H(|at|) , \tag{9.76}$$

where $K_\nu(z)$ is the modified Bessel function of the third kind and of order ν (Abramowitz and Stegun 1972). The exponent H satisfies $0 < H < 1$; a and c are positive constants. The corresponding spectral density is

$$S(\omega) = \frac{c^2\, 2^{H-1/2}\, a^{2H-1}\, \Gamma(H+1/2)}{(a^2 + \omega^2)^{H+1/2}}, \tag{9.77}$$

so that the spectral density (9.65) of the Ornstein–Uhlenbeck process is recovered in the limit $H = 1/2$. Let us also mention that the Yaglom noises are continuous in probability with a Hölder exponent H, for almost all realizations.

Applying the Kolmogorov formula (9.37)–(9.38) to this stationary Gaussian process we find its ε-entropy per unit time to be

$$h_{\rm SK}(\varepsilon) \simeq a\, \frac{2H+1}{2\pi} \left[\frac{2^{H-1/2}\Gamma(H+3/2)}{H\pi}\right]^{1/(2H)} \left(\frac{c}{\varepsilon\sqrt{a}}\right)^{1/H}, \tag{9.78}$$

in nats per unit time, so that (9.75) is recovered when $H = 1/2$. We observe that the exponent of ε^{-1} increases indefinitely as $H \to 0$. In that limit, the power spectrum given by (9.77) approaches the $1/f$-noise limit. In this sense, the $1/f$-noise may be said to have the highest degree of dynamical randomness, or to be of maximal complexity.

Fractional Brownian motions

These nonstationary time- and amplitude-continuous Gaussian processes are obtained from the Yaglom noises in the limit where $a \to 0$ (Mandelbrot and Van Ness 1968). Their power spectra behave as $\omega^{-(2H+1)}$ with a divergence at zero frequency due to their nonstationarity. Accordingly, their ε-entropy per unit time satisfies

$$h(\varepsilon) \sim \left(\frac{1}{\varepsilon}\right)^{1/H}. \tag{9.79}$$

Berger has calculated $h_{\rm SK}(\varepsilon)$ for the Brownian motion $H = 1/2$ and found the same result (9.75) as for the stationary Ornstein–Uhlenbeck process (Berger 1970). This feature may be explained by the fact that the small scales of the Ornstein–Uhlenbeck process are identical with those of the Brownian motion. We conclude this section with the comment that the stationarity or nonstationarity of these time- and amplitude-continuous Gaussian processes does not modify the divergence of the ε-entropy per unit time.

9.3.6 White noise

The white noise is a stationary Gaussian random process which is a distribution, rather than a function, with respect to time. The random variable $W(t)$ of the white noise is of zero mean and its correlation is $\langle W(0)W(t)\rangle = \delta(t)$.

We have already seen that the (ε, τ)-entropy depends on ε or on τ for processes which produce random events continuously in amplitude or in time. We may thus expect that the (ε, τ)-entropy of the white noise also has a dependency in both ε and τ.

A random process approximating the white noise can be constructed by discretization of the time axis into small intervals τ and by considering that $X_j = W(j\tau)$ are independent Gaussian variables of zero mean and unit variance. Therefore, applying the result (9.61) to the entropy per time interval τ, we find that the entropy per unit time is

$$h(\varepsilon, \tau) \simeq \frac{1}{\tau} \log \frac{1}{\varepsilon}, \qquad \text{(white noise)}, \tag{9.80}$$

which increases indefinitely as $\tau \to 0$.

Another method to generate random functions that approximate the white noise is to consider a Gaussian process with a constant spectral density in a large but finite bandwidth $|\omega| < \omega_c$ up to an ultraviolet cutoff beyond which the spectral density is zero ($\omega_c = \pi/\tau$). The SK ε-entropy (9.37)–(9.38) of this process leads to the same result as (9.80).

9.3.7 Lévy flights

Wang (1992) has studied a model of anomalous diffusion by Lévy flight, due to Klafter, Blumen and Shlesinger (1987). In the model, a particle undergoes random walks which consist of straight steps interrupted by jumps. The probability density for a single step $\mathbf{r}$ in time t is given by

$$p(\mathbf{r}, t) = \psi(\mathbf{r})\, \delta(r - t^{\nu}), \tag{9.81}$$

and the steps are independent of each other. The function ψ presents a power-law decay $\psi(\mathbf{r}) \sim r^{-\mu}$ as $r \to \infty$ and the process is assumed to occur in a d-dimensional space. The realization of these processes is a trajectory in space which is continuous and piecewise linear. The velocity is then piecewise constant with discontinuities at the jumps. Different regimes exist according to the exponent $\alpha = \nu(\mu - d + 1) - 1$.

The ε-entropy $H(\varepsilon, T)$ for the velocity process during a time interval T behaves as

$$H(\varepsilon, T) = \begin{cases} T\ d\ \log \frac{1}{\varepsilon} & \text{for } \alpha > 1\,, \\ \frac{T}{\log T}\ d\ \log \frac{1}{\varepsilon} & \text{for } \alpha = 1\,, \\ T^{\alpha}\ d\ \log \frac{1}{\varepsilon} & \text{for } \alpha < 1\,. \end{cases} \tag{9.82}$$

Accordingly, the process is regularly random for $\alpha > 1$ with a positive ε-entropy per unit time, but it is sporadically random for $\alpha \le 1$ where the ε-entropy per unit time vanishes

$$h(\varepsilon) = \begin{cases} d\ \log \frac{1}{\varepsilon} & \text{for } \alpha > 1\,, \\ 0 & \text{for } \alpha \le 1\,. \end{cases} \tag{9.83}$$

We see that when $\alpha > 1$ the process has the same degree of randomness in $\log(1/\varepsilon)$ as for time-discrete amplitude-continuous processes. We remark also that, according to (9.81), the choice of a random vector $\mathbf{r}$ in $\mathbb{R}^d$ is equivalent to a random choice of a time interval t and to a random direction on the $(d-1)$-sphere. Therefore $\varepsilon^d \sim \tau\ \delta^{d-1}$ where τ is an infinitesimal time interval while δ is the diameter of an infinitesimal cell of the $(d-1)$-sphere. We have equivalence with birth-and-death processes like the Poisson process when $d = 1$ and $\alpha > 1$.

9.3.8 Classification of the time random processes

In the preceding sections, we have obtained the (ε, τ)-entropy per unit time for a variety of stochastic processes. We observe that there exist two ways to classify the processes using the general (ε, τ)-entropy $H(\varepsilon, \tau, T)$ over a time interval T.

First, there is the dependency on the time T. We have seen that $H(\varepsilon, \tau, T)$ is extensive in T for most processes, a property we called regular randomness [cf. (9.12)]. In contrast, for some processes that we called sporadically random, $H(\varepsilon, \tau, T)$ grows more slowly than the time T according to $H(\varepsilon, \tau, T) \sim T^{\alpha}$ where $0 < \alpha < 1$ is the exponent of sporadicity [cf. (9.11)]. Let us here mention the work by Ebeling and Nicolis (1991, 1992), and others who investigated the property of extensivity of the block entropy (9.7) in language texts, in music, as well as in some DNA sequences (Ebeling et al. 1987, Grassberger 1989, Li and Kaneko 1992). These studies have revealed that such informational processes are often on or near this sporadicity or 'edge of chaos' between periodic and chaotic behaviours.[2]

2. In time evolution processes, the nonextensivity of the block entropy can be interpreted by the occurrence of new and broader comprehensive schemes emerging on longer and longer periods of time. When such a new comprehensive scheme emerges a sudden data compression of the previous information sequence becomes possible. In history, some

Secondly, there is the dependency of $H(\varepsilon, \tau, T)$ on ε and τ. This dependency disappears for the chaotic deterministic processes because their trajectories are smooth. On the contrary, the dependency remains for processes which have non-smooth trajectories. For instance, the trajectories of a birth-and-death process give the population numbers which vary discontinuously in time. The Ornstein–Uhlenbeck, the Yaglom, and the Brownian processes are continuous but nondifferentiable so that the dependency in ε remains for the envelope of $h(\varepsilon, \tau)$ and for the SK ε-entropy. Hence, all these noises have an (ε, τ)-entropy per unit time which diverges as ε or $\tau \to 0$. As a consequence, they are much more random than the chaotic processes. Furthermore, we can compare one process with another, as we do with Yaglom noises of different exponents H, in order to show that some of them are more random than others. In this way, we can establish equivalences between the degrees of randomness of the processes.

Such conclusions are based on the principle of entropy invariance by Kolmogorov, Sinai, Ornstein, and others, which states that the KS entropy per unit time is invariant under an isomorphism which maps a process onto another (Cornfeld et al. 1982, Ornstein 1974). For instance, we have constructed an isomorphism between a time-discrete Markov chain and a deterministic chaotic map in Chapter 4. The KS entropy is the same for both processes showing that information on the time evolution of the processes is strictly preserved going from one description to the other. The invariance of the KS entropy generalizes to the (ε, τ)-entropy per unit time if we admit some differences due to the different definitions of (ε, τ) in the two processes between which we want to establish an isomorphism.

The (ε, τ)-entropy can thus be used to establish the distinction between chaos and noise because it gives a quantitative measure of the randomness of a system. In a chaotic system, the ε-entropy per unit time presents a plateau so that

$$H(\varepsilon, \tau, T) \simeq T\, h_{\mathrm{KS}} + d_{\mathrm{I}} \log \frac{1}{\varepsilon}\,, \tag{9.84}$$

(Grassberger et al. 1991). The dependency on the overall time T provides the value of the KS entropy which is well defined as $\varepsilon \to 0$. However, the next-to-leading term diverges as $\varepsilon \to 0$, which provides the information dimension of the

analogy may be drawn between the above comprehensive schemes and the Kantian syntheses. Indeed, Kant has described history as a process in which theses and antitheses develop themselves during successive periods, each of which is followed by another period during which a synthesis occurs between the previous thesis and antithesis. Along similar lines of thought, recent works on the control of chaos have shown that the ability to control a system requires that the system evolves in a state of marginal stability. This suggests that adaptative evolutionary systems tend therefore to evolve toward the edge of chaos which is a marginal regime between complete periodicity and complete dynamical disorder.

chaotic attractor. In contrast, if the term which is proportional to T diverges as $\varepsilon \to 0$ or $\tau \to 0$, the conclusion would be that the system is noisy:

$$H(\varepsilon, \tau, T) \simeq T \, h(\varepsilon, \tau) . \tag{9.85}$$

In that case, the appropriate stochastic model could in principle be inferred from the type of divergence of the ε-entropy per unit time. This method has been applied, in particular, to turbulence (Wang and Gaspard 1992).

We have compiled in Table 9.1 the entropies $H(\varepsilon, \tau, T)$ for various periodic and sporadic processes. In Table 9.2, we gathered the entropies per unit time $h(\varepsilon, \tau)$ or $h_{SK}(\varepsilon)$ for the random processes considered in this chapter. Since we

Table 9.1 Periodic and sporadically random processes.

Process	$H(\varepsilon, \tau, T)$
Periodic	$\log T$
Feigenbaum attractor[(a)]	$\log T$
Intermittent maps[(b)] ($z \geq 2$)	T^α $[\alpha = 1/(z-1)]$
Intermittent maps[(b)] ($z = 2$)	$T/\log T$
Lévy flights ($\alpha < 1$)[(c)]	$T^\alpha \, d \log(1/\varepsilon)$
Lévy flights ($\alpha = 1$)[(c)]	$(T/\log T) \, d \log(1/\varepsilon)$

Table 9.2 Time random processes.

Process	$h(\varepsilon, \tau)$ *or* $h_{SK}(\varepsilon)$
Periodic	0
Feigenbaum attractor[(a)]	0
Intermittent maps[(b)]	0
Lévy flights ($\alpha \leq 1$)[(c)]	0
Lévy flights ($\alpha > 1$)[(c)]	$d \log(1/\varepsilon)$
Deterministic chaos	h_{KS}
Bernoulli and Markov	h_{KS}
Birth-and-death	$\log(1/\tau)$
Time-discrete, amplitude-continuous	$d \log(1/\varepsilon)$
Ornstein–Uhlenbeck	$(1/\varepsilon)^2$
Yaglom	$(1/\varepsilon)^{1/H}$
Brownian	$(1/\varepsilon)^2$
Fractional Brownian	$(1/\varepsilon)^{1/H}$
White noise	$(1/\tau)\log(1/\varepsilon)$
Boltzmann–Lorentz (1 particle)	$\log(1/\tau\Delta^3 v \Delta^2 \Omega)$

(a) Grassberger 1986a.
(b) The intermittent maps are defined in the unit interval according to $x_{n+1} = x_n + c x_n^z$ modulo 1 (Manneville 1980, Pomeau and Manneville 1980, Gaspard and Wang 1988).
(c) Wang 1992.

have only considered a limited number of known random processes, the list is far from being exhaustive.

9.4 Spacetime random processes

9.4.1 (ε, τ)-entropy per unit time and volume

For processes evolving in time and in space, dynamical randomness can occur not only at each time step but also at each space point. Let us consider a simple example of a chain of spins 1/2 which evolves in time according to a pure Bernoulli process. Each spin takes the values $\pm 1/2$ independently of the state of its neighbours and independently of its own previous state. The amount of data produced by this system will be proportional to the time T of observation but also to the space volume L under observation. Since the process is of Bernoulli type no compression of data is possible and the record of the spacetime diagram will require a total of TL bits. This quantity defines the entropy $H(T, L)$ of the process over a time T and a volume L.

The previous results for time random processes can be generalized by replacing the group of time translations by the group of translations in time and in a space of dimension d. In this way, we can define the entropy $H(\varepsilon, \tau, T, L)$ for spacetime processes in the time interval T and in a cubic domain of side L in space. This entropy grows at most like TL^d so that we can define the (ε, τ)-entropy per unit time and volume according to

$$h^{(\text{time, space})}(\varepsilon, \tau) = \lim_{T,L\to\infty} \frac{1}{TL^d} H(\varepsilon, \tau, T, L) , \tag{9.86}$$

as a generalization of $h^{(\text{time})}(\varepsilon, \tau)$ defined with (9.13)–(9.15) or (9.31). We shall speak of spacetime chaos if $h^{(\text{time, space})}(\varepsilon, \tau)$ is positive, and bounded as ε and τ go to zero.

Let us now review several different spacetime processes.

9.4.2 Deterministic cellular automata

These spacetime processes have been reactualized by Wolfram (1983, 1984). These systems are defined by a set of discrete states at discrete positions in a space. The dynamics is given by logic rules that reset the configuration of the system from one time step to the next. These processes can be viewed as degenerate Markov chains, where states are spatial configurations, and the matrix of transition probabilities (4.105) contains only zeros and ones. As a consequence, no dynamical randomness is produced with time. The dynamics simply propagates the initial configuration of the system which can eventually be random in space. The randomness is due either to the initial conditions or to information coming from outside the observed volume L^d.

Let us consider for instance the elementary cellular automata where the coupling is set between nearest neighbouring sites. We have $H(T, L) \sim L^d \log T$ for Wolfram's rule 132 (Grassberger 1986b). For other rules like 0, 32, 250, or 254, the initial state is attracted toward a spatially uniform or periodic configuration, in which case the entropy $H(T, L)$ increases like $\log(TL^d)$ at most. Besides, there is numerical evidence that the rule 22 has a sporadic behaviour in time like $H(T, L) \sim T^{0.82}$ without dependency on $L \geq 2$ (Grassberger 1986b). In all cases, the entropy per unit time and unit volume (9.86) vanishes. The same conclusion applies to the game of life (Conway 1982) as well as to the automata of Nowak and May (1992).

9.4.3 Lattice gas automata

These processes are discrete in time and in space, and physical quantities like the velocity take discrete amplitudes (Dab et al. 1991, Lawniczak et al. 1991, Kapral et al. 1992, Ernst 1992). Contrary to the deterministic cellular automata, however, the lattice gas automata are nondegenerate Markov chains with transition probabilities between zero and one. Accordingly, randomness is generated at each spacetime point and the entropy is proportional to the spacetime hypervolume TL^d as in the above example of a chain of randomly flipped spins. These processes have in general a positive entropy per unit time and volume which can be calculated using Eq. (9.40) and which does not depend on either ε or τ due to the discreteness of these models. The same conclusion applies to *probabilistic* cellular automata.

9.4.4 Coupled map lattices

These dynamical systems are defined by deterministic mappings (Oppo and Kapral 1987, Kaneko 1989, Kapral 1991). A continuous variable is assigned to each space position on a lattice. Therefore, the process is discrete in time and in space but has continuous amplitudes. For a finite number of coupled maps, the entropy will then be given by

$$H(T, L) = T \sum_{\bar{\lambda}_i > 0} \bar{\lambda}_i \,, \tag{9.87}$$

according to Pesin's theorem (cf. Chapter 4). The number of positive Lyapunov exponents is typically proportional to the number of degrees of freedom which itself grows proportionally with the volume (see the Introduction). As a consequence, we have that $H(T,L) \sim TL^d$. This rapid increase of the entropy is the feature of the dynamical regimes of spacetime chaos.

9.4.5 Nonlinear partial differential equations

Such equations define dynamical systems of infinite-dimensional phase spaces, with continuous spacetime and observables. Well-known examples are the Navier–Stokes equations (Ruelle 1982), the Brusselator reaction–diffusion equations (DeWit et al. 1992, 1993, 1996), or the Kuramoto–Sivashinsky equation (Pomeau et al. 1984, Manneville 1985). For partial differential equations (PDE) or the previous coupled map lattices, it is possible to introduce the density $g(\lambda)$ of Lyapunov exponents as the number of exponents having their value in the range $[\lambda, \lambda + d\lambda[$ for a system of unit volume. The entropy is then given by

$$H(T,L) \simeq TL^d \int_0^{\bar{\lambda}_{\max}} \lambda \, g(\lambda) \, d\lambda \, , \tag{9.88}$$

where $\bar{\lambda}_{\max}$ is the largest Lyapunov exponent. This result holds for every deterministic system of arbitarily large spatial extension. In a regime of spacetime chaos the entropy per unit time and volume (9.86) is then positive.

9.4.6 Stochastic spin dynamics

Glauber (1963) and Kawasaki (1966) introduced kinetic models describing the stochastic dynamics of spins in a solid. These models are defined as birth-and-death processes of large spatial extension. According to the results of Subsection 9.3.3, their entropy diverges with $\tau \to 0$ according to

$$H(T,L) \sim TL^d \log \frac{1}{\tau} \, . \tag{9.89}$$

The extra randomness with respect to the lattice gas automata or the coupled map lattices has its origin in the assumption that the time is continuous. If we rather adopt a version of the dynamics which is discrete in time the spin dynamics reduce to Markov chains as in lattice gas automata.

9.4.7 Spacetime Gaussian fields

These processes are the spacetime generalizations of the Brownian motions of Sec. 9.3. Pinsker and Sofman (1983) have generalized the Kolmogorov formula (9.37)–(9.38) to those stationary Gaussian processes. If the spectral density is

$$S(\mathbf{k}, \omega) = \int_{\mathbb{R}^{d+1}} e^{-i(\mathbf{k}\cdot\mathbf{r}+\omega t)} \, \langle X(\mathbf{r},t)X(0,0)\rangle \, d\mathbf{r} \, dt \, , \tag{9.90}$$

the SK ε-entropy per unit time and volume is given by

$$\varepsilon^2 = \frac{1}{(2\pi)^{d+1}} \int_{\mathbb{R}^{d+1}} \min[\theta \, , \, S(\mathbf{k},\omega)] \, d\mathbf{k} \, d\omega \, , \tag{9.91}$$

$$h_{\rm SK}^{(\text{time, space})}(\varepsilon) = \frac{1}{2(2\pi)^{d+1}} \int_{\mathbb{R}^{d+1}} \max\left[0 \, , \, \log \frac{S(\mathbf{k},\omega)}{\theta}\right] d\mathbf{k} \, d\omega \, . \tag{9.92}$$

For instance, let us suppose that the spectral density decreases at large frequencies and wavenumbers like

$$S(\mathbf{k},\omega) \sim (\omega^\alpha + k^\beta)^{-\nu} \tag{9.93}$$

with $(1/\alpha + d/\beta) < \nu$, so that the integral of the power spectrum is bounded. The ε-entropy then diverges as

$$h_{\rm SK}^{(\text{time, space})}(\varepsilon) \sim \left(\frac{1}{\varepsilon}\right)^\gamma , \qquad \gamma = \frac{2(1/\alpha + d/\beta)}{\nu - (1/\alpha + d/\beta)} \, . \tag{9.94}$$

We notice that these continuous random fields may develop a very high degree of randomness compared with the other examples. A possible spacetime generalization of Yaglom noises is defined with $\alpha = \beta = 2$, and $\nu = H + (d+1)/2$, so that the exponent of the ε-entropy (9.94) becomes $\gamma = (d+1)/H$.

9.4.8 Sporadic spacetime random processes

Wang (1995) has further extended the above considerations and introduced several concepts of sporadicity in space or time, in which the dynamical entropy $H(\varepsilon, \tau, T, L)$ has anomalous scalings in L or T. The general situation is the case of spatio-temporal sporadicity where the entropy scales anomalously with both L and T

$$H(\varepsilon, \tau, T, L) \sim T^\alpha \, L^{d_f} \, , \qquad \text{with } 0 \le \alpha \le 1 \quad \text{and} \quad 0 \le d_f \le d \, . \tag{9.95}$$

Moreover, Wang (1995) has given examples of several processes where the entropy has an anomalous scaling with the volume. This situation may be called *spatially sporadic chaos* in which

$$H(\varepsilon, \tau, T, L) \sim \begin{cases} T \, L^{d_f} \, , & (0 < d_f < d) \, , \\ T \, \frac{L^d}{(\log L)^\beta} \, , & (\beta > 0) \, . \end{cases} \tag{9.96}$$

This is the case in directed percolation processes at criticality as well as in the Chaté-Manneville coupled map lattice model also at criticality (Wang 1995).

9.4.9 Classification of spacetime random processes

A classification can be carried out for spacetime processes in analogy with the previous classification of the time random processes. The behaviour of the (ε, τ)-entropy $H(\varepsilon, \tau, T, L)$ versus its different parameters may lead to different classifications according to the extensivity in T, L, or some combination of T and L, or according to the divergence as $\varepsilon \to 0$ or $\tau \to 0$. Table 9.3 gives these behaviours for several spacetime processes.

9.5 Random processes of statistical mechanics

In the above discussion, we have given a survey of several time and spacetime random processes in terms of the (ε, τ)-entropy per unit time or per unit time and volume. This quantity provides a quantitative characterization of dynamical randomness, for instance, in digits per seconds produced on certain scales ε. We are now turning to microscopic and mesoscopic processes of physics and chemistry, like Brownian motion, where randomness is produced by thermal fluctuations or noises. The number of active degrees of freedom is extremely high in a mole of gas (10^{23}). Therefore, although the motion of atoms or molecules in a gas or a liquid is ultimately described by the deterministic Newton equations, the extremely large number of particles renders a stochastic description desirable.

We gave in the Introduction an estimation of the KS entropy per unit time in a

Table 9.3 Spacetime random processes.

Process	$H(\varepsilon, \tau, T, L)$
Cellular automata	$< TL^d$
Conway's game of life	$< TL^d$
Lattice gas automata	TL^d
Coupled maps in spacetime chaos	TL^d
Nonlinear PDE's in spacetime chaos	TL^d
Glauber or Kawasaki spin dynamics	$TL^d \log(1/\tau)$
$d+1$-dimensional Yaglom fields	$TL^d \, (1/\varepsilon)^{(d+1)/H}$
d-dimensional ideal gas (many particles)	$TL^{d-1} \log(1/\Delta^d q \Delta^d p)$
Lorentz gas (fixed scatterers)	$TL^d + c\, TL^{d-1} \log(1/\Delta^d q \Delta^d p)$
Hard-sphere gas	TL^d
Boltzmann–Lorentz (many particles)	$TL^d \log(1/\tau \Delta^3 v \Delta^2 \Omega)$
Spatially sporadic chaos ($0 < d_f < d$)[(a)]	TL^{d_f}
Spatially sporadic chaos ($\beta > 0$)[(a)]	$TL^d/(\log L)^{\beta}$

[(a)] Wang 1995.

gas of interacting particles. This quantity is an upper bound on the information produced in the motion of particles in a fluid. On the other hand, we have seen in this chapter that stochastic models can also be characterized by the information they produce per unit time. A comparison is possible in view of the principle of equivalence by Kolmogorov. Indeed, the stochastic models are supposed to provide a faithful representation of the dynamics of a certain subset of degrees of freedom in a large system like a fluid, at least in a certain domain of validity. In this domain, the model is derived by some (approximate) isomorphism between the process induced by Newton's equations and some stochastic equation like the Langevin equations. According to the principle of equivalence, the entropy per unit time must be (approximately) invariant under this (approximate) isomorphism. Here, the entropy should be defined for a certain partition of phase space which corresponds to the subsystem modelled by the stochastic equation. Therefore, a comparison can be established between the original deterministic process and the stochastic model thanks to their (ε, τ)-entropy per unit time. This comparison should reveal the domain where the stochastic process provides a faithful model. However, we have seen before that most of the random processes have a diverging (ε, τ)-entropy as $\varepsilon \to 0$ or $\tau \to 0$, although the (ε, τ)-entropy is expected to saturate at the KS entropy per unit time of the original process which is deterministic. This is the case, in particular, for the Brownian motion of a colloidal particle in a fluid which is modelled by a Langevin process for which the ε-entropy diverges like ε^{-2} although the fluid is at most chaotic (see Fig. 9.9). Therefore, we notice that the validity of the stochastic models is in general limited to intermediate scales in ε or τ.

We shall estimate the (ε, τ)-entropy per unit time of different stochastic models of statistical mechanics to compare their dynamical randomness. Most of these models have a large spatial extension so that they should be considered as spacetime processes. There exist different classes of models as illustrated in Fig. 9.10.

9.5.1 Ideal gases

The simplest models are the *ideal gases* where the particles are not interacting with each other (Cornfeld et al. 1982). As a consequence, there is no local dynamical randomness generated in these systems and the entropy per unit time and volume vanishes since all the Lyapunov exponents are zero. However, the ideal gases are not completely devoid of randomness since new particles continuously arrive from large distances with arbitrary positions and velocities within the Maxwellian distribution. Therefore, they constitute examples of systems which are dynamically random but which do not show any local sensitivity to initial conditions. Other models of the same class are the harmonic solids and the Toda lattice, each of them having an infinite spatial extension and a finite energy density (van Hemmen 1980, Toda 1989). These infinite systems have the properties of ergodicity, mixing, and Kolmogorov, as well as dynamical randomness without sensitivity to initial conditions.

Because of the continuous arrival of new particles from infinity, the entropy $H(T, L)$ is expected to be proportional to the surface L^{d-1} of the observed volume $V = L^d$ as well as to the time T. At each new arrival, a measurement is carried out of the positions and velocities at the intersection of the trajectory with the surface of the volume V. These positions and velocities are therefore random variables with continuous distributions so that the entropy should diverge as $\log(1/\varepsilon)$ where $\varepsilon = \Delta^d q \Delta^d v$.

This is confirmed by a calculation based on a standard counting of the number of complexions à la Boltzmann (Gaspard 1991b, 1992e, 1994). The particles are assumed to be noninteracting and to follow straight lines, $\mathbf{q}_k = \mathbf{q}_{0k} + \mathbf{v}_k t$. The particles are uniformly distributed in space with a density ρ and the velocity distribution has a normalized density $f(\mathbf{v})$. We suppose that there is a surface Σ in the three-dimensional space. For simplicity, we assume that the surface Σ is the plane $z = 0$. This plane is partitioned into small cells of area $\Delta^2 A = \Delta x \Delta y$. Let $\mathbf{n}$ be a unit vector normal to Σ at the point (x, y). The number of particles

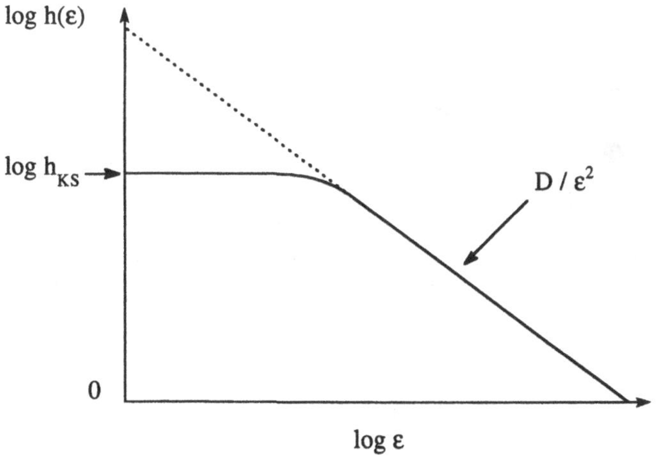

Figure 9.9 Comparison between the behaviour of the ε-entropy per unit time expected for a deterministic diffusion process (for instance in the periodic Lorentz gas or the periodic multibaker) and that for a stochastic Langevin model of this process (Gaspard 1997a).

crossing the cell $[x, x + \Delta x[\otimes[y, y + \Delta y[$ of the surface Σ with a velocity in $[\mathbf{v}_i, \mathbf{v}_i + \Delta\mathbf{v}[$ and $\mathbf{n} \cdot \mathbf{v}_i > 0$ during the time interval $[0, T]$ is equal to

$$N_i = \rho \ (\mathbf{n} \cdot \mathbf{v}_i) \ T \ \Delta^2 A \ f(\mathbf{v}_i) \, \Delta^3 v \, , \tag{9.97}$$

where the index i labels the different cells in the velocity space. The initial positions $\mathbf{q}_{0k}$ of the N_i trajectories are distributed uniformly in the space volume $(\mathbf{n} \cdot \mathbf{v}_i) T \Delta^2 A$ where the number of possible positions is equal to

$$M_i = \frac{(\mathbf{n} \cdot \mathbf{v}_i) \ T \ \Delta^2 A}{\Delta x \Delta y \Delta z} = \frac{(\mathbf{n} \cdot \mathbf{v}_i) \ T}{\Delta z} \, . \tag{9.98}$$

The combinatorial problem is therefore to place N_i particles in M_i different

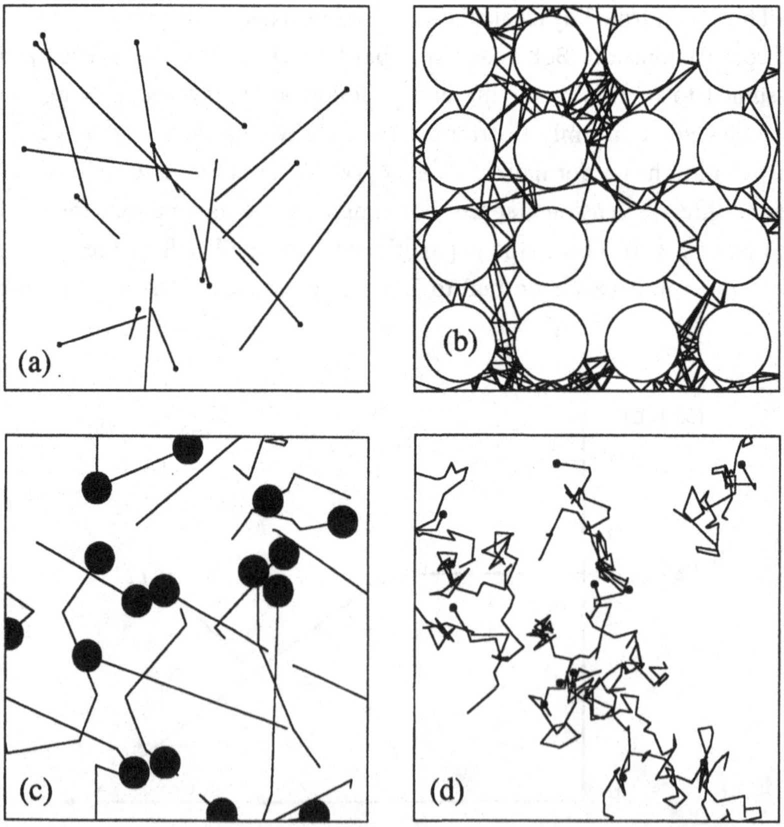

Figure 9.10 Schematic representation of four examples of models of spacetime processes used in statistical mechanics: (a) the ideal gas; (b) the finite Lorentz gas; (c) the hard-sphere gas; (d) the Boltzmann–Lorentz gas where each collision is a random process contrary to the gases a, b, and c which are deterministic (Gaspard and Wang 1993).

positions for each velocity $\mathbf{v}_i$ so that the total number of complexions is equal to

$$\prod_i \frac{1}{N_i!} M_i^{N_i} . \tag{9.99}$$

The entropy per unit time is then given by taking the logarithm of the number of complexions

$$\begin{aligned} h^{(\text{time})} &= \lim_{T\to\infty} \frac{1}{T} \ln \prod_i \frac{1}{N_i!} M_i^{N_i} \\ &= \rho \int_\Sigma d^2A \int d^3v \, (\mathbf{n}\cdot\mathbf{v}) \, f(\mathbf{v}) \, \ln \frac{e}{\rho \, f(\mathbf{v}) \, \Delta^3 q \, \Delta^3 v} , \end{aligned} \tag{9.100}$$

where we used Stirling's formula. The standard thermodynamic entropy per unit volume can be calculated in a similar way as

$$s^{(\text{space})} = \rho \int d^3v \, f(\mathbf{v}) \, \ln \frac{e}{\rho \, f(\mathbf{v}) \, \Delta^3 q \, \Delta^3 v} , \tag{9.101}$$

where we see that both entropies (9.100)–(9.101) are comparable. In both cases, the entropies are calculated for a partition into phase space cells of volume $\Delta^3 q \Delta^3 v$. Hence, the entropies diverge in the limit $\Delta^3 q \Delta^3 v \to 0$ as expected. Indeed, the KS entropy which is obtained in this limit is known to be infinite for the ideal gas (Goldstein et al. 1975).

The entropy per unit time contains a surface integral over Σ which can be interpreted as the surface of a detector. The entropy per unit time is the average number of nats (bits or digits) obtained by recording with a precision $\Delta^3 q \Delta^3 v$ the positions and velocities of the particles hitting the detector surface Σ per unit time. Dynamical randomness therefore has its origin in the flux of particles coming from large distance with arbitrary positions and velocities. For a gas at equilibrium where the velocity distribution is Maxwellian, both the volume and time entropies are functions of the density ρ and of the temperature of the gas. The preceding formulas generalize to stationary beams of particles where the phase space density $\rho f(\mathbf{v})$ may be nonuniform in space also (Gaspard 1991b, 1992e, 1994).

In summary, the dynamical entropy for an ideal gas in a domain of size L of $\mathbb{R}^d$ and over a time interval T is given by

$$H(\Delta^d q \Delta^d p, T, L) \sim T \, L^{d-1} \log(1/\Delta^d q \Delta^d p) , \tag{9.102}$$

with $\Delta^d p = m^d \Delta^d v$. The entropy per unit time of quantum ideal gases can be evaluated with the definition of quantum dynamical entropy by Connes, Narnhofer, and Thirring (1987). The quantum entropies turn out to be comparable to the above classical entropies with the standard correspondence that $\Delta q \Delta p \simeq 2\pi\hbar$, where $\hbar$ is the Planck constant. Therefore, the quantal properties remove the divergence in the above entropies because the smallest scales in phase space are inactive in quantum mechanics. Quantum mechanically, the ideal gases thus have the same power of dynamical randomness as classical chaotic systems, as discussed elsewhere (Gaspard 1991b, 1992e, 1994; Hudetz 1990).

9.5.2 The Lorentz gases and the hard-sphere gases

In contrast to ideal gases, particles interact with each other in a real gas. This interaction has nearly no effect on the estimation of the entropy per unit volume (9.101) if the gas is diluted. However, the interaction is crucial for transport properties and also for the entropy per unit time and volume, which may turn to be positive as explained in the Introduction.

Before facing the full complexity of the dynamics of a gas where all the particles move, we may consider the Lorentz gases where the particles interact with scatterers which are immobile while the particles are mutually independent. In such models, the entropy per unit time and volume is already positive.

In a Lorentz gas, the system decomposes into independent particles in elastic collisions on disks or spheres so that the dynamical instability can be studied on the energy shell of individual particles. Let $\bar{\lambda}_i$ be the positive Lyapunov exponents per unit length on the energy shell corresponding to the unit velocity $v = 1$. The KS entropy per unit time of the motion of one particle on this shell is then given by $h_{\text{KS}} = v \sum_{\bar{\lambda}_i > 0} \bar{\lambda}_i$. Let us suppose that the gas of independent particles has a uniform density of ρ particles per unit volume and a velocity distribution with the density $f(v)$ defined for $v \in [0, \infty[$. The average KS entropy per unit time and per unit volume would be

$$h^{(\text{time, space})} = \rho \int_0^\infty dv \; v \; f(v) \sum_{\bar{\lambda}_i > 0} \bar{\lambda}_i \; . \tag{9.103}$$

Therefore, the dynamical randomness in a spacetime domain TL^d is then given by

$$H(T, L) \simeq h^{(\text{time, space})} \; T \; L^d, \tag{9.104}$$

which is extensive in time and space in constrast to the ideal gases. Equation (6.32) shows how the Lyapunov exponent vanishes with the radius of the disks of the Lorentz gas so that the entropy (9.103) also vanishes in this ideal-gas limit.

The transition between the ideal gases and the Lorentz gases can be further illustrated if we consider a gas of independent particles which are scattered by a finite array of hard spheres fixed in space and occupying only a finite volume $V = L^d$. Since the collisional dynamics of the particles is defocusing in the region V of the lattice of scatterers the entropy $H(T, L)$ contains a term which is proportional to TL^d and to the sum of positive Lyapunov exponents of the Lorentz gas. However, particles are continuously coming from large distances and enter the scatterer on the surface of the lattice. This shower of new particles is another source of randomness which is now proportional to the time T, to the surface L^{d-1} of the scatterer, and to a factor with a logarithmic divergence as in the ideal gases. Therefore, the entropy $H(T, L)$ contains two terms,

$$H(T,L) \sim T\,L^d + c\,T\,L^{d-1} \log(1/\Delta^d q \Delta^d p) \tag{9.105}$$

where c is a positive constant. When the volume is large the first term dominates and the system is at the stage of spacetime chaos as in the hard-sphere gas. However, when there is no scatterer in the region V the first term vanishes and we recover the term (9.102) characteristic of the ideal gas.

The case of the hard-sphere gas has been treated in the Introduction where an estimation of the entropy per unit time and unit volume was given by Eq. (0.10). This calculation shows that both the hard-sphere gases and the Lorentz gases have the dynamical randomness of spacetime chaos.

9.5.3 The Boltzmann–Lorentz process

The Boltzmann–Lorentz process is a model for the relaxation of the velocity distribution function toward the Maxwellian equilibrium distribution as described by the Boltzmann equation (Résibois and De Leener 1977). This model has been developed into a method of simulation of a fluid of particles in molecular dynamics, called the Bird method (Bird 1963, 1990).

For simplicity, we assume that the relaxation occurs uniformly in space and that the variation of the particle density around its mean value is small enough to justify the linearization of the Boltzmann equation. The linear Boltzmann equation is then

$$\frac{\partial f(\mathbf{v}_1)}{\partial t} = \rho \int d^3v_2\, d^2\Omega\, |\mathbf{v}_1 - \mathbf{v}_2|\, \sigma(\theta, \varphi, \mathbf{v}_1, \mathbf{v}_2)\, f_e(v_2)\, [f(\mathbf{v}_1') - f(\mathbf{v}_1)]\,, \tag{9.106}$$

where ρ is the particle density, σ is the differential cross-section of the binary collision, and

$$f_e(v) = \left(\frac{m\beta}{2\pi}\right)^{3/2} \exp\left(-\frac{\beta}{2}\, m\, v^2\right). \tag{9.107}$$

Equation (9.106) describes the time evolution of the probability density of the velocity of a test particle No. 1 undergoing multiple collisions with other particles No. 2 in the gas, which form a bath (or reservoir). The outcoming velocities $\mathbf{v}_1'$ and $\mathbf{v}_2'$ after each binary collision are uniquely determined by the velocities $\mathbf{v}_1$ and $\mathbf{v}_2$ of the two particles entering the collision together with the impact unit vector locating the relative positions of the particles No. 1 and No. 2 at the point of closest approach (Fig. 9.11a).

The Boltzmann–Lorentz equation describes the successive collisions as successive random events where the velocity $\mathbf{v}_2$ of the bath particle as well as the solid angle $\Omega = (\theta, \phi)$ of the aforementioned impact unit vector are random variables (Fig. 9.11b). Besides, the Boltzmann–Lorentz equation has the form of a birth-and-death process in the continuous velocity and solid-angle variables. Discretizing the velocity and the solid angle, it has the form

$$\frac{d}{dt}\, p_{\mathbf{v}_1} = \sum_{\mathbf{v}_1'} W_{\mathbf{v}_1\mathbf{v}_1'}\, p_{\mathbf{v}_1'} \,, \tag{9.108}$$

where

$$p_{\mathbf{v}_1} = f(\mathbf{v}_1)\, \Delta^3 v \,. \tag{9.109}$$

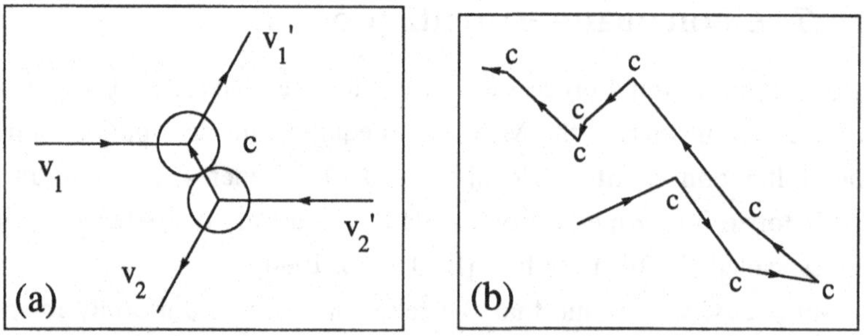

Figure 9.11 (a) Schematic representation of a typical collision between two hard spheres of ingoing velocities $\mathbf{v}_1$ and $\mathbf{v}_2$, and of outgoing velocities $\mathbf{v}_1'$ and $\mathbf{v}_2'$. $\mathbf{c}$ is the impact unit vector along the line joining the two discontinuities in the trajectories of the particles No. 1 and No. 2. (b) Schematic representation of a typical trajectory of the Boltzmann–Lorentz random process. The trajectory changes its direction at each collision c. From collision to collision, the process uses six random variables which are continuously distributed:
(1) the intercollisional time to which corresponds the infinitesimal Δt;
(2) the velocity $\mathbf{v}_2$ of the particle No. 2 coming and returning to the thermal bath with a Maxwell distribution and corresponding to the infinitesimal $\Delta^3 v_2$;
(3) the angles $\Omega = (\theta, \phi)$ of the impact unit vector to which corresponds the infinitesimal $\Delta^2\Omega$. Each continuous random variable contributes to the ε-entropy by a logarithmic divergence of its corresponding infinitesimal (Gaspard and Wang 1993).

Moreover,

$$W_{\mathbf{v}_1\mathbf{v}_1'} = \rho\, \Delta^3 v\, \Delta^2\Omega\, |\mathbf{v}_1 - \mathbf{v}_2|\, \sigma(\Omega, \mathbf{v}_1, \mathbf{v}_2)\, f_{\mathrm{e}}(v_2)\,, \tag{9.110}$$

where $\mathbf{v}_1'$ is a function of $\mathbf{v}_2$ and Ω. We shall now estimate the ε-entropy per unit time for the stochastic process of the Boltzmann–Lorentz equation in the case of a hard-sphere gas. The differential cross-section is then constant $\sigma = d^2/16$ where d is the diameter of the particles. The ε-entropy per unit time for a birth-and-death process is given by (9.52). In the present case, the role of ε is played by $\Delta^3 v$, $\Delta^2\Omega$, as well as $\tau = \Delta t$. We find

$$h(\Delta t \Delta^3 v \Delta^2 \Omega) = \left(- \sum_{\mathbf{v}_1} p^{\mathrm{e}}_{\mathbf{v}_1}\, W_{\mathbf{v}_1\mathbf{v}_1}\right)\left(\ln \frac{e}{\Delta t}\right) - \sum_{\mathbf{v}_1 \neq \mathbf{v}_1'} p^{\mathrm{e}}_{\mathbf{v}_1}\, W_{\mathbf{v}_1\mathbf{v}_1'}\, \ln\, W_{\mathbf{v}_1\mathbf{v}_1'} + \mathcal{O}(\Delta t)\,, \tag{9.111}$$

where

$$- \sum_{\mathbf{v}_1} p^{\mathrm{e}}_{\mathbf{v}_1}\, W_{\mathbf{v}_1\mathbf{v}_1} = \frac{\rho d^2}{16} \int d^2\Omega\, d^3 v_1\, d^3 v_2\, |\mathbf{v}_1 - \mathbf{v}_2|\, f_{\mathrm{e}}(v_1)\, f_{\mathrm{e}}(v_2)\,, \tag{9.112}$$

while

$$\sum_{\mathbf{v}_1 \neq \mathbf{v}_1'} p^{\mathrm{e}}_{\mathbf{v}_1}\, W_{\mathbf{v}_1\mathbf{v}_1'}\, \ln\, W_{\mathbf{v}_1\mathbf{v}_1'} = \frac{\rho d^2}{16} \int d^2\Omega\, d^3 v_1\, d^3 v_2\, |\mathbf{v}_1 - \mathbf{v}_2|\, f_{\mathrm{e}}(v_1)\, f_{\mathrm{e}}(v_2) \times \ln\left[\frac{\rho d^2}{16}\, \Delta^2\Omega\, \Delta^3 v\, |\mathbf{v}_1 - \mathbf{v}_2|\, f_{\mathrm{e}}(v_2)\right]. \tag{9.113}$$

The evaluation of these integrals gives

$$h(\Delta t \Delta^3 v \Delta^2 \Omega) = \rho\, d^2 \sqrt{\frac{\pi}{m\beta}}\, \ln\left(\frac{1595.2}{\rho\, d^2\, m\, \beta\, \Delta t\, \Delta^3 v \Delta^2 \Omega}\right) + \mathcal{O}(\Delta t)\,, \tag{9.114}$$

for the one-particle process (Gaspard and Wang 1993). We see that this (ε, τ)-entropy per unit time has the form of the ε-entropy per unit time corresponding to the random choice of six continuous random variables at time intervals separated by the mean intercollisional time, $\mathcal{T}_{\mathrm{intercoll}} \sim \ell/\bar{v}$. $\ell \sim 1/(\rho\pi d^2)$ is the mean free path while $\bar{v} \sim (m\beta)^{-1/2}$ is the average velocity. The six continuous random variables are the three velocity components $\mathbf{v}_2$ of the particles of the bath, the two angles $\Omega = (\theta, \phi)$ of the impact unit vector, and the random intercollisional time. As a consequence, ε is here the product of the ε's for each one of these random variables: $\varepsilon = \Delta t \Delta^3 v \Delta^2 \Omega$.

The preceding reasoning was done for the linear Boltzmann–Lorentz equation for simplicity but it can also be applied to the Boltzmann equation under the condition that we first map the Boltzmann equation onto a stochastic process. Indeed, the Boltzmann equation cannot be the master equation of a stochastic process because it is nonlinear whereas a master equation must be linear in order to have an interpretation in terms of a random process. Such a master equation for the Boltzmann equation has been obtained by several authors, it has the form of a birth-and-death process where the random variables are the numbers N_i of particles having the velocity $\mathbf{v}_i$. The average values of N_i are related to the probability density of the velocity according to $\bar{N}_i = \rho f(\mathbf{v}_i)\Delta^3 v$. The problem is now mapped onto a birth-and-death process for the binary collisions (see van Kampen 1981). The ε-entropy per unit time can then in principle be calculated as before with a similar logarithmic divergence in ε.

The above calculation concerns the random flight of a single particle in collision with other particles issued from the bath. In the case of a fluid, we may consider that the particles form a bath for themselves, which is at the basis of Boltzmann's arguments. The above entropy per unit time must therefore be multiplied by the number of particles so that the entropy over a time interval T in a volume L^d for such a *Boltzmann–Lorentz process with many particles* is given by

$$H(\Delta t \Delta^d v \Delta^{d-1}\Omega, T, L) \sim T\, L^d \log(1/\Delta t \Delta^d v \Delta^{d-1}\Omega)\,, \tag{9.115}$$

where d is here the spatial dimension. As a consequence of the logarithmic divergence, the Boltzmann–Lorentz gas with many particles has a higher degree of dynamical randomness than spacetime chaos.

The entropy (9.114) can be compared with the maximum Lyapunov exponent (0.6). The entropy of the full process is obtained by multiplication with the number of particles in the fluid. In this case, comparison can be made with the entropy (0.10). We see that the randomness of the Boltzmann–Lorentz process may exceed the upper bound given by the KS entropy if the ε quantities become too small. We can estimate the minimum value of ε allowed by the microscopic chaos of Newton's equations from the comparison of (9.114) with (0.6). We find

$$\Delta t\ \Delta^3 v\ \Delta^2\Omega\ >\ d\,\frac{k_B T}{m}\,, \tag{9.116}$$

where d is the particle diameter. The equation (9.116) gives the limit of applicability of the stochastic process. Crossing this limit would lead to suppose that the process is more random than it is actually according to its deterministic evolution.

9.6 Brownian motion and microscopic chaos

The comparison between the ε-entropies of a deterministic chaotic process and one of its stochastic models moreover suggests that both functions should be comparable except at small values of ε as depicted in Fig. 9.9. Therefore, the ε-entropy per unit time of a Brownian motion may provide evidence for the underlying microscopic chaos, as we now argue (Gaspard 1997a).

9.6.1 Hamiltonian and Langevin models of Brownian motion

Brownian motion is the irregular motion of a colloidal particle submitted to incessant collisions with the atoms or molecules of a gas or liquid. This motion was first observed for small pollen grains by the botanist Robert Brown in 1827. At the microscopic level, it is described by the classical Hamiltonian

$$H = \frac{\mathbf{P}^2}{2M} + \sum_{i=1}^{N} \frac{\mathbf{p}_i^2}{2m} + \sum_{1\leq i<j\leq N} V(|\mathbf{r}_i - \mathbf{r}_j|) + \sum_{i=1}^{N} \tilde{V}(|\mathbf{R} - \mathbf{r}_i|) , \tag{9.117}$$

where M is the mass of the colloidal particle, m the mass of the atoms or molecules, V the interparticle potential (for instance the Lennard–Jones interaction in rare gases), and $\tilde{V}$ the interaction potential between the atoms or molecules and the colloidal particle (Résibois and De Leener 1977). A Fokker–Planck equation for the position and velocity of the colloidal particle, $\mathbf{X} = (\mathbf{R}, \mathbf{V} = \mathbf{P}/M)$, has been derived from Hamiltonian (9.117) in the limit of large M/m ratios (Lebowitz and Rubin 1963, Résibois and Davis 1964). In the hydrodynamic limit of large spatial distances, the Fokker–Planck process reduces to the Langevin equation for the lone position of the colloidal particle

$$\dot{\mathbf{R}} = \mathbf{W} , \tag{9.118}$$

with

$$\langle W_i(t)\rangle = 0 , \qquad \langle W_i(t) W_j(t')\rangle = 2\, D\, \delta_{ij}\, \delta(t-t') , \tag{9.119}$$

(Wax 1954). The diffusion coefficient of a colloidal particle of radius a is known to be given by

$$D = \frac{k_{\mathrm{B}} T}{6\pi a \eta} , \tag{9.120}$$

where η is the fluid viscosity (for water, $\eta = 0.01$ g/cm sec; for air or argon, $\eta \simeq 2\times 10^{-4}$ g/cm sec; at room temperature and pressure) (Landau and Lifshitz

1963). In the classical experiments on Brownian motion by Perrin (1970), the radii of the colloidal particles were in the range 0.1 – 0.5 μm.

9.6.2 A lower bound on the positive Lyapunov exponents

For Brownian motion or Langevin processes like the Ornstein–Uhlenbeck process, the ε-entropy behaves as

$$h(\varepsilon) \simeq A \frac{D}{\varepsilon^2} , \tag{9.121}$$

in digits per unit time where D is the diffusion coefficient, ε is the diameter of cells in the physical position space, and A is a positive constant [cf. Eqs. (9.73) and (9.75)]. The increase of the entropy as $\varepsilon \to 0$ shows that randomness exists on arbitrarily small scales in Brownian motion as we see in Fig. 9.2. Nevertheless, on very small scales below the mean free path, the deterministic dynamics of the fluid particles becomes apparent and the ε-entropy (9.121) cannot grow indefinitely since the scaling property of the trajectory is only valid on the scales which are larger than the mean free path according to kinetic theory. Therefore, the ε-entropy saturates at a value given by the KS entropy because the KS entropy is the supremum of all the partition entropies according to (4.9) and (9.16). This behaviour is depicted in Fig. 9.9 for a system like the Lorentz gas (see also Fig. 9.8).

We now reach the conclusion of our argument. The ε-entropy (9.121) is an estimation of the entropy per unit time of a partition $\mathscr{C}_\varepsilon$ where the colloidal particle belongs to cells of radius ε in the space $\mathbf{R}$ while all the other variables, $\mathbf{P}, \mathbf{r}_1, \ldots, \mathbf{r}_N, \mathbf{p}_1, \ldots, \mathbf{p}_N$, remain arbitrary. In view of the inequality that the entropy per unit time with respect to an arbitrary partition $\mathscr{C}$ is always lower than its supremum which is the KS entropy [cf. Eqs. (4.9), (9.15), (9.16)] and because of Pesin's equality (4.41) of the KS entropy per unit time with the sum of the positive Lyapunov exponents, we get

$$0 \leq h(\mathscr{C}) \leq h_{\text{KS}} = \sum_{\bar{\lambda}_i > 0} \bar{\lambda}_i . \tag{9.122}$$

Therefore, we can conclude that the partition entropy $h(\mathscr{C}_\varepsilon) = h(\varepsilon)$ gives us a lower bound on the sum of positive Lyapunov exponents:

$$h(\mathscr{C}_\varepsilon) \simeq A \frac{D}{\varepsilon^2} \leq \sum_{\bar{\lambda}_i > 0} \bar{\lambda}_i . \tag{9.123}$$

Brownian motion can be observed with video cameras and optical microscopes which can reach resolutions of the order of $\varepsilon \simeq 0.5\mu$m (Perrin 1970). Recording the motion of colloidal particles, we can imagine measuring the ε-entropy per unit time. Table 9.4 contains several possible values for the ε-entropy corresponding to the preceding experimental conditions. Lower bounds of the order of 1–600 digits/sec may be expected. Although minute with respect to the theoretical value (0.7) for the maximal Lyapunov exponent in a fluid, the suggested experiment can nevertheless show that the sum of positive Lyapunov exponents is nonvanishing (Gaspard 1997a).

A similar reasoning can be applied to other random processes like the erratic oscillations of a torsion balance in a dilute gas or to Nyquist electric noise of thermodynamic origin (Pathria 1972).

9.6.3 Some conclusions

In this section, we developed in detail the argument that the dynamical randomness of Brownian motion or of other thermodynamic noises can give us a lower bound on the sum of Lyapunov exponents for the microscopic dynamics of the system. The argument is based on several hypotheses, in particular, the existence of positive Lyapunov exponents in Hamiltonian systems like (9.117). Such an existence can be proved in hard-sphere systems (Sinai 1979, Sinai and Chernov 1987, Chernov 1997) and there is strong numerical evidence of the validity of such an assumption in Hamiltonians like (9.117) at high energies and low densities corresponding to room temperatures and pressures (Posch and Hoover 1988, 1989; Dellago et al. 1995, 1996; van Beijeren et al. 1997).

The measurement of dynamical randomness of Brownian motion as described above would give us experimental evidence that microscopic chaos is fundamental for a transport process like diffusion. The evidence would be given by the

Table 9.4 For colloidal particles of radius a in suspension in a fluid at room temperature (300°K) and pressure (1 atm.), values of diffusion coefficient D given by (9.120), of the ε-entropy per unit time $h(\varepsilon) \simeq AD/\varepsilon^2$ for a resolution $\varepsilon = 0.5\ \mu$m, which is the lower bound on the sum of positive Lyapunov exponents. We have here taken the value $A \simeq 1.34$ digit/sec of Eq. (9.73). For comparison, we give an estimation of the maximum Lyapunov exponent as well as of the sum of positive Lyapunov exponents for 1 cm^3 of the corresponding fluid.

fluid	a [μm]	D [cm^2/sec]	$h(\varepsilon)$ [digits/sec]	$\bar{\lambda}_{\max}$ [digits/sec]	$\sum_{\bar{\lambda}_i>0} \bar{\lambda}_i$ [digits/sec·cm^3]
water	0.5	4.4×10^{-9}	2.4	10^{12}	$10^{34} - 10^{35}$
	0.1	2.2×10^{-8}	12	"	"
air or Ar	0.5	2.2×10^{-7}	120	10^{10}	$10^{29} - 10^{30}$
	0.1	1.1×10^{-6}	600	"	"

observation of a positive entropy per unit time. As a consequence, such an observation would show that a mechanism of stretching and folding exists in the phase space dynamics of the system. The observation of microscopic chaos in the Brownian motion would thus support this chaotic hypothesis.

In summary, we have developed our argument by showing that quantitative relations exist between apparently disconnected features. In particular, it is often thought that deterministic chaos has nothing to do with stochastic noises. Certainly, a distinction should be made between macroscopic and microscopic chaos as discussed in the Introduction. In the present chapter, we have shown that macroscopic chaos and stochastic noises are distinct phenomena (which may nevertheless be superposed in the same system). On the other hand, stochastic noises and microscopic chaos are two aspects of the very same phenomenon which is the motion of atoms or molecules in matter. The entropy per unit time is the concept which helps us to establish quantitatively the connection.

We should remark that microscopic chaos does not seem to be completely accessible to observation because Newtonian dynamics and chaos theory predicts values of the Lyapunov exponents and the KS entropy which are extremely large with respect to what can be practically measured. In this regard, we may also wonder to what extent the extreme dynamical randomness of a particle fluid as given by Eq. (0.10) is a property of the system or of the classical description. The answer to this question should come from a better understanding of dynamical randomness in a quantum description. We have summarized elsewhere several results known on this problem (Gaspard 1991b, 1992e, 1994).

We conclude with the remark that the property of chaos is concerned with the behaviour of the n-time correlation functions as we mentioned in Subsection 4.1.3. Multi-time properties have not been much explored till now in nonequilibrium statistical mechanics which has mainly been focused on 2-time correlation functions. Nevertheless, we think that modern multi-time techniques in nuclear magnetic resonance and in nonlinear quantum optics may give access to general or particular n-time statistical properties. In this context, the exponential decay (4.10) of the Shannon–McMillan–Breiman theorem could be used for more direct evaluations of the entropies per unit time.

Chapter 10

Conclusions and perspectives

The present work has been devoted to the study of the consequences of dynamical chaos on transport and reaction-rate phenomena in classical nonequilibrium statistical mechanics, in the light of recent results in dynamical systems theory.

10.1 Overview of the results

10.1.1 From dynamical instability to statistical ensembles

The starting point of our development has been to assume a classical dynamics of trajectories in phase space and a central role has been attributed to the properties of linear stability. During the last decade, the importance of linear stability has been emphasized in many works which have shown that trajectories are often unstable in nonlinear dynamical systems. This observation has shifted the paradigm used in mechanics from the harmonic oscillator

$$H = \frac{1}{2}(p^2 + \omega^2 q^2) , \qquad (10.1)$$

to the inverted harmonic potential

$$H = \frac{1}{2}(p^2 - \lambda^2 q^2) . \qquad (10.2)$$

Indeed, the harmonic oscillator (10.1) has been used since Galileo as the paradigm of mechanical systems. Certainly, its great interest lies in the fact that it is a model of recurrent phenomena with some period $T = 2\pi/\omega$. For such stable motions, the trajectory itself can be used to make predictions on future time evolutions.

In contrast, all the trajectories of the inverted harmonic potential (10.2) are unstable. This dynamical instability isolates each trajectory into an individualistic behaviour. In such unstable systems, long-time predictions only become possible for statistical ensembles of individual trajectories. One such prediction is the exponential decay with a lifetime given by the inverse of the Lyapunov exponent: $\tau = 1/\lambda$. This conclusion can be reached by an extension of the conventional methods of classical mechanics. First, a canonical change of coordinates

$$\begin{cases} q = \frac{1}{\sqrt{2\lambda}}(x-y) , \\ p = \sqrt{\frac{\lambda}{2}}(x+y) , \end{cases} \tag{10.3}$$

such that $dq \wedge dp = dx \wedge dy$ transforms the Hamiltonian (10.2) into the form

$$H = \lambda \, x \, y \, . \tag{10.4}$$

In these new coordinates, the equations of motion become separated

$$\begin{cases} \dot{x} = +\,\partial_y H = +\,\lambda \, x \, , \\ \dot{y} = -\,\partial_x H = -\,\lambda \, y \, , \end{cases} \tag{10.5}$$

so that the flow is

$$\mathbf{\Phi}^t(x,y) = \left(e^{+\lambda t}x, e^{-\lambda t}y\right) . \tag{10.6}$$

The time evolution of the statistical average of an observable $A(x, y)$ over the initial density $f(x, y)$ can be expanded in asymptotic series of either $\exp(-\lambda t)$ for $t > 0$ as

$$\begin{aligned} \langle A\rangle_t &= e^{-\lambda t} \int dx\, dy_0\, f\left(e^{-\lambda t}x, y_0\right) A\left(x, e^{-\lambda t}y_0\right) \\ &= \sum_{l,m=0}^{\infty} e^{-\lambda(l+m+1)t} \left[\int dx\, x^l\, \partial_y^m A(x,0)\right] \left[\int dy_0\, y_0^m\, \partial_{x_0}^l f(0,y_0)\right] , \end{aligned} \tag{10.7}$$

or of $\exp(+\lambda t)$ for $t < 0$ as

$$\begin{aligned} \langle A\rangle_t &= e^{+\lambda t} \int dx_0\, dy\, f\left(x_0, e^{+\lambda t}y\right) A\left(e^{+\lambda t}x_0, y\right) \\ &= \sum_{l,m=0}^{\infty} e^{+\lambda(l+m+1)t} \left[\int dx_0\, x_0^l\, \partial_{y_0}^m f(x_0,0)\right] \left[\int dy\, y^m\, \partial_x^l A(0,y)\right] , \end{aligned} \tag{10.8}$$

which define the forward and backward semigroups.

Contrary to the trajectories of (10.2) which are unbounded because of the instability, the time averages (10.7) and (10.8) are bounded in general and they even decay in such a dynamically unstable system. We emphasize that, in spite of the fact that long-time predictions are only possible at the level of statistical ensembles, the decay over the lifetime $\tau = 1/\lambda$ is nevertheless as certain as the periodicity of the harmonic oscillator. These dynamically unstable systems are therefore models of transient processes, which is a feature of so many natural phenomena.

In transient processes, asymptotic expansions like (10.7) or (10.8) for $t \to \pm\infty$ generate a spectrum of discrete rates of exponential relaxation, which may be interpreted as classical resonances or as generalized eigenvalues of the Liouville operator $\hat{L} = \{H, \cdot\}$. The formation of a discrete spectrum of such complex eigenfrequencies has its origin in the analyticity of the observable and the initial density as assumed in the Taylor expansions (10.7) and (10.8). In this perspective, the above construction of the semigroups of the Liouvillian dynamics is based on asymptotic expansions of the statistical averages around the point at infinity in the complex plane of the time t. The present work has been devoted to the development of such asymptotic expansions in time for nonequilibrium statistical mechanics.

In Chapter 1, we present the tools to define the linear dynamical instability in terms of the Lyapunov exponents. The local dynamical instability in phase space is characterized in terms of the local decomposition of the fundamental matrix of linear stability, called the homological decomposition. Accordingly, a scalar field of stretching rates and associated vector fields are introduced, which give the stable and unstable directions at each point in phase space, as well as the local rates of instability. These vector fields form a general Lie algebra based on the Poisson bracket. Thanks to this fundamental structure, the positive, negative, and zero Lyapunov exponents are defined locally in the phase space of the system. In Hamiltonian systems and billiards, the sum of the positive Lyapunov exponents is consequently given in terms of the local curvature of wavefronts of trajectories.

On the other hand, we show that the presence of continuous symmetries implies that a Lie group is formed by the subset of vector fields associated with the zero Lyapunov exponents. In this respect, the vector fields of the homological decomposition extend the structure of the Lie group of the constants of motion into a larger structure which has not yet been emphasized until now.

In Chapter 2, we summarize several important topological properties of trajectories in phase space. In particular, the concept of a saddle-type repeller is defined in comparison with the well-known concept of an attractor. Moreover, we show how the escape-time functions can give evidence for repellers. We introduce a new distribution and its cumulative function in order to characterize the spectrum of periodic orbits. The Laplace transform of this distribution gives the topological zeta function. The problem of counting periodic orbits is presented and the difficulty due to the overlap of cells in Markov partitions in closed systems is pointed out.

Chapter 3 describes the principles and the modern methods of Liouvillian dynamics. The concept of resonance is introduced by analytic continuation of the spectral functions toward complex frequencies, which are known as the Pollicott–Ruelle resonances in deterministic systems. The resonances appear as the eigenvalues of the Frobenius–Perron operator at complex frequencies, which complement the standard Koopman eigenvalues at real frequencies when the system is dynamically unstable. The associated eigenstates turn out to be given by (Schwartz) distributions instead of regular functions. When the system is mixing, the spectrum of Koopman eigenvalues contains only one eigenvalue corresponding to the invariant state in closed systems and it contains no eigenvalue at all in open systems with a repeller (see Fig. 10.1). The relevance of the Koopman spectral theory appears therefore to be restricted to nonmixing systems among which are the integrable classical systems like the harmonic oscillators and the pseudointegrable billiards. When the system has a continuous spectrum of real frequencies, Pollicott–Ruelle resonances may exist as generalized complex eigenfrequencies of the Liouvillian dynamics beside the Koopman real eigenfrequencies, which allows a much more precise description of the time evolution of statistical ensembles.[1]

The calculation of the Pollicott–Ruelle resonances differs depending on whether the nonwandering invariant set is a hyperbolic set with a stationary point or with periodic orbits, which are the two types of critical elements in a flow. We therefore develop in parallel the theory of classical resonances for both types of critical elements. We construct the full spectral decomposition of the Liouvillian dynamics near hyperbolic points with incommensurable stability eigenvalues. We argue that Jordan-block structures may appear in the presence of commensurabilities, which we explicitly show for a generic Hamiltonian potential with a maximum.

For hyperbolic sets with dense unstable periodic orbits, the characteristic determinant of the Frobenius–Perron operator is given by a Selberg–Smale Zeta

1. We also introduce in Subsection 3.5.3 generalized eigenstates corresponding to the *real* frequencies of the real continuous spectrum. These generalized eigenstates of the real continuous spectrum do not seem to have been introduced before. They allow a simple definition of Stone's projection operators, of the shift states, of the time operator, and a simple derivation of the Wiener–Khinchin theorem.

function which is a product involving the periods and the stability eigenvalues of the periodic orbits. Moreover, we show how the eigenvalue problem for a flow can be reduced to the eigenvalue problem of the corresponding Poincaré–Birkhoff map. The eigenstates can in principle be obtained using the Grothendieck–Ruelle generalization of the Fredholm theory. A method is described which gives the eigenprojections onto the left and right eigenstates in terms of periodic orbits.

These results show that a direct connection exists between the properties of the linear stability of the stationary points or of the periodic orbits and the time evolution of the statistical ensembles. Accordingly, the complex eigenfrequencies of the Liouvillian dynamics appear as intrinsic properties of the system as

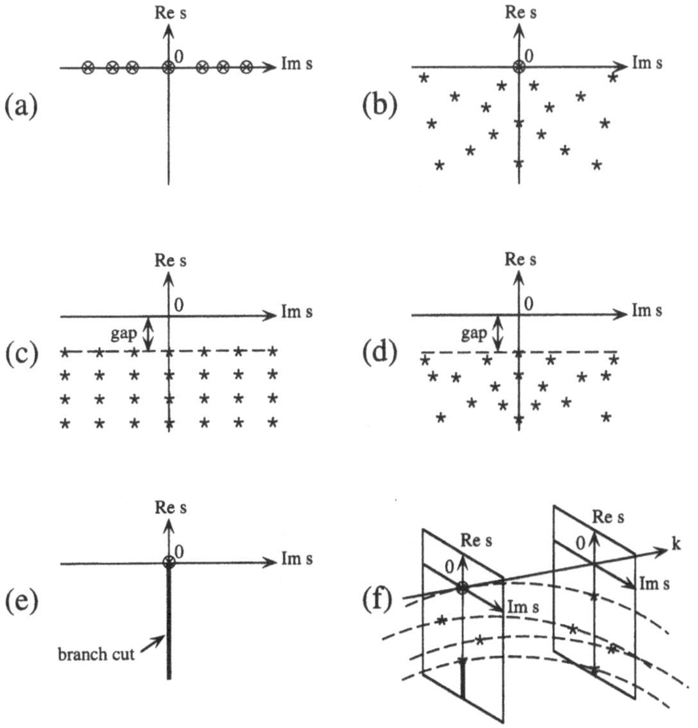

Figure 10.1 The resonance spectrum of different types of systems in the complex plane of the rate $s = \mathrm{Re}s + i\mathrm{Im}s$: (a) closed, multiperiodic, and integrable system; (b) closed, mixing, and chaotic system; (c) open, unstable, and periodic system (like the two-disk scatterer); (d) open, unstable, and chaotic system (like the three- and four-disk scatterers); (e) critical bifurcating system (like the pitchfork bifurcation); (f) spatially extended, mixing, and chaotic system (here, in the product of the complex plane of the rate s with the Brillouin zone of wavenumbers k).

illustrated with the inverted harmonic potential. In this sense, the Liouvillian dynamics can be viewed as a genuine part of classical mechanics.

Furthermore, we describe the metamorphoses of the spectrum of resonances at simple bifurcations like the pitchfork, the Hopf, and the saddle-loop homoclinic bifurcations. We show that the critical slowing down results in branch cuts formed by the accumulation of many resonances at bifurcations. Simple spectral decompositions in terms of the Pollicott–Ruelle resonances are explicitly given for the r-adic maps, where a connection is established with the Euler–Maclaurin expansions in terms of Bernoulli polynomials.

10.1.2 Dynamical chaos

The inverted harmonic potential (10.2) is an extreme case in which trajectories are of exponential character rather than harmonic oscillations. Chaos constitutes an intermediate behaviour in which the trajectories may remain bounded. Dynamical instability occurs between infinitesimally nearby trajectories, which is enough to develop a sensitivity to initial conditions and, hence, unpredictability. In turn, the local dynamical instability has consequences on the global topology of trajectories. For instance, if the system is closed the stretching of the phase-space volumes has to be followed by the folding of these volumes, which offers multiple choices for the time evolution of possible trajectories. As a consequence of folding, a branching process is generated in the space of trajectories, which results in dynamical randomness characterized by the topological or the KS entropies per unit time. These quantities are introduced by partitioning the phase space into cells labelled by integers, which is commonly carried out in the record of experimental data. In this way, *a dynamics which is deterministic on the real numbers may become nondeterministic on the integer numbers.* The dynamical randomness is a further intrinsic reason why statistical ensembles should be considered to make predictions. After long times, the stretching and folding dynamics of closed chaotic systems tends to mix and homogenize the statistical ensembles of trajectories over phase space.

To describe the chaotic properties, we introduce a model of observation in Chapter 4. With this model, we give an operational interpretation of the KS entropy in relation to the algorithmic complexity, which defines randomness according to Chaitin (1987) and Kolmogorov (1983). In this framework, dynamical randomness appears as a property of the uncountable set of almost all the trajectories of a system, to the exclusion of the countable set of regular trajectories which may be generated by a code or program with a binary length growing like the logarithm of the time. As explained by Chaitin, this definition of randomness is based on a formally undecidable proposition in the sense of Gödel, concerning an individual trajectory [namely, Eq. (4.2)]. In this way, Gödel's incompleteness theorem provides a basis for the introduction of a probabilistic description of chaotic dynamical systems.[2] The algorithmic complexity allows us to distinguish between different types of randomness. Regular randomness arises when the algorithmic complexity grows linearly with time as defined by Chaitin and Kolmogorov. Between regular randomness and quasiperiodicity, an intermediate randomness called sporadicity is possible in which the algorithmic complexity grows with time more slowly than linearly but faster than logarithmically (Gaspard and Wang 1988).

Dynamical randomness is a property which is concerned with the repetitive measurement of a property of the system. Such repetitive measurements are described in terms of n-time correlation functions. It is therefore natural to find that the KS entropy actually controls the decay of the n-time correlation functions according to the Shannon–McMillan–Breiman theorem. It is remarkable that the n-time correlation functions can have such well-defined properties. These high-order correlation functions have not yet been much explored since nonequilibrium statistical mechanics is traditionally focused on the 2-time correlation functions which give the transport coefficients according to the Green–Kubo relations. The 2-time correlation functions are at the top of the hierarchy of the n-time correlation functions. We notice that the same structure exists in thermal quantum field theories. As we show in Subsection 9.5.3, stochastic assumptions like Boltzmann's Stosszahlansatz destroy the finiteness of the KS entropy on small scales in phase space. Therefore, kinetic theory usually makes too strong assumptions on the properties of the hierarchy of the n-time correlation functions.

More general properties of the n-time correlation functions are characterized in the large-deviation formalism we present in Chapter 4. The 2-time correlation functions and their Fourier transforms – which are the spectral functions – can be used to study the statistical correlation of an observable with another observable taken at a later time. In mixing systems, two such observables become statistically independent after a long time, which suggests that time averages may statistically fluctuate around their mean asymptotic value. If the time correlations decay fast enough, these dynamical fluctuations can behave like Gaussian random variables according to the central limit theorem. This famous theorem of probability theory is only concerned with the majority of the dynamical fluctuations centered around the mean value. In contrast, the

2. In open systems, a probabilistic description is already required because almost all the trajectories are unbounded, i.e., because they escape to infinity.

large-deviation formalism is concerned with the large and rare fluctuations far from the mean value. These large fluctuations are studied at the level of the n-time correlation functions, which allows the definition of quantities like the KS entropy, the topological pressure, various generating functions and their Legendre transforms.

The probabilistic description of chaotic dynamics is built on top of the topological dynamics as illustrated in Fig. 10.2. In this regard, the probabilistic dynamics is certainly much richer than the topological dynamics due to the extra assumptions introduced concerning the observables. The large-deviation formalism is based on the definition of the topological pressure which is the dynamical analogue of the free energy of equilibrium statistical mechanics. A different topological pressure may be defined for different observables. When the observable vanishes, the topological pressure reduces to the topological entropy. In this sense, the large-deviation formalism extends beyond the topological properties as schematically depicted in Fig. 10.2. This formalism allows the definition of uncountably many invariant measures based on the different possible observables. However, there is only a unique invariant measure which is selected by the time and/or ensemble averaging which is carried out in a specific numerical experiment. This unique invariant measure is called the natural invariant measure and corresponds to probability weights given according to the local instability in phase space. The topological pressure can be used to evaluate the different characteristic quantities of chaos like the mean Lyapunov exponents, the dynamical entropies, the escape rate, the generalized fractal dimensions,.... The topological pressure can also be calculated in terms of the unstable periodic orbits of the system through generalized zeta functions which are given by the characteristic determinant of some Frobenius–Perron operators.

A very interesting aspect of the large-deviation formalism is the possibility of dynamical phase transitions where the uniqueness in the selection of the natural invariant measure fails. This is the case in nonhyperbolic systems where one invariant measure has the unstable trajectories for support while another invariant measure has the neutral trajectories for support. Therefore, the large-deviation formalism is a powerful tool for the analysis of chaos in nonhyperbolic systems.

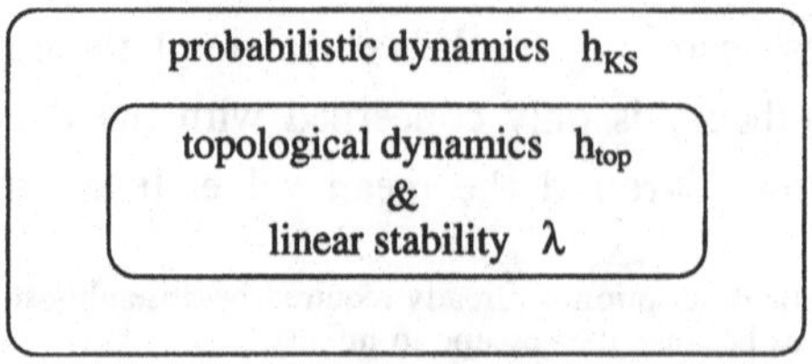

Figure 10.2 Schematic representation of the probabilistic properties defined with respect to the topological and stability properties of the phase-space trajectories.

In particular, the invariant measures can be extended by analytic continuation beyond criticality into supercritical invariant measures, which allows the effective definition of exponential transients even if the long-time decay is algebraic.

10.1.3 Fractal repellers and chaotic scattering

In open hyperbolic systems, a set of unstable trajectories may exist which is called a repeller. This is the case in most classical scattering systems. The repeller controls the dynamics of the trajectories bouncing off the scatterer and, in particular, of the trajectories escaping from the scatterer. Chapter 5 is devoted to this phenomenon which we illustrate with several examples like the disk scatterers and some models of unimolecular reactions.

In such scattering systems, the dynamical chaos manifests itself in the geometry of the repeller in phase space. Because of the escape process, the stretching and folding mechanism shapes the repeller as a fractal object. This fractal is characterized by the generalized fractal dimensions. On the other hand, the repeller is naturally characterized by its escape rate which turns out to be related to the difference between the sum of positive Lyapunov exponents and the KS entropy. The large-deviation formalism is thus a very powerful method for the study of classical scattering processes. A more detailed description of the escape process can be given in terms of the different Pollicott–Ruelle resonances of the Liouvillian dynamics. The leading Pollicott–Ruelle resonance already gives the long-time escape rate but the other resonances are associated with shorter transient features of the escape process as we show in Chapter 5. Such results nicely illustrate the importance of the classical resonances in the understanding of the time evolution of statistical ensembles.

In particular, unimolecular reactions can be conceived as the time evolution of such statistical ensembles in a scattering system. Therefore, the classical reaction rate of a unimolecular reaction is given as the escape rate of this process. Several time scales are often observed in numerical simulations of unimolecular reactions, which can be further characterized in terms of the other Pollicott–Ruelle resonances. In nonhyperbolic systems, an effective escape rate – as well as effective Pollicott–Ruelle resonances – can be defined by analytic continuation of these quantities from the chaotic phase to the ordered phase at the phase transition. With these methods, decay processes with multiple exponentials can be justified as discussed in Chapters 4 and 5. The theory presented in this work therefore provides the natural framework for the study of chemical reactions in classical mechanics.

10.1.4 Scattering theory of transport

When the scatterer has a large spatial extension the Liouvillian dynamics inside the scatterer becomes dominated by the slow transport processes such as deterministic diffusion, for instance. In particular, the escape process at the boundary of the system becomes controlled by the transport processes. In Chapter 6, we use this observation in order to relate the transport properties to the chaos properties. In this way, we obtain the escape-rate formulas which give the diffusion coefficient in terms of the characteristic quantities of chaos like the Lyapunov exponents, the KS entropy, or the Hausdorff dimension of the fractal repeller.

The escape-rate formulas are generalized to the other transport and reaction-rate coefficients by considering suitable absorbing boundaries in phase space. These absorbing boundaries are expressed in terms of the Helfand moments associated with each transport or reaction-rate process.

Accordingly, in large open systems, the large-scale transport properties determine fractal structures on the smallest scales in phase space. We may therefore understand that extremely singular objects actually control the dynamics of transport. Indeed, at the phenomenological level, transport is described in terms of the hydrodynamic modes which have the strange property that they relax exponentially. This property is in apparent contradiction with classical mechanics if we have in mind the harmonic oscillator (10.1) as the paradigm. However, the escape process of the inverted harmonic potential (10.2) shows that exponential behaviours are perfectly compatible with classical mechanics. Large scatterers from which particles only escape at the borders are weakly open systems, in which the huge dynamical instability cannot manifest itself by a massive escape of trajectories. In such systems, the dynamical instability mainly manifests itself by dynamical randomness in the bulk of the scatterer. Only a small part of the dynamical instability produces escape at the borders. The escape rate is associated with a certain mode of decay, which is nothing other than the dominant hydrodynamic mode. In this regard, we find that the exponential damping of the hydrodynamic mode becomes possible by fractalization on the finest scales in phase space.

10.1.5 Relaxation to equilibrium

The above results shed new light on the mechanism of approach to thermodynamic equilibrium, which can now be understood thanks to the concept of Pollicott–Ruelle resonance associated with the fractal-like hydrodynamic modes. This is the case in both open and closed systems of large spatial extension.

One aspect of this problem can be nicely illustrated with the baker map (Lebowitz and Penrose 1973):

$$\phi(x, y) = \begin{cases} \left(2x, \frac{y}{2}\right), & 0 \leq x \leq \frac{1}{2}, \\ \left(2x - 1, \frac{y+1}{2}\right), & \frac{1}{2} < x \leq 1. \end{cases} \tag{10.9}$$

One of the sure predictions on the fate of a typical trajectory is that it is uniformly distributed over phase space after a long enough time. The Liouvillian dynamics allows further predictions concerning the rate of approach to equilibrium for averages of some smooth observable $A(y)$ for instance. Let us consider an ensemble of N trajectories which are initially distributed uniformly in the lower half of the phase space ($y < 1/2$) as depicted in Fig. 10.3a. A corresponding probability density is obtained by the accumulation of infinitely many copies in the statistical ensemble as $N \to \infty$

$$f(x, y) = f(y) = \begin{cases} 1, & 0 \leq y < \frac{1}{2}, \\ 0, & \frac{1}{2} \leq y \leq 1, \end{cases} \tag{10.10}$$

which is depicted in Fig. 10.3b. The trajectories evolve in time under the baker map (10.9). After a finite number of iterations, the N trajectories appear uniformly distributed over phase space. However, the evolved probability density, i.e. $f[\phi^{-t}(x, y)]$, never becomes the uniform density but has jumps between 0 and 1 over smaller and smaller scales $\Delta y \sim 2^{-t}$, as time t increases (see Fig. 10.3b). Nevertheless, a smooth asymptotic time evolution can be predicted for the ensemble average $\langle A \rangle_t$ thanks to the spectral decomposition in terms of the Pollicott–Ruelle resonances. Indeed, for observables and densities which depend only on the variable y, we have the identity

$$\begin{aligned} \langle A \rangle_t &= \int_0^1 dx \int_0^1 dy\; A(x, y)\, f[\phi^{-t}(x, y)] \\ &= \int_0^1 dy\; f(y)\, (\hat{P}^t A)(y), \end{aligned} \tag{10.11}$$

where $\hat{P}$ is the Frobenius–Perron operator (3.218) of the dyadic map ($r = 2$). As

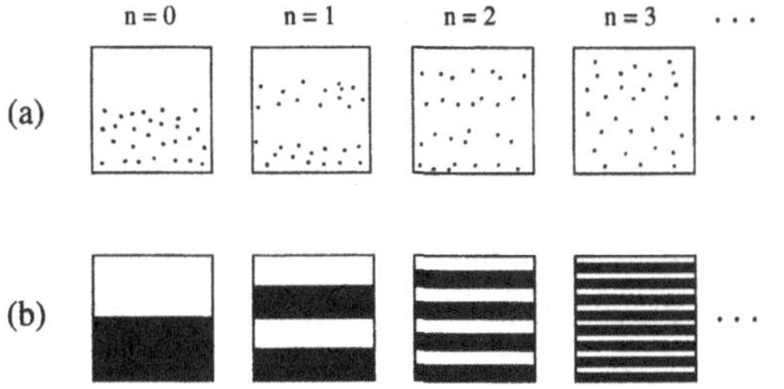

Figure 10.3 (a) Time evolution of a statistical ensemble of N trajectories under the baker map (10.9). (b) Time evolution of the corresponding probability density in the limit $N \to \infty$.

a consequence of the Euler–Maclaurin expansion (3.221)–(3.223), we find that

$$\begin{aligned} \langle A \rangle_t &= \langle f | \hat{P}^t | A \rangle = \sum_{m=0}^{\infty} \frac{A^{(m-1)}(1) - A^{(m-1)}(0)}{m!\, 2^{mt}} \int_0^{1/2} B_m(y)\, dy \,, \\ &= \frac{1}{2} \int_0^1 A(y)\, dy - \frac{A(1) - A(0)}{2^{t+3}} + \mathcal{O}(2^{-3t}) \,, \end{aligned} \tag{10.12}$$

where $A^{(n)}(y) = d^n A/dy^n$ and $B_m(y)$ is the mth Bernoulli polynomial. The first term is nothing other than the product $\langle f \rangle \langle A \rangle$ which is the asymptotic value expected for a mixing system. The other terms represent exponential decays of the time average. The coefficients are given by the values of the function $A(y)$ and of its derivatives at the borders of the phase space. This strange feature of such expansions has its origin in the fact that the evolved density $(\hat{P}^{\dagger t} f)(y)$ is a function alternating between the values 0 and 1 over 2^{t+1} small equal intervals. This function has the effect of taking derivatives of the averaged function in integrals over $0 \leq y \leq 1$. A similar asymptotic expansion can be obtained for $t \to -\infty$.

We can therefore express the approach to equilibrium in terms of well-defined exponential decays even in a system like the baker map which is time-reversal symmetric and area preserving. However, the baker map does not have the spatial extensivity which is the feature of systems sustaining transport processes. Therefore, the previous discussion has to be reconsidered in spatially extended systems.

With this motivation, we have introduced the multibaker model which is a deterministic realization of a symmetric random walk constructed as the spatial extension of the baker map (Gaspard 1992a). This model shares several properties with the Lorentz gas. In particular, both are models of deterministic diffusion and the multibaker map is area preserving like the Poincaré–Birkhoff map of the Lorentz gas. Both maps are of hyperbolic character with a positive mean Lyapunov exponent. The spatial extension and homogeneity of these models have the consequence that Pollicott–Ruelle resonances accumulate near the origin. These resonances correspond to relaxation modes with a wavelength smaller than the full size of the system but larger than the size of the phase-space structures producing the chaotic dynamics (like the disks of the Lorentz gas or the squares of the multibaker chain). On the other hand, the fast transients on the shorter phase-space scales may be controlled by branch cuts rather than Pollicott–Ruelle resonances because the time-correlation functions entering the Green–Kubo relations may have algebraic decays as often observed in numerical simulations. In systems of large spatial extensions, we should thus expect that branch cuts could coexist with the leading Pollicott–Ruelle resonances in the plane of complex frequencies.

In large but finite systems, the leading Pollicott–Ruelle resonances therefore mimic the discrete spectrum of the eigenvalues of the phenomenological diffusion equation (Gaspard 1992a). In infinite translationally-invariant systems, the accumulation of the leading Pollicott–Ruelle resonances creates a continuum of resonances labelled by the wavenumber $\mathbf{k}$ which is introduced by spatial Fourier transforms or, equivalently, by quasiperiodic boundary conditions. We show in Chapter 7 that the Liouvillian dynamics consequently decomposes into a continuum of $\mathbf{k}$-subspaces. In each $\mathbf{k}$-subspace, the Liouvillian dynamics is therefore ruled by a $\mathbf{k}$-dependent Frobenius–Perron operator which has Pollicott–Ruelle resonances depending on the wavenumber $\mathbf{k}$. In this way, the leading resonance provides the dispersion relation of deterministic diffusion. The leading resonance is the continuation toward $\mathbf{k} \neq 0$ of the Koopman eigenvalue at the origin associated with the equilibrium invariant measure. The isolation of this Koopman eigenvalue implies that, under appropriate conditions, an isolated leading resonance is generated when $\mathbf{k} \neq 0$.

As we show in Chapter 7, the Green–Kubo relation of diffusion can be deduced from the leading Pollicott–Ruelle resonance. Higher-diffusion coefficients like the Burnett and super-Burnett coefficients can in principle also be deduced from the leading Pollicott–Ruelle resonances. The hydrodynamic mode of diffusion is then given as the eigenstate associated with the leading resonance. This eigenstate is shown to be a (Schwartz) distribution rather than a function. This is in contrast with the construction of the hydrodynamic modes based on kinetic equations like the linearized Boltzmann equation or the Fokker–Planck equation, where the modes are given by the eigenfunctions of the collision operator. Here, eigenfunctions are found because of the stochastic approximation at the basis of these kinetic equations. In the Liouvillian description, the point-like character of the dynamics of trajectories in phase space prevents the eigenstates from being functions. This property already appears for the eigenstates of the Liouvillian dynamics of the inverted harmonic potential (10.2). Therefore, the origin of this property lies in the dynamical instability but not in the dynamical chaos. Dynamical chaos further complicates the eigenstates which acquire fractal-like properties. This is explicitly shown for the multibaker in which the distributions of the eigenstates can be represented by deRham-type functions under certain conditions, as discussed in Chapter 7. The logical role of the multibaker is thus to disprove the validity of the assumption that hydrodynamic modes are eigenfunctions in general. A single case, namely the multibaker, is sufficient for this purpose.

In Chapter 7, we develop the construction of the hydrodynamic modes of diffusion for general suspended flows, which can be reduced to a map with a Poincaré surface of section. This condition is not restrictive in itself because the results obtained with a Poincaré surface of section can be translated back to the original phase-space variables as done in particular for the Green–Kubo relations.

The hydrodynamic modes are constructed by assuming that the domain of definition of the Frobenius–Perron operator is much larger than a space of functions. Indeed, since the Frobenius–Perron operator is a substitution operator it acts, in particular, on the Dirac distributions and its derivatives. In this regard, we notice that the Dirac distributions represent points in phase space so that the consideration of trajectories immediately extends the domain of definition of the Frobenius–Perron operators to include the Dirac distributions in particular. If we now consider an ensemble of trajectories which is statistically regular, the domain of definition of the Frobenius–Perron operator may be restricted to functions such as (3.3) but this restriction turns out to be inadequate to study the asymptotic time evolutions even of averages over statistically regular ensembles. Indeed, the root and eigenstates of the forward and backward semigroups are found to be Schwartz distributions. Most of these distributions are not even singular measures because they involve derivatives of Dirac distributions. These distributions are defined as linear functionals over spaces of test functions which should be smooth enough. It is only for observables and initial densities in these spaces of test functions that the exponential decays can be justified. In this regard, we should emphasize that classical systems can sustain a very large variety of time evolutions, ranging from the evolution of individual trajectories if the initial distribution is Dirac's, to exponential decays for ensembles with entire analytic densities, passing by uncharacterized general decays for ensembles with only square-integrable densities. All these time evolutions are possible in principle as long as arbitrary initial conditions can be given to the different particles of the system, which is a fundamental property assumed since Galileo and Newton. However, the exponential decays observed in hydrodynamics can only be justified for ensembles of trajectories and for observables, which are represented by smooth enough functions. From this viewpoint, we can understand that very different time evolutions are possible and compatible for individual trajectories or ensembles of trajectories. Figure 10.4 gives a schematic representation of the state space of the Liouvillian dynamics. We notice that the polarization into forward and backward semigroups only concerns a special subset of the possible states on which the Frobenius–Perron operator may act.

10.1.6 Nonequilibrium steady states and entropy production

As a continuation of the previous results on the hydrodynamic modes of diffusion, a general form is obtained for the nonequilibrium steady states corresponding to a linear gradient of concentration. These steady states of the thermodynamic branch are deduced in different ways. First, we proceed to a direct construction starting from the flux boundary conditions. This construction is based on the general idea that nonequilibrium systems can be modelled in terms of the Liouvillian dynamics by imposing boundary conditions which express the nonequilibrium constraints at the level of the phase-space probability density. To each nonequilibrium constraint, there corresponds an invariant measure which has the plain phase space for support and which represents the nonequilibrium steady state.

When flux boundary conditions are imposed at large distances for a given nonequilibrium gradient, the invariant measure may become singular. The resulting singular steady state can be deduced from the construction of the hydrodynamic modes based on the Liouvillian eigenvalue problem. The steady-state

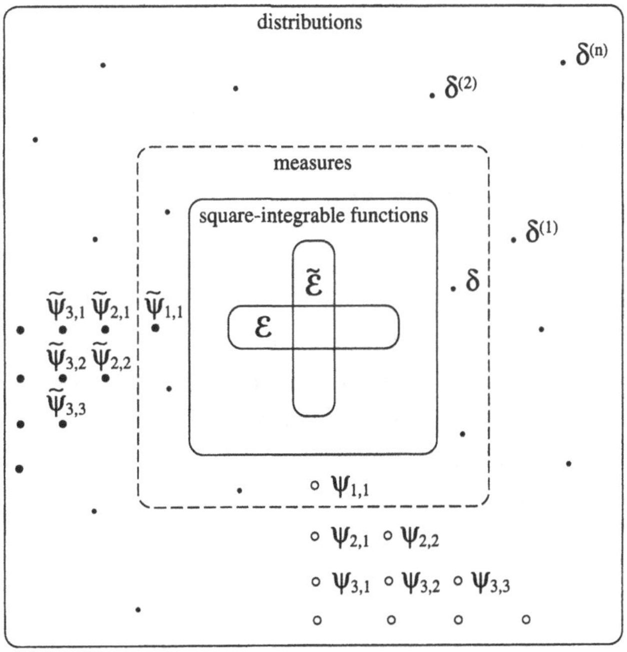

Figure 10.4 Schematic representation of the state space of a typical reversible Liouvillian dynamics. The dots represent the Dirac distributions and their derivatives. The blank circles are the root vectors $\psi_{j,i}$ of the forward semigroup while the filled circles are the root vectors $\tilde{\psi}_{j,i}$ of the backward semigroup.

invariant measure is composed of a term which is expected from the phenomenological macroscopic description and of an extra term which describes fluctuations around the mean phenomenological term due to the deterministic dynamics. A remarkable result is that this deduction from the Frobenius–Perron operator actually leads to the steady-state measures introduced by Lebowitz and MacLennan in 1959. These steady states are also equivalent to local integrals of motion introduced by Zubarev in 1962. Since then, these nonequilibrium steady states have been used in several works on nonequilibrium steady states (Procaccia et al. 1979a, 1979b; Kirkpatrick et al. 1982; Eyink et al. 1996), which shows that the extra term has an important role in the properties of nonequilibrium fluids.

For deterministic diffusion in the multibaker, we furthermore show that the singular character of the steady-state measures in large systems is at the origin of the entropy production expected from irreversible thermodynamics. In the multibaker, the nonequilibrium steady states of diffusion are given in terms of the Takagi function which is nondifferentiable because the corresponding measure is singular (Tasaki and Gaspard 1994, 1995). It turns out that the Takagi function has precisely the appropriate nondifferentiability to explain the phenomenological entropy production, which is most remarkable. We carry out a deduction of this form of the second law of irreversible thermodynamics under the assumption that the nonequilibrium constraint is maintained at arbitrarily large distances. In this regard, the second law appears as an asymptotic behaviour emerging in the bulk of out-of-equilibrium systems far from the boundaries. Because the second law appears as an emerging property there is no contradiction with the constancy of the fined-grained entropy. For a general transport process, the corresponding nonequilibrium steady state is also given by a singular measure in the large-system limit, as shown in Chapter 8. The singular part of the steady-state measure is directly related to the transport coefficient by the Green–Kubo formula as demonstrated by Eq. (8.69). Thanks to this complete generality, we may expect that the entropy production of irreversible thermodynamics can be deduced by the same argument as in the special case of diffusion in the multibaker map.

10.1.7 Irreversibility

Nonequilibrium thermodynamics gives a particular meaning to irreversibility in the realm of macroscopic phenomena in low-energy physico-chemical processes. A difficulty arises from the apparent paradox that a phenomenological equation like the diffusion equation is not invariant under the time reversal $t \to -t$

although Newton's equations are (Prigogine 1980). A related problem comes from the Poincaré recurrence theorem which precludes the use of individual trajectories to explain relaxation phenomena in bounded volume-preserving systems.

The way out of these well-known difficulties goes by Boltzmann's observation that macroscopic laws are not directly concerned with individual trajectories but by statistical ensembles of trajectories. In this way, Gibbs could explain the statistical independence of events separated by a long time, which is at the basis of the mixing property. The main point is that statistical ensembles do not behave the same way as individual trajectories in dynamically unstable systems. In particular, the time evolution of statistical ensembles with infinitely many trajectories is no longer plagued by Poincaré recurrences in typical systems. Hence, the statistical ensembles may asymptotically approach the thermodynamic equilibrium in the weak limits $t \to \pm\infty$. Moreover, the spectral decomposition of the Liouvillian dynamics in terms of the Pollicott–Ruelle resonances provides the asymptotic expansion of the statistical averages as $t \to \pm\infty$. This long-time limit appears as a highly singular limit.

According to the time-reversal symmetry of Newton's equation, the statistical comparison of a present event with a future event is equivalent to the comparison with a similar past event. However, the asymptotics $t \to \pm\infty$ is studied by analytic continuation to complex frequencies, which splits the group of evolution into two semigroups. In unstable systems, the forward semigroup is associated with eigenstates which are smooth along the unstable manifolds in phase space although the eigenstates of the backward semigroup are smooth along the stable manifolds. Since the stable and unstable manifolds are distinct objects in phase space, there is a nonequivalence under time-reversal symmetry which is introduced in the asymptotic description. This nonequivalence can manifest itself for certain observables which are not time-reversal symmetric.[3] This nonequivalence has the consequence that the eigenstates of the forward semigroup are not defined on the same space of test functions as the eigenstates of the backward semigroup.

Therefore, when the hydrodynamic modes of diffusion are derived from the Liouvillian dynamics, they have a domain of validity restricted to *either* positive times ($t > 0$) *or* negative times ($t < 0$). The construction of the asymptotic expansion at positive times *may not* be prolongated toward negative times. Accordingly, the phenomenological diffusion equation is only valid as an

3. $C(t) = C(-t)$ for a whole class of time correlation functions. But there is another class for which $C(t) = -C(-t)$ in which we find $C(t) = \langle \mathbf{q}_0 \cdot \dot{\mathbf{q}}_t \rangle$ and still another one for which $C(t)$ is not equal to $C(-t)$ or $-C(-t)$.

approximation of the forward semigroup for positive times. We may equally well be interested by the backward decay of the correlation functions in the same diffusive system. This decay under the backward semigroup would be approximated by the anti-diffusion equation. We may summarize the situation as

$$\begin{cases} \partial_t \rho \simeq +D\,\nabla^2 \rho\,, & \text{for} \quad t > 0\,, \\ \partial_t \rho \simeq -D\,\nabla^2 \rho\,, & \text{for} \quad t < 0\,. \end{cases} \tag{10.13}$$

Therefore, the aforementioned paradox of irreversibility is avoided since the diffusion and the anti-diffusion equations are restricted to different time domains.[4]

A similar explanation holds in open systems with escape. Here, any localized density of particles asymptotically approaches zero as $t \to \pm\infty$ because all the particles escape to infinity. By time-reversal symmetry, the forward escape rate has the same value as the backward escape rate. For instance, $\gamma^+ = \gamma^- = \lambda$ in the case of the inverted harmonic potential (10.2). However, the decay under the forward time evolution is asymptotically described by the leading eigenstate Ψ^+ which has the unstable manifold W_u for support. Hence, Ψ^+ is distinct from the leading eigenstate of the backward semigroup Ψ^- because this latter has the stable manifold W_s for support. They can nevertheless be mapped onto each other by the time-reversal transformation: $\Psi^+(q,p) = \Psi^-[\mathbf{I}(q,p)] = \Psi^-(q,-p)$. The time-reversal symmetry is here broken by the condition at infinity that no particle enters under forward or backward time evolution.

In a similar way, the problem of irreversibility disappears in systems maintained out of equilibrium with boundary conditions imposed on the phase-space probability density. In particular, the flux boundary condition breaks the time-reversal symmetry as discussed in Chapter 8. Indeed, the ingoing particles are assumed to be distributed with uniform densities although the density becomes highly complicated for the outgoing particles. Therefore, the invariant measure of such nonequilibrium systems breaks the time-reversal symmetry in general: $\mu(V) \neq \mu(\mathbf{I}V)$. Dynamical systems are defined in ergodic theory as the triplet composed of the phase space, the flow itself, and the invariant measure: $(\mathcal{M}, \mathbf{\Phi}^t, \mu)$ (Arnold and Avez 1968). This definition of dynamical system is Liouvillian because the definition refers to an invariant measure and, thus, to some statistical ensembles. A dynamical system is time-reversal symmetric only if $\mathbf{I}\mathcal{M} = \mathcal{M}$, $\mathbf{\Phi}^t \circ \mathbf{I} = \mathbf{I} \circ \mathbf{\Phi}^{-t}$, *and* $\mu(V) = \mu(\mathbf{I}V)$. Therefore, the dynamical

4. We can imagine a general dynamical system nonsymmetric under time reversal in which the forward diffusion coefficient differs from the backward diffusion coefficient: $D^+ \neq D^-$. In such systems, the forward semigroup cannot be mapped onto the backward semigroup by time reversal. In time-reversal symmetric systems, the equality $D^+ = D^- = D$ holds and the forward and backward semigroups can be mapped on each other.

system is not time-reversal symmetric as soon as one of the three conditions fails. Accordingly, we should expect that the time-reversal symmetry is broken by the boundary conditions imposed on out-of-equilibrium systems. A similar argument has been discussed for quantum systems by Frensley (1990).

This discussion shows that the breaking of the time-reversal symmetry can be explained in terms of boundary conditions at the level of the time evolution of statistical ensembles. These results tend to shift the emphasis toward the dynamical instability which appears as the crucial property. The role of dynamical instability has for long been neglected in physics and the other natural sciences. Only recently, dynamical instability has been more systematically considered in view of its constructive role in many phenomena, especially in far-from-equilibrium systems.

10.1.8 Possible experimental support for the hypothesis of microscopic chaos

In the last chapter, Chapter 9, the problem of dynamical randomness is placed in a broader context than deterministic dynamical systems with the aim of establishing connections with the rich phenomenology of physico-chemical processes.

First, we establish a distinction between random processes in which randomness is distributed regularly in time with a positive entropy per unit time and other processes in which random events occur sporadically with a vanishing entropy per unit time. Several examples of sporadicity are listed.

Moreover, a comparison is carried out between different stochastic models like deterministic chaos, Langevin processes, birth-and-death processes,... with a regular randomness. This comparison is based on the concept of ε-entropy per unit time which provides a quantitative measure of dynamical randomness on some scale ε. The different stochastic processes can be distinguished by the behaviour of the ε-entropy as $\varepsilon \to 0$. This entropy saturates for deterministic processes like chaos while it displays a characteristic divergence in stochastic processes like the Langevin processes. This difference allows us to make a distinction between noises and chaos.

In Chapter 9, we also use the concept of ε-entropy to compare different models of statistical mechanics and, in particular, of Brownian motion. Brownian motion can be viewed as the random process induced on a colloidal particle due to the collisions with the particles of a fluid. This large dynamical system is chaotic in its microscopic phase space. Therefore, the dynamical randomness displayed by the Brownian particle turns out to be part of the dynamical randomness produced by the underlying microscopic chaos. The ε-entropy per unit time of the Brownian motion can thus provide evidence for the microscopic chaos. We argue that a lower bound on the positive Lyapunov exponents of the fluid can be obtained thanks to the ε-entropy. In this way, evidence for dynamical instability of the microscopic dynamics could be obtained experimentally. This would provide an experimental support of the hypothesis that typical Hamiltonian systems of statistical mechanics can be well approximated by hyperbolic dynamical systems like hard-sphere gases. This hypothesis of microscopic chaos is the basis of the present work in which hyperbolic chaotic systems are used as prototypes of the systems of statistical mechanics. This hypothesis is supported by many numerical simulations which reveal a spectrum of positive Lyapunov exponents in typical many-particle systems (Posch and Hoover 1988, 1989; Dellago et al. 1995, 1996). The measurement of a lower bound on the positive Lyapunov exponents in Brownian motion would thus bring experimental support to this hypothesis of microscopic chaos.

The idea that gases are chaotic systems is not new as we discussed in the Introduction. Krylov (1944) has addressed the question of the disorder in time of systems of statistical mechanics, which is today characterized by the KS entropy per unit time. Sinai (1979) and Chernov (1991) have furthermore considered the disorder in space and in time by introducing the entropy per unit time and unit volume. In this perspective, we see the importance of experimental support for microscopic dynamical chaos.

10.2 Perspectives and open questions

The novel concepts of chaos, of fractal, and of classical resonances open new perspectives in physics and in the natural sciences in general. Determinism acquires a very different meaning than previously envisaged. The dynamical instability limits the possible prediction of the time evolution of a dynamically unstable system within the Lyapunov horizon at the level of trajectories and to statistical statements for longer times. On the other hand, the discovery that deterministic systems may have random time evolutions is of particular importance because many random phenomena are today found to be described by deterministic dynamical systems although they were previously thought to belong to different classes of systems. Similarly, dynamical instability shows that exponential relaxations can be accomodated within statistical mechanics although such behaviour previously appeared incompatible with mechanics. Therefore, the new results considerably enlarge the domain of applicability of dynamical systems theory and of classical mechanics to the natural sciences in general.

Many questions have been explicitly or implicitly open in several chapters of the present work. They concern several aspects of dynamical systems theory, with functional analysis as the basis of spectral decompositions in terms of the Pollicott–Ruelle resonances, with nonequilibrium statistical mechanics and the experimental observation of microscopic chaos, as well as with extensions to quantum mechanical systems.

10.2.1 Extensions to general and dissipative dynamical systems

In this work, we mainly focus the discussion on symplectic dynamical systems. However, most of the properties we discuss are of general validity since they actually find their origin in dissipative dynamical systems. Indeed, in recent decades, fractal objects in phase space have been introduced with the Smale horseshoe and the strange attractors of the dissipative dynamical systems. Many results of this work are based on the concept of chaotic repeller which remarkably extends without modification from dissipative systems to Hamiltonian systems. Chaotic scattering has also been studied in dissipative dynamical systems as in fluid mechanics where it provides a useful understanding of some advection phenomena (Jung et al. 1993, Ott and Tél 1993). In a similar context, the multibaker is a simple model for mixing of a passive scalar in a convecting fluid. In this perspective, we formulate the Liouvillian theory of deterministic diffusion in full generality. The theory can be used to study transport by diffusion and even anomalous diffusion in general spatially extended systems whether the vector field is dissipative or not.

Important open questions concern the extension to nonhyperbolic and nonergodic systems. For nonhyperbolic systems, we make several suggestions in particular in Chapter 4 where it is shown that the nonhyperbolicity manifests itself by dynamical phase transitions. In a recent work by Alonso et al. (1996), some consequences of nonhyperbolicity on the spectrum of the Pollicott–Ruelle resonances have also been studied. This work shows that the universality of the Feigenbaum period-doubling cascade can induce universal properties in the resonance spectrum. Extensions of such a result to other universal bifurcation cascades may be expected. Nonhyperbolicity also manifests itself at bifurcations of dynamical systems by the formation of branch cuts at complex frequencies as shown for the pitchfork bifurcation, in particular (cf. Chapter 3). The extensions of these results to more general bifurcations is an open problem of great interest because typical nonhyperbolic chaotic attractors are known to be highly sensitive to parametric perturbations due to incessant bifurcations. The Liouvillian approach to bifurcations of Hamiltonian systems is also of great interest in relation to the typical nonhyperbolicity of Hamiltonian systems. The Liouvillian dynamics of bifurcations in billiards when trajectories are tangent to the walls is also largely unexplored.

In nonbifurcating systems, many questions are open concerning the spectral decomposition even in the apparently simple case of stationary points. In Chapter 3, we give the general decomposition for stationary points without commensurabilities between the stability eigenvalues. We morever show with a simple example that Jordan blocks should in general be expected when the stability eigenvalues are commensurable. Here, the general construction of the spectral decomposition is still lacking although we may expect that it is based on the same methods as used for the reduction of a vector field to its Birkhoff normal form.

The general construction of the spectral decomposition in systems with dense periodic orbits has been developed greatly in recent years but many questions remain open. There exist several formulations and, in particular, the one based on the isomorphism to a symbolic dynamics (Ruelle 1987). The Jordan-block structure expected in area-preserving maps does not seem to have been discussed in this formulation. Moreover, the Fredholm theory has not been developed much for the direct construction of the eigenstates and many questions are open concerning the mathematical justifications of the formal algebraic manipulations. Indeed, the fact that the eigenstates are distributions instead of functions as in quantum mechanics requires considerable work in functional analysis to properly define the spaces on which these distributions are defined. The long time taken to reach a mathematically rigorous definition of Dirac's delta distribution should suggest that patience is required in such mathematical developments.

Furthermore, few studies have been devoted to the construction of the homological decomposition of the fundamental matrix of linear stability as discussed in Chapter 1.

10.2.2 Extensions in nonequilibrium statistical mechanics

The scattering theory of transport is only at a very early stage of development in our opinion. If deterministic diffusion has been analyzed in detail in systems with two degrees of freedom like the multibaker and the periodic Lorentz gas, many problems can be investigated in larger and more complex systems.

Recent works by van Beijeren and Dorfman (1995) and by Appert el al. (1996)

have been devoted to random Lorentz gases which are either deterministic or probabilistic automata. The work by Appert et al. (1996) has shown that such random Lorentz gases may present some kind of dynamical phase transitions. On the other hand, Klages and Dorfman (1995) have discovered that the diffusion coefficient can vary in a highly irregular manner under variations in the system parameters due to the structural instability of the topology of trajectories and in relation to the singular character of the steady-state measures of Chapter 8. This irregularity implies that the diffusion coefficient is not a differentiable function of the parameters, which suggests some connection with the nonanalyticity of density expansions of the transport coefficients as observed since Bogoliubov's work (Balescu 1975).

Many open questions concern the systems with many degrees of freedom. In particular, the characterization of high-dimensional fractal objects in phase space is still in its infancy. Very few results exist on the partial dimensions of such high-dimensional fractals which could be used to extend the escape-rate formula expressed in terms of the partial Hausdorff codimension from two-degree-of-freedom systems to larger systems. The extension of the escape-rate formulas to conservative systems with external fields is also very interesting, as well as to thermostatted dynamical systems. Another current open problem is the precise calculation of the KS entropy per unit time and unit volume in dilute gases of hard spheres or of particles interacting with a potential. In this regard, let us mention that kinetic-type methods have been developed for the systematic calculation of the KS entropy of such systems by van Beijeren, Dorfman, and coworkers (1995, 1997).

The construction of the hydrodynamic modes for general transport processes remains a major open problem in the prolongation of the results of Chapter 7. Such a construction should be based on the understanding of the way the hydrodynamic modes are interrelated to the conservation laws expressed either locally or globally. The work by Bunimovich and Spohn (1996) suggests that such a construction would be possible in two-particle fluids but the distinction between the number of particles in the system and the number of copies of the system in the statistical ensembles requires further understanding. Similarly, the formation of hydrodynamic Goldstone modes by spontaneous symmetry breaking is a fascinating problem which seems to be related to the breaking of ergodicity in a dynamical system like the hard-disk system at high densities. The explicit contruction of the chemio-hydrodynamic modes in deterministic models of reaction-diffusion systems is still open. In Chapter 8, we also mention the possibility of modelling far-from-equilibrium systems with appropriate boundary conditions in the Liouvillian dynamics. The resonances of the Liouvillian dynamics can also provide a characterization of the special decays of the time-correlation functions in far-from-equilibrium systems.

Many results of the present work rest on the assumption of a microscopic chaos. The experimental test of this chaotic hypothesis by the observation of the dynamical randomness of Brownian motion as we explain in Chapter 9 is of particular importance.

10.2.3 Extensions to quantum-mechanical systems

In the present work, we have restricted ourselves to classical systems. The extension of the previous results to quantum systems would require further work with the specific methods of quantum mechanics. The main reason is that quantum mechanics induces time evolutions which are much more coherent than in classical mechanics. In particular, the spectrum of a closed quantum systems is always discrete as given by the Bohr frequencies $\omega_{mn} = (E_m - E_n)/\hbar$ where $\hbar$ is the Planck constant. Continuous spectra are only expected in scattering quantum systems and in many-body quantum systems. This result opens three possible lines of investigation.

In a first approach, closed quantum systems are studied in the semiclassical limit, which formally corresponds to the limit $\hbar \to 0$. Many works have recently addressed this problem and Gutzwiller has developed a periodic-orbit theory in order to semiclassically quantize classically chaotic systems (Gutzwiller 1990). The aim of Gutzwiller's theory is to obtain the eigenvalues of the Hamiltonian operator in terms of the classical periodic orbits. In this context, important efforts are devoted to the understanding of level statistics in semiclassical terms (Andreev et al. 1996). On the other hand, the time-correlation functions are expected to approach the classical behaviour described in the present work in the limit $\hbar \to 0$. As shown elsewhere, the semiclassics of the time-correlation functions involves both the Liouvillian periodic-orbit theory of Chapter 3 for a dominant quasiclassical part and Gutzwiller periodic-orbit theory for quantum coherent corrections (Gaspard and Burghardt 1997, Gaspard and Jain 1997).

A similar behaviour is expected in quantum scattering systems. Here, Gutzwiller's theory has been used in order to obtain semiclassically the scattering resonances of the Hamiltonian operator, i.e., the complex energies which are the poles of the quantum scattering matrix (Gaspard and Rice 1989a, Smilansky 1991). On the other hand, Liouvillian resonances are also defined in the context of the dynamics of quantum statistical ensembles described by a density matrix in which incoherent superpositions of wavefunctions are considered. We have discussed elsewhere the connection with the results of the present work (Gaspard and Burghardt 1997).

Certainly, many-body quantum systems should be expected to have the dynamical properties required to sustain relaxation and transport processes which are at the centre of the preoccupations of the present work. Indeed, it is only in such systems that we may expect that the time-correlation functions decay fast enough for the transport coefficients to be well defined. In this context, we mention the work by Jakšić and Pillet (1996) who have recently defined resonances of the quantum Liouvillian dynamics of the spin-boson system. Moreover, Connes, Narnhofer, and Thirring (1987) have defined a dynamical entropy for quantum systems, which is in correspondence with the classical ε-entropy per unit time of Chapter 9 for the ideal gases (Gaspard 1991b, 1992e, 1994; Hudetz 1990). More recently, Alicki and Fannes (1994) have proposed another definition which is promising for further applications. Methods are thus also developed for the definition of dynamical randomness in large quantum systems. In this perspective, it appears that thermal many-body quantum systems and thermal quantum fields can also display dynamical randomness. But the description of these fascinating problems is beyond the present work.

References

Abraham, R. & Marsden, J. G. (1978). *Foundations of Mechanics*. Reading, Massachusetts: Benjamin-Cumming.

Abramowitz, M. & Stegun, I. A. (1972). *Handbook of Mathematical Functions*. New York: Dover.

Adler, R. L. & Weiss, B. (1970). *Mem. Am. Math. Soc.* **98**. Providence: Am. Math. Society.

Alder, B. J. & Wainwright, T. E. (1969). *Phys. Rev. Lett.* **18**, 988.

Alekseev, V. M. & Yakobson, M. V. (1981). *Phys. Rep.* **75**(5), 287.

Alicki, R. & Fannes, M. (1994). *Lett. Math. Phys.* **32**, 75.

Alonso, D., MacKernan, D., Gaspard, P. & Nicolis, G. (1996). *Phys. Rev. E* **54**, 2474.

Anderson, P. W. (1984). *Basic Notions of Condensed Matter Physics*. Menlo Park, California: Benjamin/Cummings.

Andreev, A. V., Agam, O., Simons, B. D. & Al'tshuler, B. L. (1996). *Phys. Rev. Lett.* **76**, 3947.

Anosov, D. V. (1967). *Proc. Steklov Inst. Math.* **90**, 235; *Russian Math. Surveys* **22**, 103.

Antoniou, I. & Tasaki, S. (1992). *Physica A* **190**, 303.

Antoniou, I. & Tasaki, S. (1993). *J. Phys. A* **26**, 73.

Antoniou, I. & Qiao, B. (1996). *Phys. Lett. A* **215**, 280.

Appert, C., van Beijeren, H., Ernst, M. H. & Dorfman, J. R. (1996). *Phys. Rev. E* **54**, R1013.

Arnold, V. I. (1978). *Mathematical Methods of Classical Mechanics*. New York: Springer-Verlag.

Arnold, V. I. (1982). *Geometrical Methods in the Theory of Ordinary Differential Equations*. New York: Springer-Verlag.

Arnold, V. I., ed. (1988). *Encyclopedia of Mathematical Sciences: Dynamical Systems III*. Berlin: Springer-Verlag.

Arnold, V. I. & Novikov, S. P., ed. (1990). *Encyclopedia of Mathematical Sciences: Dynamical Systems IV*. Berlin: Springer-Verlag.

Arnold, V. I. & Avez, A. (1968). *Ergodic Problems of Classical Mechanics*. New York: W. A. Benjamin.

Artuso, R. (1991). *Phys. Lett. A* **160**, 528.

Artuso, R., Cvitanović, P. & Kenny, B. G. (1989). *Phys. Rev. A* **39**, 268.

Artuso, R., Aurell, E. & Cvitanović, P. (1990). *Nonlinearity* **3**, 325, 361.

Artuso, R., Casati, G. & Guarneri, I. (1996). *J. Stat. Phys.* **83**, 145.

Aumaître, S. (1995). *Etude de la viscosité d'un "fluide" périodique à deux particules en interaction de Yukawa à partir d'une simulation numérique.* DEA memorandum ULB (unpublished).

Baladi, V., Eckmann, J.-P. & Ruelle, D. (1989). *Nonlinearity* **2**, 119.

Balescu, R. (1960). *Phys. Fluids* **3**, 52.

Balescu, R. (1961). In: *Lectures in Theoretical Physics* (eds. Brittin, W. E., Downs, B. W. & Downs, J.), vol. III, pp. 382–444. New York: Interscience.

Balescu, R. (1975). *Equilibrium and Nonequilibrium Statistical Mechanics.* New York: Wiley.

Baranyai, A., Evans, D. J. & Cohen, E. G. D. (1993). *J. Stat. Phys.* **70**, 1085.

Baras, F. & Malek Mansour, M. (1997). *Adv. Chem. Phys.* **100**, 393.

Bauer, L. & Reiss, E. L. (1978). *SIAM J. Appl. Math.* **35**, 508.

Beck, C. & Schlögl, F. (1993). *Thermodynamics of Chaotic Systems.* Cambridge, England: Cambridge University Press.

Beenakker, C. W. & van Houten, H. (1989). *Phys. Rev. Lett.* **63**, 1857.

Beijeren, H., van (1982). *Rev. Mod. Phys.* **54**, 195.

Beijeren, H., van & Dorfman, J. R. (1995). *Phys. Rev. Lett.* **74**, 4412. (1996). *ibid.* **76**, 3238(E).

Beijeren, H., van, Dorfman, J. R., Posch, H. A. & Dellago, Ch. (1997). *Phys. Rev. E* **56**, 5272.

Bensimon, D. & Kadanoff, L. P. (1984). *Physica D* **13**, 82.

Berger, T. (1970). *IEEE Trans. Inf. Theor.* **16**, 134.

Berger, T. (1971). *Rate Distortion Theory.* Englewood Cliffs: Prentice-Hall.

Bessis, D., Paladin, G., Turchetti, G. & Vaienti, S. (1988). *J. Stat. Phys.* **51**, 109.

Biham, O. & Kvale, M. (1992). *Phys. Rev. A* **46**, 6334.

Billingsley, P. (1965). *Ergodic Theory and Information.* New York: Wiley.

Bird, G. A. (1963). *Phys. Fluids* **6**, 1518.

Bird, G. A. (1990). In: *Microscopic Simulations of Complex Flows* (ed. Mareschal, M.), pp. 1–13. New York: Plenum Press.

Birkhoff, G. D. (1927). *Dynamical Systems,* Colloq. Publ. vol. IX. New York: Am. Math. Society.

Birkhoff, G. D. (1931). *Proc. Natl. Acad. Sci. USA* **17**, 656.

Bleher, S., Ott, E. & Grebogi, C. (1990). *Physica D* **46**, 87.

Blümel, R. (1993). In: *Quantum Chaos* (eds. Casati, G., Guarneri, I. & Smilansky, U.), Proceedings of the CXIX International Course of Physics "Enrico Fermi", Varenna, 1991, pp. 385–398. Amsterdam: North-Holland.

Blümel, R., Dietz, B., Jung, C. & Smilansky U. (1992). *J. Phys. A* **25**, 1483.

Boas, R. P. & Buck, R. C. (1958). *Polynomial Expansions of Analytic Functions.* Berlin: Springer-Verlag.

Bogoliubov, N. N. (1946). *J. Phys. USSR* **10**, 257, 265.

Bohr, T. & Rand, D. (1987). *Physica D* **25**, 387.

Bohr, T. & Tél, T. (1988). In: *Directions in Chaos* (ed. Hao Bai-lin, U.), vol. II, pp. 194–237. Singapore: World Scientific.

Boltzmann, L. (1896). *Vorlesungen über Gastheorie.* Leipzig: Barth. English translation (1964). *Lectures on Gas Theory.* New York: Dover.

Boon, J.-P. & Yip, S. (1980). *Molecular Hydrodynamics.* New York: Dover.

Born, M. & Green, H. S. (1946). *Proc. Roy. Soc. Lond. A* **188**, 10.

Bouchaud, J.-P. & Le Doussal, P. (1985). *J. Stat. Phys.* **41**, 225.

Bouckaert, L. P., Smoluchowski, R. & Wigner, E. P. (1936). *Phys. Rev.* **50**, 58.

Bowen, R. (1975). *Equilibrium States and the Ergodic Theory of Anosov Diffeomorphisms,* Lect. Notes in Math. Vol. 470. Berlin: Springer-Verlag.

Bowen, R. & Ruelle, D. (1975). *Invent. Math.* **29**, 181.

Breymann, W. , Tél, T. & Vollmer, J. (1996). *Phys. Rev. Lett.* **77**, 2945.

Brout, R. & Prigogine, I. (1956). *Physica* **22**, 621.

Buchleitner, A., Delande, D., Zakrzewski, J., Mantegna, R. N., Arndt, M. & Walther, H. (1995). *Phys. Rev. Lett.* **75**, 3818.

Bunimovich, L. A. (1979). *Commun. Math. Phys.* **65**, 295.

Bunimovich, L. A. (1985). *Sov. Phys. JETP* **62**, 842.

Bunimovich, L. A. & Sinai, Ya. G. (1980). *Commun. Math. Phys.* **78**, 247, 479. (1986) *ibid.* **107**, 357(E).

Bunimovich, L. A., Sinai, Ya. G. & Chernov, N. I. (1990). *Russian Math. Surveys* **45**, 105.

Bunimovich, L. A. & Spohn, H. (1996). *Comm. Math. Phys.* **176**, 661.

Burghardt, I. & Gaspard, P. (1994). *J. Chem. Phys.* **100**, 6395.

Burghardt, I. & Gaspard, P. (1995). *J. Phys. Chem.* **99**, 2732.

Callen, H. B. & Welton, T. A. (1951). *Phys. Rev.* **83**, 34.

Casati, G. & Ford, J. (1979). *Stochastic Behavior in Classical and Quantum Hamiltonian Systems.* Berlin: Springer-Verlag.

Chaitin, G. J. (1987). *Algorithmic Information Theory.* Cambridge, England: Cambridge University Press.

Chen, Q., Ding, M. & Ott, E. (1990). *Phys. Lett. A* **145**, 93.

Chernov, N. I. (1991). *Funct. Anal. Appl.* **25**, 204.

Chernov, N. I. (1994). *J. Stat. Phys.* **74**, 11.

Chernov, N. I. (1997). *J. Stat. Phys.* **88**, 1.

Chernov, N. I., Eyink, G. L., Lebowitz, J. L. & Sinai, Ya. G. (1993a). *Phys. Rev. Lett.* **70**, 2209.

Chernov, N. I., Eyink, G. L., Lebowitz, J. L. & Sinai, Ya. G. (1993b). *Commun. Math. Phys.* **154**, 569.

Chernov, N. I. & Markarian, R. (1997). *Bull. Brazil. Math. Soc.*

Chernov, N. I. & Lebowitz, J. L. (1997). *J. Stat. Phys.* **86**, 953.

Choquard, Ph. & Steiner, F. (1996). *Helv. Phys. Acta* **69**, 636.

Christiansen, F., Paladin, G. & Rugh, H. H. (1990a). *Phys. Rev. Lett.* **65**, 2087.

Christiansen, F., Cvitanović, P. & Rugh, H. H. (1990b). *J. Phys. A* **23**, L713.

Cohen, A. & Procaccia, I. (1985). *Phys. Rev. A* **31**, 1872.

Cohen, E. G. D., ed. (1962). *Fundamental Problems in Statistical Mechanics.* Amsterdam: North-Holland.

Cohen, E. G. D., de Schepper, I. M. & Zuilhof, M. J. (1984). *Physica B* **127**, 282.

Cohen, E. G. D. & de Schepper, I. M. (1990). *Il Nuovo Cimento D* **12**, 521.

Connes, A., Narnhofer, H. & Thirring, W. (1987). *Commun. Math. Phys.* **112**, 691.

Conway, J. H. (1982). In: *Winning Ways* (eds. Berkelamp, E., Conway, J. & Guy, R.), vol. 2. New York: Academic Press.

Cornfeld, I. P., Fomin, S. V. & Sinai, Ya. G. (1982). *Ergodic Theory.* Berlin: Springer-Verlag.

Cross, M. C. & Hohenberg, P. C. (1993). *Rev. Mod. Phys.* **65**, 851.

Cvitanović, P. (1988). *Phys. Rev. Lett.* **61**, 2729.

Cvitanović, P. (1991). *Physica D* **51**, 138.

Cvitanović, P. & Eckhardt, B. (1989). *Phys. Rev. Lett.* **63**, 823.

Cvitanović, P. & Eckhardt, B. (1991). *J. Phys. A* **24**, L237.

Cvitanović, P., Gaspard, P. & Schreiber, T. (1992). *Chaos* **2**, 85.

Cvitanović, P. & Eckhardt, B. (1993). *Nonlinearity* **6**, 277.

Cvitanović, P. & Vattay, G. (1993). *Phys. Rev. Lett.* **71**, 4138.

Cvitanović, P., Eckmann, J.-P. & Gaspard, P. (1995). *Chaos, Solitons, and Fractals* **6**, 113.

Dab, D., Boon, J.-P. & Li, Y.-X. (1991). *Phys. Rev. Lett.* **66**, 2535.

Dana, I., Murray, N. W. & Percival, I. C. (1989). *Phys. Rev. Lett.* **62**, 233.

Dantus, M., Bowman, R. M., Gruebele, M. & Zewail, A. H. (1989). *J. Chem. Phys.* **91**, 7437.

deGroot, S. & Mazur, P. (1962). *Nonequilibrium Thermodynamics.* Amsterdam: North-Holland.

Dekker, H. & van Kampen, N. G. (1979). *Phys. Lett. A* **73**, 374.

Dellago, Ch. & Posch, H. A. (1995). *Phys. Rev. E* **52**, 2401.

Dellago, Ch., Posch, H. A. & Hoover, W. G. (1996). *Phys. Rev. E* **53**, 1485.

deRham, G. (1957). *Rend. Sem. Mat. Torino* **16**, 101.

Dettmann, C. P. & Morris, G. P. (1996). *Phys. Rev. E* **53**, R5541.

Devaney, R. L. (1990). *Chaos, Fractals, and Dynamics*. Menlo Park, California: Addison-Wesley Publ. Co.

DeWit, A., Dewel, G., Borckmans, P. & Walgraef, D. (1992). *Physica D* **61**, 289.

DeWit, A., Dewel, G. & Borckmans, P. (1993). *Phys. Rev. E* **48**, R4191.

DeWit, A., Lima, D., Dewel, G. & Borckmans, P. (1996). *Phys. Rev. E* **54**, 261.

DeWitt-Morette, C., Maheshwari, A. & Nelson, B. (1979). *Phys. Rep.* **50**, 255.

Doob, J. (1953). *Stochastic Processes*. New York: Wiley.

Dörfle, M. (1985). *J. Stat. Phys.* **40**, 93.

Dorfman, J. R. & Cohen, E. G. D. (1970). *Phys. Rev. Lett.* **25**, 1257.

Dorfman, J. R. & Cohen, E. G. D. (1972). *Phys. Rev. A* **6**, 776.

Dorfman, J. R. & Cohen, E. G. D. (1975). *Phys. Rev. A* **12**, 292.

Dorfman, J. R. & Gaspard, P. (1995). *Phys. Rev. E* **51**, 28.

Dorfman, J. R., Ernst, M. H. & Jacobs, D. (1995). *J. Stat. Phys.* **81**, 497.

Doron, E., Smilansky, U. & Frenkel, A. (1991). *Physica D* **50**, 367.

Driebe, D. J. & Ordóñez, G. E. (1996). *Phys. Lett. A* **211**, 204.

Dunford, N. & Schwartz, J. T. (1958). *Linear Operators*, vol. I. New York: Interscience-Wiley.

Dunford, N. & Schwartz, J. T. (1963). *Linear Operators*, vol. II. New York: Interscience-Wiley.

Dunford, N. & Schwartz, J. T. (1971). *Linear Operators*, vol. III. New York: Interscience-Wiley.

Ebeling, W., Feistel, R. & Herzel, H. (1987). *Physica Scripta* **35**, 761.

Ebeling, W. & Nicolis, G. (1991). *Europhys. Lett.* **14**, 191.

Ebeling, W. & Nicolis, G. (1992). *Chaos, Solitons and Fractals* **2**, 635.

Eckhardt, B. (1987). *J. Phys. A* **20**, 5971.

Eckhardt, B. (1988). *Physica D* **33**, 89.

Eckhardt, B. (1993). In: *Quantum Chaos* (eds. Casati, G., Guarneri, I. & Smilansky, U.), Proceedings of the CXIX International Course of Physics "Enrico Fermi", Varenna, 1991, pp. 77–111. Amsterdam: North-Holland.

Eckhardt, B. & Wintgen, D. (1990). *J. Phys. B* **23**, 355.

Eckhardt, B. & Wintgen, D. (1991). *J. Phys. A* **24**, 4335.

Eckhardt, B. & Yao, D. (1993a). *Physica D* **65**, 100.

Eckhardt, B. & Russberg, G. (1993b). *Phys. Rev. E* **47**, 1578.

Eckhardt, B., Russberg, G., Cvitanović, P., Rosenqvist, P. E. & Scherer, P. (1995). In: *Quantum Chaos* (eds. Casati, G. & Chirikov, B.), pp. 405–433. Cambridge, England: Cambridge University Press.

Eckmann, J.-P. (1981). *Rev. Mod. Phys.* **53**, 643.

Eckmann, J.-P. (1989). *Proceedings of the IXth Int. Congr. on Mathematical Physics*, pp. 192–207. Bristol: IOP Publishing Ltd.

Eckmann, J.-P. & Ruelle, D. (1985). *Rev. Mod. Phys.* **57**, 617.

Eckmann, J.-P., Oliffson Kamphorst, S., Ruelle, D. & Ciliberto, S. (1986). *Phys. Rev. A* **34**, 4971.

Eggleston, H. G. (1949). *Q. J. Math. Oxford Ser.* **20**, 31.

Enskog, D. (1921). *Kungl. Svenska Vet. Akad. Handl.* **63**, No. 4.

Ernst, M. H. (1992). In: *Microscopic Simulations of Complex Hydrodynamic Phenomena* (eds. Mareschal, M. & Holian, B. L.), pp. 153–168. New York: Plenum Press.

Ernst, M. H., Hauge, E. H. & van Leeuwen, J. M. J. (1970). *Phys. Rev. Lett.* **25** 1254.

Ernst, M. H., Dorfman, J. R., Nix, R. & Jacobs, D. (1995). *Phys. Rev. Lett.* **74**, 4416.

Evans, D. J. (1989). *J. Stat. Phys.* **57**, 745.

Evans, D. J., Hoover, W. H., Failor, B. H., Moran, B. & Ladd A. J. C. (1983). *Phys. Rev. A* **28**, 1016.

Evans, D. J., Cohen, E. G. D. & Morriss, G. P. (1990a). *Phys. Rev. A* **42**, 5990.

Evans, D. J. & Morriss, G. P. (1990b). *Statistical Mechanics of Nonequilibrium Liquids.* New York: Academic.

Evans, D. J., Cohen, E. G. D. & Morriss, G. P. (1993). *Phys. Rev. Lett.* **71**, 2401.

Ewing, G. (1978). *Chem. Phys.* **29**, 253.

Eyink, G. L., Lebowitz, J. L. & Spohn, H. (1996). *J. Stat. Phys.* **83**, 385.

Farber, E., ed. (1961). *Great Chemists.* New York: Interscience Publishers.

Feller, W. (1968). *An Introduction to Probability Theory and Its Applications,* Vol. I. New York: Wiley.

Feller, W. (1971). *An Introduction to Probability Theory and Its Applications,* Vol. II. New York: Wiley.

Fermi, E. (1936). *Thermodynamics.* New York: Dover.

Forster, D. (1975). *Hydrodynamic Fluctuations, Broken Symmetry, and Correlation Functions.* Reading, Massachusetts: Benjamin/Cummings.

Fox, R. F. (1995). *Chaos* **5**, 619.

Fox, R. F. (1997). *Chaos* **7**, 254.

Fraser, A. M. & Swinney, H. L. (1986). *Phys. Rev. A* **33**, 1134.

Frensley, W. R. (1990). *Rev. Mod. Phys.* **62**, 745.

Friedman, B. (1956). *Principles and Techniques of Applied Mathematics.* New York: Wiley.

Friedman, B., Oono, Y. & Kubo, I. (1984). *Phys. Rev. Lett.* **52**, 9.

Gallavotti, G. & Ornstein, D. S. (1974). *Commun. Math. Phys.* **38**, 83.

Gallavotti, G. & Cohen, E. G. D. (1995a). *Phys. Rev. Lett.* **74**, 2694.

Gallavotti, G. & Cohen, E. G. D. (1995b). *J. Stat. Phys.* **80**, 931.

Gantmacher, F. R. (1959). *Applications of the Theory of Matrices.* New York: Interscience.

Gaspard, P. (1990). *J. Phys. Chem.* **94**, 1.

Gaspard, P. (1991a). In: *Solitons and Chaos* (eds. Antoniou, I. & Lambert, F.), pp. 46–57. Berlin: Springer-Verlag.

Gaspard, P. (1991b). In: *Quantum Chaos* (eds. Cerdeira, H. A., Gutzwiller, M. C., Ramaswamy, R. & Casati, G.), pp. 348–380. Singapore: World Scientific.

Gaspard, P. (1992a). *J. Stat. Phys.* **68**, 673.

Gaspard, P. (1992b). *J. Phys. A* **25**, L483.

Gaspard, P. (1992c). *Phys. Lett. A* **168**, 13.

Gaspard, P. (1992d) In: *From Phase Transitions to Chaos* (eds. Györgyi, G., Kondor, I., Sasvári, L. & Tél, T.), pp. 322–334. Singapore: World Scientific.

Gaspard, P. (1992e). In: *Quantum Chaos – Quantum Measurement* (eds. Cvitanović, P., Percival, I. & Wirzba, A.), pp. 19–42. Dordrecht: Kluwer.

Gaspard, P. (1993a). *Chaos* **3**, 427.

Gaspard, P. (1993b). In: *Quantum Chaos* (eds. Casati, G., Guarneri, I. & Smilansky, U.), Proceedings of the CXIX International Course of Physics "Enrico Fermi", Varenna, 1991, pp. 307–383. Amsterdam: North-Holland.

Gaspard, P. (1994). *Prog. Theor. Phys. Suppl.* **116**, 369.

Gaspard, P. (1995). In: *Dynamical Systems and Chaos* (eds. Aizawa, Y., Saito, S. & Shiraiwa, K.), vol. 2, pp. 55–68. Singapore: World Scientific.

Gaspard, P. (1996). *Phys. Rev. E* **53**, 4379.

Gaspard, P. (1997a). *Adv. Chem. Phys.* **XCIX**, 369.

Gaspard, P. (1997b). *J. Stat. Phys.* **88**, 1215.

Gaspard, P. (1997c). *Physica A* **240**, 54.

Gaspard, P. & Nicolis, G. (1983). *J. Stat. Phys.* **31**, 499.

Gaspard, P. & Nicolis, G. (1985). *Physicalia Magazine (J. Belg. Phys. Soc.)* **7**, 151.

Gaspard, P. & Wang, X.-J. (1988). *Proc. Natl. Acad. Sci. (USA)* **85**, 4591.

Gaspard, P. & Rice, S. A. (1989a). *J. Chem. Phys.* **90**, 2225, 2242, 2255. **91**, E3279.

Gaspard, P. & Rice, S. A. (1989b). *J. Phys. Chem.* **93**, 6947.

Gaspard, P. & Nicolis, G. (1990). *Phys. Rev. Lett.* **65**, 1693.

Gaspard, P. & Alonso Ramirez, D. (1992). *Phys. Rev. A* **45**, 8383.

Gaspard, P. & Baras, F. (1992). In: *Microscopic Simulations of Complex Hydrodynamic Phenomena* (eds. Maréschal, M. & Holian, B. L.), pp. 301–322. New York: Plenum Press.

Gaspard, P. & Rice, S. A. (1993). *Phys. Rev. A* **48**, 54.

Gaspard, P. & Wang, X.-J. (1993). *Phys. Rep.* **235** (6), 321.

Gaspard, P., Alonso, D., Okuda, T. & Nakamura, K. (1994). *Phys. Rev. E* **50**, 2591.

Gaspard, P. & Baras, F. (1995). *Phys. Rev. E* **51**, 5332.

Gaspard, P. & Dorfman, J. R. (1995). *Phys. Rev. E* **52**, 3525.

Gaspard, P., Nicolis, G., Provata, A. & Tasaki, S. (1995a). *Phys. Rev. E* **51**, 74.

Gaspard, P., Alonso, D. & Burghardt, I. (1995b). *Adv. Chem. Phys.* **XC**, 105.

Gaspard, P. & Burghardt, I. (1997). *Adv. Chem. Phys.* **101**, 491.

Gaspard, P. & Jain, S. R. (1997). *Pramana – J. of Phys.* **48**, 503.

Geisel, T., Nierwetberg, J. & Zacherl, A. (1985). *Phys. Rev. Lett.* **54**, 616.

Gelfand, I. M. (1961). *Lecture on Linear Algebra.* New York: Dover.

Gelfand, I. M. & Shilov, G. (1968). *Generalized Functions*, vol. 2. New York: Academic Press.

Gibbs, J. W. (1902). *Elementary Principles in Statistical Mechanics.* New York: Scribner.

Glansdorff, P. & Prigogine, I. (1971). *Thermodynamics of Structure, Stability, and Fluctuations.* New York: Wiley.

Glauber, R. J. (1963). *J. Math. Phys.* **4**, 294.

Gohberg, I. C. & Klein, M. G. (1971). *Introduction à la théorie des opérateurs linéaires non auto-adjoint dans un espace hilbertien.* Paris: Dunod.

Goldstein, H. (1950). *Classical Mechanics.* Reading, Massachusetts: Addison-Wesley Publ. Co.

Goldstein, S., Lebowitz, J. L. & Aizenman, M. (1975). In: *Dynamical Systems, Theory and Applications* (ed. Moser, J.), p. 112. Berlin: Springer-Verlag.

Gomez Llorente, J. M. & Taylor, H. S. (1989). *J. Chem. Phys.* **91**, 953.

Grassberger, P. (1986a). *J. Theor. Phys.* **25**, 907.

Grassberger, P. (1986b). *J. Stat. Phys.* **45**, 27.

Grassberger, P. (1989). *IEEE Trans. Inf. Theory* **35**, 669.

Grassberger, P. & Procaccia, I. (1983). *Phys. Rev. A* **28**, 259.

Grassberger, P., Badii, R. & Politi, A. (1988). *J. Stat. Phys.* **51**, 135.

Grassberger, P., Schreiber, T. & Schaffrath, C. (1991). *Int. J. of Bif. and Chaos* **1**, 521.

Gray, S. K., Rice, S. A. & Noid, D. W. (1986). *J. Phys. Chem.* **81**, 1083.

Grebogi, C., Ott, E. & Yorke, J. A. (1983). *Phys. Rev. Lett.* **50**, 935.

Grebogi, C., Ott, E. & Yorke, J. A. (1988). *Phys. Rev. A* **37**, 1711.

Green, M. S. (1951). *J. Chem. Phys.* **19**, 1036.

Green, M. S. (1952). *J. Chem. Phys.* **20**, 1281.

Green, M. S. (1954). *J. Chem. Phys.* **22**, 398.

Green, M. S. (1960). *Phys. Rev.* **119**, 829.

Grossmann, S. & Thomae, S. (1977). *Z. Naturforsch. A* **32**, 1353.

Grothendieck, A. (1955). *Produits tensoriels topologiques et espaces nucléaires*, Memoir **16**. Providence, R. I.: Amer. Math. Society.

Guckenheimer, J. & Holmes, P. (1983). *Nonlinear Oscillations, Dynamical Systems, and Bifurcations of Vector Fields.* New York: Springer-Verlag.

Gustavson, F. G. (1966). *Astron. J.* **71**, 670.

Gutzwiller, M. C. (1990). *Chaos in Classical and Quantum Mechanics.* New York: Springer-Verlag.

Hadamard, J. (1898). *J. Math. Pur. Appl.* **4**, 27.

Hamermesh, M. (1962). *Group Theory and its Application to Physical Problems.* New York: Dover.

Hannay, J. H. & Ozorio de Almeida, A. M. (1984). *J. Phys. A* **17**, 3429.

Hansen, K. T. (1993). *Nonlinearity* **6**, 753, 771.

Hao Bai-Lin, U., ed. (1984). *Chaos I: a reprint collection.* Singapore: World Scientific.

Hao Bai-Lin, U., ed. (1990). *Chaos II: a reprint collection.* Singapore: World Scientific.

Harrison, W. A. (1980). *Solid State Theory.* New York: Dover.

Hasegawa, H. H. & Saphir, W. C. (1992a). *Phys. Lett. A* **161**, 471, 477.

Hasegawa, H. H. & Saphir, W. C. (1992b). *Phys. Rev. A* **46**, 7401.

Hasegawa, H. H. & Driebe, D. J. (1992). *Phys. Lett. A* **168**, 18.

Hasegawa, H. H. & Driebe, D. J. (1993). *Phys. Lett. A* **176**, 193.

Hasegawa, H. H. & Driebe, D. J. (1994). *Phys. Rev. E* **50**, 1781.

Hasley, T. C., Jensen, M. H., Kadanoff, L. P., Procaccia, I. & Shraiman, B. I. (1986). *Phys. Rev. A* **33**, 1141.

Hata, M. (1986). In: *Patterns and Waves* (eds. Nishida, T., Mimura, M. & Fujii, H.), pp. 259–278. Tokyo: Kinokuniya & Amsterdam: North-Holland.

Hata M. & Yamaguti, M. (1984). *Jpn. J. Appl. Math.* **1**, 183.

Hedlund, G. A. (1939). *Bull. Amer. Math. Soc.* **45**, 241.

Helfand, E. (1960). *Phys. Rev.* **119**, 1.

Helfand, E. (1961). *Phys. Fluids* **4**, 681.

Hemmen, J. L., van (1980). *Phys. Rep.* **65**, 43.

Hénon, M. (1969). *Quart. Appl. Math.* **27**, 291.

Hénon, M. (1976). *Commun. Math. Phys.* **50**, 69.

Hénon, M. (1988). *Physica D* **33**, 132.

Hénon, M. & Heiles, C. (1964). *Astron. J.* **69**, 73.

Holbrook, K. A., Pilling, M. J. & Robertson, S. H. (1996). *Unimolecular Reactions*, 2nd edition. Chichester: Wiley.

Hoover, W. G. (1985). *Phys. Rev. A* **31**, 1695.

Hoover, W. G. (1991). *Computational Statistical Mechanics.* Amsterdam: Elsevier.

Hopf, E. (1937). *Ergodentheorie.* Berlin: Springer-Verlag.

Hopf, E. (1939). *Ber. Verh. Sächs. Akad. Wiss. Leipzig* **91**, 261.

Hudetz, T. (1990). Diploma Thesis, Vienna (unpublished).

Isola, S. (1988). *Commun. Math. Phys.* **116**, 343.

Jakšić, V. & Pillet, C.-A. (1996). *Commun. Math. Phys.* **176**, 619; **178**, 627.

Jalabert, R. A., Baranger, H. U. & Stone, A. D. (1990). *Phys. Rev. Lett.* **65**, 2442.

Joachain, C. J. (1975). *Quantum Collision Theory.* Amsterdam: North-Holland.

John, W., Milek, B., Schanz, H. & Seba, P. (1991). *Phys. Rev. Lett.* **67**, 1949.

Jung, C. (1986). *J. Phys. A* **19**, 1345.

Jung, C. (1987). *J. Phys. A* **20**, 1719.

Jung, C. &. Scholz, H. J (1987). *J. Phys. A* **20**, 3607.

Jung, C. & Tél, T. (1991). *J. Phys. A* **24**, 2793.

Jung, C., Tél, T. & Ziemniak, E. (1993). *Chaos* **3**, 555.

Kadanoff, L. P. & Swift, J. (1968). *Phys. Rev.* **166**, 89.

Kadanoff, L. P. & Tang, C. (1984). *Proc. Natl. Acad. Sci. U.S.A.* **81**, 1276.

Kamgar-Parsi, B., Cohen, E. G. D. & de Schepper, I. M. (1987). *Phys. Rev. A* **35**, 4781.

Kampen, N. G., van (1977). *J. Stat. Phys.* **17**, 71.

Kampen, N. G., van (1978). *Suppl. Prog. Theor. Phys.* **64**, 389.

Kampen, N. G., van (1981). *Stochastic Processes in Physics and Chemistry.* Amsterdam: North-Holland.

Kaneko, K. (1989). In: *Formation, Dynamics and Statics of Patterns* (ed. Kawasaki, K.). Singapore: World Scientific.

Kantz, H. & Grassberger, P. (1985). *Physica D* **17**, 75.

Kaplan, J. L. & Yorke, J. A. (1979). *Commun. Math. Phys.* **67**, 93.

Kapral, R. (1991). *J. Math. Chem.* **6**, 113.

Kapral, R., Lawniczak, A. & Masiar, P. (1992). *J. Chem. Phys.* **96**, 2762.

Kawasaki, K. (1966). *Phys. Rev.* **145**, 224.

Kirkpatrick, T. R., Cohen, E. G. D. & Dorfman, J. R. (1982). *Phys. Rev. A* **26**, 950, 972, 995.

Kirkwood, J. G. (1946). *J. Chem. Phys.* **14**, 180.

Kittel, C. (1976). *Introduction to Solid State Physics*. New York: Wiley.

Klafter, J., Blumen, A. & Shlesinger, M. F. (1987). *Phys. Rev. A* **35**, 3081.

Klages, R. (1996). *Deterministic Diffusion in One-Dimensional Chaotic Dynamical Systems*. Berlin: Wissenschaft und Technik Verlag.

Klages, R. & Dorfman, J. R. (1995). *Phys. Rev. Lett.* **74**, 387.

Klein, M. & Knauf, A. (1992). *Classical Planar Scattering by Coulombic Potentials*. Berlin: Springer-Verlag.

Knauf, A. (1987). *Commun. Math. Phys.* **110**, 89.

Knauf, A. (1989). *Ann. Phys. (N. Y.)* **191**, 205.

Knuth, D. E. (1969). *The Art of Computer Programming*, vol. 2, Chap. 3. Reading, Massachusetts: Addison-Wesley.

Kolmogorov, A. N. (1954). Reprinted in: Casati & Ford (1979).

Kolmogorov, A. N. (1956). *IRE Trans. Inform. Theory* **1**, 102.

Kolmogorov, A. N. (1958). *Dokl. Acad. Sci. USSR* **119**(5), 861.

Kolmogorov, A. N. (1959). *Dokl. Acad. Sci. USSR* **124**(4), 754.

Kolmogorov, A. N. (1983). *Russian Math. Surveys* **38**, 29.

Kolmogorov, A. N. & Tikhomirov, V. M. (1959). *Uspekhi Mat. Nauk* **14**, 3.

Koopman, B. O. (1931). *Proc. Natl. Acad. Sci. U.S.A.* **17**, 315.

Korsch, H. J. & Wagner, A. (1991). *Computers in Phys.* **5**, 497.

Kovács, Z. & Tél, T. (1990). *Phys. Rev. Lett.* **64**, 1617.

Kramers, H. A. (1940). *Physica* **7**, 284.

Krámli, A., Simányi, N. & Szász, D. (1992). *Commun. Math. Phys.* **144**, 107.

Kreyszig, E. (1991). *Differential Geometry*. New York: Dover.

Krylov, N. S. (1944). *Nature* **153**, 709.

Krylov, N. S. (1979). *Works on the Foundations of Statistical Physics*. Princeton: Princeton University Press.

Kubo, R. (1957). *J. Phys. Soc. Jpn.* **12**, 570.

Landau, L. D. (1936). *Phys. Z. Sowj. Union* **10**, 154.

Landau, L. D. & Lifshitz, E. M. (1963). *Fluid Mechanics*. New York: Pergamon Press.

Landauer, R. (1970). *Philos. Mag.* **21**, 863.

Lanford, O. E. (1973). In: *Statistical Mechanics and Mathematical Problems* (ed. Lenard, A.). Berlin: Springer-Verlag.

Laskar, J. (1989). *Nature* **338**, 237.

Lasota, A. & Mackey, M. C. (1985). *Probabilistic Properties of Deterministic Systems*. Cambridge, England: Cambridge University Press.

Lauritzen, B. (1990). *Phys. Rev. A* **43**, 603.

Lawniczak, A., Dab, D., Kapral, R. & Boon, J.-P. (1991). *Physica D* **47**, 132.

Lax, P. D. & Phillips, R. S. (1967). *Scattering Theory*. New York: Academic Press.

Lebowitz, J. L. (1959). *Phys. Rev.* **114**, 1192.

Lebowitz, J. L. & Rubin, E. (1963). *Phys. Rev.* **131**, 2381.

Lebowitz, J. L. & Penrose, O. (1973). *Physics Today*, February, p. 23.

Lenard, A. (1960). *Ann. Phys. (N. Y.)* **3**, 390.

Li, W. & Kaneko, K. (1992). *Europhys. Lett.* **17**, 655.

Lichtenberg, A. J. & Lieberman, M. A. (1983). *Regular and Stochastic Motion*. New York: Springer-Verlag.

Lindenberg, K., Shuler, K. E., Freeman, J. & Lie, T. L. (1975). *J. Stat. Phys.* **12**, 217.

Livi, R., Politi, A. & Ruffo, S. (1986). *J. Phys. A* **19**, 2033.

Lloyd, J., Niemeyer, M., Rondoni, L. & Morris, G. P. (1995). *Chaos* **5**, 536.

Lorenz, E. (1963). *J. Atm. Sci.* **20**, 130.

Lorenz, E. (1993). *The Essence of Chaos*. London: UCL Press.

Lorentz, H. A. (1905). *Proc. Roy. Acad. Amst.* **7**, 438, 585, 684.

Lützen, J. (1990). *Joseph Liouville 1809–1882: Master of Pure and Applied Mathematics*. New York: Springer-Verlag.

Lyapunov, M. A. (1907). *Problème Général de la Stabilité du Mouvement*, Ann. Facult. Sci. Toulouse, 2nd series, vol. 9. Reprinted (1947). Princeton: The University Press.

MacDonald, S. W., Grebogi, C., Ott, E. & Yorke, J. A. (1985). *Physica D* **17**, 125.

Machta, J. & Zwanzig, R. (1983). *Phys. Rev. Lett.* **50**, 1959.

MacKay, R. S. (1993a). In: *Quantum Chaos* (eds. Casati, G., Guarneri, I. & Smilansky, U.), Proceedings of the CXIX International Course of Physics "Enrico Fermi", Varenna, 1991, pp. 1–50. Amsterdam: North-Holland.

MacKay, R. S. (1993b). *Renormalisation in Area-Preserving Maps*. Singapore: World Scientific.

MacKay, R. S., Meiss, J. D. & Percival, I. C. (1984). *Physica D* **13**, 55.

MacKay, R. S. & Meiss, J. D. (1987). *Hamiltonian Dynamical Systems: A Reprint Selection*. Bristol: Adam Hilger.

MacKernan, D. & Nicolis, G. (1994). *Phys. Rev. E* **50**, 988.

Mackey, M. C. (1989). *Rev. Mod. Phys.* **61**, 981.

MacLennan, J. A., Jr. (1959). *Phys. Rev.* **115**, 1405.

Malek Mansour, M., Van Den Broeck, C., Nicolis, G. & Turner, J. W. (1981). *Ann. Phys.* **131**, 283.

Mandelbrot, B. B. (1982). *The Fractal Geometry of Nature*. San Francisco: Freeman.

Mandelbrot, B. B. & Van Ness, J. W. (1968). *SIAM Review* **10**, 422.

Manneville, P. (1980). *J. de Phys.* **41**, 1235.

Manneville, P. (1985). In: *Macroscopic Modeling of Turbulent Flows* (ed. Pironneau, O.), Lecture Notes in Physics, vol. 230, p. 319. New York: Springer-Verlag.

Manning, A. (1971). *Bull. London Math. Soc.* **3**, 215.

Marcus, C. M., Westervelt, R. M., Hopkins, P. F. & Gossard, A. C. (1993). *Chaos* **3**, 643.

Marcus, M. & Minc, H. (1964). *A Survey of Matrix Theory and Matrix Inequalities*. New York: Dover.

Mareschal, M., Malek Mansour, M., Puhl, A. & Kestemont, E. (1988). *Phys. Rev. Lett.* **61**, 2550.

Maxwell, J. C. (1890). *The Scientific Papers of James Clerk Maxwell*. Cambridge: The University Press. Reprinted (1952). New York: Dover.

Mayer, D. H. (1991). In: *Ergodic Theory, Symbolic Dynamics and Hyperbolic Spaces* (eds. Bedford, T., Keane, M. & Series, C.). Oxford, England: Oxford University Press.

Mayer, D. H. & Roepstorff, G. (1987). *J. Stat. Phys.* **47**, 149.

Mayer, D. H. & Roepstorff, G. (1988). *J. Stat. Phys.* **50**, 331.

Meiss, J. D. (1992). *Rev. Mod. Phys.* **64**, 795.

Meiss, J. D. & Ott, E. (1985). *Phys. Rev. Lett.* **55**, 2741.

Misra, B. (1978). *Proc. Natl. Acad. Sci. USA* **75**, 1627.

Morette, C. (1951). *Phys. Rev.* **81**, 848.

Mori, H. (1958). *Phys. Rev.* **112**, 1829.

Morriss, G. P. (1987). *Phys. Lett. A* **122**, 236.

Morriss, G. P. (1989). *Phys. Lett. A* **143**, 307.

Morriss, G. P. & Rondoni, L. (1994). *J. Stat. Phys.* **75**, 553.

Moser, J. (1958). *Astron. J.* November, 439.

Moser, J. (1973). *Stable and Random Motions in Dynamical Systems*. Princeton, New Jersey: Princeton University Press.

Naisse, J. (1980). *Introduction à la physique quantique*. Lecture Notes ULB (unpublished).

Narnhofer, H. (1980). *Phys. Rev. D* **22**, 2387.

Narnhofer, H. & Thirring, W. (1981). *Phys. Rev. A* **23**, 1688.

Neumann, J., von (1932). *Proc. Natl. Acad. Sci. USA* **18**, 70.

Nicolis, G. (1995). *Introduction to Nonlinear Science*. Cambridge, England: Cambridge University Press.

Nicolis, G. & Prigogine, I. (1977). *Self-Organization in Nonequilibrium Systems*. New York: Wiley.

Nicolis, G. & Malek Mansour, M. (1978). *Prog. Theor. Phys. Suppl.* **64**, 249.

Nicolis, G. & Nicolis, C. (1988). *Phys. Rev. A* **38**, 427.

Nicolis, G. & Prigogine, I. (1989). *Exploring Complexity*. New York: Freeman.

Nicolis, G. & Gaspard, P. (1994). *Chaos, Solitons and Fractals* **4**, 41.

Nicolis, G. & Daems, D. (1996). *J. Phys. Chem.*, **100**, 10187.

Noid, D. W., Gray, S. K. & Rice, S. A. (1986). *J. Chem. Phys.* **84**, 2649.

Nosé, S. (1984a). *J. Chem. Phys.* **81**, 511.

Nosé, S. (1984b). *Mol. Phys.* **52**, 255.

Nowak, M. A. & May, R. (1992). *Nature* **359**, 826.

Onsager, L. (1931). *Phys. Rev.* **37**, 405; **38**, 2265.

Oono, Y. (1993). *Prog. Theor. Phys.* **89**, 973.

Oppo, G.-L. & Kapral, R. (1987). *Phys. Rev. A* **36**, 5820.

Ornstein, D. S. (1974). *Ergodic Theory, Randomness, and Dynamical Systems.* New Haven: Yale University Press.

Oseledets, V. I. (1968). *Trans. Moscow Math. Soc.* **19**, 197. Reprinted in: Sinai (1991).

Ott, E. (1993). *Chaos in Dynamical Systems.* Cambridge, England: Cambridge University Press.

Ott, E. & Tél, T., eds. (1993). *Chaos* **3** (4). Focus issue on "Chaotic Scattering".

Ozorio de Almeida, A. M. (1988). *Hamiltonian Systems: Chaos and Quantization.* Cambridge, England: Cambridge University Press.

Parry, W. & Pollicott, M. (1983). *Ann. Math.* **118**, 573.

Parry, W. & Pollicott, M. (1990). *Astérisque* **187–188**, 1.

Pathria, R. K. (1972). *Statistical Mechanics.* Oxford: Pergamon.

Pauli, W. (1928). *Festschrift zum 60. Geburtstage A. Sommerfelds*, p. 30. Leipzig: Hirzel.

Perrin, J. (1970). *Les atomes.* Paris: Presses Universitaires de France.

Pesin, Ya. B. (1976). *Math. USSR Izv.* **10**(6), 1261.

Pesin, Ya. B. (1977). *Russian Math. Surveys* **32**(4), 55.

Pesin, Ya. B. (1989). In: *Dynamical Systems II* (ed. Sinai, Ya. G.), pp. 108–151. Berlin: Springer-Verlag.

Petit, J. M. & Hénon, M. (1986). *Icarus* **60**, 536.

Pettini, M. (1993). *Phys. Rev. E* **47**, 828.

Pinsker, M. S. & Sofman, L. B. (1983). *Problems of Inf. Transm.* **19**(3), 214.

Poincaré, H. (1892). *Les Méthodes Nouvelles de la Mécanique Céleste.* Paris: Gauthier-Villars.

Pollicott, M. (1985). *Invent. Math.* **81**, 413.

Pollicott, M. (1986). *Invent. Math.* **85**, 147.

Pollner, P. & Vattay, G. (1996). *Phys. Rev. Lett.* **76**, 4155.

Pomeau, Y. & Résibois, P. (1975). *Phys. Rep.* **19**, 63.

Pomeau, Y. & Manneville, P. (1980). *Commun. Math. Phys.* **74**, 189.

Pomeau, Y., Pumir, A. & Pelce, P. (1984). *J. Stat. Phys.* **37**, 39.

Posch, H. A. & Hoover, W. G. (1988). *Phys. Rev. A* **38**, 473.

Posch, H. A. & Hoover, W. G. (1989). *Phys. Rev. A* **39**, 2175.

Press, W. H., Flannery, B. P., Teukolsky, S. A. & Vetterling, W. T. (1986). *Numerical Recipies.* Cambridge, England: Cambridge University Press.

Prigogine, I. (1949). *Physica* **15**, 272.

Prigogine, I. (1961). *Introduction to Thermodynamics of Irreversible Processes.* New York: Wiley.

Prigogine, I. (1962). *Nonequilibrium Statistical Mechanics.* New York: Wiley.

Prigogine, I. (1980). *From Being to Becoming.* San Francisco: Freeman.

Procaccia, I., Ronis, D., Collins, M. A., Ross, J. & Oppenheim, I. (1979a). *Phys. Rev. A* **19**, 1290.

Procaccia, I., Ronis, D. & Oppenheim, I. (1979b). *Phys. Rev. Lett.* **42**, 287.

Rand, D. (1989). *Ergod. Theory Dyn. Syst.* **9**, 527.

Ravani, M., Derigletti, B., Broggi, G., Brun, E. & Badii, R. (1988). *J. Opt. Soc. Am. B* **5**, 1029.

Résibois, P. & Davis, H. T. (1964). *Physica* **30**, 1077.

Résibois, P. & De Leener, M. (1977). *Classical Kinetic Theory of Fluids.* New York: Wiley.

Rice, S. A., Gaspard, P. & Nakamura, K. (1992). *Adv. Class. Traj. Methods* **1**, 215.

Risken, H. (1984). *The Fokker–Planck Equation.* Berlin: Springer-Verlag.

Robbins J. M. (1989). *Phys. Rev. A* **40**, 2128.

Roepstorff, G. (1987). *On the Exponential Decay of Correlations in Exact Dynamical Systems*, preprint. Princeton: Institute for Advanced Study (unpublished).

Rössler, O. (1979). *Ann. N. Y. Acad. Sci.* **316**, 376.

Roukes, M. L. & Alerhand, O. L. (1989). *Phys. Rev. Lett.* **65**, 1857.

Roux, J.-C., Simoyi, R. H. & Swinney, H. L. (1983). *Physica D* **8**, 257.

Ruelle, D. (1976). *Invent. Math.* **34**, 231.

Ruelle, D. (1978). *Thermodynamic Formalism.* Reading, Massachusetts: Addison-Wesley.

Ruelle, D. (1982). *Commun. Math. Phys.* **87**, 287.

Ruelle, D. (1983). *C. R. Acad. Sci. Paris* **296**, Série I, 191.

Ruelle, D. (1986a). *Phys. Rev. Lett.* **56**, 405.

Ruelle, D. (1986b). *J. Stat. Phys.* **44**, 281.

Ruelle, D. (1987). *J. Differ. Geom.* **25**, 99, 117.

Ruelle, D. (1989a). *Commun. Math. Phys.* **125**, 239.

Ruelle, D. (1989b). *Chaotic Evolution and Strange Attractors.* Cambridge, England: Cambridge University Press.

Ruelle, D. (1989c). *Elements of Differentiable Dynamics and Bifurcation Theory.* Boston: Academic Press.

Ruelle, D. (1992). In: *Mathematical Physics X* (ed. Schmüdgen, K.), pp. 43–51. Berlin: Springer-Verlag.

Ruelle, D. (1994). *Dynamical Zeta Functions for Piecewise Maps of the Interval.* Providence: Amer. Math. Society.

Ruelle, D. (1996). *J. Stat. Phys.* **85**, 1.

Ruelle, D. (1997). *J. Stat. Phys.* **86**, 935.

Rugh, H. H. (1992). *Nonlinearity* **5**, 1237.

Sano, M. M. (1994). *J. Phys. A* **27**, 4791.

Schell, M., Fraser, S. & Kapral, R. (1982). *Phys. Rev. A* **26**, 504.

Schnakenberg, J. (1976). *Rev. Mod. Phys.* **48**, 571.

Seidel, W. (1933). *Proc. Natl. Acad. Sci. USA* **19**, 453.

Shannon, C. E. (1960). In: *Information and Decision Processes* (ed. Machol, R.), p. 93. McGraw-Hill.

Shannon, C. E. & Weaver, W. (1949). *The Mathematical Theory of Communication.* Urbana: The University of Illinois Press.

Shilnikov, L. P. (1970). *Math. USSR Sbornik* **10**, 91.

Sinai, Ya. G. (1959). *Dokl. Acad. Sci. USSR* **124**(4), 768.

Sinai, Ya. G. (1961). *Izv. Acad. Sci. USSR* **25**(6), 899.

Sinai, Ya. G. (1966). *Izv. Acad. Sci. USSR* **30**(1), 15.

Sinai, Ya. G. (1968). *Funct. Anal. Appl.* **2**, 61, 245. Reprinted in: Sinai (1991).

Sinai, Ya. G. (1970). *Russian Math. Surveys* **25**, 137.

Sinai, Ya. G. (1972). *Russian Math. Surveys* **27**, 21.

Sinai, Ya. G. (1979). In: Krylov (1979), p. 239.

Sinai, Ya. G. & Chernov, N. I. (1987). *Russian Math. Surveys* **42**(3), 181.

Sinai, Ya. G. ed. (1991). *Dynamical Systems: Collection of Papers.* Singapore: World Scientific.

Sinai, Ya. G. (1996). *Int. J. Bifur. Chaos* **6**, 1137.

Smale, S. (1980a). *Ann. N. Y. Acad. Sci.* **357**, 260. Reprinted in: Smale (1980b).

Smale, S. (1980b). *The Mathematics of Time.* New York: Springer-Verlag.

Smilansky, U. (1991). In: *Chaos and Quantum Physics* (eds. Giannoni, M.-J., Voros, A. & Zinn-Justin, J), pp. 371–442. Amsterdam: Elsevier Science Publishers.

Sommerer, J. C., Ku H.-C. & Gilreath, H. E. (1996). *Phys. Rev. Lett.* **77**, 5055.

Spohn, H. (1980). *Rev. Mod. Phys.* **53**, 569.

Spohn, H. (1991). *Large Scale Dynamics of Interacting Particles.* Berlin: Springer-Verlag.

Sussman, G. J. & Wisdom, J. (1988). *Science* **241**, 433.

Szász, D. (1996). *Studia Scientiarum Mathematicarum Hungarica* **31**, 299.

Szepfalusy, P. & Tél, T. (1986). *Phys. Rev. A* **34**, 2520.

Takagi, T. (1903). *Proc. Phys. Math. Soc. Japan Ser. II* **1**, 176.

Tasaki, S., Antoniou, I. & Suchanecki, Z. (1993). *Phys. Lett. A* **179**, 97.

Tasaki, S., Antoniou, I. & Suchanecki, Z. (1994). *Chaos, Solitons, and Fractals* **4**, 227.

Tasaki, S. & Gaspard, P. (1994). In: *Towards the Harnessing of Chaos* (ed. Yamaguti, M.), pp. 273–288. Amsterdam: Elsevier.

Tasaki, S. & Gaspard, P. (1995). *J. Stat. Phys.* **81**, 935.

Tél, T. (1986). *Phys. Lett. A* **119**, 65.

Tél, T. (1987). *Phys. Rev. A* **36**, 1502.

Tél, T. (1989). *J. Phys. A* **22**, L691.

Tél, T. (1990). In: *Directions in Chaos* (ed. Hao Bai-Lin, U.), vol. 3. Singapore: World Scientific.

Tél, T., Vollmer, J. & Breymann, W. (1996). *EuroPhys. Lett.* **35**, 659.

Tersigni, S. H., Gaspard, P. & Rice, S. A. (1990). *J. Chem. Phys.* **92**, 1775.

Tikhomirov, V. M. (1963). *Russian Math. Surveys* **18**, 51.

Toda, M. (1989). *Theory of Nonlinear Lattices.* Berlin: Springer-Verlag.

Tolman, R. C. (1938). *The Principles of Statistical Mechanics.* New York: Oxford University Press.

Touma, J. & Wisdom, J. (1993). *Science* **259**, 1294.

Troll, G. & Smilansky, U. (1989). *Physica D* **35**, 34.

Uehling, E. A. & Uhlenbeck, G. E. (1933). *Phys. Rev.* **43**, 552.

Ulam, S. & von Neumann, J. (1947). *Amer. Math. Soc. Bull.* **53**, 1120, Abstract 403.

Van Hove, L. (1951). *Bull. Cl. Sci. Acad. Roy. Belg.* Tome **XXVI**, Fascicule 6 (exemplaire hors commerce).

Van Hove, L. (1954). *Phys. Rev.* **95**, 249.

Van Hove, L. (1955). *Physica* **21**, 517.

Van Hove, L. (1957). *Physica* **23**, 441.

Van Hove, L. (1959). *Physica* **25**, 268.

Vattay, G. (1994). *Prog. Theor. Phys. Suppl.* **116**, 251.

Vernon-Lovitt, W. (1950). *Linear Integral Equations.* New York: Dover.

Vlassov, A. (1938). *Zh. Eksp. Teor. Fiz.* **8**, 291.

Walker, J. (1988). *Sci. Am.* **259**, 112.

Walters, P. (1981). *An Introduction to Ergodic Theory.* Berlin: Springer-Verlag.

Wang, X.-J. (1989). *Phys. Rev. A* **40**, 6647.

Wang, X.-J. (1990). *Commun. Math. Phys.* **131**, 317.

Wang, X.-J. (1992). *Phys. Rev. A* **45**, 8407.

Wang, X.-J. (1995). *Phys. Rev. E* **52**, 1318.

Wang, X.-J. & Gaspard, P. (1992). *Phys. Rev. A* **46**, R3000.

Wang, X.-J. & Hu (1993). *Phys. Rev. E* **48**, 728.

Wax, N., ed. (1954). *Selected Papers on Noise and Stochastic Processes.* New York: Dover.

Weiss, G. H. (1986). *J. Stat. Phys.* **42**, 3.

Wiener, N. (1948). *Cybernetics.* Cambridge, Massachusetts: The MIT Press.

Wiggins, S. (1992). *Chaotic Transport in Dynamical Systems.* New York: Springer-Verlag.

Wintgen, D., Richter, K. & Tanner, G. (1992). *Chaos* **2**, 19.

Wintgen, D., Bürgers, A., Richter, K. & Tanner, G. (1994). *Prog. Theor. Phys.* **116**, 121.

Wisdom, J., Peale, S. J. & Mignard, F. (1984). *Icarus* **58**, 137.

Wolfram, S. (1983). *Rev. Mod. Phys.* **55**, 601.

Wolfram, S. (1984). *Physica D* **10**, 1.

Yaglom, A. M. (1962). *An Introduction to the Theory of Stationary Random Functions.* New York: Dover.

Yaglom, A. M. (1987). *Correlation Theory of Stationary and Related Random Functions.* Berlin: Springer-Verlag.

Yamamoto, T. (1960). *J. Chem. Phys.* **33**, 281.

Young, L. S. (1982). *Ergod. Th. and Dyn. Syst.* **2**, 109.

Yvon, J. (1935). *La Théorie Statistique des Fluides et l'Equation d'Etat.* Paris: Hermann.

Zewail, A. H. (1991). *Faraday Discuss. Chem. Soc.* **91**, 207.

Zubarev, D. N. (1962). *Sov. Phys. Dokl.* **6**, 776.

Zubarev, D. N. (1974). *Nonequilibrium Statistical Thermodynamics.* New York: Consultants.

Zvonkin, A. K. & Levin, L. A. (1970). *Russian Math. Surveys* **25**, 83.

Zwanzig, R. (1965). *Ann. Rev. Phys. Chem.* **16**, 67.

Index

Page numbers in bold type indicate the most relevant pages.